MW00476122

HIGH-SPEED SEMICONDUCTOR DEVICES

HIGH-SPEED SEMICONDUCTOR DEVICES

Edited by

S.M. Sze

AT&T Bell Laboratories
Murray Hill, New Jersey

National Chiao Tung University
Hsinchu, Taiwan

A Wiley-Interscience Publication
John Wiley & Sons, Inc.
New York / Chichester / Brisbane / Toronto / Singapore

Library of Congress Cataloging in Publication Data:

High-speed semiconductor devices / edited by S.M. Sze.

    p.   cm.

    Based on notes for lectures and talks given at the 1988
International Electron Devices and Materials Symposium, held at the
National Sun Yat-sen University, Kaohsiung, Taiwan, R.O.C.

    "A Wiley-Interscience publication."
    Includes bibliographical references.

    1. Semiconductors. 2. Diodes, Semiconductor. 3. Transistors.
I. Sze, S. M., 1936–
TK7871.85.H52   1990
621.381'52—dc20   90–38458
                                             CIP

ISBN 0–471–62307–5

Contents

Contributors

P. M. Asbeck
Rockwell International Science Center
Thousand Oaks, California

J. C. Bean
AT&T Bell Laboratories
Murray Hill, New Jersey

F. Beltram
AT&T Bell Laboratories
Murray Hill, New Jersey

J. R. Brews
AT&T Bell Laboratories
Murray Hill, New Jersey

F. Capasso
AT&T Bell Laboratories
Murray Hill, New Jersey

M. A. Hollis
Lincoln Laboratory
Massachusetts Institute of Technology
Lexington, Massachusetts

S. Luryi
AT&T Bell Laboratories
Murray Hill, New Jersey

R. A. Murphy
Lincoln Laboratory
Massachusetts Institute of Technology
Lexington, Massachusetts

S. J. Pearton
AT&T Bell Laboratories
Murray Hill, New Jersey

S. Sen*
AT&T Bell Laboratories
Murray Hill, New Jersey

N. J. Shah
AT&T Bell Laboratories
Murray Hill, New Jersey

S. M. Sze†
AT&T Bell Laboratories
Murray Hill, New Jersey

W. T. Tsang
AT&T Bell Laboratories
Murray Hill, New Jersey

---

*Present Address: University of Calcutta, Calcutta, India.
†Present Address: National Chiao Tung University, Hsinchu, Taiwan, Republic of China.

Preface

High-speed semiconductor devices are key components for advanced electronic systems that process digital data at rates higher than 1 Gb/s (i.e., $> 10^9$ *bits* per second) or handle analog signals at frequencies above 1 GHz (i.e., $> 10^9$ *Hz*). These devices with their ever increasing speed are vital to the continued growth of the electronic industry. The purpose of this book is to provide an introduction to the physical principles and operational characteristics of these devices.

The book is intended for use by senior undergraduate or first-year graduate students in applied physics, electrical engineering, and materials science, and as a reference for engineers and scientists involved in semiconductor device research and development. It is assumed that the reader has already acquired a basic understanding of device operations such as those given in *Physics of Semiconductor Devices, 2nd ed.* (Wiley, 1981). With this as a basis, the present book elaborates on the high-speed aspect of device performances.

The text began as a set of lecture notes for two short courses and two invited talks given by the contributing authors at the 1988 International Electron Devices and Materials Symposium held at the National Sun Yat-sen University, Kaohsiung, Taiwan, R.O.C. We have substantially expanded and updated the lecture notes to include the most advanced and important topics in high-speed semiconductor devices. In Chapters 1 and 2, material properties, advanced technologies, and novel device building blocks are presented. This information is the basis for understanding and analyzing devices in the subsequent chapters. In Chapter 3 through 10, each chapter considers a group of closely related devices and is presented in such a way that it is self-contained and essentially independent of the other chapters. More than 35,000 papers in the literature have provided the background. For each chapter a brief historical review, as well as a general outline, is given in its introduction. The physics of devices and their mathematical formulation are then presented and are generally arranged in logical sequence without heavy reliance on the original papers. The last section of each chapter presents a summary and a discussion of future device trends. The problem set at the end of each chapter forms an integral part of the development of the topics.

In the course of the writing many people have assisted us and offered their support. We would first like to express our appreciation to the management of AT&T Bell Laboratories, MIT Lincoln Laboratory, Rockwell International Science Center, and the National Chiao Tung University for providing the environments in which we worked on the book. We have benefited from suggestions made by the reviewers: Dr. E. R. Brown and Prof. A. L. McWhorter of MIT Lincoln Laboratory, Prof. C.Y. Chang of the National Chiao Tung University, Profs. J. J. Coleman and K. Hess of University of Illinois, Profs. L. F. Eastman, J. W. Mayor and P. J. Tasker of Cornell University, Prof. J. F. Gibbons of Stanford University, Drs. A. A. Grinberg, A. F. J. Levi, W. T. Lynch, K. K. Ng and R. K. Watts of AT&T Bell Laboratories, Prof. G. I. Haddad of University of Michigan, Prof. C. M. Hu of University of California, Dr. G. J. Iafrate of U.S. Army Electronics Technology and Devices Laboratory, Prof. H. Kroemer of University of California, Dr. T. P. Lee of Bell Communication Research Inc., Dr. A. G. Lewis of Xerox Corporation, Profs. U. K. Mishra and R. J. Trew of North Carolina State University, Prof. M. S. Shur of University of Minnesota, and Drs. M. Heiblum and D. D. Tang of IBM Watson Research Center.

We are further indebted to Mr. N. Erdos for technical editing of the entire manuscript and Messrs. E. Labate and R. A. Matula for their literature search. Thanks are also due Ms. L. Levy and Ms. A. M. McGowan for providing thousands of technical papers on high-speed devices, Ms. M. H. Baker, Ms. L. L. Freund, Ms. E. C. Hung, Ms. J. Marotta, Ms. H. F. Womack, and the members of Text Processing Centers who typed the initial drafts and the final manuscripts, and Mr. D. A. Spranza, Ms. D. Sowinska-Khan, and members of the Art Departments who furnished the hundreds of technical illustrations used in the book. At our publishers, John Wiley and Sons, we wish to acknowledge Mr. G. Telecki who encouraged us to undertake this book project. Finally, we wish to thank Ms. T. W. Sze for preparing the Appendixes and Ms. F. F. Fang, Ms. B. L. Huang, Ms. S. L. Hsiau, Mr. Y. H. Hsu, Mr. C. S. Ren, Mr. M. Y. Tzeng, Ms. S. I. Tzeng for preparing the Index.

S. M. SZE

Hsinchu, Taiwan
April 1990

Introduction

Organization of the Book

The book is organized into three parts:

Part 1: Materials, technologies, and device building blocks
Part 2: Field-effect and potential-effect devices
Part 3: Quantum-effect, microwave, and photonic devices

Part 1, Chapters 1 and 2, is intended as a summary of material properties (e.g., lattice constant, bandgap, and drift velocity), advanced processing technologies (e.g., molecular beam epitaxy, ion implantation, and electron beam lithography), and device building blocks (e.g., heterojunction, planar-doped barrier, and resonant-tunneling double barriers). Since we assume that the reader has already acquired an introductory understanding of semiconductor devices such as those given in *Physics of Semiconductor Devices*,[1] the operational principles of the basic devices (e.g., *p-n* junction, Schottky contact, metal-oxide-semiconductor capacitor, etc.) will not be repeated here. The information presented in Part 1 will be used as a basis for understanding and analyzing high-speed devices considered in subsequent chapters.

Part 2, Chapter 3 through 7, considers classic semiconductor devices such as the bipolar transistor and the MOSFET (metal-oxide-semiconductor field-effect transistor). These devices are the workhorses for the present-day electronic systems.

Part 3, Chapters 8 through 10, deals with quantum-effect, microwave, and photonic devices. Quantum-effect devices can have dimensions an order of magnitude smaller than field-effect and potential-effect devices. Therefore, they can be operated at higher speed than those discussed in Part 2. In addition, they are functional devices, that is, they can perform complicated electronic functions with much fewer components than devices discussed in Part 2. Microwave and photonic diodes are two-terminal devices. They have many unique and novel properties; specifically, they are used in millimeter-wave (30 to 300 GHz) and optoelectronic applications.

High-Speed Semiconductor Devices, Edited by S.M. Sze. ISBN 0-471-62307-5
© 1990 John Wiley & Sons, Inc.

Silicon and Gallium Arsenide

The most important semiconductors for high-speed devices are silicon (Si) and gallium arsenide (GaAs) and its related III-V compounds and solid solutions (e.g., the binary compound InP, ternary compounds $Al_{0.3}Ga_{0.7}As$, and quaternary compound $Ga_{0.13}In_{0.87}As_{0.37}P_{0.63}$).

In addition, we have the superlattice semiconductor, which is an artificial one-dimensional periodic structure constituted by different semiconductor materials with a period of about 100 Å. The superlattice semiconductors include the Si-based materials (e.g., GeSi/Si) and GaAs-based materials (e.g., GaAlAs/GaAs).

Figure 1 shows the world production of polished single-crystal Si wafers and III-V compound semiconductor wafers.[2,3] Because of the pre-eminent position in very large scale integrated (VLSI) circuit applications, Si-based devices now constitute about 95% of all semiconductor devices sold worldwide. However, for high-speed applications III-V compounds have certain speed advantages over silicon because of their higher carrier mobilities and higher effective carrier velocities. If the production trend continues, by year 2000 the silicon wafer area will reach 10^{11} cm^2 (or about 200 million wafers with 250-mm diameter) while the III-V compound wafer area will be about 10^9 cm^2 (or about 10 million wafers with 100-mm diameter).

We anticipate that hybrid material systems (e.g., based on a heteroepitaxial process) will become increasingly important; thus the distinction between Si and III-V compound materials may becomes less well defined. For example, one can envisage an advanced integrated circuit architecture in which super high

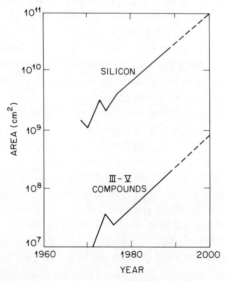

Fig. 1 World production of single-crystal silicon and III-V compound semiconductors. (After Refs. 2 and 3)

speed GaAs/AlGaAs devices are located on heteroepitaxially grown islands on a silicon wafer, integrated with the silicon VLSI circuits by a suitable metallization process. Furthermore, RCL delays due to on-chip interconnect lines cause severe signal attenuation. There is an enormous potential in the development of optical communications within a silicon wafer by using, for example, electronically triggered compound-semiconductor lasers or light-emitting diodes made in aforementioned heteroepitaxial materials.[4]

Field-Effect and Potential-Effect Transistors

Field-effect transistors (FETs) are voltage-controlled devices. The control electrode is capacitively coupled to the active region of the device; and the charge carriers are separated spatially by an insulator or a depletion layer. Figure 2 shows the family tree of field-effect transistors. It includes three device groups: the MOSFET (Chapter 3), the homogeneous FET (Chapter 4), and the heterostructure FET (Chapter 5). We shall now briefly describe these three groups.

The first MOSFET was fabricated in 1960 using a thermally oxidized silicon substrate. Although present-day MOSFETs have been scaled down to submicron region, the choice of silicon and thermally grown silicon dioxide used in the first MOSFET remains to be the most important combination. Figure 3a shows the scaling of MOSFET gate length. Since the early 1960s the annual rate of the reduction for commercial integrated circuits has been 13%. If the trend continues, the minimum device length will shrink to 0.2 μm in the year 2000. As the gate is scaled down, other device dimensions are also reduced at comparable rates. Figures 3b and c show the reductions of the source/drain junction depth and the gate oxide thickness, respectively.

The NMOS (n-channel MOSFET) has higher speed than PMOS (p-channel MOSFET) because the charge carriers in the n-channel are electrons, which have higher mobility. The CMOS (complementary MOSFET) provides both NMOS and PMOS on the same chip. CMOS circuits consume substantially less power than NMOS circuits because significant current is conducted through the CMOS circuit only during switching. SOI (silicon-on-insulator) is a non-

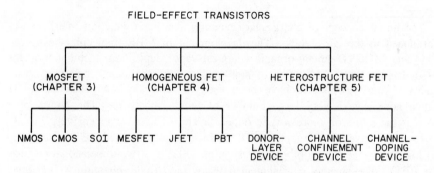

Fig. 2 Family tree of field-effect transistors.

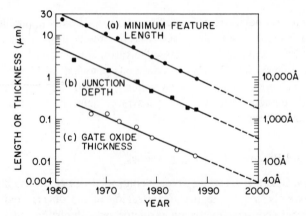

Fig. 3 Scaling of MOSFET dimensions.

epitaxial approach to providing single-crystal silicon. The advantages of making ICs in SOI substrates include low parasitic capacitance for high-speed operation, latch-up free CMOS circuits, and stacked three-dimensional ICs by repeating the SOI process.

The homogeneous FET group includes the JFET (junction FET) and the MESFET (metal-semiconductor FET). These devices were proposed in 1952 and 1966, respectively, and both devices are made on homogeneous semiconductor material such as the n-type GaAs. The JFET is basically a voltage-controlled resistor employs a p-n junction as a "gate" to control the resistance and thus the current flows between two ohmic contacts (i.e., the source and drain contacts). The MESFET is similar to a JFET; however, it uses a metal-semiconductor rectifying contact (Schottky contact) instead of a p-n junction for the gate electrode. Both JFET and MESFET offer many attractive features for high-speed ICs, because they can be made from semiconductors that have high mobilities and high velocities (e.g., GaAs). Another homogeneous FET is the PBT (permeable-base transistor), which has fine metal grids with epitaxial overgrowth of a semiconductor covering the grids. It can be operated at high-current density with high transconductance; it is thus an ultra-high-speed power device.

The initial concept of charge accumulation at a heterojunction interface was introduced in the late 1960s. The development of MBE (molecular-beam epitaxy) and MOCVD (metal-organic chemical vapor deposition) techniques in the 1970s has made the fabrication of high-quality semiconductor heterostructures a reality. These techniques have been used to fabricate heterostructure FETs which now have a very large number of family members. The donor layer devices have one or more n-type doped layers. The most extensively investigated donor layer device is the MODFET (modulation-doped FET) also called HEMT (high electron-mobility transistor) or TEGFET (two-dimensional electron gas FET). The channel confinement devices, such as the quantum-well-channel transistor, are designed specifically to eliminate the transfer of carriers from the channel to the higher bandgap confining layers. The channel-doping devices use methods that distribute doping to allow charge to populate the device. One such device is the MISFET (metal-insulator-semiconductor FET) in which the "insulator layer" is a higher bandgap semiconductor (e.g., $Al_{0.3}Ga_{0.7}As$ with $E_g = 1.8$ eV) while the channel region is in a doped lower bandgap semiconductor (e.g., GaAs with $E_g = 1.4$ eV).

To compare the performances of various field-effect transistors, a key figure of merit is the cutoff frequency f_T, which is related to the carrier transit time across the device active region. The cutoff frequency versus the device length is shown in Fig. 4 for silicon NMOS and compound-semiconductor MESFET and MODFET. The length represents the minimum critical dimension in the device (i.e., the channel length L for NMOS, and the gate length L_G for MESFET and MODFET). Note that for the devices shown in Fig. 4, f_T varies approximately as the reciprocal of the device length.

The smallest channel length in the figure is 0.1 μm. This length is considered to be a lower practical limit for MOSFETs operated at room temperature (see Chapter 3). Silicon NMOS, which has the most advanced processing technology, has an extrapolated f_T of 50 GHz at 0.1 μm. Because of parasitics, fan-out, and interconnect delays, real circuits will operate at frequencies lower than f_T (e.g., a maximum operating frequency of 10 to 20 GHz is expected for a Si NMOS circuit with 0.1-μm channel lengths). For a given length, the MESFETs and MODFETs exhibit a factor of 3 to 8 higher cutoff frequencies than silicon NMOS, mainly due to their higher mobilities and higher velocities. However, compound-semiconductor technology is relatively primitive as compared to silicon technology. We have to improve the compound-semiconductor technology to fabricate such devices with high yield and high reliability.

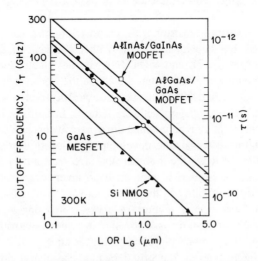

Fig. 4 Cutoff frequency versus channel length (L) or gate length (L_G) of various high-speed field-effect devices.

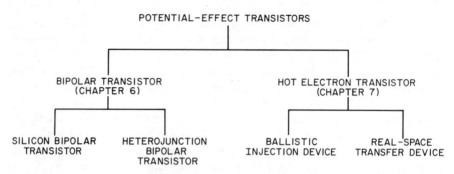

Fig. 5   Family tree of potential-effect transistors.

Figure 5 shows the family tree of potential-effect devices. It includes two groups: the bipolar transistor (Chapter 6) and the hot-electron transistor (Chapter 7). Potential-effect devices are current-controlled. The control electrode is resistively coupled to the active device region, and the charge carriers are separated energetically by an energy barrier.

The invention of the bipolar transistor in 1947 was the beginning of the modern electronics era. This device remains a key component in high-speed computers and many modern communication systems. Although the heterojunction bipolar transistor (HBT) was envisioned in late 1950s to offer substantial improvements in performance over the silicon bipolar transistor, the implementation of HBT was delayed by decades due to the technological difficulties of obtaining a perfect interface between dissimilar semiconductors. At present, the material systems most extensively studied are those involving semiconductors that have identical lattice constants such as in the GaAlAs/GaAs and InGaAs/InP system.

The first hot-electron transistor was proposed in the early 1960s; its full potential, however, has become realizable with the advent of the advanced epitaxial technology (e.g., MBE and MOCVD). By hot electron we mean an electron having an energy more than a few kT above the Fermi level, where k and T are Bolzmann constant and lattice temperature, respectively, and thus the electron is not in thermal equilibrium with the lattice. In ballistic injection devices, the injected hot carriers are maintained in a narrow energy range; this group includes the metal-base transistor and the planar-doped barrier transistor. In real-space transfer devices, the hot carriers are transferred from a narrow-bandgap semiconductor layer to a wide-bandgap layer where they may have different mobility. If the mobility in the wide-bandgap layer is lower, negative differential resistance will occur. Therefore, there is a strong analog to the Gunn effect based on the momentum-space transfer in which electrons are transferred from a low-energy, high-mobility valley to higher-energy, low-mobility satellite valleys. Important devices in the group include the charge injection transistor and the negative-resistance FET. Strictly speaking, the real-space transfer devices are temperature-effect devices; that is, we modulate the effective temperature of the charge carriers, which are then transferred over the barrier.

The cutoff frequency for potential-effect transistors is not uniquely tied to the emitter stripe width as is the channel or gate length in the case of field-effect transistors. For example, for silicon bipolar transistors, f_T is a weak function of the emitter stripe width S and varies approximately as $50\,(0.1/S)^{1/4}$ GHz, where S is in micrometers. For all potential-effect transistors, the base thickness, and the resistances and capacitances of the emitter and collector are also important in determination of f_T. Values of f_T of the order of 200 GHz have been obtained from GaAlAs/GaAs HBTs.

It is important to point out that the MOSFET of the field-effect transistor family and the bipolar transistor of the potential-effect transistor family are not totally unrelated devices; they can be considered as extremes in a continuum of device configurations.[5] To illustrate this continuum, Fig. 6 shows the step-by-step transformation from a MOSFET to a bipolar transistor. When we replace Si by GaAs and the insulator (SiO$_2$) by a wide-bandgap semiconductor (AlGaAs), we obtain the MODFET. If the wide-bandgap semiconductor is replaced by a homogeneous semiconductor identical to the substrate material, we obtain the MESFET (or JFET depending on the gate). We can stack MESFETs together to form a multichannel FET. Turning the device 90° and reducing the gate length gives rise to the vertical-channel FET. When the thickness of the metal grids is reduced to the submicron region, we have a permeable-base transistor. If the metal grids are connected together, to form a thin metal layer, we obtain the metal-base transistor. We can replace the thin metal base by a heavily doped semiconductor layer to form the δ-doped bipolar transistor. By increasing the base layer thickness, we obtain the classic bipolar transistor or the heterojunction bipolar transistor depending on the semiconductor material used for the emitter layer. Of course, the transformation can be done in reverse order.

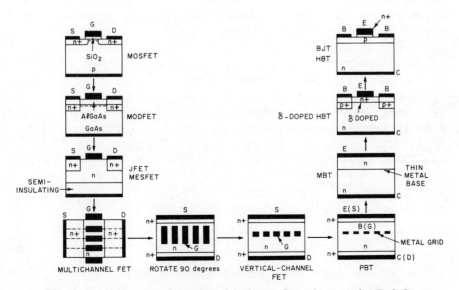

Fig. 6 Continuum transformation of device configurations. (After Ref. 5)

In addition to the transformation of one device configuration to another, we can also combine them to achieve improved circuit performances. For example, the BiCMOS (bipolar-CMOS) may have the advantages of high cutoff frequency and high-current drive in silicon bipolar transistors and the high-density capability and low-power dissipation in silicon CMOS devices. This hybrid device system can also be coupled with the previously mentioned hybrid material system to form monolithically integrated circuits that will enhance system performances and create new system architectures. Since the lower practical device length for MOSFET is about 0.1 μm, the continuum of device configurations discussed above implies that for other field-effect transistors as well as potential-effect transistors, the lower practical device length may also be limited to about 0.1 μm. From Fig. 4, we expect that for those devices shown, the maximum cutoff frequency is about 300 GHz.

Quantum-Effect, Microwave, and Photonic Devices

To increase the cutoff frequency above 300 GHz and to reduce the device dimension below 0.1 μm, many novel device structures have been proposed. These include the molecular electronic devices and the neural devices; however, they generally suffer from low response speeds. The most promising high-speed devices are the quantum-effect devices considered in Chapter 8. A quantum-effect device uses resonant tunneling to provide controlled transport. In such devices, the operational distance is comparable to de Broglie wavelength, which is about 200 Å at room temperature. These small dimensions give rise to a quantum size effect that alters the band structures and density of states and enhances device transport properties.

Figure 7 shows the family tree of quantum-effect devices. The basic device building block is the resonant tunneling diode, which is a double-barrier structure having four heterojunctions and one quantum well. By increasing the num-

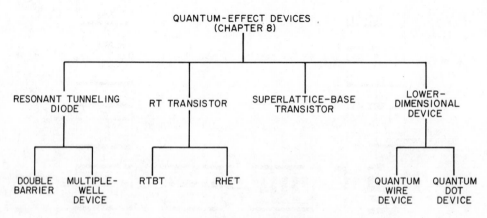

Fig. 7 Family tree of quantum-effect devices.

ber of barriers in series, we have multiple-well devices. Many novel current-voltage characteristics can be obtained by inserting a resonant-tunneling structure into a potential-effect or field-effect device, such as the RTBT (resonant-tunneling bipolar transistor) and RHET (resonant-tunneling hot-electron transistor). These devices can be used as functional devices; that is, they can perform relatively complex circuit functions at high speed with reduced component counts. It is expected that by moving toward lower dimensional devices (e.g., quantum-wire for one-dimensional structure or quantum dot for zero-dimensional structure) very high component density ($>10^{10}$ components/cm^2) and very high speed performance ($f_T > 1000$ GHz) are possible.

The microwave and high-speed photonic diodes are covered in Chapters 9 and 10, respectively (Fig. 8). The microwave diodes are two-terminal devices that can generate, amplify, or detect signals at microwave frequencies, especially in the millimeter wave region (30 to 300 GHz). The advantages of millimeter waves over lower microwave and infrared systems include small size, light weight, and operation in adverse weather (e.g., transmitting through fog and dust). These devices are based on physical principles considered in Chapters 2 to 8, that is, quantum tunneling for tunnel devices, potential effect for transit-time devices, and field effect for transferred-electron devices.

High-speed photonic devices are those that can convert electrical energy into optical radiation (light-emitting diode and semiconductor laser) and those that can detect optical signals through electronic processes (photodetector). These devices must be able to be modulated at high frequencies; they are critically important for many electronic systems, especially in optical-fiber communication systems.

It is conceivable that by using the hybrid material/device technology, we can integrate the field-effect, potential-effect, quantum-effect, microwave, as well as photonic devices monolithically to form super integrated circuits to meet the future demand for sophisticated electronic systems.

A remark on notation: To keep the notation simple, it is necessary to use the simple symbols more than once, with different meanings for different devices.

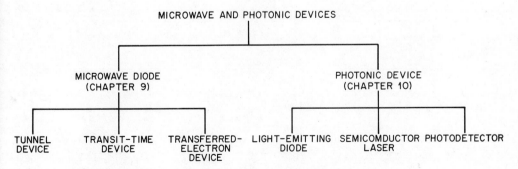

Fig. 8 Family tree of microwave and photonic devices.

However, within each chapter, each symbol is used with only one meaning and is defined the first time it appears. Many symbols do have the same meanings consistently throughout this book; they are summarized in Appendix A for convenient reference. Also listed in the appendixes are the International System of Units (Appendix B) and the Physical Constants (Appendix C).

The research and development of high-speed semiconductor devices are moving at a rapid pace. Many of the devices, such as the MODFET, the HBT, and the quantum-effect devices, are under intensive investigation. Their ultimate performance is not fully understood at the present time. The material presented in this book is intended to serve as a foundation. The references listed at the end of each chapter can supply more information.

REFERENCES

1. S. M. Sze, *Physics of Semiconductor Devices,* 2nd ed., Wiley, New York, 1981.

2. C. W. Pearce, "Crystal Growth and Wafer Preparation," in *VLSI Technology,* 2nd ed., S. M. Sze, Ed., McGraw-Hill, New York, 1988.

3. J. D. Meindl, "Ultra-Large Scale Integration," *IEEE Trans. Electron Dev.,* **ED-31,** 1555 (1984).

4. S. Luryi and S. M. Sze, "Possible Device Applications of Silicon Molecular Beam Epitaxy," in *Silicon Molecular Beam Epitaxy,* E. Kasper and J. C. Bean, Eds., CRC Press, Boca Raton, FL, 1988.

5. E. B. Stoneham, "The Search for the Fastest Three-Terminal Device," *Microwaves,* February 1982, p. 55.

I MATERIALS, TECHNOLOGIES, AND DEVICE BUILDING BLOCKS

1 Materials and Technologies

J. C. Bean
AT&T Bell Laboratories
Murray Hill, New Jersey

1.1 INTRODUCTION

With the tremendous strides being made in superconductor, polymer, and amorphous material research, it is somewhat surprising that semiconductors continue as the basis for virtually all commercial high-speed electronic devices. The semiconductor's tenacity can be attributed to several factors. Most important, semiconductors can be produced as densely packed, highly perfect crystals. Crystals are, by definition, intensely uniform and this uniformity provides a natural basis for the fabrication of large numbers of closely matched devices. Dense crystals have the additional advantage that they offer few attractive sites for foreign atoms. A solidifying crystal thus excludes many impurities, and repeated solidification (zone refining) can produce semiconductors with impurity concentrations of less than one tenth part per billion. This compares with polycrystalline metal or insulator purities of parts per thousand and non-crystalline polymer purities of parts per hundred.

The structural perfection and purity of a semiconductor crystal are manifested in a number of important ways. Mechanical properties are often enhanced, and indeed, certain semiconductors have a hardness and yield resistance greater than steels and refractory metals. This strength facilitates reduction in component size and enhances production yields. Crystalline order can also increase thermal conductivity, allowing the dissipation of waste energy and the dense packing of devices. Finally, perfection and purity reveal the key electrical property of a semiconductor: it is an insulator on the verge of becoming a conductor. This delicate balance means that the deliberate, localized introduction of electrically active impurities will have a large effect on electrical properties, giving the semiconductor its tremendous versatility.

High-Speed Semiconductor Devices, Edited by S.M. Sze. ISBN 0-471-62307-5
© 1990 John Wiley & Sons, Inc.

With this portfolio of virtues, can we be assured that semiconductors will remain the ultimate high-speed electronic material? Of course not. History is replete with examples of unforeseen obsolescence and, indeed, semiconductor supremacists may have created just the tools for unseating the semiconductor. Techniques such as ion implantation and rapid thermal processing give us the ability to synthesize materials that occur nowhere in nature. Other techniques, such as molecular beam epitaxy, allow us to tailor layer structure down to the atomic level. As these tools are applied to non-semiconductor materials, we can safely assume that novel properties will be discovered and that the semiconductor will eventually encounter serious competition. For the moment, however, discussion of high-speed materials can be confined to semiconductors and materials related to semiconductors by either chemistry or structure.

1.2 THE IDEAL SEMICONDUCTOR: A WISH LIST

Semiconductors can be presented in a tabular form based on their position in the periodic table (i.e., column IV, III-V, or II-VI) and the structural, thermal, electronic, and optical properties enumerated for each material. For the purposes of this book, it makes more sense to turn the problem around, to take the device designer's perspective and attempt to define the "ideal semiconductor." The emphasis is then on those generic properties that contribute to high-speed device performance, rather than on minute and perhaps irrelevant differences in physical parameters. Moreover, this engineering perspective reflects the fact that we are no longer strictly bound by naturally occurring material properties. Devices need not be tailored to materials; rather, to an increasing degree, materials can be tailored to the device application.

1.2.1 Carrier Transport

The ideal high-speed semiconductor should have at least one type of charge carrier that responds strongly and rapidly to changes in the applied electric field. In a constant field, $\mathcal{E}$, a charged Newtonian particle, would accelerate at a rate $q\mathcal{E}/m$ until it suffered a collision that randomized its velocity vector. The average velocity would then be zero and the process would repeat itself. Averaged over many cycles, the mean velocity would then be $q\tau\mathcal{E}/2m$, where τ is the average time between collisions (or "relaxation time"). The ratio of velocity to field, $q\tau/2m$ is then termed the mobility. In a more realistic quantum-mechanical picture, carrier motion must be described on the basis of Bloch functions sharing the lattice periodicity.[1] One nevertheless arrives at the same relationship if an effective mass, m^*, is substituted for the free-electron mass. The effective mass is inversely proportional to the k-space curvature of relevant bandedge:

$$\mu = \frac{\text{velocity}}{\mathcal{E}} = \frac{q\tau}{2m^*} = \left[\frac{q\tau}{2\hbar^2}\right]\frac{d^2E}{dk^2}, \tag{1}$$

where $\hbar$ is Plank's constant over 2π, and the second derivative of the band-edge energy is evaluated in the k-space direction of carrier motion. In general, the directionality of this derivative is unimportant, and we can treat the effective mass as a simple scalar quantity. This is possible because either the band is symmetric or there are degenerate band edges symmetrically arrayed in k-space, and the derivative over the sum is dominated by a similar term. (Note, however, that this symmetry may be broken in the strained- or two-dimensional layered devices discussed later in this book.) Based on this expression, we conclude that the band structure of the ideal semiconductor should have very sharply curved valleys and peaks.

This model explains mobility but does not provide a rule for selecting one semiconductor over another (i.e., how does one predict sharply peaked band extrema?). As the underlying physics is based on scattering, we might expect that mobility would be a complex and seemingly random function of semiconductor purity, crystalline structure, and atomic size. Figures 1 and 2 plot measured electron and hole mobilities as a function of minimum bandgap for the well-characterized semiconductors. It is strikingly apparent that the variation is not random and that, especially for electrons, mobility varies inversely with minimum bandgap. The answer can be found if, instead of considering a single bandedge, we calculate the effect of band interaction. This can be done using

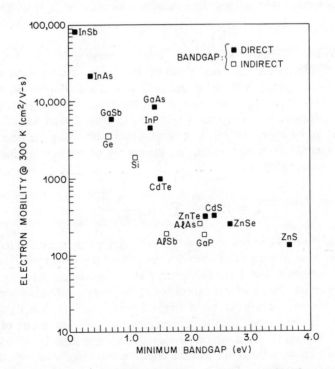

Fig. 1 Low-field electron mobilities of the common semiconductors (at room temperature).

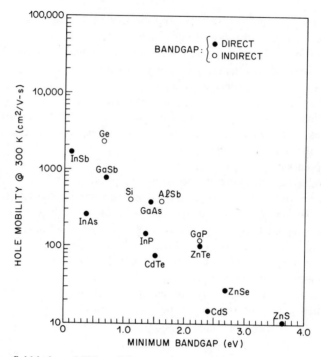

Fig. 2 Low-field hole mobilities of the common semiconductors (at room temperature).

the so-called **k·p** technique in which the effective mass is derived on the basis of a simple perturbation of the band structure at $k = 0$ (or other local extrema).[2] Both the wave function and energy are expanded in a Taylor series, substituted into the Schrödinger equation, and the higher-order terms discarded. After some manipulation, it can be shown that at these critical band points, the effective mass can be approximated by a sum of terms involving the energy difference between the band in question and the energy of neighboring bands evaluated at the same value of k:

$$\frac{m}{m_n^*} = 1 + \frac{2}{m} \sum_{j \neq n} \frac{|p_{n,j}|^2}{E_n(k_0) - E_j(k_0)]}, \tag{2}$$

where m is the free-electron mass, m^* the effective mass, n a fixed index referring to the band of interest, j an index identifying the neighboring bands, p a momentum operator, and E the band energy at the momentum k_0.

The important feature of this expression is the energy difference appearing in the denominators. If a band has a single nearest neighbor, the complex sum may be dominated by one such term, involving only the interaction of this pair of bands. Moreover, for a typical semiconductor, this term is generally much larger than one, with the result that the effective mass increases with the difference in band energies. This broadening of conduction band minima, with increasing bandgap, is clearly evident in the band structures of Fig. 3. Expressed in terms of mobility, Eq. 2 yields the inverse scaling observed in the data of Figs. 1 and 2. (Note: because $E_g \neq \Delta E$ for indirect materials, a rigorous comparison with Eq. 2 demands that we replot these points.)

This perturbation model serves as the basis for a generalization that the ideal high-speed semiconductor should have a small, direct energy gap. As in most engineering problems a competing consideration prevents us from going to extremes. In this case, the problem is thermally generated carriers. If the bandgap becomes too small, electron-hole pairs may form spontaneously, creating a leakage current that is manifested as a "dark current" in an optical detector or premature breakdown in a high-field device.

If scattering rates were fixed, mobility would remain constant and carrier velocity would be limited only by the magnitude of the applied field. In fact, as carrier energies increase, new scattering mechanisms come into play, and the carrier relaxation times decrease. At medium fields (e.g., 10^4 V/cm), carriers may acquire sufficient energy to excite acoustic phonons within the semiconductor lattice. Such scattering is seen as a roll-off in the velocity-field characteristic (or a decease in the ac mobility). This effect is particularly pronounced in indirect-bandgap materials where the multiplicity of conduction-band minima provides a large variety of scattering paths. As the electric field increases to above 10^5 V/cm, carriers have enough energy to excite optical phonons. For further increases in field, energy is immediately transferred from the carriers to lattice heating, and velocities saturate, generally at a value of about 10^7 cm/s.

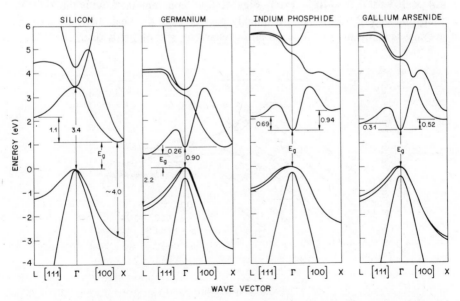

Fig. 3 Energy-band structure of Si, Ge, InP, and GaAs derived from pseudopotential calculations. (After Ref. 3)

Between the initial roll-off and final velocity saturation, certain semiconductors display an anomalous peak in drift velocity, as shown in Fig. 4. To understand this, one must delve somewhat deeper into semiconductor band structure. Figure 3 shows that the conduction bands of many semiconductors, such as GaAs, have a number of secondary minima at energies only slightly above the principal valley. At either very low fields or very high fields, the carriers will behave as described above. At intermediate energies, however, electrons may acquire just enough energy to be transferred into one of the secondary minima. These minima have much broader valleys corresponding to heavier effective masses and lower mobilities. This transfer process can thus produce the downturn in mobility observed in the GaAs data of Fig. 4. Although this result is sometimes termed velocity overshoot, we will reserve that term for the transient effect, described below. Instead, this steady-state, band structure phenomenon will be termed the transferred-electron effect. As discussed in Chapter 9, transferred-electron devices exploit the breakup of carriers into fast and slow moving domains to produce negative resistance and microwave oscillation.

It has also been suggested that we could employ this velocity peaking in a bipolar or field-effect transistor. This would be sufficient ground for including such an effect in our list of ideal semiconductor attributes. However, more careful consideration shows that it may not be possible to exploit this phenomenon. Those materials displaying the maximum transferred-electron effects have bandgaps of about 1.5 eV. A typical junction will thus drop about 1.4 V in a distance ranging from 100 to 10,000 Å (for doping of 1×10^{19} and $1 \times 10^{15}/cm^3$, respectively). The built-in fields will thus fall in the range from 1.4×10^4 to 1.4×10^6. From Fig. 4, it is apparent that, with the possible exception of InP, electrons will quickly pass through the peak and may have an average velocity of 6 to $8 \times 10^6 cm/s$ (characteristic of fields over 10^5 V/cm).

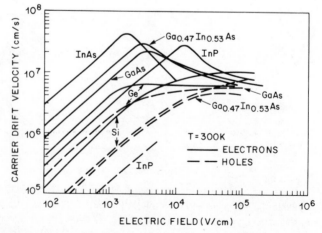

Fig. 4 Compilation of velocity-field characteristics for carriers in the most heavily utilized semiconductors. (After Ref. 4)

At typical device operating voltage of 5 V, a more detailed calculation shows that the advantage of even indium phosphide fades and the indirect material, silicon, may actually come out slightly ahead with its high-field velocity of $1 \times 10^7 \text{cm/s}$.

A fundamentally different, and transient, velocity overshoot effect has been observed in certain submicron research devices produced in the late 1970s.[5,6] In the Newtonian model, if a device is small enough, a carrier may pass through a high-field region so quickly that few, if any, scattering events occur, and motion is termed ballistic. The velocity is no longer the average of a series of velocity ramps and scattering events but in certain situations may be closer to a single velocity ramp. In the absence of multiple scattering, we can no longer assume that the carrier and lattice are in thermal equilibrium. Given that scattering rates increase with lattice temperature, strange things can begin to occur.

A complete explanation of this phenomenon requires detailed quantum mechanics and is extensively simulated by Monte Carlo computer modeling of carrier trajectories. However, clear descriptions are still not readily available. In the end, the degree of velocity overshoot depends on how long the carrier can avoid scattering, which is, in turn, a function of the variety of possible scattering mechanisms. Again, the multiplicity of conduction-band minima put the indirect-bandgap semiconductor at a disadvantage by providing many scattering paths. The net result is that while overshoot is predicted to occur in many

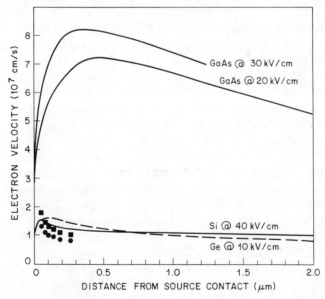

Fig. 5 Solid and dashed lines: calculated electron velocity in submicron field-effect transistors as a function of carrier position within the channel. Plots are shown for those fields producing the strongest overshoot. (After Ref. 7) Data points: average channel mobilities inferred from transconductance data for short channel NMOS silicon transistors at 300 K (squares) and 77 K (circles). (After Ref. 8)

materials, ballistic behavior can be maintained over distances of a micron in a direct-bandgap III-V semiconductor, but only 1000 Å in an indirect-bandgap column IV semiconductor. Moreover, as indicated in Fig. 5, the magnitude of the overshoot is very much larger in the direct-bandgap materials. Thus, on grounds of transient velocity overshoot (but not steady-state transferred-electron effects), we are once again led to favor a direct-bandgap material in our search for the ideal semiconductor.

1.2.2 Heterostructures

If carrier speed were the only consideration, we could stop now and select the material with the smallest direct bandgap that is consistent with a tolerable level of thermal carrier generation. We must, however, create a device in real space where carriers are confined to move in well-defined regions. In early semiconductor devices, localization was accomplished strictly by use of the *p-n* junctions created by electrically active impurities. More recently, this has been supplemented by the use of band structure to control carrier position.[9,10] Not only are additional degrees of freedom provided but the electrical and physical design of the device are often decoupled in a strongly advantageous manner.

If two semiconductors of differing bandgap are brought together, their band edges will have one of the alignments shown in Fig. 6. In case a, both electrons and holes near the junction tend to migrate to the smaller bandgap semiconductor. In case b, electrons and holes tend to separate, each going into a different material. In case c, one carrier is unaffected while the other favors one member of the pair. While these alignments could be easily described in words, specialists refer to these alignments as Type I, Type II, and Type III, respectively. Depending upon the type of the alignment and dopant, the migration of carriers across the heterojunction may produce thin intrinsic regions or charge accumulations, with corresponding spikes or dips in the band edge. The width of the disturbances increases at lower doping levels. In lightly doped structures, where tunneling is not likely, the spikes may behave as barriers to carrier transport and the dips act as two-dimensional charge wells. In terms of our ideal semiconductor, we cannot specify one single alignment as being preferred. In a laser, we want to accumulate both carriers in the same region to enhance recombination. A Type I alignment is, therefore, favored. In a heterojunction bipolar transistor, we want to inhibit the flow of one carrier with a large barrier while leaving the other carrier unaffected. A Type III alignment is thus favored, and the list goes on, device by device.

We can, however, invoke heterostructures to make one generalization about our ideal semiconductor: it cannot stand alone. To capitalize on the power of heterostructures, we must combine at least two semiconductors in such a way that an idealized interfacial structure is maintained. To do this, atomic bonding must be continued, without interruption, across the interface. Therefore, the two materials must either have the same atomic spacing or be able to deform to adopt a common spacing. These two situations are described as lattice-matched

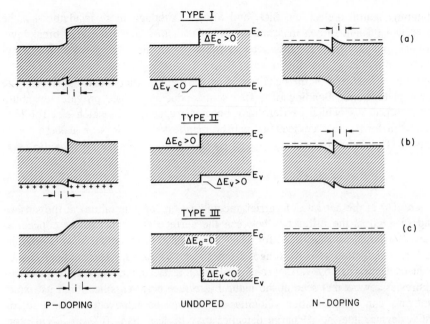

Fig. 6 Possible heterostructure band alignments. Type I: smaller bandgap is nested within the larger. Type II: staggered band alignment. Type III: continuous conduction (or valence) band edge with all of bandgap difference appearing at other band edge. Note that, as a function of doping, the band edge discontinuities may produce barriers and localized depletion of one carrier at the heterointerface.

epitaxy and strained-layer epitaxy. We can improve upon this situation by two other approaches. If the materials are closely related, such as GaAs and AlAs, heterojunctions need not be confined to the two pure materials but can contain alloys such as $Al_xGa_{1-x}As$. These alloys not only provide a continuous range of band-edge discontinuities at the interface but can also involve some change in the type of the alignment. The second approach involves strained layers where the strain itself affects the bandgap and alignment. This strain can often be tuned at the designer's request. Both of these topics are detailed later in the chapter.

1.2.3 Dielectrics and Surfaces

We have identified the ideal semiconductor as a crystalline, direct, narrow-bandgap material, lattice matched to at least one other semiconductor. In other words, virtually any semiconductor except silicon. Given silicon's dominant position in electronics, there are clearly other factors that must be considered. First, there is the issue of dielectrics. Not only are dielectrics the basis of metal-oxide-semiconductor (MOS) devices but they are crucial in virtually any integrated device. Silicon is unique in that it possesses two of the most outstanding natural dielectrics, SiO_2 and Si_3N_4. Although these insulators can be deposited on non-silicon materials, it is difficult to achieve the high-breakdown voltage and electrical stability seen in films grown directly by reaction of oxygen or ammonia with a silicon wafer.

Equally important is the behavior of carriers at a semiconductor surface. The incomplete atomic bonding of a free semiconductor surface provides an abundance of sites at which carriers may become trapped or at which electron-hole pairs can form. In a device, this can be manifested as a loss or leakage mechanism. This behavior is characterized by a surface recombination velocity:

$$s = \frac{J}{q(n-n_0)},\tag{3}$$

where J/q is the net rate of carrier recombination (or generation) at the surface and $(n-n_0)$ is the difference between the surface and bulk carrier concentrations (for holes, substitute p's for n's). In a typical semiconductor, this undesirable recombination velocity ranges from 10^3 to 10^6 cm/s. This number can be reduced somewhat by various chemical treatments, but no treatment produces figures as low as that seen at the natural interface between silicon and its oxide. Not only can recombination velocities of 10 cm/s be achieved, but in a silicon MOS device, interfacial charge densities may be less than 10^{10}/cm^2. This corresponds to less than one charge site in 10^5 interfacial atoms.

To these electrical benefits, we must add the fact that a dense, grown, chemically resistant oxide, such as SiO_2, is a natural masking material for lithographic processing. This factor, however, is becoming less important. As device sizes become smaller and more exotic materials are used, it is less and less desirable to repeatedly subject a silicon wafer to the 1000°C (or higher) temperatures necessary for thick thermal oxidation. Oxidation is increasingly reserved for the growth of the MOS gate oxide or for the reduction of surface recombination velocity (i.e., passivation). For other, less critical steps, oxides or nitrides are deposited by gas-phase reactions rather than by thermal oxidation of the surface. Thus, at least in terms of lithography, this trend will tend to put silicon on a more equal footing with other semiconductors for which processing, based on deposited dielectrics, has long been the rule.

1.2.4   Mechanics and Economics

To complete our picture of the ideal semiconductor, we must move beyond electronics into the realm of mechanics and economics. These topics are related. As a semiconductor is converted from a wafer to a complex electronic device, it passes through hundreds of processing steps. If the device is to be affordable, these steps must be completed rapidly, largely by automated means. A loss due to breakage of one device or wafer in a thousand at each step may appear trivial, but if breakage occurs at each of several hundred steps, almost one-half of the starting material will be lost. To this material wastage is added the huge overhead of an integrated circuit line that may have an annual operating cost in the millions of dollars and a setup cost of $20 to over $100 million that must be amortized over a useful lifetime of as little as 5 years.

Beyond simple breakage, there is the fact that an affordable device will depend on the repeated overlay of lithographic patterns over wafers 75 to 200 mm in diameter. For these overlying patterns to line up with a micron or sub-micron linewidth tolerance, not only is wafer stability critical, but for optical lithography, wafers must be manufactured to flatnesses of order 1 μm over at least 1 cm^2.

All of these factors suggest the use of an extremely hard, unyielding semiconductor material. These parameters are a reflection of the semiconductor's atomic bonding strength and correlate roughly with the semiconductor's melting point. Melting points are tabulated in Table 1 for a number of semiconductors along with two common measures of hardness and mechanical strength. Data for stainless steel and certain refractory metals are included for comparison.

Both measures of mechanical strength strongly favor the covalent bonding of column IV materials over the polar bonding of III-V and II-VI semiconductors.[11] The strength of the covalent bond is also reflected in the high thermal conductivities of these materials. In high scales of device integration, power dissipation can be the factor-limiting device-packing densities, and these conductivities also put the column IV materials at an advantage.

TABLE 1 Mechanical Properties of Common Semiconductors and Metals

	Melting Point (° C)	Knoop Hardness (kg/mm^2)	Yield Strength (10^{10} dynes/cm^2)	Thermal Conductivity (W/cm-°C)	Thermal Expansion ($10^{-6}\delta L/L$-C°)
Diamond	>3550	7000	53	20	1.0
Tungsten	3410	485	4.0	1.78	4.5
Silicon carbide	2700	2480	21	3.5	3.3
Molybdenum	2610	275	2.1	1.38	5.0
Aluminum arsenide	1740	481		0.8	5.20
Zinc selenide	1520	150		0.13	7
Gallium phosphide	1467	945		0.97	5.8
Silicon	1415	1150	7.0	1.57	2.56
Gallium arsenide	1238	750		0.54	6.8
Stainless steel	1000–1400	600	2.1	0.32	12
Cadmium telluride	1098	100		0.07	5.5
Indium phosphide	1070	535		0.68	4.5
Indium arsenide	943	381		0.26	5.19
Aluminum	660	130	0.17	2.36	25
Indium antimonide	525	223		0.18	5.04

Although it is generally far less important, the cost of the semiconductor wafer is also reflected in the cost of a device. Wafer costs are a complex function of material, purification, and crystallization costs, but we would expect costs to scale inversely with the atomic abundance of the raw materials in the earth's crust. For the constituents of the common semiconductors, these abundances are listed in Table 2 in descending order (stated as a weight fraction).[12] Silicon is strongly favored, followed by zinc sulfide followed by a loose group of germanium, GaAs, and AlAs, with many of the II-VIs far behind.

Our characterization of the ideal semiconductor thus leads to the following conclusions. If ultimate high-speed performance is paramount and one is little concerned with either cost or level of integration, the basic physics suggests the use of a narrow direct-bandgap semiconductor from columns III-V or II-VI of the periodic table. This rule is tempered, however, if surface effects or dielectrics are important to device performance. In this situation, the attraction of the III-Vs and II-VIs fades somewhat in favor of the column IV materials. Finally, in situations where cost or levels of integration are critical, mechanics and natural abundance strongly favor the use of silicon as a base material.

1.3 THE MAJOR SEMICONDUCTOR FAMILIES

Silicon is far and away the major player in the current practice of high-speed electronics. Its dominance is not so much the product of its own electronic properties (which are comparatively mediocre) but rather the result of its natural abundance, its outstanding mechanical properties, and its superb natural dielectrics. However, for the bulk of new applications, a single stand-alone semiconductor is not adequate. We must have a family of semiconductors to exploit the full potential of heterostructures or, as it has come to be called, "bandgap engineering."

Figure 7 presents the semiconductor family tree. The horizontal axes plot the crystalline lattice constant in either angstroms or as a percentage mismatch to silicon's lattice constant of 5.43 Å. The vertical axes plot the minimum bandgap in both electron volts and as an effective wavelength where

TABLE 2 Abundance of the Semiconductor Constituents in the Earth's Crust

Element	Abundance	Element	Abundance	Element	Abundance
Silicon	.283	Zinc	7×10^{-5}	Cadmium	2×10^{-7}
Aluminum	.083	Gallium	1.5×10^{-5}	Indium	1×10^{-7}
Phosphorus	.001	Germanium	5×10^{-6}	Mercury	8×10^{-8}
Sulfur	2.6×10^{-4}	Arsenic	1.8×10^{-6}	Selenium	5×10^{-8}
Carbon	2×10^{-4}	Antimony	2×10^{-7}	Tellurium	1×10^{-9}

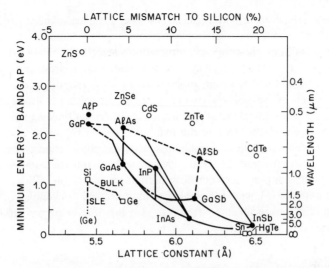

Fig. 7 Compilation of minimum bandgap versus lattice constant data for the column IV, III-V, and II-VI semiconductors. Right axis indicates the wavelengths of light that would be emitted by a laser or LED for a material of the corresponding bandgap. For the column IV and III-V materials, connecting lines give information for alloys of the materials at the endpoints of a given line segment. Solid lines indicate a direct bandgap and dashed lines an indirect bandgap. For Ge-Si, the line denoted BULK corresponds to unstrained, lattice-mismatched growth and the line SLE to strained layer epitaxy of GeSi on unstrained Si. (After Ref. 13)

λ (in μm) $= 1.24/E_g$ (in eV). Below this wavelength a semiconductor will begin to absorb light and a direct-bandgap material to emit light. If different III-V materials can be combined as an alloy, the points are connected by lines indicating the bandgap of the alloy with a solid line denoting a direct bandgap and a dashed line an indirect bandgap. However, these alloy lines should be used with a bit of caution: they indicate the bandgap an alloy should have but do not indicate whether that alloy can be synthesized. Many compositions may be thermodynamically unstable and will spontaneously form precipitates or regions of differing composition. These thermodynamically unstable regions are known as miscibility gaps and are identified in the thermodynamic literature.[14] Note that, in the literature, an alloy such as GaAs-AlAs is generally denoted by $Al_xGa_{1-x}As$, where the column III (or II) specie(s) are on the left in alphabetical order, the column V (or VI) specie(s) on the right, and x or y denote the composition fraction of the chemically interchangeable atoms.

The semiconductor families are then defined by those materials that share the same lattice constant (i.e., on a vertical line) or by materials that may be combined, as alloys, to lattice match a third material. For such alloys, the lattice constant can be approximated by a linear interpolation between the values of the constituents (e.g., an alloy of 47% gallium arsenide and 53% indium arsenide will lattice match indium phosphide). The one added qualification is that families are currently confined to materials of the same columns of the periodic table. The problem is that, although materials such as column IV silicon and column III-V gallium phosphide are crystallographically compatible, as impurities they dope one another. During any high-temperature processing, especially crystal growth, there is a strong probability of such interdiffusion. This may not be a problem if the active area of the device is well away from the crystallographic junction. However, in that case, the device will not exploit the potential of the heterojunction. One of the materials is used simply as a substrate, or if devices are fabricated in both materials, it is simply a scheme (generally not cost effective) to combine disparate devices without the need to wire together two chips. Cross-doping may not be a permanent limitation. Processing is, in general, moving to lower temperatures and we may eventually overcome thermal-interdiffusion problems. However, for the time being we can define the following major semiconductor families as follows.

1.3.1   Aluminum Gallium Arsenide ($Al_xGa_{1-x}As$)

The aluminum arsenide–gallium arsenide system is one of only two instances where related semiconductors share virtually the same lattice constant. In addition, both materials have high melting points and good mechanical properties: they share one constituent, simplifying processing, and over much of the composition range, their bandgap is direct and produces light emission at wavelengths visible to the human eye. Further, unlike many compound semiconductors, all compositions from AlAs to GaAs can be synthesized without the formation of secondary phases or precipitates (i.e., there is no miscibility gap). For all of these reasons, ternary $Al_xGa_{1-x}As$ was the first and remains the foremost, practical semiconductor family.[15,16] Along with these more obvious strengths, $Al_xGa_{1-x}As$ has a number of more subtle advantages. Over much of the composition range, it has a Type I heterostructure band alignment that concentrates carriers in a GaAs layer clad on both sides by AlAs (or $Al_xGa_{1-x}As$). This confinement is the key to achieving the so-called carrier inversion necessary for the operation of a solid-state laser (and the first such lasers were fabricated in this system in the late 1960s). The direct bandgap of this system also falls in a range that produces high electron mobilities while being large enough that thermally generated current (dark noise) is not a significant problem.

On the down side, aluminum has an extremely high affinity for atmospheric oxygen. Oxygen, especially at a heterostructure interface, can degrade electrical properties as manifested in reduced mobilities and carrier trapping. This oxygen affinity may also account for the fact that $Al_xGa_{1-x}As$ has a very high surface recombination velocity of order $10^6 cm/s$. Combined with the lack of a good passivating dielectric, this means that surface effects can be especially damaging, particularly as devices are scaled to submicron dimensions and total surface areas become comparable to active device junction areas. In this situation, the intended "intrinsic" device performance may be overwhelmed by the surface-dominated "extrinsic" device.

While the presence of highly toxic arsenic might make these materials less attractive, the comparative disadvantage is offset by the fact that compounds of non-toxic silicon, such as silane, can be spontaneously flammable or explosive. As such, the safe handling of virtually any common semiconductor requires expensive and carefully designed facilities.

1.3.2 Gallium Indium Arsenide Phosphide on Indium Phosphide ($Ga_xIn_{1-x}As_yP_{1-y}$/InP)

In heterostructure electronics, the supremacy of $Al_xGa_{1-x}As$ is being challenged by growth of $Ga_xIn_{1-x}As_yP_{1-y}$, lattice matched to InP.[17]. This shift can be ascribed to three major factors. For one, this system has proven to be electrically more benign, probably because of the absence of highly reactive aluminum. It appears that material of high quality can be achieved with less stringent processing and, perhaps, with less concern about oxygen impurities in the starting materials. Once produced, the material has at least one vastly improved electrical characteristic: a surface recombination velocity of order 10^3cm/s. While not to the 10 cm/s standard of thermally oxidized silicon, this represents a vast improvement over the $Al_xGa_{1-x}As$ value of 10^6. Improved surfaces have already been shown in scaling experiments demonstrating sustained high performance as indium-based devices were reduced to sizes well below 1 μm.

The second advantage of InP materials is that there is a greater difference between the secondary conduction-band minima and the primary minimum than in the GaAs materials (0.94 and 0.69 eV versus 0.31 and 0.52 eV, respectively, as shown in Fig. 3). This increased separation means that higher fields can be tolerated without the onset of mobility-degrading, transferred-electron effects. In Fig. 4, this is evident in the rightward position of InP's electron mobility peak (at 1.2×10^4V/cm). If an InP device can be designed and operated at voltages low enough that such fields are not exceeded, there is the potential of achieving a two- to threefold enhancement in electron velocity.

The final advantage to the quaternary $Ga_xIn_{1-x}As_yP_{1-y}$/InP system is in the area of optoelectronics. While lasers can be fabricated in this and the $Al_xGa_{1-x}As$ system, the indium-based lasers emit in the 1.3 to 1.5 μm range at which quartz fibers have a minimum in loss. The compatibility with fibers suggests that this family may be the natural system in which to pursue the integration of electronic and optical processing for communication purposes.

The indium materials have several shortcomings. The need to deal with high phosphorus overpressures and the current small scales of production make the substrate crystals significantly more expensive than even GaAs. Also, as indicated in Table 1, the material is somewhat weaker and may suffer higher breakage rates during processing. Although breakage can be reduced by increasing wafer thickness, this only exacerbates the cost issue. Finally, InP will evaporate as a molecule (i.e., congruently) at temperatures of only 360^oC, sharply limiting the range of acceptable device-processing temperatures.

1.3.3   Gallium Indium Arsenide Antimonide ($Ga_xIn_{1-x}As_ySb_{1-y}$)

Indium antimonide can be combined with gallium arsenide to produce an alloy system spanning lattice constants from 5.65 to 6.47 Å, with direct bandgaps from 1.43 to 0.18 eV. Depending on the alloy composition, InP, InAs, GaSb, or AlSb can serve as a suitable substrate. Figure 1 shows that the highest mobilities are achieved at the larger lattice constant end of this system where the small 0.70-, 0.35-, and 0.18-eV bandgaps of GaSb, InAs, and InSb yield extraordinary room temperature mobilities of 6000, 22,000, and 80,000 cm^2/V-s, respectively. These mobilities have attracted some research attention, but in general, the system suffers from the relative immaturity of substrate-manufacturing technologies and the shortcoming of small bandgaps. At room temperature, these bandgaps are only several kT wide. Thus, there is a strong probability of spontaneous thermal generation of electron-hole pairs across the bandgap, resulting in exceedingly high junction leakage currents. Although such leakage puts these materials out of the current mainstream of high-speed device work, the situation could change if predicted trends toward refrigerated device operation are borne out.

1.3.4   Cadmium Mercury Telluride ($Cd_xHg_{1-x}Te$)

The final heavily investigated heterostructure family consists of $Cd_xHg_{1-x}Te$ on CdTe (or other) substrates.[18] This system is studied because, at the mercury-rich end, the bandgap diminishes to zero and the material becomes metallic. This property can be of interest in basic physics investigations and has one principal commercial application: infrared imaging at multi-micron wavelengths. However, outside of this area, the attractions of $Cd_xHg_{1-x}Te$ are offset by the need for refrigeration to suppress thermal current, the strongly inferior mechanical properties, and the difficulty in producing high-quality CdTe substrates. It is unlikely that this materials system will become a major player in the area of high-speed electronics

1.3.5   Germanium Silicide on Silicon and Strained-Layer Epitaxy

In the early 1980s, a new and unorthodox player emerged on the heterostructure scene: Ge_xSi_{1-x} on Si.[19] The lattice constant of germanium is over 4% larger than that of silicon (Fig. 7). That may not sound like much, but at the atomic level, it suggests that at least 1 in 25 interfacial atoms would be improperly bonded, providing an abundance of potential trapping sites. The technique for overcoming these problems, strained-layer epitaxy, requires a somewhat lengthy discussion. This is justified, however, by the fact that Ge_xSi_{1-x} is becoming a prototype for the application of strained-layer epitaxy throughout the semiconductor families.

Figure 8 gives a simplified view of the alternatives presented when two lattice-mismatched semiconductors are combined. If both lattices are rigid, they will retain their fundamental lattice spacings, and the interface will contain rows

STRAINED LAYER EPITAXY

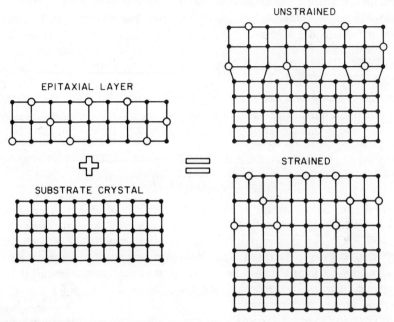

Fig. 8 Schematic representation of possible means of growing lattice-mismatched materials. In the unstrained case, both materials retain their bulk structure resulting in misbonded atoms at the interface. In the strained case, the thinner, or more elastic, layer compresses (dilates) in the plane of growth to match the substrate. Elastic forces then compel it to dilate (compress) in a direction perpendicular to the interface. (After Ref. 20)

of misbonded atoms described as misfit or edge dislocations. Alternatively, if the epitaxial layer is flexible (and much thinner than the substrate), it may distort itself in the plane of growth to conform to the substrate spacing. One can then compare the energetics of dislocations and volume strain to decide which will be the favored structure.

This model is fine for a "Gedanken" experiment, but it bears only a distant relation to the realities of epitaxial crystal growth. In real life, we start with a substrate crystal and lay down planes of the epitaxial material, one atomic layer at a time. If there is a strong chemical dissimilarity between the layers or a very large lattice mismatch, the epitaxial material balls up to form microscopic islands. This occurs in the growth of GaAs on Si and is analogous to the formation of water droplets on a waxed surface. As the islands grow, they will eventually coalesce into a continuous sheet. If the islands are not lattice matched to the substrate, it is likely that there will be misbonding at the coalescing island edges and a heavily dislocated epitaxial layer will form.

Another situation arises if the two materials are chemically similar (e.g., from the same columns of the periodic table) and not too strongly mismatched (e.g., 0 to 4%). It is then very likely that the first atomic plane will grow as a continuous, strained, dislocation-free layer. This occurs because the strain energy of a single atomic layer is almost always less than that of an array of misbonded, misfit dislocations. The question is then, what will happen as the second, third, and fourth layers are deposited? As the strained-layer thickness increases, the total number of distorted atomic bonds grows, and at some point the total energy exceeds that of an array of interfacial dislocations. It is then energetically desirable to begin forming such an array; this thickness is referred to as the equilibrium-critical-layer thickness. In the case of a thin epitaxial layer on a thick and comparatively rigid substrate, this equilibrium thickness is governed by equations of the form [21]

$$L_c = \frac{b}{8\pi f (1 + \nu) \, [\ln(L_c/b) + 1]},$$ (4)

where b is the Burger vector of the stress-relieving dislocation (a measure of atomic displacement around the dislocation, generally of order one lattice parameter), ν is the Poisson ratio for the epitaxial material (a measure of its elasticity), and f is the fractional lattice mismatch between the two materials.

Two important points should be noted. First, even in equilibrium, it is not desirable to relax all of the strain instantaneously by forming a fully developed dislocation array. A minimization of energy demands that, as the equilibrium-critical-layer thickness is exceeded, there is a rather gradual trade-off between strain relaxation and dislocation growth. Second, equilibrium can be very difficult to reach. A close examination of Fig. 8 shows that the formation of an interfacial misfit dislocation is atomically equivalent to the addition or removal of an entire atomic plane running from the top of the epitaxial layer down to the substrate interface. This cannot happen instantaneously. Under the influence of strain, the process is believed to begin at the surface, with the plane gradually growing downward toward the interface. Growth can occur either by the migration of interstitial atoms to, or atom vacancies from, the edge of the growing plane. This process will be a strong function of temperature, and at low temperatures it may be possible to postpone dislocation formation to growth thicknesses well above the equilibrium-critical-layer thickness. Complete theories of kinetically limited strained-layer growth are still under development,[22] but it has been shown that such critical thicknesses tend toward the form [23,24]

$$L_c = \frac{b^2(1-\nu)}{8\pi w f^2} \, [\ln(L_c/b)],$$ (5)

where w is a semi-empirical parameter determined by fitting to one experimental point. The most significant difference between Eqs. 4 and 5 is the $1/f$ versus $1/f^2$ dependence on misfit. The latter, kinetic, case yields a much steeper dependence of critical thickness upon strain.

Such a strong kinetic delay in dislocation formation occurs in the Ge_xSi_{1-x}/Si system (Fig. 9). The equilibrium calculation suggests that dislocation-free growth will be maintained for thickness diminishing from a few hundred ang-

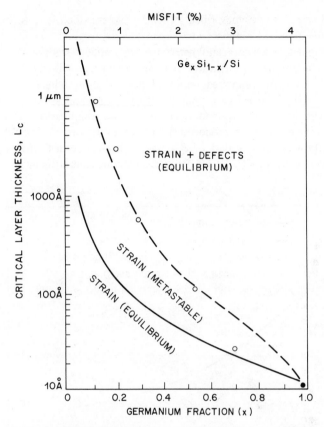

Fig. 9 Solid line: calculated thickness above which it becomes energetically favorable to form misfit dislocations in strained layer GeSi grown on Si. (After Ref. 21) Points: experimental data for low temperature MBE growth. (After Ref. 19) Dashed line: trend calculated for simple model of kinetically limited defect formation. (After Refs. 23 and 24)

stroms at $Ge_{0.1}Si_{0.9}$ (0.4 % mismatch) to 10 Å for pure germanium on silicon (4.2% mismatch). However, if such layers are grown by a low-temperature technique, such as molecular beam epitaxy, dislocation formation is inhibited up to a much greater thickness, called the metastable-critical-layer thickness. These values are "metastable" because between the two curves the crystal would like to form dislocations to lower its energy but has not been given enough time at high temperature. (In fact, dislocations begin to form as soon as the equilibrium limit is exceeded, but in this system the numbers remain small until the metastable limit is reached. The process is analogous to avalanche breakdown in a diode where instead of electrons multiplying, dislocations suddenly multiply.)

Equilibrium strained-layer epitaxy, and its extension to much thicker metastable layer growth, effectively adds silicon to the list of heterostructure semiconductor families. And, as described in later chapters, Ge_xSi_{1-x}/Si has already been employed in a number of classic heterostructure devices including modulation-doped transistors and heterojunction bipolar transistors. There is, however, an additional ramification of strained-layer growth that is making its use attractive even in compound semiconductor systems, where lattice-matched alternatives exist. The effect is that of strain upon band structure. When an epitaxial layer grows in strained-layer form, the compression or dilation of bonds is equivalent to that produced by pressures of order 100,000 atm. As typified by the Ge_xSi_{1-x}/Si data of Fig. 10, such huge forces affect interatomic bonding and can be manifested by gross changes in band structure. In the ger-

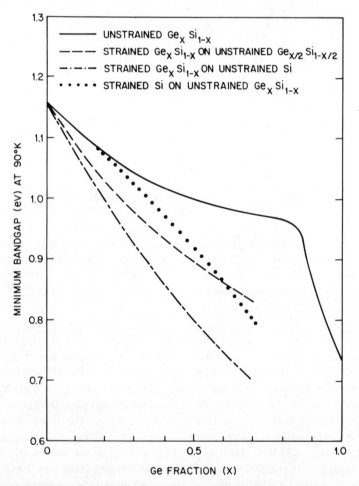

Fig. 10 Calculations showing the dramatic effect of strain upon semiconductor bandgaps. Solid line: bulk (unstrained) GeSi. Dotted line: silicon strained to match the lattice constant of Ge_xSi_{1-x}/Si. Dashed line: Ge_xSi_{1-x} strained to lattice match $Ge_{x/2}Si_{1-x/2}/Si$. Dash-dot line: Ge_xSi_{1-x} strained to lattice match silicon. (After Ref. 13)

manium silicide system, the effect is to reduce the bandgap well below that of the unstrained alloy. This means that a given difference in bandgaps can be achieved by the use of much more dilute Ge_xSi_{1-x} layers on silicon. Because they are more dilute, layers can be grown to greater thicknesses, increasing their resistance to dislocation formation.

A second, possibly generic, benefit of strained-layer epitaxy derives from its effect on band alignment at a semiconductor heterointerface. In our earlier discussion of Type I-III band alignment, we implicitly assumed that the alignment is fixed for each pair of semiconductors. This need not be the case if strain is present. Strain affects any semiconductor bandgap. If one semiconductor of a heterostructure pair is affected more strongly than the other, even a uniform strain may produce an alteration in band alignment and possibly a change in type. This is even more likely in a strained-layer case where the layers will have a different level of strain (possibly even of a different sign). The effect has been clearly documented for Ge_xSi_{1-x}, where growth on silicon produces a nested Type I alignment but an alternation of two Ge_xSi_{1-x} layers with differing strain can yield a staggered Type II alignment. Although it is too early to say whether strained-layer epitaxy will find commercial application, investigators in virtually every major semiconductor laboratory are actively pursuing the enhanced manipulation of band structure in a variety of semiconductor systems.

1.4 SYNTHESIS AND FABRICATION TECHNIQUES

To complete our overview of high-speed materials we must now consider the techniques for synthesizing the starting crystals and fabricating those crystals into completed devices. As before, the details are left to reference books[25-27] and the emphasis will be on basic principles and those factors that tend to differentiate materials. The discussion is divided into sections on bulk and epitaxial crystal growth, diffusion and oxidation, ion implantation and rapid thermal processing, wet and dry etching, and lithography. Because of their critical role in heterostructure device fabrication, the techniques of crystal growth and ion implantation will receive particular attention.

1.4.1 Bulk Crystal Growth

Bulk semiconductor crystals are generally fabricated in one of two ways.[28] In one approach, a charge of material is loaded in a container (generally of quartz) with a small single-crystal seed placed at one end. The charge may consist of measured amounts of the unreacted constituents or pieces of polycrystal synthesized in another apparatus. The container, or "boat," is heated until the charge melts and wets the seed crystal. The temperature is then decreased from the seed end, so that the charge crystallizes, using the seed as a template. In its simplest implementation, the heating and cooling are accomplished by pushing the boat into a furnace and then gradually withdrawing it. Alternatively, the boat may remain fixed within the furnace and the furnace temperature profile ramped up and down electronically. In the first case, the process is termed gradient-freeze crystal growth; in the latter, Bridgman crystal growth.

Wetting of the seed and mounting of the furnace are generally easier if the boat lies horizontally, and until recently this was the practice with virtually all III-V and II-VI materials. The shortcoming is apparent in the cross section of the resultant crystal. The combination of a generally cylindrical boat and a planar surface to the melt result in a D-shaped cross section that not only varies with the filling of the boat but is incompatible with well-developed, round-wafer handling equipment. For this reason, vertical variations of these techniques have been developed for both GaAs and InP.

The alternative, Czochralski (CZ), bulk crystal growth technique consists of holding a container of liquid charge material in a vertical crucible so that its top surface is just barely above the melting point (Fig. 11). A seed crystal is lowered into the center of the melt from above and withdrawn. Heat conduction up the seed cools the liquid, and crystal begins to grow. The crystal is forced to a roughly circular cross section by the rotation of the seed about its axis and the

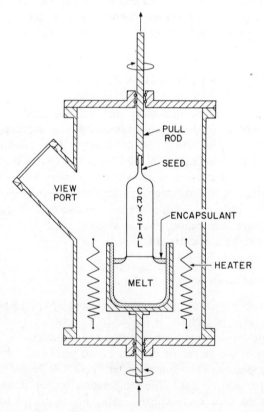

Fig. 11 Schematic of Czochralski-style crystal grower used to produce substrate ingots of Si, GaAs, and GaP.

counter rotation of the melt, which counters the crystal's natural tendency toward faceted growth. The diameter of the crystal can be closely controlled by adjusting the rate at which the seed is raised. This diameter control is generally accomplished by monitoring the weight of the crystal hanging from the seed, although optical and X-ray techniques have also been employed.

Czochralski growth has the advantage of producing long ingots with tightly regulated, nearly circular, cross sections and near total utilization of charge materials. It does, however, place a premium on the mechanical strength of the material in that the entire crystal must hang from the seed as it is pulled from the melt. For this reason, the Czochralski technique is ideal for silicon where it is possible to suspend a 2-m long, 15-cm diameter, 100-kg crystal from a seed a few millimeters on a side (Fig. 12).

Czochralski growth is also used for the relatively strong GaAs and GaP materials, but here its use is complicated by the high vapor pressures of the column V component: approximately 2 and 30 atm for As and P, respectively, at the compound melting points. This is the first of many vapor pressure concerns, and for that reason, elemental vapor pressures are plotted for the semiconductor constituents in Fig. 13. The Czochralski growth chamber must not only withstand these pressures but allow for the multiple mechanical feedthroughs shown in Fig. 11. The rod holding the seed both rotates and slides. To keep the top

Fig. 12 Photograph of 200- and 100-mm diameter ingots of silicon grown by the Czochralski technique. Note the remnant of the seed crystal visible as a point to the larger ingot. During growth this rod, less than 5 mm in diameter, supported the entire weight of the rotating crystal. (Photo courtesy of K. L. Bean, Texas Instruments)

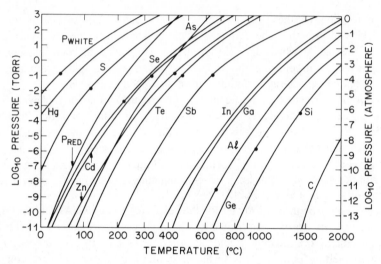

Fig. 13 Elemental vapor pressures of the semiconductor constituents as a function of temperature. Note that the vapor pressure of an element above a compound will be lower and these curves, therefore, define an upper bound to pressures encountered in semiconductor processing. (Data from Ref. 29)

of the charge at the melting point as its level in the crucible falls, the crucible is also gradually raised as well as rotated. Finally, because vapors of arsenic or phosphorus will try to leave the melt and condense on the walls of the apparatus, the melt is generally sealed by a molten layer of a second material (such as boron oxide) floating on its surface. This is then referred to as liquid encapsulated Czochralski, or LEC, crystal growth.

The more sophisticated Czochralski technique gives silicon and, to a somewhat lesser extent, GaAs and GaP a comparative advantage in terms of bulk crystal growth. In certain devices, however, bulk silicon crystals may have one important weakness. The resistivity of a silicon crystal may be increased only by purification. Although heroic efforts may produce resistivities of over 10,000 Ω-cm, common Czochralski silicon generally peaks at about 100 Ω-cm. In contrast, the resistivity of many compound semiconductors may be greatly increased through use of carrier-trapping impurities such as chrome or iron. Substrates doped with these species can have resistivities of over 10^8 Ω-cm. The use of these semi-insulating (SI) substrates can eliminate speed-degrading stray capacitances and grossly simplify the electrical isolation of devices. There is, however, evidence of these species, especially chrome, diffusing from substrates to degrade the heterointerfaces of epitaxial layers.

1.4.2 Epitaxial Crystal Growth

Although very early semiconductor devices were produced by actually changing the doping of a melt as a bulk crystal was grown, this sufficed for only the

crudest structures. Today, the substrate crystal is almost always a simple plat-
form for an epitaxial layer. With epitaxy ("epi") it is much easier to grow
material that has the requisite uniformity both across a wafer and in depth; also,
the dopant profile or semiconductor composition can be controllably varied with
depth. This vertical control is the natural complement to the lateral structuring
produced by lithographic processes, and together they achieve the three-
dimensional device structure. For these reasons, epitaxy is the rule in virtually
all devices and materials described in this book.

Certain early epitaxial structures were formed by sliding a substrate wafer
over successive molten puddles of semiconductor with differing composition and
doping. Although this liquid-phase epitaxy (LPE) is still used for certain
devices, it is completely overshadowed by vapor-phase deposition procedures.
The best developed of these is chemical vapor deposition (CVD), where a flow
of molecules containing the semiconductor and doping species passes over a
heated substrate crystal.[28] Those molecules striking the heated crystal release
the desired species, yielding crystal growth. The flux of arriving reactant atoms
is related to the gas pressure by the expression [30]

$$F = \frac{P}{\sqrt{2\pi mkT}}, \tag{6}$$

where F is the flux in molecules per unit area per time, P is the reactant partial
pressure, m is the gas molecular weight, k is the Boltzmann constant, and T is
the gas temperature. In conventional units, this can be rewritten as

$$F \text{ (Mol./cm}^2 - \text{s)} = 3.5 \times 10^{22} \, P \text{(torr)}/\sqrt{m \text{(g)} \, T \text{(K)}}. \tag{7}$$

The deposition rate is then the product of this flux multiplied by the net probability of condensation or reaction. In many instances, CVD makes use of
hydrides; for example, doped silicon may be produced by flows of SiH_4 in combination with either B_2H_6, PH_3, or AsH_3. Alternatively, halogen-containing
species such as $SiCl_4$ may be employed; for compound semiconductors, organometallic species such as $As(CH_3)_3$ or $Ga(CH_3)_3$, referred to as trimethyl arsenic
and gallium, are used in a procedure known as metal-organic CVD or MOCVD
(Fig. 14).

The key requirements for chemical vapor deposition are that the reacting
molecule release the desired species at an acceptable substrate temperature and
that the remaining species exit the system without adversely affecting the epitaxy. Adverse reactions consist, for example, of the introduction of carbon that
could dope a compound semiconductor or nucleate an incompatible SiC crystal
phase on silicon. If suitable chemistries can be identified, chemical vapor deposition has a fundamental strength based upon its use of reversible chemical reactions. This is important because the most difficult thing in epitaxial crystal
growth is the initial creation of an atomically clean and ordered substrate surface
to serve as a template for the subsequent layer. Using a reaction such as

$$SiCl_4 \leftrightarrow Si + 2Cl_2 \tag{8}$$

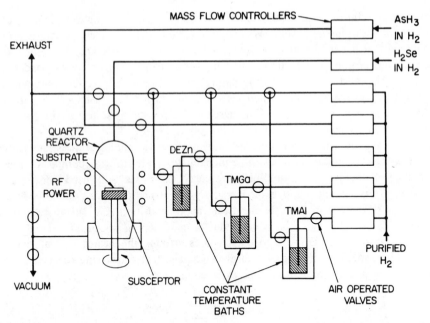

Fig. 14 Schematic of an atmospheric-pressure metal-organic CVD system used in the growth of doped AlGaAs semiconductor layers. As and Se dopant species are supplied as hydride molecules regulated by mass flow controllers. Metallic column III species and Zn are generated by vapors of diethylzinc (DEZn), trimethylgallium (TMGa), and trimethylaluminum (TMAl). Gases are mixed in quartz reaction vessel containing RF heated substrate crystal. (Courtesy of R. D. Dupuis, AT&T Bell Labs)

a clean, ordered substrate may be produced by initially flooding the reactor with the reaction product, Cl_2, and reversing the reaction to back etch the substrate surface. The Cl_2 pressure is then reduced, $SiCl_4$ pressure increased, and growth commences. The use of a slight Cl_2 pressure during growth can have the further effect of removing any misbonded atoms or impurities that would otherwise nucleate dislocations. Analogous effects can be exploited with other chemistries and other semiconductors.

It can also be an advantage if the molecular reaction probability is low enough that the reactant species strike a surface many times before decomposing. The precise gas flow is then less important, and not only may deposition uniformity increase, but it is frequently possible to achieve uniform growth over closely packed batches of 50 or more wafers. There is the further possibility that the reactant may decompose on the semiconductor, but not the insulator-coated semiconductor. Deliberate oxide patterning may then be used to produce well-defined localized or "selective area" epitaxial growth that can be readily exploited in many device structures.

Chemical vapor deposition has two principal disadvantages. First, the need to decompose a molecule generally entails the use of fairly high growth temperatures (generally two thirds or more of the crystal's melting point in degrees kelvin). At these temperatures, there will be substantial diffusive rearrangement of both the dopants and, in heterostructures, the semiconductor species. This rearrangement produces a blurring of the intended composition profiles or even the out-diffusion of species from the back of a wafer into the vapor and back into the growing epi (referred to as autodoping). Second, in chemical vapor deposition, we control what enters the reactor but not what arrives at the growing semiconductor surface. A change in the incoming gas chemistry does not produce a corresponding change at the surface. There will be a delay and a time averaging. In addition, from atmospheric pressure down to below 1 torr, the reactant gases form a semi-stagnant boundary layer at the epi surface, introducing additional delays and averaging. The net result of both the temperature and gas transport effects is an inability to produce abrupt vertical changes in epitaxial composition. In silicon, for example, several thousand angstroms may be needed to change the composition or doping by an order of magnitude, despite a near instantaneous adjustment of valves at the gas-mixing manifold.

Two research thrusts address the weaknesses of chemical vapor deposition. First, the use of low-pressure chemical vapor deposition (LPCVD) is increasing. LPCVD improves the abruptness of compositional transitions by at least an order of magnitude. Second, many investigators now employ an alternative vapor-phase technique that uses either simple atoms or molecules composed only of semiconductor or dopant atoms. Reactant decomposition is then much easier, and growth temperatures can generally be lowered to the point where solid-state diffusion and autodoping are insignificant. This technique is called molecular beam epitaxy, or MBE.[31,32]

In its earliest successful form, MBE consisted of a vacuum chamber containing a heated GaAs substrate and small crucibles containing gallium, aluminum, and arsenic (see Fig. 15a). The crucibles were heated to temperatures producing evaporation of Ga, Al, and As_4 vapors, and the arrival of these atoms at the substrate produced crystal growth. For MBE of low vapor pressure materials, such as Ge_xSi_{1-x}, resistively heated crucibles are often replaced by electron beam evaporation sources and ion beams (Fig. 15b).

Several crucial points make the technique successful and desirable. First, as gas pressure decreases, the probability of molecular collisions drops sharply. This probability can be characterized by the average distance between collisions, or mean free path, given by [30]

$$L = \frac{kT}{\sqrt{2}\,\pi r^2 P},$$
(9)

where πr^2 is the molecular collision cross section and other symbols are as in Eq. 6. For a room temperature gas with air-like molecules, this can be reduced to

$$L \text{ (cm)} \cong \frac{5 \times 10^{-3}}{P \text{ (torr)}}.$$
(10)

Thus, at pressures below 10^{-5} torr, mean free paths exceed the vacuum chamber dimensions.

COMPOUND SEMICONDUCTORS

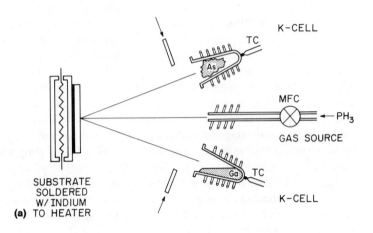

SILICON + SILICIDES

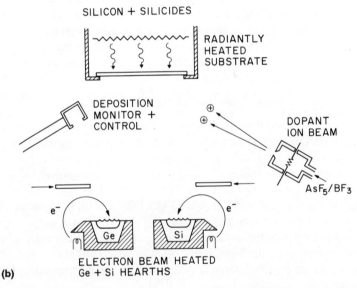

Fig. 15 Schematic of MBE systems for growth of compound semiconductors and silicon materials. (a) Compound semiconductors: sample is mounted vertically by indium soldering on a heated molybdenum block. As and Ga are evaporated from resistively heated K-cells with temperature controlled by a thermocouple (TC). Phosphorus flux is provided by a mass flow controlled (MFC) gas jet with a heater to produce breakup (cracking) of the PH_3 molecule. (b) Silicon compounds: sample mounted horizontally and heated by radiation from a filament. Ge and Si evaporated from liquid pools produced by electron beam heating of solid charge material. Evaporation rates are monitored and controlled by flux sensors. Dopants are supplied by low-energy ion implantation.

Second, at room temperature, virtually all semiconductor constituents have negligible vapor pressures (P, Hg, and S are the exceptions, as shown in Fig. 13). Consequently, charges may be heated to produce evaporation, but the resultant vapor condenses completely on the first cool surface it encounters. If this is combined with the use of long mean free paths, a simple door placed between the source crucible and the substrate will halt deposition of that species on the substrate instantaneously. (The doors are referred to as shutters and the crucibles as Knudsen cells, or simply K-cells.) It is thus possible, and indeed straightforward, to produce modulations in doping and composition down to the scale of single atomic layers. Further, because tightly bound, multiple-species reactants need not be decomposed, substrate temperatures are low enough that there is generally no measurable rearrangement of atoms during subsequent growth.

Finally, in the growth of compound semiconductors, the third crucial point to the success of molecular beam epitaxy is in the area of stoichiometry control. A compound, such as GaAs, must have a precise 1:1 ratio of each atom. If this is not the case, either there will be vacancies on one atomic site or atoms will be forced into interstitial positions. In both situations, these improperly bonded sites may be electrically active, degrading the electronic properties of the material. If the deviation from the proper 1:1 stoichiometric ratio becomes as "large" as one part per million, precipitates of a second phase (e.g., solid arsenic) nucleate with potentially disastrous effects on the perfection of the surrounding crystal. A crystal solidifying from a melt spontaneously adsorbs the proper ratio of atoms at its surface. At the low temperatures of MBE, a substrate tends to adsorb all atoms. Knudsen-cell temperature cannot be regulated with enough accuracy to produce part per million control of arrival rates, apparently precluding growth of compound materials.

The solution comes from the observation that low-substrate-temperature ranges can be found over which the higher vapor pressure arsenic adheres to a previously deposited gallium or aluminum layer but not to a completed monolayer of arsenic (because the bond strength of As to As is much weaker than that of As to Ga or Al). Stoichiometric growth is assured as long as there is a slight oversupply of As and the growth rate is determined strictly by the metal arrival rate. Analogous regimes have been found for virtually all III-V and II-VI semiconductors.

Molecular beam epitaxy has the advantage of near-perfect vertical control of composition and doping through an epi layer. It is also conducted at low enough temperatures that disparate materials such as silicon and $CoSi_2$ can be epitaxially combined without adverse interdiffusion, or metastable structures such as thick strained-layer Ge_xSi_{1-x} may be grown. However, its strengths are also its weaknesses. The immediate condensation of reactants means that MBE is intrinsically a line-of-sight process. While schemes are developing for uniform deposition on multiple substrates, one does not have the freedom of chemical vapor deposition to simply stack 50 or more wafers in a row.

Throughput is thus a hotly debated issue. In addition, the use of essentially non-reversible reactions means that one cannot beneficially back etch the starting substrate nor count on etching of impurities or misbonded atoms during growth. Extremely high vacuums are therefore employed, and it is possible that the purity or crystalline perfection of epi layers may not match the very high standards of chemical vapor deposition.

Ironically, the common solution to the weaknesses of both molecular beam epitaxy and chemical vapor deposition may be the increasing use of CVD-like gas chemistries in MBE-like vacuum systems. These hybrid techniques are receiving substantial research attention, and early results have already combined many of the advantages of easily replaceable gaseous sources with much of the composition control usually associated with solid evaporation source molecular beam epitaxy. The feasibility of these more exotic techniques has been enhanced by new generations of vacuum pumps, such as closed-cycle cryopumps and turbomolecular pumps, that greatly simplify achievement of ultra-high-vacuum conditions (i.e., $< 10^{-8}$ torr). In their variations, these new techniques go by a variety of names, with practitioners of chemical vapor deposition referring to ultra-low-pressure or ultra-high-vacuum CVD (i.e., ULP CVD or UHV CVD) and users of molecular beam epitaxy discussing gas source MBE or metallorganic MBE or chemical beam epitaxy (i.e., GSMBE or MOMBE or CBE).

1.4.3 Diffusion and Oxidation

Historically, diffusion and oxidation provided the basis for modern integrated circuits and could be considered as important as crystal growth. Although oxidation remains the key to fabrication of MOS gates, the use of oxidation and diffusion is otherwise declining. The reason is that both processes depend on thermal rearrangement of a crystal at very high temperatures. This rearrangement also blurs doping and composition modulations, leads to undesirable chemical interactions, and prohibits use of metastable materials. Outside of silicon, there is the additional problem that at the requisite temperatures at least one component of the semiconductor has a very high pressure. An attempt to do an open-tube-furnace diffusion on GaAs will end with a puddle of gallium and an arsenic-contaminated laboratory. Therefore, high-temperature processing often requires that compound semiconductors be laboriously sealed within a quartz ampoule or otherwise placed in an environment that provides a rich overpressure of their volatile species.

In the area of oxidation, none of the other semiconductors have a thermal oxide that either approaches the dielectric quality of SiO_2 or even justifies the trouble or thermal degradation of high-temperature processing. For these and a variety of other more subtle reasons, oxidation and diffusion are increasingly used only in specific Si MOS steps or in the creation of shallow contact layers. Indeed, we could argue that these techniques are almost fundamentally incompatible with the heterostructure devices that are the major topic of this book.

1.4.4   Ion Implantation and Rapid Thermal Annealing

The successor to solid-state diffusion, in its ability to produce local introduction of dopants, is ion implantation.[33,34] This technology grew out of the nuclear industry's use of ion beam mass spectrometers for isotope separation. A gas or vaporized solid is ionized, focused into a beam, and accelerated across a fixed potential into a magnetic mass spectrometer (Fig. 16). The selected beam is then accelerated across a second, variable potential and directed into the semiconductor crystal. A fixed, comparatively low, pre-magnet accelerating potential is used to minimize the magnetic field strength (and thus magnet size) and ensures a one-to-one correspondence between magnet current and the selected ion mass. The variable post-magnet potential is used to control the total ion energy, and in certain research machines, its polarity can be switched to either accelerate or decelerate the ions.

The dopant penetration is determined by the rate at which the ion loses energy to electronic excitation and nuclear scattering. Although first-principle derivations of these mechanisms exist, it is general practice to use one of a number of tabulations listing the popular semiconductors and the penetration of each dopant ion as a function of its energy.[35] It is, however, fundamentally a

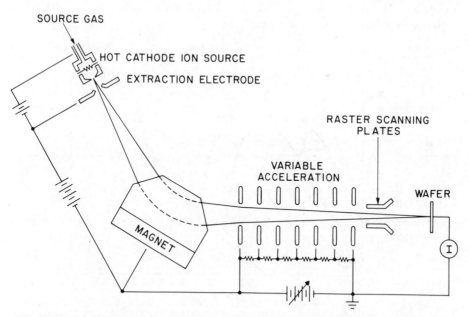

Fig. 16 Schematic of a modern high-current ion implanter. An ion beam is generated by the electron discharge from a heated cathode. An electromagnet both focuses the beam and selects the ion species with the desired charge-to-mass ratio. Selected beam is then accelerated and scanned across the wafer to incorporate the dopant uniformly. Scanning may be accomplished by electronic movement of the beam or physical displacement of the sample.

billiard-ball process. For a fixed energy, a dopant penetrates less deeply the heavier it is. For a fixed dopant and energy, the penetration increases as the average mass of the substrate atoms decreases.

The range of available energies is determined by implanter design characteristics. If too low an accelerating potential is used, it can be difficult to focus the ion beam, and this will degrade the mass spectrometer's ability to select only the desired ion species. Further, at very low beam energies, the field from the charge within the beam can overwhelm the applied accelerating potential, leading to the so-called space-charge limitation of current. For these reasons, minimum operating energies are generally in the range of 20 to 30 keV. On the high end, one is limited by insulator and atmospheric breakdown that forces one to go to longer and longer accelerating columns. Common commercial equipment now reaches several hundred keV and laboratory units may reach over 1 MeV. In the commercial 20- to 400-keV energy range, dopant penetration is generally from a few hundred to a few thousand angstroms. The statistics of the collision process dictate that the dopant atoms do not all come to rest at the same depth but are distributed with a more or less Gaussian profile. The width of the Gaussian profile increases with the range such that low-energy implantations produce shallow, sharply peaked profiles and high energies yield deep, broad distributions (Fig. 17). For applications that demand non-Gaussian dopant distributions, multiple implantations may be conducted with different ion energies and total ion fluences (or doses).

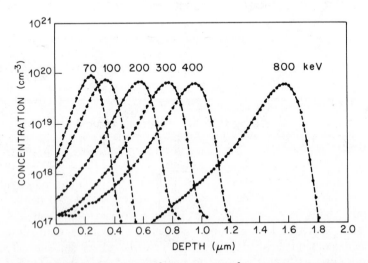

Fig. 17 Implantation profiles for 10^{15} boron ions/cm^2 implanted into polycrystalline Si at different energies. Shallower implants can be represented by Gaussian distributions, but deeper implants display a skewed distribution modeled by more complex functions. For implantation in single crystals, distributions can also display deeply penetrating tails produced by easy movement of ions down the channels between atomic rows. (After Ref. 36)

In its basic form, ion implantation can be used with any semiconductor. The differences are primarily in the methods for dealing with the damage produced by the incoming ions. A beam of only 30 to 50 eV has sufficient energy to knock a semiconductor atom off its proper atomic site. Since all practical ion implanters greatly exceed this energy, a large number of such displacements are expected for each incoming ion. One might think that the number of displacements would decrease steadily as the ion lost energy and thus expect most of the damage to occur at a depth shallower than the dopant range and very close to the semiconductor surface. This result would be desirable because electrically active junctions generally form on the deep side of the implantation distribution, and surface damage not only might anneal more easily but could be etched away.

Unfortunately, this is not the case. The chance of a direct nuclear collision between the ion and a semiconductor atom is small. Most of the energy transfer occurs through the interaction of the nuclei's charges. At high energies, ions are moving so quickly that they do not have enough time to kick a given nucleus hard enough to knock it from its crystalline site. Most of the displacement occurs near the end of an ion's range. Conservation of momentum dictates that most of the displaced atoms are kicked deeper and that the final damage distribution closely coincides with the dopant distribution or is slightly deeper. Device performance depends on the near complete annealing of this displacement damage.

Two opposing strategies have proven effective in silicon and the compound semiconductors. Because silicon crystals are very strongly bound, there is a strong driving force toward recrystallization. In addition, all atomic sites are geometrically equivalent and the crystallization process unambiguous. It has proven most effective to maximize ion implantation damage by use of heavy ion masses and room temperature implantation so that the crystal is converted to a completely amorphous state down to below the ion's penetration range. With suitable annealing, the amorphous layer then seeds on the underlying crystal and regrows toward the surface in the aptly named process of solid-phase epitaxy (SPE). The ions assume proper substitutional positions and the quality of the final layer is actually superior to that achieved by annealing an implanted layer that was not fully amorphized.

This solid-phase-epitaxy process does not work well with the compound semiconductors. Not only are the crystals more weakly bound, but the lattice has two geometrically non-equivalent sites. Using GaAs as an example, gallium atoms must recrystallize onto gallium sites and arsenic atoms onto arsenic sites. A gallium atom on an arsenic site is referred to as an anti-site defect, and because it does not have the proper number of bonding electrons for this site, it acts just like a dopant atom and produces electrical activity, trapping, or both. Thus, although solid-phase recrystallization occurs in compound semiconductors, the resultant quality is vastly inferior to that of the starting crystal.

In compound semiconductors, the successful strategy is to minimize damage accumulation by implanting at elevated temperatures. This prevents amorphization and makes it likely that the damage from one ion will anneal before a second ion creates damage in the same vicinity. The absence of overlapping damage clusters simplifies the rearrangement process and maximizes the chance of restoring crystal quality. To cope with the tendency of compound semiconductors to decompose when they are heated, these semiconductors are generally coated with a thin sealing layer of Si_3N_4 or SiO_2 during the heated implantation.

Even with continuous damage annealing, it may still be desirable to subject an implanted compound semiconductor to a post-implantation heating step. For silicon, such a heating step is the key to damage restoration. However, if the implanted wafer is subjected to conventional, slow, furnace annealing, the sharp ion implantation profiles may be broadened by diffusion and the compound semiconductors may decompose. Indeed, in a modern low-temperature process, the temperature required for complete damage removal may be exceeded only by the original bulk crystal growth step. It is imperative that the high-temperature anneal time be minimized, and to this end, techniques have been developed for rapid thermal annealing (RTA).[37,38] Rather than insert samples into furnaces, anneals for ion implantation and metallization are accomplished by means of rapidly cycling banks of infrared lamps (Fig. 18). RTA heats samples uniformly, often reaching the annealing temperature in seconds or fraction of a second. The sample is then usually held at peak temperature for less than 1 min, and when the lamps are extinguished, it will cool in a matter of seconds. Almost arbitrary time and temperature combinations are accessible, and it is frequently possible to define an anneal that will result in neither dopant migration or material decomposition.

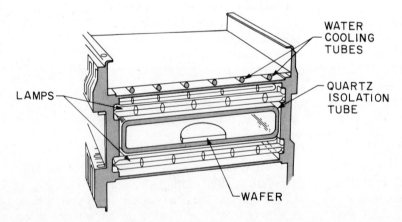

Fig. 18 Schematic of the rapid-thermal-annealing furnace used to remove damage from a ion-implanted sample or to sinter a metal to a semiconductor contact. Samples sit upon a silicon wafer. An inert gas flows through a water-cooled quartz tube and short pulses of heat are provided by external banks of infrared lamps. Temperature is monitored and controlled either by an optical pyrometer aimed at the sample or by a thermocouple welded to the silicon stage. (Courtesy of A. G Associates).

1.4.5 Wet and Dry Chemical Processing

More than any other area, chemical processing of semiconductors is the realm of secret recipes and black magic.[39] This may be attributed in part to the diversity of semiconductors and to the fact that the main practitioners are engineers and physicists rather than chemists. Although systematic trends are more difficult to discern, a few statements can be made. In terms of wet chemistry, silicon again stands alone. The reason is obvious when we look at any shelf of chemicals: with only one major exception, these chemicals may all be stored in containers of silicon dioxide. Oxidized silicon can withstand virtually any acid, base, solvent, or combination thereof. This chemical resistance can be a great advantage in cleaning silicon or patterning a dielectric, metal, polymer, or other material deposited on silicon. Virtually any chemistry may be tried and the process customized to the materials being used.

Hydrofluoric acid is the one chemical that attacks silicon dioxide. Although dangerous to the skin, it is otherwise relatively well behaved. Along with thermal oxidation, lithography, and diffusion, it provides the basis for standard silicon IC processing. Etching silicon itself can be a mixed blessing. If the goal is to etch silicon uniformly, with no orientation dependences (i.e., isotropically), the only common etchant is a combination of hydrofluoric acid and nitric acid (sometimes buffered with glacial acetic acid to produce a so-called CP4 etch). This etch works by a two-step process with the nitric acid oxidizing the silicon and the hydrofluoric acid removing the oxide. It is a fast, highly exothermic, and sometimes violent etch, making it difficult to produce smooth surfaces, particularly at low etch rates.

Silicon is much better behaved if one is trying to produce non-uniform etch features bounded by certain crystallographic planes.[11,40] Chemicals such as potassium hydroxide and ethylene diamine pyrocatechcol (EDP) will attack <111> planes as much as 100 times more slowly than they will etch other facets. If an oxide masked <100> silicon wafer is etched in either solution, material will be removed rapidly up until the <111> planes bounded by the mask are reached. This produces the inverted square pyramidal pits indicated in Fig. 19a. On masked <110> wafers, where <111> planes lie perpendicular to the surface, deep narrow trenches may be produced with depth-to-width aspect ratios of almost 100:1 (Figs. 19c and 20). EDP has the further property that it stops etching if it encounters silicon strained by the implantation of an undersized dopant atom such as boron. It is thus possible to completely etch away an underlying substrate and leave a free-standing, 1000-Å thick film many square centimeters in area. These and other "silicon micromachining" tricks have been the basis for non-electronic devices ranging from ink-jet print heads to an on-chip solid-state gas chromatograph (complete with metering valves) to at least one attempt at an actual gearbox.

Non-silicon semiconductors do not have the same wet-chemical resistance nor the very well-developed tendencies toward anisotropic etching. Solutions are therefore milder and include mixtures of hydrochloric acid, sulfuric acid, or

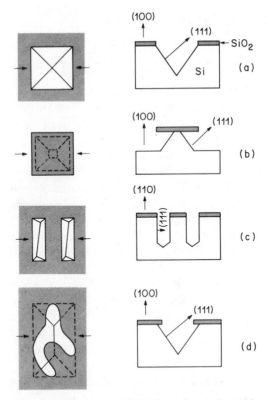

Fig. 19 Schematic of various features produced by anisotropic etching of silicon. (a) Square hole in oxide masked (100) wafer etches rapidly until bounding (111) facets are reached. (b) Square oxide mask on (100) wafer protects Si to produce square pyramid. Continued etching undercuts mask leaving it at the top of a narrowed pyramid. (c) Slots in oxide masked (110) wafer etch to produce trenches with vertical (111) walls (trench bottom has complex shape because terminating (111) planes are not parallel with wall (111) planes). (d) Irregular hole in oxide masked (100) wafer is undercut until etch reaches (111) planes bounding its outer edges. The etch can thereby improve upon irregular lithography or be used to determine facet direction. (After Ref. 11)

bromine methanol solutions. This lower resistance can, however, be of benefit in the etching of non-planar structures. In contrast to silicon, on these structures it is relatively easy to produce localized etching with smooth surfaces, free of residues. Mesa structures are straightforward and form the basis for most research and many production compound semiconductor devices. Compound semiconductors have another advantage: despite their crystallographic similarities, different members of a semiconductor family almost always etch in somewhat different chemistries. Wet chemistry can often be used to remove or undercut one layer selectively or to mask another with a chemically resistant layer. Again, this reinforces the tendency to use non-planar device technologies with these materials. Although non-planar processing can hinder large-scale

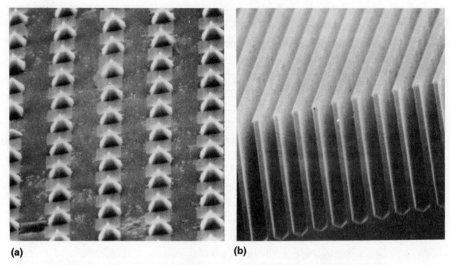

(a)   (b)

Fig. 20 Scanning electron micrographs of anisotropically etched silicon. (a) Features corresponding to Fig. 19b. Note faintly visible oxide squares suspended at top of pyramids. (b) Features corresponding to Fig. 19c. (After Ref. 40)

integration, the range of morphologies and the ease of device isolation can grossly simplify not only preliminary device research but also production of certain high-performance devices and circuits.

Dry chemical processing consists of the use of plasmas, or ion beams, to produce localized etching of a semiconductor surface. Plasmas may be operated with an unbiased substrate to produce generally isotropic etching or with specialized chemistries and bias to produce anisotropic material removal (in which case they are referred to as reactive ion etching, or RIE). Ion beam systems use special large-area, low-energy (50 to 2000 eV) ion sources to produce what is generally referred to as ion milling. Because the ions are formed into beams, this milling is inherently anisotropic and masking layers can be used to produce trench features with deep, well-defined, vertical walls.

In general, an advantage of dry chemical processes is the ability to produce uniform, reproducible wide-area etching. In vacuum plasmas, gas flow tends be even, thus avoiding the problem of localized depletion of a wet chemical that would produce reduced etch rates at the center of a wafer or center of an etch feature. The beams in an ion miller may be focused and the wafers scanned mechanically to ensure uniform etch rates. In plasmas, the color of the discharge glow can be monitored to detect when a given feature is being etched and the desired endpoint reached. Similarly, the number of atoms removed by an ion miller varies directly as the integral of the incoming ion flux, facilitating accurate calibration and real-time monitoring. With both techniques, the use of energetic ions means also that they will be largely insensitive to any resist residues inadvertently left in masking windows, further enhancing reproducibility.

For all of these reasons, dry chemical processing is beginning to dominate silicon processing and to make strong inroads in compound semiconductor lines.

There is, however, a downside to the use of energetic ions in that they can produce the damage problems encountered in ion implantation. As with ion implantation, this is of particular concern in compound semiconductor processing. In current silicon processing, difficulties are avoided if the etching is done well away from the active device volume. However, as device dimensions continue to shrink, the situation may change. Because of these concerns, the late 1980s have seen a major upsurge in research on an alternate technique known as electron cyclotron resonance (ECR) etching. In this variation of plasma processing, ionization of gases is induced by microwave excitation in a magnetic field. This ionization occurs some distance from the semiconductor surface, and the resultant ions arrive with energies well below the 40- to 50-eV range necessary for atomic displacement (or sputter removal). Etching occurs more by a wet chemical-like reaction process but with many of the potential uniformity and reproducibility advantages of a vacuum process.

1.4.6   Lithography

All semiconductor devices are fabricated through the repeated exposure of photoactive resist layers to light, X-ray, electron, or ion beams.[41,42] Where they are exposed, these generally polymer materials undergo bond linking (or breaking) such that the exposed material becomes insoluble (or soluble) in a developing solution. The patterned layers are then used to mask subsequent chemical processing or ion implantation steps. Research in this field centers on several areas (listed from the more immediate to more speculative): (1) extension of optical lithography equipment and resists to the shorter wavelengths of the far ultraviolet region of the spectrum ($\lambda \leq 0.3$ μm); (2) use of electron beam machines for writing the original photomask; (3) ion beam repair of photomasks; (4) the development of affordable, ultra-short-wavelength X-ray lithography equipment; and (5) use of direct, maskless writing of patterns on wafers by electron or ion beam lithography.

In general, lithographic technologies are not material specific and thus confer no particular advantage or disadvantage on a particular semiconductor. There are, however, at least three indirect arguments that are relevant to particular semiconductor systems. First, in order to achieve fine, submicron linewidths, optical lithography equipment is constrained to a very narrow depth of focus. Optimal focus can be maintained only over the near planar features of a silicon style device architecture. Even in this situation, the application of the active resist is often preceded by the deposition of a non-photoactive polymer layer to smooth the surface (the composite is referred to as a multilevel resist).

Second, to maximize feature control, older "contact" lithography cameras clamp the glass masking plate tightly against the semiconductor during exposure. To flatten the semiconductor wafer, substantial forces are exerted and most equipment is designed for the greater mechanical tolerances of silicon. When this equipment is used with the softer of the III-V semiconductors and most of the II-VI semiconductors, substantial wafer breakage can occur. Breakage not only wastes the particular wafer but may ruin the much more expensive mask. Newer, non-contacting cameras eliminate this problem, but they can be substantially more expensive and have not yet made their way into many experimental processing lines.

There is, however, a third point that returns to the basic physics and works against silicon. The generally higher carrier velocities and longer scattering lengths of a compound semiconductor often mean that comparable high-speed device performance can be achieved without being forced to submicron linewidths. In silicon, high-speed performance is dependent on submicron channel lengths or emitter stripe widths, and the eventual exploitation of hot carrier or velocity overshoot phenomenon may depend upon subtenth micron features. Thus, although silicon can be more amenable to leading-edge lithography, its high-speed performance may depend on this capability, which is afforded only by use of some of the most expensive and delicate semiconductor processing equipment.

1.5   TOTAL THERMAL BUDGET

To this point, we have discussed the various semiconductor fabrication and processing techniques as if they were independent of one another. This suggests that a given step in the complex device fabrication sequence could be altered with no consideration of the preceding or following steps. Nothing could be further from the truth. It comes down to a question of metastability. We have discussed metastability in the context of strained layers and, indirectly, in questions of metalization, thermal diffusion, and decomposition. However, at their heart, all semiconductor devices depend on one form of metastability or another. If they were allowed to reach a truly equilibrium configuration, compositional gradients would disappear, dopants distribute themselves uniformly, metals interdiffuse, and volatile species disappear. This is the crux of the third law of thermodynamics.

Each of the semiconductor fabrication steps can be thought of as a push away from an equilibrium configuration that will then be actively restored by subsequent thermal processing. The key parameter is diffusion length:

$$L_D = \sqrt{D(T)t}, \tag{11}$$

where $D(T)$ is the thermal diffusivity at temperature T and t is the time. The diffusion length describes the characteristic distance over which a species will move in a given processing step. Not only is D a strong function of temperature, but for processes such as substitutional diffusion, it can be enhanced by ion beam processes that increase the population of vacant lattice sites. In the extreme cases of disordered or polycrystalline material, diffusivities increase by many orders of magnitude. Such diffusion may easily obliterate the desired device configuration.

The general strategy is therefore to permit steps at the highest temperature and longest duration only at the early stages of a process and only at those points, such as crystal growth, where the configuration is intrinsically uniform. For each step, the time-temperature product must then move steadily downward as the process proceeds. Further, the ordering induced at a given step must tolerate not only the thermal processing of that step but the integral of all subsequent diffusion lengths. This introduces the concept of total thermal budget, or total thermal exposure. The net result is that a powerful, low-temperature processing step may be useless if it must be placed early in the device fabrication process, and the results cannot tolerate subsequent processing. Similarly, an exciting high-temperature innovation may be of benefit only if it can be situated early in the fabrication sequence or if its use eliminates multiple steps, thus maintaining the thermal budget. Given the increased use of disparate heterostructure materials and fine dimensions inherent in this book's device discussions, this concept (and warning) should be kept in mind.

1.6 SUMMARY AND FUTURE TRENDS

The main theme of this chapter is that despite the often confusing details, there is a fundamental unity to both semiconductor materials and semiconductor processing. The materials are linked by a common crystal structure and bonding that yield a common physical foundation. The resulting devices may have a wide variation in parameters, such as mobility or bandgap, but these are really the product of rather subtle shifts in band structure. In a similar manner, processing steps, although themselves unrelated, must be assembled into a total sequence that creates and maintains a specific, ordered, and thus metastable device configuration, despite the forces of entropy. An important ramification of this commonality is that there is seldom, if ever, a right material or process but only a preferred approach. And, indeed, current preferences may be easily offset by subsequent innovations in material synthesis and processing.

PROBLEMS

1. Glass fibers have a minimum loss window of about 1.5 μm, making this wavelength attractive for optical communications circuits. List the materials that might be used for light sources and detectors in such a system.

2. List the ternary (three component) semiconductor alloys that can be lattice matched to an InP substrate. Give the approximate composition and bandgap for each alloy.

3. For one of the alloys in the previous problem, assume that the ternary is perfectly lattice matched to the InP substrate at an epitaxial growth temperature of 500°C. When the sample is cooled to room temperature, what will be the lattice mismatch between the layers (assume the properties of the ternary can be interpolated from its binary constituents). Note that this mismatch must be accommodated by strain or by defects with a spacing of the lattice constant/mismatch.

4. A silicon device will be fabricated using lithographic masks with a 100-mm field of view. If subsequent masks have features with a 1-μm linewidth that must be aligned to a 1/4-μm tolerance, mask level to mask level, what variation in room temperature can be tolerated at the lithography apparatus (assume the glass mask has a thermal expansion coefficient of 1×10^{-6}/K).

5. Epitaxial layers of AlAs and Si will be grown by MBE at a rate of one atomic monolayer per second. Assume that at the MBE growth temperature an oxygen molecule has a 100% probability of binding to an aluminum atom and a 0.1% probability of binding to a silicon atom. For each material, how low must one keep the residual oxygen pressure in the MBE system in order to grow layers containing no more than 0.01% oxygen? (Note, both semiconductors have a density of about 5×10^{22} atoms/cm^3.)

6. In silicon (as in all semiconductors) ion implantation damage anneals gradually. Quality improves steadily over the temperature range 600 to 1000°C, making it desirable to use the highest anneal temperature the device will tolerate. Assume the implanted dopant has a thermal diffusivity with an activation energy of 4 eV. If an 800°C, 5-min furnace anneal is to be replaced by a 30-s RTA, what RTA temperature will produce equivalent dopant displacement?

REFERENCES

1. R. A. Smith, *Semiconductors,* Cambridge University Press, Cambridge (1978).

2. J. Callaway, *Quantum Theory of the Solid State,* Academic, New York (1974), Chap, 4.

3. J. R. Chelikowski and M. L. Cohen, "Nonlocal Pseudopotential Calculations for the Electronic Structure of Eleven Diamond and Zinc-blende Semiconductors," *Phys. Rev.* **B14,** 556–582 (1976).

4. T. Y. Chang, private communication.

5. L. Reggiani, Ed., *Hot-Electron Transport in Semiconductors, Topics in Applied Physics,* Vol. 58, Springer-Verlag, Berlin (1985).

6. J. Ruch, "Electron Dynamics in Short Channel Field-Effect Structures," *IEEE Trans. on Electron Dev.* **ED19,** 652–654 (1972).

7.   R. S. Huang and P. H. Ladbrooke, "The Physics of Excess Electron Velocity in Submicron-Channel FET's," *J. Appl. Phys.* **48,** 4791–4798 (1977).

8.   G. A. Sai-Halasz, M. R. Wordeman, D. P. Kern, S. Rishton, and G. Ganin, "High Transconductance and Velocity Overshoot in NMOS Devices at the 0.1-μm Gate-Length Level," *IEEE Electron Dev. Lett.* **EDL9,** 464–446 (1988).

9.   V. Narayanamurti, *Physics of Semiconductors,* Vol. 1, World Scientific, Singapore (1987), p. 3.

10.   G. Bauer, F. Kuchar, and H. Heinrich, Eds., *Two Dimensional Systems, Heterojunctions and Superlattices,* Springer-Verlag, Berlin (1984).

11.   K. E. Peterson, "Silicon as a Mechanical Material," *Proc. IEEE* **70,** 420 (1982).

12.   *Handbook of Chemistry and Physics,* CRC, Boca Raton, FL.

13.   J. C. Bean, "The Growth of Novel Silicon Materials," *Phys. Today,* October 1986, pp 2–8.

14.   M. Hansen, *Constitution of Binary Alloys,* McGraw-Hill, New York (1958).

15.   S. Adachi, "GaAs, AlAs, and $Al_xGa_{1-x}As$: Material Parameters for Use in Research and Device Applications," *J. Appl. Phys.* **58,** R1-29 (1985).

16.   M. Shur, *GaAs Devices and Circuits,* Plenum, New York (1987).

17.   T. P. Pearsall, Ed. *GaInAsP Alloy Semiconductors,* Wiley, New York (1982).

18.   R. Dornhaus and G. Nimtz, *Properties and Applications of the $Hg_{1-x}Cd_xTe$ Alloy System in Narrow-Gap Semiconductors,* Springer Tracts in Modern Physics, Vol. 98, Springer-Verlag, Berlin (1985).

19.   J. C. Bean, L. C. Feldman, A. T. Fiory, S. Nakahara, I. K. Robinson, "Ge_xSi_{1-x}/Si Strained-Layer Superlattice Grown by Molecular Beam Epitaxy," *J. Vac. Sci. and Technol.* **A2,** 436 (1984).

20.   J. C. Bean, "Strained Layer Epitaxy of Germanium-Silicon Alloys," *Science* **230,** 127–131 (1985).

21.   J. W. Matthews, "Defects Associated with the Accommodation of Misfit between Crystals," *J. Vac. Sci. and Technol.* **12,** 126 (1975) and references therein.

22.   J. Y. Tsao, B. W. Dodson, S. T. Picraux and D. M. Cornelison, "Critical Stresses for Si_xGe_{1-x} Strained Layer Plasticity," *Phys. Rev. Lett.* **59,** 2455 (1987).

23.   R. People and J. C. Bean, "Calculation of Critical Layer Thickness versus Mismatch for Ge_xSi_{1-x}/Si Strained-Layer Heterostructures," *Appl. Phys. Lett.* **47,** 322–324 (1985).

24.   R. People and J. C. Bean, "Erratum: Calculation of Critical Layer Thickness versus Mismatch for Ge_xSi_{1-x}/Si Strained-Layer Heterostructures," *Appl. Phys. Lett.* **49,** 229 (1986).

25.   W. S. Ruska, *Microelectronic Processing,* McGraw-Hill, New York (1987).

26. S. Wolf and R. N. Tauber, *Silicon Processing for the VLSI Era,* Lattice, Sunset Beach, CA (1987).

27. S. M. Sze, Ed., *VLSI Technology,* McGraw-Hill, New York, 1988.

28. C. W. Pearce, "Crystal Growth and Wafer Preparation," in *VLSI Technology,* S. M. Sze, Ed., McGraw-Hill, New York (1988), Chaps. 1 and 2.

29. R. E. Honig and D. A. Kramer, "Vapor Pressure Data for the Solid and Liquid Elements," *RCA Rev.* **30,** 285 (1969).

30. A. Roth, *Vacuum Technology,* North-Holland, New York (1979).

31. E. H. C. Parker, Ed., *The Technology and Physics of Molecular Beam Epitaxy,* Plenum, New York (1985).

32. E. Kasper and J. C. Bean, Eds., *Silicon Molecular Beam Epitaxy,* CRC, Boca Raton, FL (1988).

33. R. G. Wilson and G. R. Brewer, *Ion Beams with Applications to Ion Implantation,* Wiley, New York (1973).

34. M. D. Giles, "Ion Implantation," in *VLSI Technology,* S. M. Sze, Ed., McGraw-Hill, New York (1988), Chap. 8.

35. J. F. Gibbons, W. S. Johnson, and S.,W. Mylroie, *Projected Range Statistics,* Wiley, New York (1975).

36. W. K. Hofker, "Implantation of Boron in Silicon," *Phillips Res. Rep. Suppl.* **38** (1975).

37. T. O. Sedgwick, "Short Time Annealing," *J. Electrochem. Soc.* **130,** 484 (1983).

38. G. K. Celler and T. E. Seidel, "Transient Thermal Processing of Silicon," in *Silicon Integrated Circuits,* Supplement 2C to *Applied Solid State Science,* D. Kahng, Ed., Academic, New York (1985).

39. W. Kern and C. A. Deckert, "Chemical Etching," in *Thin Film Processes,* J. L. Vossen and W. Kern, Eds., Academic, New York (1978), Chap. 5.

40. K. E. Bean, "Anisotropic Etching of Silicon," *IEEE Trans. on Electron Dev.* **ED-25,** 1185 (1978).

41. R. K. Watts, "Lithography," in *VLSI Technology,* S. M. Sze, Ed., McGraw-Hill, New York (1988), Chap. 4.

42. R. K. Watts, Ed., *Submicron Integrated Circuits,* Wiley, New York (1989).

2 Device Building Blocks

S. Luryi
AT&T Bell Laboratories
Murray Hill, New Jersey

2.1 INTRODUCTION

In describing the structure elements of modern devices, we shall make every effort to avoid repetition of the material covered in Ref. 1. For example, although the basic *p-n* junction is still the most important element of many semiconductor devices, we shall omit any description of the *p-n* junction theory here and assume the reader is well familiar with it. Of course, to a varying degree of detail, it is decribed in virtually all texts on semiconductors and semiconductor devices and readers wishing to refresh their memory are advised to read the classical treatment[2] or consult some of the more modern references.

We shall assume that the reader is familiar with the standard equations governing the carrier transport (the drift-diffusion equation) and the electrostatic potential distribution (Poisson equation) in semiconductors. Familiarity with the equilibrium semiconductor statistics will also be assumed.

Exercise 1: Fill in the steps leading from the general expression for the electron density in the conduction band,

$$n = \int_{E_C}^{\infty} g_C(E - E_C) \left[e^{(E - E_F)/kT} + 1 \right]^{-1} dE , \qquad (1)$$

in terms of the conduction-band density of levels $g_C(E)$, the electron chemical potential E_F, and the temperature, to the standard formula, valid for nondegenerate semiconductors:

$$n = N_C(T) e^{-(E_C - E_F)/kT} , \qquad (2)$$

where N_C is the effective density of states in the conduction band. Expres-

High-Speed Semiconductor Devices, Edited by S.M. Sze. ISBN 0-471-62307-5
© 1990 John Wiley & Sons, Inc.

sions 1 and 2 are valid both for intrinsic and doped semiconductors in equilibrium; of course, the Fermi level E_F at a given temperature is controlled by the number and the type of impurities. In equilibrium both E_F and T are constant, while n and E_C can be coordinate-dependent.

This chapter consists of two major parts. The first part, Sections 2.2 to 2.5, discusses "classical" elements, that is, those whose description does not explicitly invoke quantum-mechanical concepts,† the second (Sections 2.6 and 2.7) concentrates precisely on quantum-mechanical phenomena. In the first part, we shall be mainly considering unipolar transport in structures with abrupt doping profiles.

2.2   ELEMENTS OF CARRIER TRANSPORT IN SEMICONDUCTOR DEVICES

The most widely used approximation to describe carrier transport in semiconductor devices is the drift-diffusion equation,

$$\mathbf{J} = q\,(n\mu\mathcal{E} + D\,\boldsymbol{\nabla}\,n)\,, \tag{3}$$

where $\mathbf{J}$ is the current density, $\mathcal{E}$ is the field acting on electrons, and μ and D are, respectively, the mobility and the diffusion coefficients for electrons. These coefficients are defined in a completely independent way‡ and so it may be surprising that there exists a relation between them:

$$qD \,=\, \mu kT, \qquad \text{Einstein relation for non-degenerate case}. \tag{4}$$

This relation follows directly from Eq. 3 and the equilibrium Boltzmann statistics: only if Eq. 4 is obeyed, will Eq. 3 yield zero current in equilibrium.

Exercise 2: Show that if the equilibrium carrier concentration is given by Eq. 1 and the transport equation by Eq. 3, then the following equation must hold:

$$\mu \,=\, qD\,\frac{d\,(\ln n)}{dE_F}\,. \tag{4a}$$

This generalized Einstein relation is valid in both degenerate and non-degenerate cases. For Boltzmann statistics it automatically reduces to Eq. 4.

The Einstein relation can be expected to be valid only to the extent that the drift-diffusion equation with constant coefficients μ and D is an adequate

---

†Implicitly, of course, quantum mechanics provides the foundation for semiconductor physics as described in any solid-state physics text.

‡The mobility is proportional to the inverse drag force coefficient b, as in $\mathcal{E} = bv$, whereas the diffusion coefficient is defined from Fick's law.

description of the carrier transport and that is the case only sufficiently close to equilibrium.

Away but not too far from equilibrium, it is very convenient to define an auxiliary quantity, called the quasi-Fermi level or "imref." The imref is a position-dependent $E_F(\mathbf{r})$, defined by extending Eq. 1 or 2 to the non-equilibrium case—when the local electron concentration $n(\mathbf{r})$ is affected by carrier injection due to the current flow. By *adjusting* (locally redefining) $E_F(\mathbf{r})$,† one can keep both the form of Eq. 1 or 2 and the correct local value of n.

The imref is a useful concept for qualitative analysis of semiconductor devices under bias. Consider, for example, the case of a forward-biased Schottky diode. Sketching the imref variation, one can pictorially distinguish between the thermionic and the diffusion theories,[1] and identify their regimes of applicability (Fig. 1). In most cases (except at very high currents), the net current is a small difference between much larger diffusion and drift components, hence the deviation from the equilibrium in the semiconductor is small and the thermionic theory is nearly correct.

Using the concept of the imref $E_F(\mathbf{r})$ and the Einstein relation (in the general form, Eq. 4a), one can re-write the drift-diffusion equation in the form

$$\mathbf{J} = \mu n \; \nabla E_F(\mathbf{r}) \; . \qquad (5)$$

Far away from equilibrium the drift-diffusion equation breaks down. Diffusion at high concentration gradients deviates from Fick's law,‡ mobility at high fields becomes field-dependent, (worse—non-local!), and even when both D and μ are well-defined, far from equilibrium, they are no longer related in a simple way. Field dependence of the mobility is briefly discussed in the next section.

2.2.1   Nonlinear Velocity-Field Characteristics

At high electric fields the carrier mobility is no longer constant. Velocity-field characteristics of typical semiconductors are shown in Chapter 1. Curves dis-

---

†Although one can always determine a definite $E_F(\mathbf{r})$ with this procedure, it is useful only when the temperature is a well-defined and constant quantity. Far from the equilibrium, the carrier statistics is not described by the Fermi distribution function underlying Eq. 1 or by the Boltzmann distribution function as in Eq. 2. In this case, there is no reason to expect that an $E_F(\mathbf{r})$, defined by the concentration, which is in turn controlled by the current flow, should vary monotonically along the current lines, as it would follow from Eq. 5. The problem is that besides the drift and diffusion terms in the current, there are also thermodiffusive fluxes driven by gradients in the effective carrier temperature. One can extend the validity of Eq. 3 by assigning the electronic system an effective temperature T_e, which does not have to be uniform over a device, and include thermodiffusive currents through terms proportional to ∇T_e.

‡Otherwise the particle velocity in a diffusion flux, $n^{-1}D \nabla n$, could become arbitrarily large, which would be unphysical.

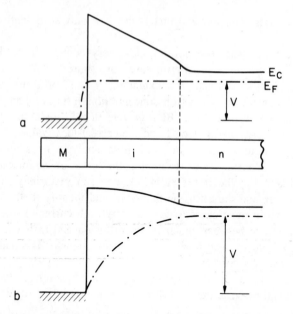

Fig. 1 Variation of the quasi-Fermi level (imref) in a forward-biased Schottky diode. Example chosen corresponds to the so-called Mott diode—a special case of the Schottky diode in which semiconductor layer (i) adjacent to the metal is undoped (intrinsic). (a) Thermionic theory: the imref E_F is assumed to be flat in the semiconductor, as if the carrier concentration on the top of the barrier were in equilibrium with that in the bulk semiconductor, (b) Diffusion theory: the imref E_F continuously joins the metal Fermi level, as if the carrier concentration on top of the barrier were in equilibrium with that in the metal. The concave band bending in the i layer is due to the uncompensated charge of mobile electrons diffusing from the semiconductor n layer.

played there represent the interpolation formulas commonly used in device modeling:

$$\text{Si:} \quad v(\mathcal{E}) = \frac{\mu\mathcal{E}}{[1 + (\mu\mathcal{E}/v_S)^2]^{1/2}} \; ; \tag{6a}$$

$$\text{GaAs:} \quad v(\mathcal{E}) = \frac{\mu\mathcal{E} + v_S(\mathcal{E}/\mathcal{E}_0)^4}{1 + (\mathcal{E}/\mathcal{E}_0)^4} \; , \tag{6b}$$

where μ, v_S, and $\mathcal{E}_0$ are parameters. Typically, one takes $\mu = 1390$ cm^2/V-s and $v_S = 10^7$ cm/s for Si and $\mu = 8000$ cm^2/V-s, $v_S = 7.7 \times 10^6$ cm/s, and $\mathcal{E}_0 = 4 \times 10^3$ V/cm for GaAs. With these parameters, the above analytic expressions give a reasonable fit to the experimentally measured drift velocities.[3, 4]

We see that at low fields $\mathcal{E}$ the carrier drift velocity v is proportional to the field, $v \equiv \mu\mathcal{E}$. One can interpret the drift as a result of carrier acceleration by the field in the presence of a drag force, $-bv$:

$$\frac{d(mv)}{dt} = q\mathcal{E} - bv \quad \rightarrow \quad v = q\frac{\mathcal{E}}{b} \quad \text{(in steady state)} , \tag{7}$$

where the meaning of μ is the inverse friction coefficient $\mu = q/b$. Alternatively, one can assume a view that transport of individual carriers represents a series of acceleration steps of random duration τ, each ending with a collision, which randomizes the direction of carrier velocity. In each step an electron acquires a velocity $v_0 - q\mathcal{E}\tau/m$. Assuming that collisions produce an isotropic velocity distribution, $< v_0 > = 0$, this implies that the average electron velocity is $- q\mathcal{E}\tau_m/m$, where $\tau_m \equiv <\tau>$. This gives the coefficient μ a meaning ($\mu = q\tau_m/m$) in terms of the characteristic time τ_m, called the momentum relaxation time. Taking this argument literally, one would conclude that τ_m is a mean time between collisions. A somewhat more refined argument[5] starts from the Boltzmann equation in the relaxation time approximation† and leads to τ_m being a more complicated weighted average of the relaxation time $\tau(k)$. The quantity τ, interpreted either as a relaxation time or a free time between collisions, depends on the electron energy, $E = \hbar^2 k^2/2m$ (and may also depend on the orientation of k—which we shall not be concerned with). As E rises, new mechanisms of scattering set in, which correspond to shorter relaxation times: an energetic carrier can emit higher-energy phonons, cause impact ionization, and so on. Some scattering mechanisms *decrease* their efficiency with carrier energy:‡ for example, the cross section for the scattering of electrons by charged impurities goes as E^{-2} (Rutherford scattering formula). The scattering probability is proportional to this cross section and the final density of states ($\sim E^{1/2}$); therefore, the relaxation time for scattering by charged impurities, $\tau^{(imp)} \propto E^{3/2}$.

If different scattering mechanisms are acting at the same time but can be regarded as "independent" (i.e., the presence of one mechanism does not affect the way other mechanisms operate), then their contributions to the total collision frequency $1/\tau$ are additive§:

$$\frac{1}{\tau(E)} = \frac{1}{\tau^{(1)}(E)} + \frac{1}{\tau^{(2)}(E)} + \cdots . \tag{8}$$

---

† This approximation corresponds to the collision integral replaced by $(f - f_0)/\tau$, where $f(k)$ is the distribution function in the presence of current flow; if all current-driving forces are removed, then $f(k)$ decays to the equilibrium distribution function $f_0(k)$ exponentially with the relaxation time $\tau(k)$.

‡ If such a mechanism is the dominant form of scattering, one can have a "runaway" instability, when carriers increase their energy in a constant field indefinitely. This instability has been extensively discussed for scattering by polar-optical phonons, where the rate of phonon emission (at energies far above the optical phonon energy) indeed decreases with increasing electron energy.

§ This proposition is an essential consequence of the relaxation time approximation. Experimentally, however, one deals not with τ but with the resistivity $\rho = 1/q\mu n$. It is often a good approximation to add contributions of different scattering mechanisms to the resistivity: $\rho = \rho^{(1)} + \rho^{(2)} + \cdots$ (Matthiessen's rule). It should be noted, however, that this approximate rule does not follow from Eq. 8, except in the trivial case of energy-independent τ, because $1/<\tau^{(1)}> \neq <1/\tau^{(1)}>$. For example, for impurity scattering near equilibrium one has $<\tau^{(imp)}> <1/\tau^{(imp)}> = 32/3\pi \approx 3.4$. Nevertheless, Matthiessen'rule is a helpful qualitative guide. In particular it tells us that the dominant scattering mechanism is the one that gives the shortest τ_m.

It is clear that if τ depends on the electron energy, then τ_m is determined by the energy distribution function for electrons driven from equilibrium. A commonly used approximation to describe such distributions is the electron temperature approximation. It assumes that the distribution function is of Boltzmann form but with an effective electron temperature $T_e = 2<E>/3k$. The philosophy behind this approximation is that multiple collisions (especially electron-electron interaction) often establish a quasi-equilibrium within the electronic system—effectively detached from the lattice. In the electron temperature approximation $\tau_m = \tau_m(T_e)$. The fact that τ_m depends on T_e or, equivalently, on $<E>$ is the origin of the non-linear v-F dependences.

At sufficiently high fields, the differential mobility in typical semiconductors becomes vanishingly small and the drift velocity saturates. In this regime the dominant scattering mechanism is emission of optical phonons. Velocity saturation happens when the average carrier energy becomes comparable to the phonon energy $\hbar\omega_{opt}$. By the order of magnitude, the saturated velocity equals $v_S \sim (\hbar\omega_{opt}/m)^{1/2}$.

Certain semiconductors, notably direct-gap III-V compounds, exhibit a range of field where the differential mobility is negative: see the velocity field curve for GaAs. While velocity saturation occurs at sufficiently high fields in all semiconductors, the negative differential mobility is a property associated with a special type of the band structure. Direct-gap III-V compound semiconductors have several heavy-mass satellite valleys [local minima of the energy dispersion $E(\mathbf{k})$] that are not too far separated in energy from the central Γ ($\mathbf{k} = 0$) valley, responsible for the conduction at low fields. Electrons energized by the electric field scatter into those upper valleys. Upper-valley electrons generally have a much lower mobility than those in the Γ valley. This happens for several reasons: relaxation time shortens because of the rapid scattering between equivalent satellite valleys, and the mobility ($\propto 1/m$) is further depressed by the heavy mass. The negative differential mobility occurs in the range where the fraction of electrons moving in the satellite valleys rises sharply. Because of the heavier mass and the valley degeneracy, the satellite valleys are weighted with a much larger density of states factor, and consequently when kT_e becomes comparable to the satellite valley separation, most of the electrons move in the satellite valleys, leading to a decrease in the average electron velocity.

2.2.2 Transient Effects in Carrier Transport; Velocity Overshoot

The concept of a velocity-field curve should be applied to transport problems in devices with great caution. It is clear that the momentum relaxation time τ_m depends on the field only indirectly—through the average energy. Therefore, in an inhomogeneous semiconductor the effect of the field can be non-local, because the average electron energy at a given point may depend on the field at nearby points, which supply electrons to the given point.

If a uniform electric field is suddenly turned on, the changed conditions do not instantly reflect on the electron distribution function $f(\mathbf{k})$. It takes time and several collisions for the distribution to reach its new steady-state form. The drift velocity v and the average carrier energy (electron temperature) T_e represent the first and the second moment of the distribution $f(\mathbf{k})$. These moments do not equilibrate at the same rate: typically, the momentum relaxation time τ_m is much shorter than the energy relaxation time τ_E. Indeed, elastic collisions are dominant for momentum relaxation, whereas an effective energy relaxation requires several inelastic interactions with phonons. Therefore, on the time scale of the momentum relaxation, the average electron energy can be considered quasi-static and the balance equation for the average momentum, Eq. 7, can be re-written in the form:

$$\frac{d(m\mathbf{v})}{dt} = q\mathcal{E} - \frac{m\mathbf{v}}{\tau_m(T_e)} \, , \tag{9}$$

where T_e itself is a function of time, "slowly" rising with a characteristic time τ_E. Thus at short times ($t < \tau_E$) after the imposition of a strong electric field, the carrier drift occurs with a low-field mobility and the velocity can substantially overshoot its steady-state value.[6]

Figure 2 shows the typical transient characteristics in Si and GaAs in response to a suddenly imposed electric field. The qualitative explanation given above is quite adequate for describing the overshoot in Si and Ge, where there is a strong dependence of the phonon scattering rates on the carrier energy. In GaAs and some other III-V compounds, the dominant scattering process in the Γ valley is due to polar optical phonons, and the rate of these processes at sufficiently high electron energies becomes nearly energy-independent. In such materials the mobility degradation at high energies is associated with the transfer of electrons to the lower-mobility upper valleys; consequently the overshoot is not seen below the threshold field for the negative differential mobility effect ($\mathcal{E} \geq 3.2\,\text{kV/cm}$ in GaAs and $\geq 11\,\text{kV/cm}$ in InP).

The overshoot phenomenon has become quite important in determining the speed of modern transistors with ultra-short gate lengths. For that purpose, it is more relevant to replot the carrier drift velocity as a function of the distance from the source (for simple estimates we can just scale the time axis by the saturation velocity $v_S \approx 10^7\,\text{cm/s}$).

Exercise 3: Using the data in Fig. 2 and assuming a uniform source to drain field, re-plot the carrier drift velocity $v(x)$ as a function of the distance from the source, $x(t) = \int dt'/v(t')$.

From the data of Fig. 2 it is clear that the characteristic overshoot distances in Si are of the order of several hundred angstroms, which is small compared to the transistor dimensions. Nevertheless, for a $0.25\text{-}\mu\text{m}$ n-channel Si field-effect transistor the overshoot effect contributes an $\approx 20\%$ enhancement in the speed.[7] In GaAs the characteristic overshoot length (i.e., the distance where electron travel at a velocity above the steady-state value) can be as high as $1\,\mu\text{m}$. Monte

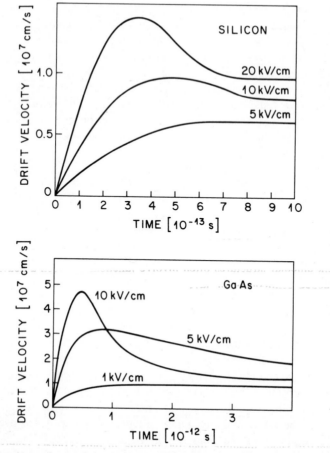

Fig. 2   Time dependence of the average velocity of electrons in Si and GaAs in response to a suddenly imposed electric field.   (After Ref. 6)

Carlo simulations of the overshoot have become a valuable tool in the design of GaAs transistors.

For times less than τ_m, that is, over distances less than the mean free path, carriers travel "ballistically." If there were no collisions at all, the carrier velocity would be determined by the conservation of energy, $mv^2/2 = q\phi(x)$, where ϕ is the electrostatic potential.   Such a situation is realized in vacuum diodes.   In the absence of a fixed charge, $\phi(x)$ itself is determined by the electron concentration and the boundary conditions.   If one assumes the boundary condition of $\mathcal{E} = 0$ and $\phi = 0$ at the cathode contact, then, solving simultaneously the Poisson equation, the current-continuity equation, and the energy conservation equation, one obtains the Child-Langmuir law of vacuum electronics:

$$ J = \frac{4}{9} \left[\frac{2q}{m} \right]^{1/2} \frac{\epsilon_s V^{3/2}}{L^2} , \qquad (10) $$

where L is the length of the diode, $V \equiv \phi(L)$, and ϵ_s is the dielectric permittivity. In recent years there was a lively discussion of the possibility of realizing a semiconductor analog of Eq. 10 in ultra-small semiconductor devices.[8] In the absence of collisions a conduction-band electron can be accelerated to velocities much higher than v_S [up to the maximum group velocity $dE/d(\hbar \mathbf{k})$ in the band; for electrons in GaAs accelerated in a $<100>$ direction this limit is $\approx 10^8$ cm/s] and such an enhancement would be important and beneficial for the performance of short semiconductor devices. Nevertheless, even though there is solid experimental evidence for the existence of collisionless transport of large groups of electrons in specially designed structures (see Chapter 7), no device structure has been demonstrated to date in which the current-voltage characteristics would convincingly conform to Eq. 10.

2.3 SYMMETRIC *n-i-n* STRUCTURE

Modern bulk unipolar structures often involve junctions between a heavily doped and an undoped semiconductor layer. In such junctions the effects of carrier injection into the undoped layer are important and one cannot use the familiar depletion approximation that works so well for understanding the properties of bipolar junctions. To illustrate these effects, let us consider the case of a symmetric *n-i-n* or *p-i-p* junction, on the assumption that the entire charge in the base (both in equilibrium and under bias) is due to mobile carriers injected into the base from the doped contact regions. This means that we neglect both the fixed charge due to any residual doping in the intrinsic layer and the mobile charge thermally generated across the forbidden gap. In other words, we consider a *unipolar* model of an *n-i-n* or *p-i-p* structure. We shall confine ourselves to the *symmetric* case, meaning that the doping in bulk contact layers on both sides is assumed the same (dropping this assumption would bring in the complication of a built-in potential difference between the two sides of the junction—without changing any issues that we would like to clarify here).

2.3.1 Equilibrium Properties

For a symmetric *n-i-n* junction, Fig. 3, the potential barrier $\phi(x)$ is due to the mobile charge diffusing into the intrinsic *i* layer from the *n* layers doped to the level N_D. The shape of this barrier, as well as the field and the charge distributions in the junction can be rather neatly expressed in a closed analytic form by solving the Poisson and the drift-diffusion equations. Introduce the dimensionless variables:

$$\xi \equiv \frac{x}{x_0}, \quad \text{(coordinate)}, \qquad \nu = \frac{n}{N_0}, \quad \text{(concentration)}, \qquad (11)$$

$$e \equiv \frac{\mathcal{E}}{\mathcal{E}_0}, \quad \text{(electric field)}, \qquad j = \frac{J}{J_0}, \quad \text{(current)},$$

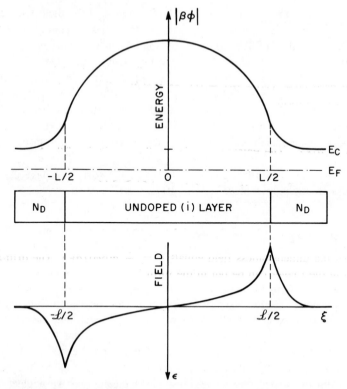

Fig. 3 Conduction-band diagram and electric field profile in a thin symmetric *n-i-n* junction in equilibrium. The solution, given by Eqs. 15 to 19, is exact provided that the intrinsic carrier concentration in the *i* layer can be neglected.

where

$$N_0 \equiv N_D \,,$$

$$x_0 \equiv L_D = \left[\frac{\epsilon_s kT}{q^2 N_D} \right]^{1/2} ,$$

$$\mathcal{E}_0 \equiv \frac{kT}{qL_D} = \left[\frac{kTN_D}{\epsilon_s} \right]^{1/2} ,$$

$$J_0 \equiv q N_D \mu \mathcal{E}_0 = \frac{q N_D D}{L_D} .$$

L_D being the Debye length in the doped layers, ϵ_s the dielectric permittivity, and μ and D are, respectively, the mobility and the diffusion constant in the *i* layer, related by Eq. 4.

Exercise 4: Estimate parameters x_0, $\mathcal{E}_0$, and J_0 for Si and GaAs at 300, 77, and 4.2 K and $N_D = 10^{18}\,\mathrm{cm}^{-3}$.

We set the origin $x = 0$ in the middle of the i layer of thickness L (in dimensionless units $\mathcal{L} \equiv L/L_D$) and choose the reference potential equal to zero far from the base layer $[\phi(\pm\infty) = 0]$, so that the dimensionless carrier concentration is of the form:

$$v(\xi) = \exp[\beta\phi(\xi)] , \tag{12}$$

where $\beta \equiv q/kT$. The Poisson equation in these units is of the form

$$\frac{de}{d\xi} = -v , \qquad |\xi| \le \mathcal{L}/2 , \tag{13a}$$

$$\frac{\partial^2(\beta\phi)}{d\xi^2} = e^{\beta\phi} - 1 , \qquad |\xi| \ge \mathcal{L}/2 . \tag{13b}$$

[Note that the dimensionless field equals $e = -\partial(\beta\phi)/\partial\xi$.] The drift-diffusion equation in the i layer can be put in the form

$$\frac{\partial}{\partial\xi}\left[\frac{1}{2}e^2 + \frac{\partial e}{\partial\xi}\right] = j . \tag{14}$$

For $j = 0$, the solution of Eq. 14, satisfying $e(0) = 0$ and the physical condition $v > 0$, is given by

$$e = -2\gamma\tan(\gamma\xi) , \qquad |\gamma\xi| < \pi/2 , \tag{15}$$

with a real γ. Integrating Eq. 15 we find the potential everywhere in the i region:

$$\beta\phi = \beta\phi(0) - \ln\cos^2(\gamma\xi), \qquad |\xi| \le \mathcal{L}/2 , \tag{16}$$

where

$$e^{\beta\phi(0)} = v(0) = 2\gamma^2 . \tag{17}$$

On the other hand, outside the i layer the potential is found by integrating Eq. 13b, namely,

$$e^{\beta\phi} - \beta\phi - 1 = \frac{e^2}{2} , \qquad |\xi| \ge \mathcal{L}/2 . \tag{18}$$

Matching Eqs. 16 and 18 at $|\xi| = \mathcal{L}/2$ and using Eq. 17, we find

$$\cos^2\left[\frac{\gamma\mathcal{L}}{2}\right] = 2\gamma^2\exp(1 - 2\gamma^2) . \tag{19}$$

This equation determines $\gamma(\mathcal{L})$ and thus completely solves the problem. The proper choice of the solution is provided by the condition in (15), namely, $|\gamma\mathcal{L}| < \pi$. It is instructive to plot $v(0)$ versus $\mathcal{L}$, Fig. 4. In the limit

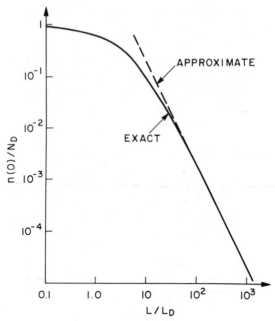

Fig. 4 Dependence of the equilibrium carrier concentration in the middle of the undoped i layer, due to electron diffusion from the adjacent n-doped layers, on the i layer thickness L and n-layer doping N_D. Solid line shows the exact solutions given by Eqs. 19 and 17, where $\nu \equiv n/N_D$ and $\mathscr{L} \equiv L/L_D$, and dashed line corresponds to the approximation of Eq. 20.

$\mathscr{L} \gg 1$ (i.e., $L \gg L_D$ and $n(0) \ll N_D$) one has $|\gamma\mathscr{L}| \to \pi$ and hence $\nu(0) \approx 2\pi^2/\mathscr{L}^2$ or, equivalently,

$$n(0) \approx \frac{2\pi^2\epsilon_s kT}{q^2 L^2}, \qquad n(0) \ll N_D. \tag{20}$$

Note that in this limit $n(0)$ is independent of N_D. Of course, the above solution is valid only if $n_i \ll n(0)$ where $n_i(T)$ is the intrinsic carrier concentration in silicon ($\sim 10^{10}$ cm^{-3} at 300 K).

2.3.2 Symmetric n-i-n Structure under Applied Bias

When an external bias is applied to an n-i-n structure, the previous equilibrium considerations do not apply. For an analytic treatment of this problem, an important issue is the choice of a tractable model of boundary conditions. A number of authors have imposed a condition on the mobile-charge concentration at the base edges—taken to be equal to the carrier concentration in the bulk of doped layers, independent of the current level. This model is rather crude and works reasonably only for long-base devices. We shall assume a more realistic boundary condition[9, 10] that consists in neglecting the variation of the quasi-

Fermi level within the doped contact layers. This means that the concentration and the field profiles in those layers are calculated "exactly" on the assumption of a quasi-equilibrium distribution, and the solutions obtained are then matched with the corresponding quantities in the base layer.

We shall again assume a symmetric diode, Fig. 5, with equal donor concentration N_D (fully ionized) in both n layers and an intrinsic (i) base of thickness L. The mobility μ and the diffusion coefficient D are assumed field-independent. Schematically, the problem is solved in the following way.[10]

The drift-diffusion equation in the i layer is given by Eq. 14 with a non-vanishing j. The first integral of Eq. 14 is of the form

$$-\frac{\partial e}{\partial \xi} - \frac{1}{2} e^2 + j\xi - \zeta = 0 , \quad (|\xi| \leq \mathcal{L}/2) \qquad (21)$$

with ζ being an integration constant† (which, of course, depends on j). Elec-

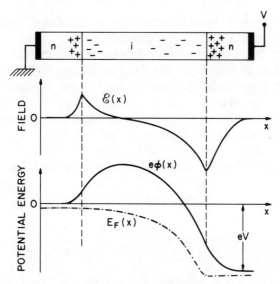

Fig. 5 Illustration of the *n-i-n* diode under applied bias. Schematically shown are profiles of the electric field and the electrostatic potential energy. The energy diagram also shows the imref $E_F(x)$ (dashed line). It is assumed that the imref is constant within the doped contact layers—resulting in the boundary conditions, Eq. 24, on the electric field in the base i.

---

†It can be shown that ζ must be a positive constant. Indeed, it is physically obvious that for low and moderate currents, $j < 1$, the potential energy must have a maximum within the base region. (The situation when the maximum moves into the doped emitter layer or disappears there entirely, falls outside our model. It corresponds to currents so high that one can no longer neglect the drop of the quasi-Fermi level in the doped regions.) At the point $\xi = \xi_{max}$ where the potential-energy maximum occurs, one has $e = 0$, and Eq. 22 implies that positive concentration can be ensured only if $\zeta > j\,\xi_{max} > 0$. From an inspection of Eq. 22, it is also clear that in the low-current limit ζ must approximately equal the dimensionless carrier concentration at $\xi = \xi_{max}$. In particular, in the limit $j \to 0$ one has $\zeta(j) \to \nu(0)$, where $\nu(0)$ is determined by Eqs. 17 and 19.

tron concentration in the intrinsic layer is then determined by the equation

$$\nu(\xi) = \frac{1}{2}e^2 - (j\xi - \zeta). \quad (|\xi| \leq \mathcal{L}/2) \tag{22}$$

In the doped contact layers (which are assumed to be in equilibrium—i.e., with a flat quasi-Fermi level) the electrostatic potential is related to the electric field, by Eq. 18 and the electron concentration is of the Boltzmann form, Eq. 12. Using these equations and matching the concentration $\nu(\pm\mathcal{L}/2)$ with that given by Eq. 22, we can express the boundary values of the potential and the field in terms of the constant ζ and the current:

$$\beta\phi_- = \zeta - 1; \tag{23a}$$

$$\beta\phi_+ = \zeta - 1 + j\mathcal{L}; \tag{23b}$$

$$e_- = -[2(\exp(\zeta - 1) - \zeta)]^{1/2}; \tag{24a}$$

$$e_+ = \{2[\exp(\zeta - 1 - j\mathcal{L}) - \zeta + j\mathcal{L}]\}^{1/2}, \tag{24b}$$

where e_- corresponds to the cathode and e_+ to the anode boundary of the i layer ($\xi = -\mathcal{L}/2$ and $\xi = \mathcal{L}/2$, respectively).

Solution of the non-linear first-order differential equation, Eq. 21, describing the electric field inside the i layer is rather cumbersome and will not be reproduced here. With the boundary relations, Eqs. 23 and 24, this solution completely determines the charge, field, and potential distributions in an n-i-n diode under bias, as well as its current-voltage characteristics. Let us confine ourselves to describing the key results in a graphical form.

First, it is convenient to picture the behavior of the parameter ζ as a function of the current density, Fig. 6. Besides helping to understand Eqs. 22 to 24, this

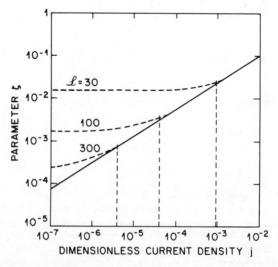

Fig. 6 Dependence of the parameter ζ on the current density. (After Ref. 10). Vertical dashes indicate points where $j\mathcal{L}/\zeta = 1$.

plot indicates the existence of two distinct regimes in the diode operation: the high-current/long-base limit ($j\mathcal{L} \gg \zeta$) and the low-current/short-base limit ($j\mathcal{L} \ll \zeta$). The calculated current-voltage characteristics of an *n-i-n* diode are displayed in Fig. 7 for several dimensionless base thicknesses. In the high-current limit, the characteristics coincide with the classical Mott-Gurney law[11] for space-charge-limited current:

$$j = \frac{9}{8} \frac{(\beta V)^2}{\mathcal{L}^3} , \quad \text{or} \quad J = \frac{9}{8} \frac{\epsilon_s \mu V^2}{L^3} . \tag{25}$$

In the low-current limit, one instead obtains a *linear* law

$$J = \frac{qn(0)\mu}{1+\delta} \frac{V}{L} \approx \frac{2\pi^2}{1+\delta} \frac{eps\,\mu V}{L^3} \frac{kT}{q} , \tag{26}$$

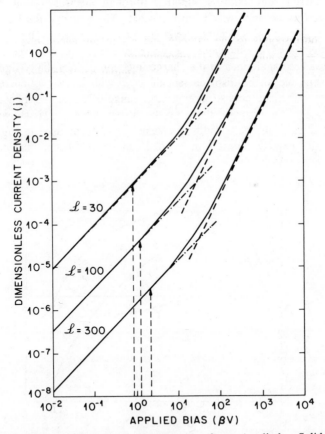

Fig. 7 The calculated current-voltage characteristics of an *n-i-n* diode. Solid lines represent the exact solution of Eq. 21 with the boundary conditions of Eq. 24. Dashed lines correspond to the Mott-Gurney regime, Eq. 25, and dash-dotted lines describe the low-current limit, Eq. 26. Vertical dashes indicate points where the parameter $j\mathcal{L}/\zeta = 1$, as in Fig. 6. (After Ref. 10).

where $\delta \equiv 2^{3/2} \mathcal{L}^{-1} \exp(\frac{1}{2})$. The transition between these two regimes is evident in Fig. 7. Interestingly, the linear regime can be seen to extend in voltage quite a bit beyond the range $\beta V \le 1$, where it can be rigorously derived, and in current beyond the "watershed" value $j\mathcal{L} \approx \varsigma$. At first sight this may appear strange: in the low-current regime electrons move over a relatively large potential barrier (see Fig. 5), and one could expect an exponential curve, typical for barrier injection. Equation 26 would then be just a first-order term in an expansion of the exponential for $V \le kT/q$ and one can hardly expect it to be valid for $V \gg kT/q$. However, the barrier in an n-i-n structure is entirely due to the mobile electrons injected into the base, and any increase in the bias not only lowers the barrier electrostatically but also moves its position toward the emitter contact, see Fig. 8. Carriers that participate in the current do not dissipate any extra energy due to this barrier. The energy required by an electron to move over the barrier maximum is supplied (through electrostatic interaction) by another electron moving away from the peak. The fact that the linear approximation, Eq. 26, remains accurate over a rather large range—up to an applied bias of order $10 kT/q$ —indicates that currents over a space-charge barrier cannot be considered analogous to the charge injection over spatially fixed barriers. It is precisely due to the displacement of x_{max} with increasing current that the linear IV characteristic persists over a large range of biases. Dependence of the position x_{max} of the virtual cathode on the current density j is shown in Fig. 9 for three values of the base width (for same $\mathcal{L}$ as those used in Fig. 7). Note that for larger base thicknesses the virtual cathode approaches the emitter con-

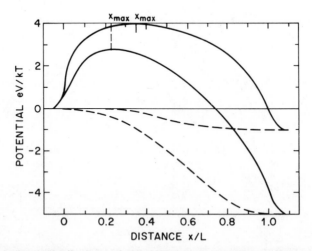

Fig. 8 Calculated distributions of the electrostatic potential (solid lines) and of the quasi-Fermi level (dashed lines) across an n-i-n diode of base width $\mathcal{L} = 30$ at two different current densities, $j = 10^{-3}$ (lower-bias curve) and $j = 5 \times 10^{-3}$ (higher-bias curve). The quasi-Fermi level in the bulk of the doped contacts to be coincident with the electrostatic potential energy, $E_F = q\phi$. (After Ref. 10).

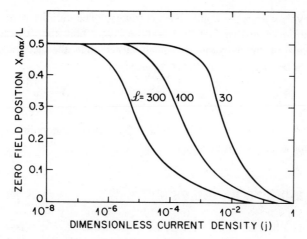

Fig. 9 Dependence of the position x_{max} of the virtual cathode on the current density, calculated for different values of the base width. (After Ref. 10).

tact faster. When x_{max} becomes small compared to L, then the Mott-Gurney law provides an accurate description.†

It is instructive to consider the behavior of the imref potential, $\eta \equiv E_F/q$, defined by the relation

$$\nu(x) = \exp[\beta(\phi - \eta)] .$$

Figure 8 also shows the position dependences of both E_F and $q\phi$, calculated for two different current densities in an *n-i-n* structure with $\mathcal{L} = 30$. Of course, the quasi-Fermi level is constant in the doped contacts—which is the basic approximation of our model. However, we see that it is also relatively flat in the vicinity of both boundaries in the base.‡ At low currents, most of the variation in E_F is concentrated near the middle of the base. With increasing current, most of the drop in E_F is shifted toward the collector contact. This is as expected: on the uphill slope of the space-charge barrier the current is a difference of two oppositely directed (diffusion and drift) fluxes. If in some region of the diode these fluxes are large compared to the net current, then this region is close to equilibrium and hence the E_F is nearly flat. The downhill slope is always further removed from equilibrium than the uphill slope and, naturally, most of the drop in the quasi-Fermi level occurs there.

Finally, let us comment on the validity of the assumed model of the boundary conditions in an *n-i-n* junction under bias. We have neglected the drop of the

---

†In Section 2.5, we shall return to this law and its generalization for space-charge-limited currents in thin films.

‡Since our purpose here is only to illustrate the behaviour of E_F inside the base, we assume the imref in the bulk of the doped contacts to be coincident with the electrostatic potential energy, $E_F = q\phi$. This implies a particular choice of the doping N_D, which may not be consistent with the non-degenerate statistics used.

quasi-Fermi levels within the doped contact layers. The validity of this approximation is conditioned on the fact that in these layers both the diffusion and the drift components of the current are much larger than the net current. This requirement can be expressed by the inequalities

$$e_{\pm} \exp(\beta\phi_{\pm}) \gg j \ , \tag{27}$$

where $\phi_{\pm}$ and $e_{\pm}$ are given by Eqs. 23 and 24. These inequalities are usually satisfied, except at very high currents.

2.4 PLANAR-DOPED BARRIERS

The planar-doped-barrier (PDB) rectifying structure (Fig. 10) was first demonstrated in GaAs molecular-beam-epitaxy (MBE) grown samples.[12] It represents an extension (the limiting case) of the camel-diode structure.[13] We begin by reviewing the theory[14] of rectification and charge injection in this important structure.

A PDB $[n\text{-}i\text{-}\delta(p^+)\text{-}i\text{-}n]$ structure represents a nearly intrinsic (i) layer of thickness L sandwiched between two n-type layers of low resistivity. In the process of epitaxial growth a p^+-doped layer of ("infinitesimal") thickness $\delta \ll L$ and doping N_A is built into the i region. Acceptors in the p^+ layer are completely ionized, that is, completely depleted of holes. A negative charge sheet of surface density $\Sigma = qN_A\delta$ gives rise to a triangular potential barrier with shoulders L_1 and L_2 and the height Φ approximately given by

$$\Phi \approx -\frac{\Sigma L_1 L_2}{\epsilon_s L} \ . \tag{28}$$

This expression corresponds to a "capacitor" model in which the fixed charge Σ induces charges only in an infinitesimally thin layer of the doped contacts at the boundaries of the i layer. In equilibrium, the barrier height is the same on both sides provided the doping in both contact layers is identical. Under an applied bias V, the height of the emitter ("uphill") barrier will decrease by the amount VL_1/L (without a loss of generality, we can assume that emitter barrier is the one that has the shoulder L_1), giving rise to an exponentially rising current

$$I \propto e^{\beta VL_1/L} \ . \tag{29}$$

Exercise 5: Consider the electrostatics of a PDB structure in the "capacitor" model and ascertain Eq. 29.

The exact shape of the potential barrier, including the effects of the mobile charge diffusion into the i layers, can be found in a manner similar to that described for the $n\text{-}i\text{-}n$ case. This will be discussed in the next two sections.

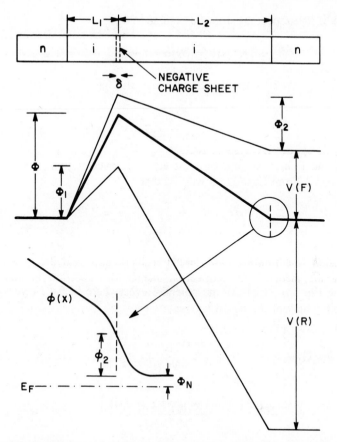

Fig. 10 Schematic illustration of the planar-doped triangular barrier. In the example drawn, the barrier shoulders are related as 3:1. Evidently, one has to apply approximately 3 times larger voltage in the reverse (R) than in the forward (F) direction to obtain the same barrier height ($\Phi_1 = \Phi_2$) for the thermionic emission, hence the rectification. Insert in the lower left corner shows the barrier shape including the effects of carrier diffusion into the i layer. Excess negative charge produces a concave shape of the potential in the (i) region, while depletion of the negative charge in the doped layer gives rise to a convex potential shape there.

2.4.1 Equilibrium Properties

The appropriate solution of the non-linear drift-diffusion equation, Eq. 14, is now of the form

$$
e = \begin{cases}
2\gamma_1 \coth[\gamma_1(\xi + \xi_1)] , & 0 \leq \xi \leq \mathcal{L}_1 \\
-2\gamma_2 \coth[\gamma_2(\mathcal{L} - \xi + \xi_2)] , & \mathcal{L}_1 \leq \xi \leq \mathcal{L} = \mathcal{L}_1 + \mathcal{L}_2
\end{cases} \quad (30)
$$

where $\mathcal{L}_i = L_i/L_D$ and ξ_i, γ_i, $i = 1,2$ are constants to be determined. At

$\xi = 0$ and $\xi = \mathcal{L}$ one has from Eqs. 18 and 30

$$e^{\beta\phi_i} - \beta\phi_i - 1 = 2\gamma_i^2\coth^2(\gamma_i\xi_i) . \qquad i = 1,2 . \qquad (31a)$$

Differentiating Eq. 30 and using Eq. 13a one also has

$$\nu_i = e^{\beta\phi_i} = \frac{2\gamma_i^2}{\sinh^2(\gamma_i\xi_i)} , \qquad i = 1,2 , \qquad (31b)$$

where ν_1 and ν_2 are the dimensionless electron concentrations at the boundaries of the i layer. From Eqs. 31a,b we find that

$$\beta\phi_i = -1 - 2\gamma_i^2 , \qquad i = 1,2 \qquad (32)$$

and

$$\sinh^2(\gamma_i\xi_i) = 2\gamma_i^2\exp(1 + 2\gamma_i^2), \qquad i = 1,2 . \qquad (33)$$

Integrating Eq. 30, we obtain the potential everywhere in the i layer relative to $\phi = 0$ in the bulk of the doped layers:

$$\beta\phi(\xi) = \beta\phi_1 - \ln\left\{\frac{\sinh[\gamma_1(\xi + \xi_1)]}{\sinh(\gamma_1\xi_1)}\right\}^2 , \qquad 0 \le \xi \le \mathcal{L}_1 \qquad (34a)$$

$$\beta\phi(\xi) = \beta\phi_2 - \ln\left\{\frac{\sinh[\gamma_2(\mathcal{L} - \xi + \xi_2)]}{\sinh(\gamma_2\xi_2)}\right\}^2 , \qquad \mathcal{L}_1 \le \xi \le \mathcal{L} . \qquad (34b)$$

From the continuity of ϕ [matching Eqs. 34a and 34b at $\xi = \mathcal{L}_1$] and using Eqs. 32 and 33, we have

$$e^{\beta\Phi} = \frac{2\gamma_1^2}{\sinh^2[\gamma_1(\xi_1 + \mathcal{L}_1)]} = \frac{2\gamma_2^2}{\sinh^2[\gamma_2(\xi_2 + \mathcal{L}_2)]} . \qquad (35)$$

On the other hand, the discontinuity in the electric field at $x = L_1$ is given by Gauss's law, $\mathcal{E}_1 - \mathcal{E}_2 = \Sigma/\epsilon_s$, where $\mathcal{E}_1 = \mathcal{E}(L_1 - \delta/2)$ and $\mathcal{E}_2 = \mathcal{E}(L_1 + \delta/2)$, whence

$$2\gamma_1\coth[\gamma_1(\mathcal{L}_1 + \xi_1)] + 2\gamma_2\coth[\gamma_2(\mathcal{L}_2 + \xi_2)] = \frac{\Sigma}{(\epsilon_s kTN_D)^{1/2}} . \qquad (36)$$

Equations 33, 35, and 36 constitute a system of 4 equations in 4 unknowns ξ_i, γ_i, which can be solved for any given PDB parameters.

The barrier height Φ is then given exactly by Eq. 35. The general expression $\Phi = \Phi(\Sigma, L_1, L_2, N_D)$ simplifies considerably if one notes that in all practical PDB in equilibrium $|e_i\mathcal{L}_i| \gg 1$, or, equivalently, $\mathcal{E}_iL_i \gg kT/q$.

Under this condition $2\gamma_i = |e_i|$, and Eq. 35 reduces to

$$\beta\Phi = \beta\Phi_i = - |e_i| \mathcal{L}_i \tag{37}$$

$$+ \ln \left[2e_i^2 \left(\left[1 + \frac{e_i^2}{2} \exp\left(1 + \frac{e_i^2}{2} \right) \right]^{1/2} - \left[\frac{e_i^2}{2} \exp\left(1 + \frac{e_i^2}{2} \right) \right]^{1/2} \right)^2 \right].$$

In the stronger limit $|e_i| \gg 1$, or $|\mathcal{E}_i| \gg 63$ kV/cm (for $N_D = 10^{18}$ cm^{-3} in Si), we can further simplify Eq. 37:

$$\beta\Phi = - |e_i| \mathcal{L}_i - \frac{e_i^2}{2} - 1 , \quad i = 1,2 . \tag{38}$$

Solving Eqs. 38 together with the Gauss law, Eq. 36, re-written in the form

$$|e_1| + |e_2| = \sigma \equiv \frac{\Sigma}{(\epsilon_s kTN_D)^{1/2}} , \tag{39}$$

where σ is the dimensionless charge-sheet density ($\sigma = 2.4$ at room temperature for $\Sigma/q = 10^{12}$ cm^{-2} and $N_D = 10^{18}$ cm^{-3}), we find

$$\beta\Phi = - \frac{\sigma\mathcal{L}_1\mathcal{L}_2}{\mathcal{L}} - \frac{\sigma(\mathcal{L}_1 - \mathcal{L}_2)^2}{2\mathcal{L}(\mathcal{L} + \sigma)} - \frac{\sigma^2}{8} \left[1 + \left(\frac{\mathcal{L}_1 - \mathcal{L}_2}{\mathcal{L}_1 + \sigma} \right)^2 \right] - \frac{2}{\mathcal{L}} \tag{40a}$$

Thus, for a symmetric PDB ($\mathcal{L}_1 = \mathcal{L}_2$) we have

$$|\Phi| = \frac{\Sigma L_1 L_2}{\epsilon_s L} + \frac{\Sigma^2}{8\epsilon_s qN_D} + \frac{2kTL_D}{qL} . \tag{40b}$$

The corrections to Eq. 28 become appreciable for large Σ and/or small N_D, that is, precisely in the limit when Eqs. 40 are strictly valid. It should be noted that even though the strong inequality $|e_i| \gg 1$ is rarely fulfilled in practice, Eqs. 40 provide a good approximation when $|e_i| \gtrsim 2$, that is, $\mathcal{E}_i^2 \gtrsim 4kTN_D/\epsilon_s$.

2.4.2 PDB under Applied Bias

It is possible to solve Eq. 14 for $J \neq 0$ on each barrier shoulder (the solution is simple if one notes that j is usually a small parameter, since J_0 is typically $> 10^7$A/cm^2) and then obtain the current-voltage (I-V) characteristic by matching the solutions for $n(x)$ at $x = L_1$. However, such a solution would be almost meaningless, because it neglects the effects of carrier acceleration and heating (e.g., the velocity saturation) on the downhill slope of the barrier.

A better approach is to use the thermionic emission theory.[1] This theory assumes that the quasi-Fermi level is constant on the uphill slope of the barrier all the way to $x = L_1$ (see Fig. 1). In the low-current limit the I-V characteristic of a PDB can be calculated by the following procedure: Assume e_1 and e_2 subject to Gauss's law, Eq. 39. Calculate Φ_1 and Φ_2. Set $\Phi_1 - \Phi_2 = V$, and

$$J = qN_D v_T e^{\beta\Phi_i} = AT^2 e^{\beta(\Phi_i + \Phi_N)} , \tag{41a}$$

where Φ_i is the injecting (uphill) barrier, $\beta\Phi_N = \ln(N_D/N_C) < 0$ is the separation between the conduction band and the Fermi level in the (non-degenerately doped) emitter, A is the effective Richardson constant, and

$$v_T = \left[\frac{kT}{2\pi m} \right]^{\frac{1}{2}} \tag{42}$$

is the mean thermal velocity of carriers in a given direction. Repeating the calculation (which can be done with a pocket calculator) for different ratios e_1/e_2, one determines the entire low-current portion of the I-V characteristic.

Figure 11 (solid line) shows the calculated I-V characteristic for an exemplary silicon PDB structure. In the limit $\mathcal{E}_i^2 \geq 4kTN_D/\epsilon_s$ one can use Eq. 38 and calculate $I(V)$ in a closed form:

$$J \doteq AT^2 e^{\beta(\Phi_0 - \epsilon_s V^2/2qL^2N_D)} \left[e^{\beta V(\tilde{L}_1/L)} - e^{-\beta V(\tilde{L}_2/L)} \right] , \tag{41b}$$

where Φ_0 is the sum of Φ given by Eq. 40a and Φ_N, and $\tilde{L}_i$ are the effective barrier shoulder lengths,

$$\tilde{L}_i = L_i + \frac{\Sigma}{qN_D} \left[1 - \frac{2L_i}{L} \right] . \tag{43}$$

We see that at low biases the PDB diode characteristics are approximately exponential with the "leverage" factors† $\lambda_i \equiv \tilde{L}_i/L < 1$; note from Eq. 43 that $\lambda_1 + \lambda_2 = 1$. A strongly asymmetric diode ($L_1 \gg L_2$) exhibits rectification with Φ_1 being the injecting and Φ_2 the blocking barrier. Most of the usefulness of planar-doped barriers derives from this property.

At high applied biases, the exponential I-V characteristic of a PDB begins to saturate and, eventually, it is replaced by a linear law (see the dashed line in Fig. 11). This happens because of (i) slowing of the effective diffusion velocity on the uphill slope, and (ii) screening of the applied field by the mobile charge dynamically stored (i.e., stored while in transit) on the downhill slope. For potential applications of the PDB concept, it is most important to understand at what current levels this saturation occurs.[14] Consider the limitation (i) first.

---

†Diode characteristics of the form $I \propto \exp(\beta V)$ are usually called "ideal." At room temperature, it takes 66 mV to increase tenfold the current of an ideal diode. Deviations from this law are commonly described by an ideality factor $n > 1$. Thus, a practical diode at room temperature requires $66n$ mV per decade of current. Our leverage factor $\lambda < 1$ describes characteristics of the form $I \propto \exp(\lambda\beta V)$ and corresponds to an inverse ideality factor.

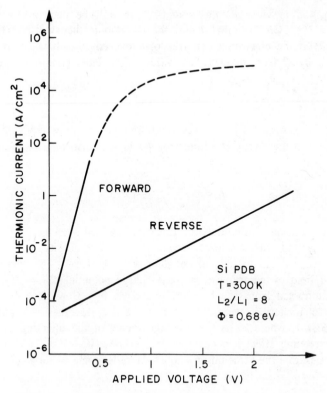

Fig. 11 Calculated *I-V* characteristics for a Si PDB diode with the following parameters: $L_1 = 250\,\text{Å}$, $L_2 = 2000\,\text{Å}$, $N_D = 10^{19}\,\text{cm}^{-3}$, $\Sigma/q = 2 \times 10^{12}\,\text{cm}^{-2}$.

As the applied bias increases, the injecting potential flattens out. Assume, for concreteness, that the uphill slope corresponds to L_1. In the limit $1/\mathcal{L}_1 \ll e_1 \ll 1$ Eq. 37 is still valid and gives $\beta\Phi_1 \approx -e_1\mathcal{L}_1 + \ln(2e_1^2)$. If we formally substitute this expression into Eq. 41a, we obtain

$$J = 2\epsilon_s\beta\, v_T\mathcal{E}_1^2 e^{-\beta\mathcal{E}_1 L_1}. \tag{44}$$

Equation 44 contains a pre-exponential factor that decreases with the bias. It formally predicts $J \to 0$ in the limit of a flat uphill slope $\mathcal{E}_1 = 0$, while it is physically clear that without a barrier we must have a space-charge-limited current flow. The curve $J(\mathcal{E})$ goes through a maximum at $\beta\mathcal{E}_1 L_1 = 2$. The reason for this unphysical behavior is twofold. On the one hand, Eq. 37 loses its validity when $\beta\mathcal{E}_1 L_1 \lesssim 2$, which may be interpreted as a restriction on the effective diffusion velocity $\mu\mathcal{E}_1$ on the uphill slope that must exceed a minimum value $2D/L_1$,

$$\mu\mathcal{E}_1 \gg \frac{D}{L_1}, \tag{45a}$$

(when Eq. 45a is violated, the potential can still be determined exactly by Eqs. 33 to 36). On the other hand, the thermionic theory itself (Eqs. 41a,b) applies only if the electron mean free path exceeds the distance in which the barrier falls by kT from its maximum value.[15] The latter condition can be cast in the form

$$\mu \mathscr{E}_1 \gg v_T . \tag{45b}$$

Restrictions, expressed by Eqs. 45a and 45b are roughly equivalent. Using Eq. 44, we can thus write the limitation (i) in the form of a restriction on the current density:

$$J \ll \frac{\epsilon_s kT}{qL_1} \frac{v_T}{L_1} \equiv J_{C1} \tag{46}$$

Next, we consider limitation (ii). Electrons emitted over the peak of the barrier drift downhill with a saturation velocity v_S. The density ρ of the injected charge is proportional to the current density, $\rho = J/v_S$. This charge screens the applied field by contributing a space-charge potential $\Delta\phi = JL_2^2/2\epsilon_s v_S$ that must be subtracted from V when one evaluates Φ_1. If without this screening effect the characteristic would be similar to Eq. 41, that is, approximately $J \propto \exp(\beta V\lambda_1)$, where λ_1 is the leverage factor of the injecting barrier, then with the screening effect it becomes $J \propto \exp[\beta\lambda_1 (V - JL_2^2/2\epsilon_s v_S)]$. We can, therefore, describe the screening by an effective leverage factor $\tilde\lambda_1 (J)$, namely,

$$\tilde\lambda_1(J) = \lambda_1 \left[1 + \frac{qJ\lambda_1}{kT} \frac{L_2^2}{2\epsilon_s v_S} \right] , \tag{47}$$

from which we can express limitation (ii) also as a restriction on the current density:

$$J \ll \frac{\epsilon_s kT}{qL_2} \frac{v_S}{L_2} \equiv J_{C2} . \tag{48}$$

When the conditions expressed by Eqs. 46 and 48 are fulfilled, we have a thermionic current and an efficient charge injection over the triangular barrier. Above J_{C1} or J_{C2} one deals with a space-charge-limited current. At room temperature, $v_T \sim v_S \sim 10^7$ cm/s and for $L_i \sim 10^{-5}$ cm ($i = 1$ or 2) one has $J_{Ci} \sim 3$ kA/cm^2. At lower T the limitation (i) becomes progressively more restrictive since $J_{C1} \propto T^{3/2}$.

Finally, let us emphasize that in the entire discussion in this section we have neglected all possible *bipolar* effects, such as an accumulation of mobile holes under the triangular barrier. The minority carriers can be injected electrically, like in a bipolar transistor, or optically as in a photodiode. An extension of the theory discussed in this section includes the effects of minority-carrier storage on the barrier height and the *IV* characteristics of a PDB diode.[16]

2.5   SPACE-CHARGE-LIMITED CURRENT IN BULK SEMICONDUCTORS AND THIN FILMS

Space-charge-limited (SCL) current in a bulk double-junction *n-i-n* diode has been discussed in Section 2.3. Conduction in such structures is due to electrons injected into the (*i*) base. The uncompensated charge of injected electrons limits the current. In this section, we shall also consider the SCL current in a thin-film diode. It will be assumed that potential barriers confine the current flow within a semiconductor layer of thickness D.

Whether one deals with the bulk or film case, the mechanism of electron transport is different in different parts of the diode. In a symmetrical structure at equilibrium, the electric field $\mathscr{E}_x$ vanishes in the middle of the diode base. As an external bias V is applied, the point $\mathscr{E}_x = 0$ moves toward the cathode (cf. Section 2.3.2). This point defines the position of the so-called virtual cathode. In the bulk case, the virtual-cathode geometry is obvious: it represents a plane perpendicular to the current flow. Only the position of this plane changes with an applied bias. For a film, the virtual cathode represents a surface defined by the condition of a vanishing normal component of the field. This surface is not necessarily equipotential, and its exact shape cannot be determined without considering the diffusion current component. The usefulness of the virtual-cathode concept lies in the fact that it allows separation of the mobile charge in the base into two groups. For charges located on the anode side of the virtual-cathode surface, the electric-field lines terminate on the positive charge at the anode electrode, whereas for charges located on the cathode side of the surface, the field lines terminate on the cathode electrode.

The transport of electrons between the cathode contact and the virtual cathode cannot be described without accounting for diffusion because in this region electrons move against the electric field. However, with increasing bias this region shrinks and, for a sufficiently high V, almost the entire base of the diode is located between the virtual cathode and the anode where the dominant transport mechanism is carrier drift in the electric field.

The virtual-cathode approximation corresponds to neglecting the variation of the quasi-Fermi level on the cathode side of the virtual cathode and the diffusion transport on the anode side. In this approximation the entire injected charge has field lines terminating on the anode. The actual field distribution depends on the shape of the electrodes. It has been rigorously shown in Section 2.3.2 that in a bulk diode the virtual-cathode approximation is quite accurate, provided that $V \geq 10kT/q$. In this section, we shall be interested mainly in the large-current limit, where this approximation is certainly justified. It will allow us to derive an analytic form of the conduction law governing the SCL current in thin films, which is analogous to the Mott-Gurney result for bulk diodes. Our results will be confirmed by a numerical simulation that does not rely on the virtual-cathode approximation and can be regarded as an exact solution of the drift-diffusion model.

2.5.1   General Form of SCL Current for Different Transport Mechanisms

The general form of the law governing SCL current in the virtual-cathode approximation can be obtained without calculations from a simple dimensional argument. Indeed, in CGS units the conductivity has the dimensions of velocity. Taking this velocity to be an effective carrier velocity v, we can write a generic expression for the SCL current in the form†

$$I \propto \frac{\epsilon}{4\pi} vV \frac{A}{L^2}, \qquad \text{(bulk)} \qquad\qquad (49)$$

$$I \propto \frac{\epsilon}{4\pi} vV \frac{W}{L}, \qquad \text{(film)}$$

where L is the length, $A = DW$ the cross-sectional area, and W the width of the diode (see Fig. 12). The relative permittivity $\epsilon \equiv \epsilon_s/\epsilon_0$ of the material enters because it scales the space-charge potential in Poisson's equation. The actual current-voltage dependence (up to a numerical coefficient) can be "derived" from Eq. 49 whenever the conduction process involves a dominant transport mechanism, which provides a unique scaling relationship between v and V. Thus, for free electron motion, the velocity scales as $v^2 \propto (q/m)V$ and one obtains laws appropriate for ballistic transport, for example, for the bulk case Child's law (Eq. 10):

$$I \propto \frac{\epsilon}{4\pi} \left[\frac{q}{m} \right]^{1/2} V^{3/2} \frac{A}{L^2}. \qquad\qquad (50)$$

For the case when electron velocity is saturated, the law is of the form of Eq. 49 with $v = v_S$, and for the case of constant mobility μ, the velocity scales as $v \propto \mu V/L$, which leads to the following expressions:

$$I = \zeta_3 \frac{\epsilon}{4\pi} \frac{\mu V^2}{L^3} A \qquad \text{(bulk)},$$

$$\qquad\qquad (51)$$

$$I = \zeta_2 \frac{\epsilon}{4\pi} \frac{\mu V^2}{L^2} W \qquad \text{(film)}.$$

The numerical coefficient ζ_3 corresponding to the bulk case is $\zeta_3 = 9/8$. For the case of a film, the corresponding coefficient ζ_2 is not universal but depends on the shape of the contacts. The SCL current in a film was first discussed long time ago[17] for a special case of a strip-contact geometry in the limit $D \to 0$. The result obtained corresponds to $\zeta_2 = 2/\pi$.

Recently, this problem was re-analyzed[18] for three representative film geometries. These geometries, illustrated in Fig. 12 (a to c), are point contacts (a),

---

†Using this dimensional argument as a cover, we shall be employing the CGS units throughout Section 2.5. Connection to formulas of Sections 2.3 and 2.4 is established by replacing the dimensionless factor $\epsilon/4\pi$ with ϵ_s in farads per meter.

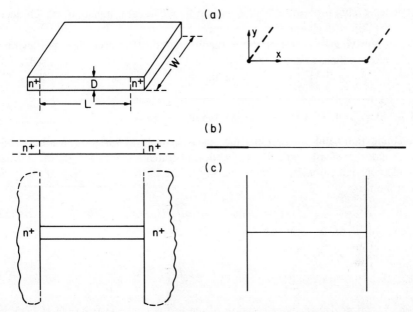

Fig. 12 Illustration of various thin-film contact geometries. The conducting contacts are provided by heavily doped semiconductor regions. The film lies in the xz plane. (a) "Edge contacts": Cross section of the contact is a square with the side equal to film thickness D. In the limit $D \to 0$, the contact is modeled by conducting filament. (b) "Strip contacts": The contacts are doped layers of the same thickness as the thin-film base. In the limit $D \to 0$, the contacts are modeled as conducting planes co-planar to the base. (c) "H-shaped contacts": The contacts represent doped bulk regions, modeled as conducting planes perpendicular to the plane of the base. (After Ref. 18).

contacts in the form of two semi-infinite strips that have the same thickness as the film (b), and contacts in the form of two infinite planes perpendicular to the film (c). Besides evaluating ζ_2 in the limit $D \to 0$ for each of these geometries, the SCL was numerically calculated for a finite film thickness, including effects of the velocity saturation.

The method of solution, employed in Ref. 18, is as follows. First, find the Green function of the Poisson equation, which describes the electric field of the injected charge and the charge induced on the electrodes. For point contacts (Fig. 12a), the location of the charge on the anode electrode is specified and the evaluation of the two-dimensional potential distribution is straightforward. For extended contacts (Figs. 12b,c) the induced charge is distributed in a non-trivial way, and the easiest way to include it is to use a Green function that satisfies the boundary condition of a fixed potential on the conducting electrode surfaces. The total potential and field distribution in the device is found next by integrating over the injected charge density. The latter is related locally to the current and the total local electric field—including that due to external charges on the electrodes. This procedure leads to a non-linear integral equation, which determines the injected charge density. Once the charge distribution is known, the SCL current-voltage characteristic is readily found. The procedure is illustrated in the next section first for the case of a bulk diode, where it is used to re-derive the classical Mott-Gurney law, and then for the case of a thin film.

2.5.2 SCL in a Bulk Diode: the Mott-Gurney Law

The conduction-band diagram and the quasi-Fermi level variation in a bulk n-i-n diode under a small applied bias were illustrated in Fig. 8. In the virtual-cathode approximation, the charge density in the base $(0 \le x \le L)$ can be written in the form

$$\rho(x) = -qn(x) + qP\,\delta(L-x) , \tag{52}$$

where

$$P \equiv \int_0^L n(x)\, dx . \tag{53}$$

In Eq. 52 we neglected the separation of the virtual cathode from the cathode by setting $x_0 = 0$. All the positive anode charge that neutralizes the charge $n(x)$, injected into the region between the virtual cathode and the anode, is placed at the plane $x = L$. The charge density ρ obeys the total neutrality condition.

The electrostatic potential in the diode base is determined by the Poisson equation

$$\frac{d^2\phi}{dx^2} = -\frac{4\pi}{\epsilon}\rho(x) , \tag{54}$$

and the boundary conditions

$$\frac{d\phi}{dx} = 0 , \quad \text{at } x \equiv x_0 = 0 \tag{55}$$

$$\phi(L) = V \tag{56}$$

The Green function $G(x, x') = -\frac{1}{2}\,|x - x'|$ of the Poisson equation, Eq. 54, allows us to write its solution in the form

$$\phi(x) = -\frac{2\pi}{\epsilon}\int_0^L |x - x'|\,\rho(x')\,dx' \tag{57}$$

$$= \frac{2\pi q}{\epsilon}\int_0^L [\,|x - x'| + |L - x|\,]\,n(x')\,dx'.$$

Note that the charge density ρ chosen, as in Eq. 52, to satisfy total neutrality, automatically guarantees that the boundary condition (Eq. 55) are obeyed. This follows from Gauss's law and can also be seen from the expression for the electric field in the base, obtained from Eq. 57:

$$\mathcal{E}(x) = -\frac{d\phi}{dx} = -\frac{4\pi q}{\epsilon} \int_0^x n(x')\, dx' . \tag{58}$$

Equation 58 combined with the constant-mobility approximation for the current density,

$$j = q\mu n(x)\mathcal{E}(x) , \tag{59}$$

leads to the following non-linear integral equation for the carrier density:

$$\frac{4\pi q^2 \mu}{-\epsilon j} n(x) \int_0^x n(x')\, dx' = 1 , \tag{60}$$

and the solution of this equation is of the form

$$n(x) = \left[-\frac{\epsilon j}{8\pi q^2 \mu x} \right]^{1/2} . \tag{61}$$

When substituted in Eq. 57 and using Eq. 56, this solution yields the Mott-Gurney law

$$I = -\frac{9\mu}{8} \frac{\epsilon}{4\pi} \frac{V^2}{L^3} A . \tag{62}$$

The method employed in this derivation of Eq. 62 differs from the original derivation.[11] It has the advantage of being convenient to generalize to the case of a thin film.

2.5.3   SCL Current in Thin Films: Analytical Results

Our analytic treatment will be confined to the limiting case of a vanishing film thickness. Physically, it is sufficient to assume that $D \ll L$. Instead of the current density used in Eq. 10, we shall use a linear current density

$$J_L \equiv \frac{I}{W} = q\mu n_s(x)\mathcal{E}_x(x) , \tag{63a}$$

where n_s is the surface electron concentration in the film and $\mathcal{E}_x$ is the x-component of the electric field in the film ($y = 0$). Of course, Eq. 63a is only valid in the approximation of constant mobility. Later, we shall also consider the SCL current in the opposite limit, corresponding to a vanishing differential mobility and constant "saturated" velocity $v = v_S$, where Eq. 63a must be replaced by

$$J_L = q\, n_s(x)\, v_S . \tag{63b}$$

An exact expression for the current-voltage characteristic in this limit will allow us to assess the range where the constant-mobility approximation is valid and to illustrate the effect of velocity saturation.

The potential distribution in the film device, determined by the two-dimensional Poisson equation,

$$\frac{d^2\phi}{dx^2} + \frac{d^2\phi}{dy^2} = -\frac{4\pi}{\epsilon}\rho(x),$$ (64)

must satisfy the boundary conditions on the electrodes and be consistent with Eq. 63. We shall only discuss the case of a film with coplanar-strip contacts (Fig. 12b), and refer the reader to the original paper[18] for the treatment of other electrode geometries. The simplest case of edge contacts is the subject of Problem 5.

Film with Two Coplanar Strip Contacts.   This case is somewhat more complicated than the bulk case because we do not know *a priori* the charge distribution induced in the extended contacts. We can circumvent this problem by using the Green function of Poisson's equation subject to the boundary condition that both electrodes are at zero potential. Potential $\tilde{\phi}(x, y)$ determined with the help of this function must then be added to the potential determined from the homogeneous (Laplace) equation that describes the field due to electrodes biased by V neglecting that due to the mobile charge.

The Green function $G(x, x'; y, y')$ of Eq. 64 can be found by the standard methods of conformal mapping, using the method of images for the source charge. For our purposes, we need only the values of $G(x, x'; y, y')$ at $y = y' = 0$. This function is of the form

$$G(\xi, \xi') = \frac{1}{2\pi}\,\text{Re}\,\ln\left[\frac{\sin[(\psi + \psi')/2]}{\sin[(\psi - \psi')/2]}\right],$$ (65)

where

$$\xi \equiv \frac{2x - L}{L}, \qquad \psi \equiv \arccos(\xi);$$

$$\xi' \equiv \frac{2x' - L}{L}, \qquad \psi' \equiv \arccos(\xi').$$

From this function the potential produced by injected electrons is obtained in the following form:

$$\tilde{\phi}(\xi) = -\frac{2\pi qL}{\epsilon}\int_{-1}^{+1} n_s(\xi')G(\xi, \xi')\,dx',$$ (66)

and the total potential is given by

$$\phi(\xi) = \tilde{\phi}(\xi) + V\left[1 - \frac{\psi(\xi)}{\pi}\right].$$ (67)

Differentiating Eq. 67, we find the electric field distribution in the film:

$$\mathcal{E}_x(\xi) \equiv -\frac{2}{L}\frac{\partial\phi}{\partial\xi} = \frac{-2}{(1-\xi^2)^{\frac{1}{2}}}\left[\frac{q}{\epsilon}\int_{-1}^{1}\frac{n_s(\xi')(1-\xi'^2)^{\frac{1}{2}}d\xi'}{\xi-\xi'} + \frac{V}{\pi L}\right]. \quad (68)$$

Expressing $\mathcal{E}_x$ with the help of Eq. 63a in terms of the surface concentration n_s and writing V in the form

$$V = -\int_0^L \mathcal{E}_x(x,0)\,dx = \left[\frac{-J_L L^2}{2\epsilon\mu}\right]^{1/2}\int_0^1\frac{d\xi}{v_s(\xi)}, \quad (69)$$

where v_s is the dimensionless surface concentration defined by

$$v_s(\xi) = \left[\frac{2q^2\mu}{-J\epsilon}\right]^{1/2} n_s(\xi), \quad (70)$$

we can bring Eq. 68 into the form of a non-linear integral equation for v_s:

$$\int_{-1}^{1}\left[\frac{1}{\pi v_s(\xi')} + \frac{v_s(\xi')(1-\xi'^2)^{1/2}}{\xi-\xi'}\right]d\xi' = \frac{(1-\xi^2)^{1/2}}{v_s(\xi)}. \quad (71)$$

A reasonably accurate approximation for the solution of this equation is achieved by assuming v_s in a form with free parameters:

$$v_s(\xi) = a(1+\xi)^\alpha(1-\xi)^\beta, \quad (72)$$

and finding the best values of the parameters ($a^2 \approx 0.18$, $\alpha \approx -0.4$, and $\beta \approx 0.4$) by minimizing the average error in Eq. 71. When Eq. 72 is substituted in Eq. 69, one obtains a current-voltage characteristic of the form of Eq. 51 with

$$\zeta_2^b \approx \frac{2^{1+2\alpha+2\beta}\pi a^2[\Gamma(2-\alpha-\beta)]^2}{[\Gamma(1-\alpha)\Gamma(1-\beta)]^2} \approx 0.70. \quad (73)$$

This result is in a good agreement with the exact expression,[17] $\zeta_2^b = 2/\pi \approx 0.64$, for the strip-contact geometry.

This analytical description of the SCL current in thin films was verified by numerical simulations of the problem[18] that took into account both the drift and the diffusion transport mechanisms, as well as the effects of field-dependent mobility. In these simulations, a material with relative permittivity $\epsilon = 10$ was assumed with doping concentration $N_{\text{background}} = 10^{12}\,\text{cm}^{-3}$ (at such low concentrations none of the calculated characteristics was sensitive to a variation in $N_{\text{background}}$). The contacts were assumed doped to $10^{19}\,\text{cm}^{-3}$ with a Gaussian tail of $0.01\,\mu\text{m}$ at the junction with the low-doped material. The low-field mobility was assumed to equal $\mu = 10^3\,\text{V-cm/s}$ and the diffusion coefficient was defined by the Einstein relation at $T = 300\,\text{K}$.

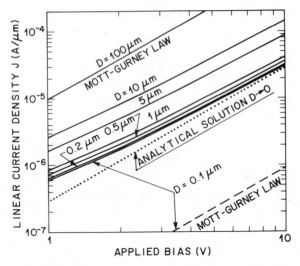

Fig. 13 Current-voltage characteristics of a thin-film diode with strip-contact geometry. Solid lines show the curves calculated numerically for different film thicknesses D with a drift-diffusion simulator, assuming a constant mobility $\mu = 1000$ cm^2/V-s and an associated diffusion coefficient determined by Einstein's relationship at room temperature. The dotted line corresponds to the analytical expression, Eqs. 51, with $\zeta_2 = 0.7$. Also shown are the predictions of the Mott-Gurney law for two values of D. For $D = 100\,\mu$m this law agrees with the numerical simulation to within the accuracy of the figure. In contrast, for a thin film the Mott-Gurney law (dashed line) gives a current density that is lower than the actual current by more than an order of magnitude. (After Ref. 18).

Figure 13 shows the current-voltage characteristics calculated using the constant-mobility approximation for different film thicknesses. For thick films $D \geq L$ under high bias ($\geq 10\,kT$), the numerically calculated IV characteristic coincides with the Mott-Gurney law. As D decreases, the simulated curves converge toward the analytical result for thin films. When $D \leq 0.2\,\mu$m there is almost no further variation of the current. This means that a further decrease in D is entirely compensated by the increasing volume density of charge in the film, so that the surface concentration $n_s(x)$ becomes independent of D.

In a thin-film diode the current density can be substantially higher (by an order of magnitude) than in the bulk case. Physically, this is due to the electric field lines spreading out of the film, which relaxes the space-charge limitation of carrier injection. Because of that, the dependence of SCL current on the diode length L differs from the bulk case. That this could be surmised "for free" by the argument leading to Eq. 51 demonstrates the power of dimensional analysis.

SCL Current in the Saturated-Velocity Limit. It is interesting to compare the results, obtained under the assumption of a constant mobility, with the opposite limiting case, when the differential mobility is zero and carrier velocity is saturated, $v = v_S$. In this case, the carrier concentration is uniform and governed by the current density—independently of the local electric field:

$$n_s = - \frac{J_L}{q v_S} . \tag{74}$$

Substituting Eq. 74 into Eq. 68 and integrating, we find the electric field:

$$\mathcal{E}_x(\xi) = - \frac{2}{(1 - \xi^2)^{1/2}} \left[\frac{\pi \xi J_L}{\epsilon v_S} - \frac{V}{\pi L} \right] . \tag{75}$$

The field must vanish at the virtual cathode position, $x = 0$, that is, $\mathcal{E}_x(\xi = -1) = 0$, whence

$$J_L = - \frac{\epsilon}{4\pi} \frac{4 v_S V}{\pi L} . \tag{76}$$

Figure 14 shows the results of a numeric simulation of a thin-film n-i-n diode. The drift-diffusion simulator assumed a velocity-field curve as in Eq. 6a. The simulated current-voltage characteristic nicely interpolates between the analytically calculated limits of constant mobility (valid when $\mu \mathcal{E}_x \ll v_S$ over most of the diode base) and constant drift velocity $v = v_S$ at high applied fields.

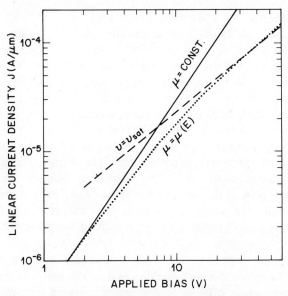

Fig. 14 Illustration of the effect of velocity saturation on the current-voltage characteristics of a thin-film diode ($L = 4.8\,\mu m$) with strip-contact geometry. Dotted line shows the results calculated for $D = 0.1\,\mu m$ by a drift-diffusion simulator, assuming a field-dependent mobility, Eq. 6a, with the low-field value $\mu = 1000\ cm^2/V\text{-s}$ and $v_S = 10^7\,cm/s$. Solid line corresponds to a simulation in which the velocity-field characteristic was assumed in the form $v = \mu \mathcal{E}_x$ with a constant $\mu = 1000\ cm^2/V\text{-s}$. The dashed line was calculated with Eq. 76, corresponding to a constant drift velocity v_S. (After Ref. 18).

Equation 76 gives a correct expression for the SCL current in the high-bias limit. Compared to the case of constant mobility, it gives a different scaling of the current with the diode length (L^{-1} instead of L^{-2}).

Whenever Eq. 51 predicts a higher current than Eq. 76, it is a clear indication that the constant-mobility approximation is invalid. For the strip-contact geometry this happens when the applied voltage exceeds the value

$$V_{\text{transition}} = \frac{4}{\pi \zeta_2^b} \frac{L v_S}{\mu} . \qquad (77)$$

2.6   ELECTRONIC SUBBANDS IN QUANTUM WELLS AND INVERSION LAYERS

In most of our discussion in the previous sections, quantum mechanics played no explicit role. This should not imply, however, that quantum effects are unimportant for bulk unipolar structures with a sharp doping profile. Consider a few examples:

Thermally assisted tunneling in ultrathin planar-doped barrier diodes. In such structures, at room temperature, the current is mainly thermionic, whereas at lower temperatures, the favored mechanisms are those that require a smaller activation energy. A similar situation occurs in forward-biased Schottky diodes.[19] One such mechanism is thermionic field emission, or thermally assisted tunneling. This mechanism is illustrated in Fig. 15 based on the experimental results in ultra-thin GaAs diodes.[20] At low temperatures, the direct thermionic current is small. The contribution of tunneling electrons at a given energy is proportional to the product of the number of electrons at this energy and the tunneling probability. The former decreases exponentially with energy according to the Boltzmann formula while the latter increases sharply with decreasing phase area of the barrier above the given energy. The product is sharply peaked at some energy E_a and the main contribution to the current is due to electrons with an energy near E_a. The quantity E_a is the activation energy in a given temperature range.

Quantum subbands in δ-doped quantum wells. The power of modern epitaxial techniques such as MBE to control doping virtually on a monolayer scale (δ-doping as the extreme case of planar doping) has been explored in creating electronic confinement in the vicinity of planar-doped layers. Figure 16 shows schematically the potential profile and electronic subbands in a δ-doped quantum well. Capacitance-voltage profiling of δ-doped GaAs shows that during the MBE growth Si impurities do not diffuse over more than two lattice constants.[21] Confined electronic systems by δ-doping can have applications in field-effect transistors.[22, 23]

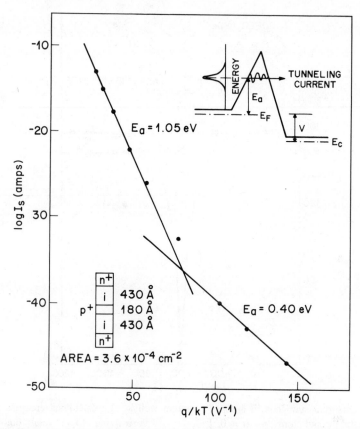

Fig. 15 Thermally assisted tunneling in ultra-thin planar-doped barrier diodes. The lower insert shows the cross section of a GaAs diode. The Arrhenius plot shows the temperature dependence of the current extrapolated to zero voltage. The upper insert illustrates the thermionic field-emission mechanism. At high T the activation energy, $E_a = 1.05\,\text{eV}$ is close to the designed barrier height. (After Ref. 20).

Doping superlattices. Implementation of periodic potentials (superlattices) by n-i-p-i doping modulation has been studied extensively in recent years.[24] These structures offer the device designer a possibility of tuning certain electrical and optical properties (e.g., the bandgap). One attractive application of n-i-p-i superlattices is the possibility of implementing infrared photodetectors in the wavelength range above the fundamental threshold for interband absorption in a given semiconductor. Figure 17 illustrates the band diagram of a *nipi* superlattice in which the n and p dopants are effectively confined to one atomic plane ("sawtooth" superlattice). The low-temperature photoluminescence spectra of this structure exhibit multiple, clearly resolved, interband transitions from quantum-confined states of the superlattice.[25] Note that the overlap of electron and hole wave functions, necessary for recombination,

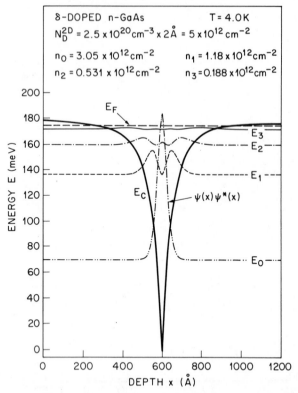

Fig. 16 Quantum subbands in δ-doped quantum wells. The calculated energies E_i and the subband populations n_i $(i = 0, 1, 2, \cdots)$ correspond to a total doping of $N_D^{2D} = 5 \times 10^{12}\,\mathrm{cm}^{-2}$. Each of the subband states is illustrated by a sketch of its probability density $|\psi(x)|^2$. (After Ref. 21).

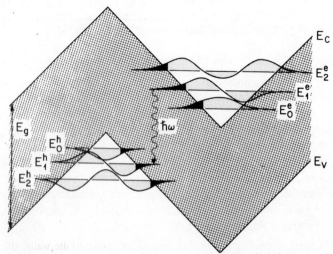

Fig. 17 Schematic band diagram of a sawtooth *n-i-p-i* superlattice by δ-doping. (After Ref. 25)

occurs in the classically forbidden region, that is, beyond the classical turning points for the carrier motion.

We shall not go further in describing these structures, referring the reader to the original papers and reviews cited above. However, the concepts of two-dimensional confinement, tunneling, and superlattices will be the main subject of the rest of this chapter. These issues will be discussed using potential barriers that represent (in first approximation) abrupt walls, as provided by heterojunctions and semiconductor-insulator interfaces. These examples were chosen because abrupt interfaces give a closer approximation to simple models with transparent solutions.

2.6.1 One-Dimensional Quantum Wells

Let us consider several one-dimensional examples of quantum wells, Fig. 18. Figure 18a shows a square-well potential (with infinite walls). The eigenfunctions (not normalized)

$$\chi_n(z) \sim \sin\left[(n+1)\pi\eta\right], \qquad n = 0, 1, 2, \cdots \tag{78a}$$

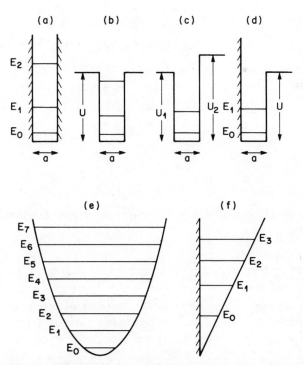

Fig. 18 Examples of quantum wells. (a) Square well will infinite walls, (b) symmetric square well with finite walls, (c) asymmetric square well with finite walls, (d) one-sided square well, (e) parabolic well, and (f) triangular well.

where $\eta = z/a$ and a is the quantum-well (QW) width, correspond to the energy eigenvalues

$$E_n = \frac{\pi}{2} \frac{\hbar^2}{ma^2} (n + 1)^2 . \tag{78b}$$

An exact solution is also available for square wells with walls of finite height (symmetric or asymmetric, Figs. 18b-d). Such wells can support only a finite number of bound states (cf. Problem 6). A symmetric well, however shallow, always has at least one bound state; an asymmetric well may have no bound states at all. For those bound states whose energies are far from the top of the confining barrier, a rough estimate of energy is obtained by assuming a well of same width but with infinite barriers. This estimate always gives an upper bound to the corresponding energy levels.

For a parabolic well (Fig. 18e), $U(z) = \frac{1}{2} m\omega^2 z^2$ and the eigenfunctions are

$$\chi_n(z) \sim e^{-\eta^2/2} H_n(\eta) , \qquad n = 0, 1, 2, \cdots \tag{79a}$$

where H_n are the Hermite polynomials, $\eta = z/a$, and $a \equiv (\hbar/m\omega)$. The energy eigenvalues are

$$E_n = \hbar\omega (n + \frac{1}{2}) = \frac{\hbar^2}{ma^2} \left[n + \frac{1}{2} \right] . \tag{79b}$$

A triangular well, confined by an electric field $\mathcal{E}$ (Fig. 18f), corresponds to the potential

$$U(z) = \begin{cases} q\mathcal{E}z & \text{if } z > 0 ; \\ \infty & \text{if } z \leq 0 . \end{cases}$$

The eigenfunctions are

$$\chi_n(z) \sim \text{Ai} \left[\eta - \frac{E_n}{q\mathcal{E}a} \right] , \qquad n = 0, 1, 2, \cdots \tag{80a}$$

where $\eta = z/a$ and $a \equiv (\hbar^2/2mq\mathcal{E})^{1/3}$, Ai$(x)$ is the Airy function, and the corresponding energy levels are

$$E_n = \frac{1}{2} \frac{\hbar^2}{ma^2} \alpha_{n+1} \approx \frac{1}{2} \left[\frac{3\pi}{2} \right]^{2/3} \frac{\hbar^2}{ma^2} \left[n + \frac{3}{4} \right]^{2/3} , \tag{80b}$$

α_i being the ith root of the Airy function in order of increasing magnitude.[†] The approximate expression on the right-hand side of Eq. (80b) is quasi-classical. Although it is supposed to be valid only asymptotically for large

---

[†]The first three roots of equation Ai$(-\alpha) = 0$ are $\alpha_1 \approx 2.338$, $\alpha_2 \approx 4.088$, and $\alpha_3 \approx 5.527$.

quantum numbers n, it provides an excellent approximation even for $n = 0$ (see Problem 7).

Note that the energy levels in these examples can be expressed in terms of a characteristic confinement energy of a quantum well, $\hbar^2/ma^2$.

Exercise 6: Evaluate this energy assuming $a = 100\,\text{Å}$ and the effective electron mass of GaAs. Do the same for the longitudinal effective electron mass of Si. Express the results in meV, in degrees K (from $E = kT$) and Hz (from $E = h\nu$). The length a is a measure of the extent of the ground-state wave function (cf. an example in Problem 7e).

2.6.2   Two-Dimensional Electron Gas

There are no truly two-dimensional (2D) systems in nature. The wave function of a quantum-well electron is, in general,†

$$\psi(x, y, z) = \sum_n \chi_n(z)\, \phi_n(x, y) , \tag{81}$$

and has a finite width in the z direction whatever the mix of states in Eq. 81, and even if only one term is present. In a truly 2D world particle *interactions* would be different.

Exercise 7: Show that in a 2D world the Coulombic potential would be logarithmic rather than $1/r$. Note the basic property that field lines begin and terminate on charges only, which is ensured by Gauss' law.

The two-dimensional electron gas (2DEG) is a *dynamically* 2D system, which means that $\chi(z)$ is fixed in a state $\chi_n(z)$ during the processes under consideration. This requires that the temperature be much lower than the energy separation to the next level in the confining potential,

$$kT << (E_{n+1} - E_n) \sim \frac{\hbar^2}{ma^2} . \tag{82}$$

It is also necessary that different χ_n should not be mixed appreciably by the perturbations, which requires that the off-diagonal matrix elements of the perturbing interaction satisfy an inequality similar to Eq. 82 for kT.

When these conditions are satisfied, it is permissible to separate the Hamiltonian into the part acting on $\chi(z)$ and that acting on $\phi(x, y)$, and the total wave function can be factored:

$$\psi(x, y, z) = \chi_n(z)\, \phi_n(x, y) . \tag{83}$$

The energy eigenstates $E_{n,\,\alpha,\,\beta}$ are then labeled by the quantum numbers n, α, β

---

†In general, for a quantum well confined by finite-height barriers, the sum in Eq. 81 runs over the states of the continuous spectrum also.

where α and β specify the state of in-plane motion. For a *free* 2DEG, the stationary in-plane states are plane waves and $(\alpha, \beta) = (k_x, k_y)$:

$$\psi_{n, k_x, k_y}(x, y, z) = \chi_n(z) \, e^{\, i(k_x x + k_y y)} \,, \tag{84}$$

and the total energy is given by

$$E_{n, k_x, k_y} = E_n + \frac{\hbar^2 k_x^2}{2m} + \frac{\hbar^2 k_y^2}{2m} \,. \tag{85}$$

Equations 84 and 85 describe the situation when the in-plane effective-mass tensor is isotropic (as is the case for the most technologically important 2DEGs: those at GaAs/AlGaAs heterojunction interfaces and those in MOS inversion layers at a Si {100} surface).

Exercise 8: Consider the more general situations. How would Eqs. 84 and 85 look for a quantum well in an anisotropic semiconductor, if the z axis is a principal symmetry axis of the effective-mass tensor, but *not* an axis of cylindrical symmetry? What if the z axis is not even a principal symmetry axis?

It should be clearly understood that the energy spectrum of a free 2DEG has no gaps above the ground-state energy E_0. Each of the levels E_n represents the seat of a parabolic spectrum that extends up to the energies that limit the free-electron approximation. If such a limit did not exist (a rather absurd supposition), then the spectrum would look like a set of ideal parabolas, extending to

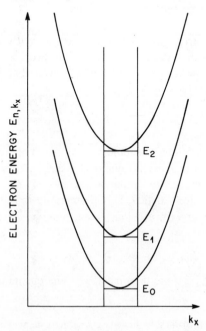

Fig. 19   Illustration of the energy spectrum of a 2D electron gas in a quantum well.

"infinity," Fig. 19. If at an energy $E > E_n$ an electron could still be regarded as free, then it could belong to any of the n subbands—from E_0 (in which case it has an extremely high kinetic energy) to E_n. It should be noted, however, that the idealization leading to the separability of the Hamiltonian and factorization of the wave function, as in Eq. 83, will, typically, break down for degenerate states belonging to different subbands. Such states are unavoidably mixed together by perturbations. Nevertheless, since the mixing force must supply or take on a finite in-plane quasi-momentum, single-subband states may be sufficiently long-lived that the mixing phenomenon can be regarded as "intersubband" scattering.

In a real 2DEG system, electrons move not only in the quantum-well potential, but also in the field of other electrons. The many-body problems that arise can be approached with techniques of varying complexity.[26] The commonly used Hartree approximation (or Hartree-Fock approximation that includes the exchange interaction) leads to a single-electron potential, which itself depends on the wave function, thus making the Schrödinger equation non-linear. A number of iterative numerical techniques have been devised to achieve a fast convergence to the self-consistent result. Useful results are often achieved by a simpler approach based on assuming a variational trial function that embodies the gross features of an expected solution.[27] This approach is illustrated in Problem 8.

2.6.3 Density of States and the Fermi Level

Consider the phase space of a single particle in d dimensions. It has $2d$ axes, corresponding to d coordinates and d momenta of the particle, Fig. 20. It is the basic tenet of quantum statistics that a hyper-volume $V^{(d)}$ (say, a hypercube $V^{(d)}$

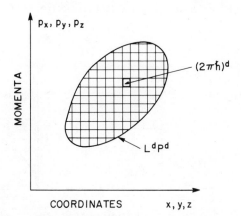

Fig. 20 Hypervolume $V^{(d)} = L^d P^d$ in the phase space contains N cells of volume $(2\pi\hbar)^d$ and twice as many electronic states.

$= L^d P^d$) in the phase space contains N distinct states, where for a spin-½ particle

$$N = \frac{2V^{(d)}}{(2\pi\hbar)^d} \; . \tag{86}$$

The factor of 2 in the numerator comes from states of different spin. For a hyper-sphere of radius R, the hyper-volume equals $R^d \pi^{d/2}/\Gamma(1+d/2)$.

For a free-electron gas, the energy dispersion relation is isotropic, $E = p^2/2m$, which allows a simple counting of states in spherical energy shells $dE = (2E/m)^{1/2} dp$:

$$1\text{D}: \quad dN = \frac{2L \, dp}{2\pi\hbar} = \left(\frac{\sqrt{m}}{\pi\hbar\sqrt{2E}}\right) L \, dE; \tag{87a}$$

$$2\text{D}: \quad dN = \frac{2L^2 2\pi p \, dp}{(2\pi\hbar)^2} = \left(\frac{m}{\pi\hbar^2}\right) L^2 \, dE; \tag{87b}$$

$$3\text{D}: \quad dN = \frac{2L^3 4\pi p^2 \, dp}{(2\pi\hbar)^3} = \left(\frac{(2m)^{3/2}\sqrt{E}}{2\pi^2\hbar^3}\right) L^3 \, dE. \tag{87c}$$

In these equations, the quantity in the brackets is the density of states $g(E) \equiv dN/dE$ (per unit length, area, or volume of the electron gas). Convenient units for $g(E)$ are $[\text{cm}^{-d} \text{ eV}^{-1}]$. Figure 21 illustrates the $g(E)$ function for 1, 2, and 3 dimensional free-electron gases.

Exercise 9: Label the axes in Fig. 21, that is, indicate the value of $g(E)$ for, say, 1, 10, or 100 meV, assuming the effective electron mass of GaAs.

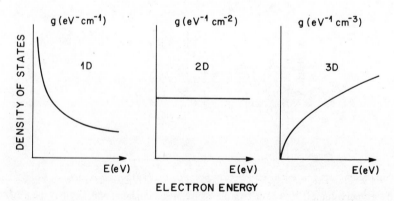

Fig. 21   Density of states of a free-electron gas in 1, 2, and 3 dimensions.

Exercise 10: Plot the total 2D density of states in a square well of width $a = 100 \text{ Å}$ (bounded by barriers of infinite height) assuming the effective electron mass of GaAs. Include contributions of the first seven subbands. On the same graph plot the 3D density of states in GaAs (multiplied by a, so that the dimensions of both curves are the same).

If there are several equivalent conduction-band minima ("valleys"), the density of states must be multiplied by their number (the valley degeneracy factor g_v), which may be different for different subbands. For example, the lowest subband in an MOS inversion layer at Si {100} surface has $g_v = 2$, with each of the two ellipsoids having isotropic dispersion relations within the xy plane, see Section 2.6.4. As long as the energy dispersion relation of the electron gas is isotropic, $E = E(\mathbf{p}^2)$, a generalization of Eqs. 87 can be obtained readily. One only has to replace $dp = (m/2E)^{1/2} dE$ by the more general $dp = (dp/dE) dE$, where dp/dE is a function of E (see, e.g., Problem 9).

When the dispersion relation is anisotropic, the problem becomes mathematically more cumbersome. It is easy to give a general definition of $g_n(E)$ for a given subband n of a 2DEG and a prescription for its evaluation:

$$g_n(E) \equiv 2 \int \frac{d^2p}{(2\pi\hbar)^2} \, \delta[E - E_n(\mathbf{p})] = \int_{C_n(E)} \frac{dC}{2\pi^2} \frac{1}{|\nabla_k E_n(\mathbf{k})|}, \quad (88)$$

where $E_n(\mathbf{p})$ is the dispersion relation in the subband n, $\mathbf{p} = \hbar\mathbf{k}$, $C_n(E)$ is the contour of constant energy in the momentum space, and ∇_k is the k-gradient of $E_n(\mathbf{k})$, (see Fig. 22); but the actual evaluation of $g_n(E)$ sometimes can be done only numerically. The simplest anisotropic case of an elliptical dispersion relation is considered in Problem 10.

The whole concept of the density of states pre-supposes that single-particle energies can be defined meaningfully. This is not always the case in a system

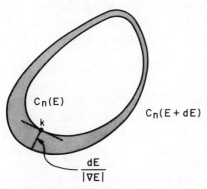

Fig. 22 Illustration of Eq. 88. The density of states $D(E)$ is proportional to the area in the momentum space enclosed between two isoenergetic contours $C_n(E)$ and $C_n(E + dE)$. To calculate this area, we integrate the normal clearance $dE / |\nabla_k E_n(\mathbf{k})|$ along the contour C_n.

of interacting electrons. When the density of states is defined, one can also determine the Fermi level E_F in terms of the occupational probability of single-electron energies[†]:

$$n_s = \int\limits_{E_0}^{\infty} dE \; g(E) \, f(E - E_F) \,, \tag{89}$$

where $f(E)$ is the Fermi function,

$$f(E) = (e^{E/kT} + 1)^{-1} \,. \tag{90}$$

Note that E_F does not have to correspond to any single-electron energy level; for example, it can lie in the forbidden gap below E_0. If $E_0 < E_F < E_1$, then, at zero temperature, only one subband is occupied. For a 2DEG with an isotropic spectrum, as in Eq. 85, the number of electrons per unit area is then given by

$$n_s = \frac{g_v m \, (E_F - E_0)}{\pi \hbar^2} \,. \tag{91}$$

The locus of **k** vectors corresponding to $E(\mathbf{k}) = E_F$ defines the Fermi "surface" (in 2D it is a contour). In the simplest case, corresponding to Eq. 91, the Fermi surface is a circle of radius

$$k_F = \sqrt{2\pi n_s / g_v} \,. \tag{92}$$

The case of more than one occupied subband is a straightforward generalization. Note, however, that for multi-valley semiconductors higher subbands may correspond to differently oriented ellipsoids, and consequently both g_v and m may vary, as is the case for an inversion layer in a silicon MOS structure, considered in the next section.

A two-dimensional electron gas at a GaAs/AlGaAs heterojunction interface (Fig. 23) is, in many respects, simpler than that in a Si MOS inversion layer. Except at high electron energies, there is only one valley (Γ) to consider ($g_v = 1$), and this valley is isotropic to a high degree. On the other hand, since the effective electronic mass in GaAs is lighter than that in Si, "quantum effects" become important under less extreme conditions, for example, at higher temperatures.

Exercise 11: Generalize Eq. 91 to the case of multiple-subband occupation for a single-valley semiconductor. Plot the Fermi function for $T = 0$, $T = 75\,\text{K}$, and $T = 300\,\text{K} \ll E_F$ with the same energy scale as that used in Exercise 10. Copy these plots onto a transparency and slide it with respect to the plots of

---

[†]The term Fermi level in semiconductor physics is synonymous with "chemical potential." It may be worthwhile to mention that the chemical potential (Gibbs free energy per particle) is well defined for any interacting system at equilibrium—independent of how well defined are the single-particle states and the $g(E)$. On the other hand, there is no guarantee for the existence of a Fermi surface, which is a single-particle idealization and may or may not be adequate to the real interacting-electron problem.

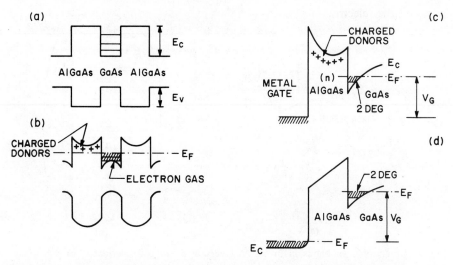

Fig. 23   Two-dimensional electron gas at a GaAs/AlGaAs heterojunction inteface. (a) Band edges and electron states in an undoped multilayer system. The band offsets in a GaAs/Al$_x$Ga$_{1-x}$As heterojunction are currently believed to be split in proportion $\Delta E_C{:}\Delta E_V \approx 3{:}2$. (b) Modulation-doped multilayer heterostructure. Energy levels of the 2D electron gas are well below the donor levels in AlGaAs, and so electrons find it energetically favorable to be spatially separated from the parent donors. A narrow (a few nanometers) layer of AlGaAs, immediately adjacent GaAs, is usually left undoped to supress scattering of 2DEG electrons by the ionized donors and, thus, enhance the mobility. (c) Conduction band diagram of the gate structure of a modulation-doped high electron-mobility transistor. (d) Conduction band diagram of a high electron-mobility transistor with an undoped gate barrier and a heavily doped GaAs gate layer.

Exercise 10 (graphically changing the Fermi level) to gain a feel for the quantitative relationship between n_s and E_F in GaAs quantum wells. Discuss the similarity between a 100-Å well at 75 K and a 50-Å well at 300 K.

2.6.4   Inversion Layer in a Silicon MOS Structure

The schematic cross section of a silicon MOS structure is illustrated in Fig. 24 along with the band-bending near the Si/SiO$_2$ interface under a sufficiently large positive gate bias. Let us look in more detail at the band structure of the 2DEG in an inversion layer on the {100} surface. First we have to recall certain basic facts about the bulk Si conduction-band structure.†

---

†Notation: equivalent crystallographic planes, e.g., (100), (0$\bar{1}$0), etc., are collectively denoted by {100}. Similarly, equivalent (symmetry-related) directions in the reciprocal lattice, e.g., [100], [0$\bar{1}$0], etc., are collectively referred to as the <·100> direction.

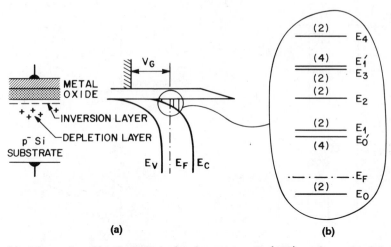

(a)   **(b)**

Fig. 24 Electronic subbands in inversion layers at a Si {100} surface. (a) Schematic cross section of a silicon MOS structure and the conduction band bending near the Si/SO$_2$ interface. (b) Energies of the seven lowest subbands for an inversion layer with $n_s = 10^{12} \text{cm}^{-2}$ in a lightly doped ($N_A = 10^{15} \text{cm}^{-3}$) p-type material at room temperature. Valley degeneracy of each level is shown in parentheses.

The conduction-band bottom edge E_C corresponds to six symmetry-related points $\mathbf{k}_\Delta$ in the reciprocal lattice located on the $<100>$ axes about 80% of the way toward the Brillouin zone boundary. The dependence $E(\mathbf{k} - \mathbf{k}_\Delta)$, plotted along the same axis (say, [100]) on which a particular $\mathbf{k}_\Delta$ is located, is approximated at low energies by a parabola that defines the *longitudinal* effective mass $m_l \approx 0.2 m_0$. Plotting $E(\mathbf{k} - \mathbf{k}_\Delta)$ from the same edge point along any direction perpendicular to [100] (e.g., [010] or [001]) results in a faster-rising parabola, which defines the *transverse* effective mass $Mt \approx m_0$. If we plot the iso-energetic surface at some energy close to $E_C \equiv E(\mathbf{k} - \mathbf{k}_\Delta)$, we obtain six ellipsoids of revolution, each elongated in its own $<100>$ direction, Fig. 25.

In a (roughly triangular) quantum well formed near the Si/SiO$_2$ interface under a positive gate bias, ellipsoids oriented differently with respect to the surface give rise to a quite different subband structure. Let us specify the actual {100} Si surface as a (100) crystal plane. Electrons in the two ellipsoids elongated in [100] direction possess the heavy mass m_l in z direction and an isotropic light mass Mt in any direction lying in the (100) plane. These electrons give rise to the subbands whose bottom-edge energies are denoted by E_n. The other four ellipsoids, whose longitudinal axes lie in the (100) plane, correspond to the light mass Mt in the [100] direction, and their subbands are denoted by E_n'. The dispersion relationship in these subbands is elliptical (cf. Problem 10).

Because the quantum-well energy levels scale with $1/m$, one has $E_0 < E_0'$, and so the inversion-layer electrons in their ground subband have an isotropic light mass. The order of the higher-lying subbands can be established only on the basis of self-consistent numerical calculations. The subband energies depend

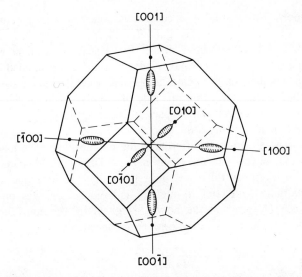

Fig. 25 Isoenergetic surfaces near the conduction-band minima in silicon. Also shown are the Brillouin zone boundaries and the reciprocal lattice directions of the conduction-band minima.

not only on the field and the background doping but also on the temperature, which affects the relative population of higher-lying subbands and, hence, the self-consistent field.

Exercise 12: Discuss whether the energy separation of the higher-lying subbands from E_0 should increase or decrease with increasing T. Consider the variation of $<z>$ with T and how this should change the effective field seen by electrons in the higher-lying subbands.

The order of the seven lowest subbands, calculated[28] for an inversion layer at a Si-{100} surface with $n_s = 10^{12}\,\mathrm{cm}^{-2}$ in a lightly doped ($N_A = 10^{15}\,\mathrm{cm}^{-3}$) p-type material at room temperature, is illustrated in Fig. 24b. The levels E_0 and E_1 are close in energy and, in fact, change their order at lower T and/or n_s.

2.6.5   Quantum Capacitance of a 2D Electron Gas

The capacitance-voltage (*CV*) characteristic of an MOS capacitor is amply discussed in Ref. 1. For capacitor on a p-type substrate, increasing the gate bias progressively brings the device into the regimes of hole accumulation (at negative V_G), depletion, weak inversion, and then strong inversion (at high positive V_G). In the strong inversion regime, the "bands are pinned," that is, the additional gate field does not penetrate beyond the inversion layer (see Problem 4).

As will be shown below, this absence of field penetration is only an approximation, which, although exceedingly good for Si MOS, is sometimes inadequate for 2DEGs with a lower effective mass. If we neglect its quantum-mechanical properties, the inversion layer behaves, electrically, as a grounded metal plate.[†] It is well known that such a plate completely shields the quasi-static electric fields emanating from charges on one side of the plate from penetrating into the other side. Thus, in a three-plate capacitor illustrated in Fig. 26a, application of a voltage to the node 1 changes the electric field only in the space filled with the dielectric ϵ_1. The situation is different if the middle plate Q is made of a two-dimensional (2D) metal, like the electron gas (2DEG) in a quantum well (QW) or an inversion layer. In this case, quite generally, the field due to charges on the plate 1 partially penetrates through Q and induces charges on the plate 2. The capacitance C_{tot} seen at the node 1 is given by the equivalent circuit of Fig. 26b, where C_1 and C_2 are the geometric capacitances,

$$C_i = \frac{\epsilon}{4\pi d_i} , \qquad i = 1, 2 \tag{93}$$

C_Q the "quantum capacitance" per unit area,

$$C_Q = \frac{g_v m e^2}{\pi \hbar^2} = g_v \frac{m}{m_0} \times 6.00 \times 10^7 \, \text{cm}^{-1} , \tag{94}$$

and m is the effective electron mass in the direction perpendicular to the QW plane. Proof of the equivalence of this circuit is left to the reader (Problem 12).

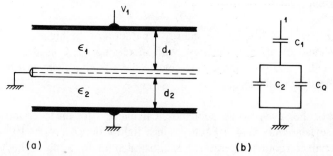

Fig. 26 Quantum capacitance. (a) Schematic illustration of a three-plate capacitor in which the middle plate represents a two-dimensional metal. The space between the plates is assumed filled with dielectrics of permittivity ϵ_1 and ϵ_2. (b) Equivalent circuit for the capacitance seen at node 1. (After Ref. 30).

[†]We are now concerned with equilibrium properties, corresponding to the "low-frequency" *CV* curves. At quasi-static frequencies (lower than the inverse generation time $1/\tau$ in the substrate) the inversion layer can be regarded as a grounded plate even without a contact. If there is a contact (e.g., through a diffused source and, possibly, through a battery) to the ground, then the high-frequency and low-frequency *CV* curves coincide and the inversion layer works as a grounded metal plate at all but microwave frequencies.

The quantum capacitance is a consequence of the Pauli principle, which requires extra energy for filling a QW with electrons. In the classical limit, $\hbar \to 0$ or $m \to \infty$, when $C_Q \to \infty$, the capacitances C_2 and C_Q drop out. If $N > 1$ subbands are occupied, the expression for C_Q must be multiplied by N; thus transition to the 3D case can be viewed as corresponding to $C_Q \to \infty$ because of $N \to \infty$. For MOS structures on a Si-{100} surface, one has $g_v = 2$ and $m = m_l \approx m_0$, so that $C_Q \gg C_1 \equiv C_{\text{oxide}}$ at all realistic oxide thicknesses. On the other hand, for a small m one can expect interesting effects when C_Q becomes comparable to the geometric capacitances. In particular the charge induced on the ground metal plate equals

$$\sigma_2 = -\sigma_1 \frac{C_2}{C_2 + C_Q}. \tag{95}$$

Partial penetration of an external field through a highly conducting 2DEG allows the implementation of novel devices.[29, 30] An important consideration is by how much one needs to vary the voltage on electrode 1 to cause a required variation of the electrostatic potential, Φ_Q, of the QW at fixed voltages‡ on the QW and on the electrode 2 (Fig. 27). This can be described by a "leverage factor"

$$\lambda \equiv \left(\frac{\partial \Phi_Q}{\partial V_1} \right)_{V_Q = \text{const}} = \frac{C_1}{C_1 + C_Q + C_2}. \tag{96}$$

The nature of quantum capacitance is different from a similar effect arising due to the field penetration through a conducting inversion layer in the weak-inversion (subthreshold) regime of a field-effect transistor. The latter effect is

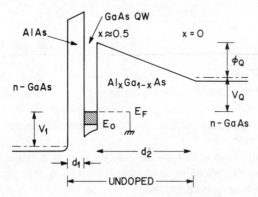

Fig. 27 Because of the field-penetration effect, the bias applied to electrode 1 changes the electrostatic potential of the well relative to its own Fermi level and that of electrode 2. (After Ref. 30)

---

‡Terminology: "voltage" is something that a voltmeter is measuring—which is the Fermi level difference (Ref. 2, p. 305). Electrostatic potential differences are much harder to measure, especially in microstructures.

associated with partial occupation of the inversion layer states at a finite temperature and disappears at $T = 0$. Motion of the depletion boundary in a subthreshold MOS system corresponds to a trivial penetration of an electric field through an "insulator," whereas the quantum-capacitance effect is a property of a highly conducting 2D Fermi system.

2.7 TUNNELING

The term "tunneling" in quantum mechanics describes a host of phenomena associated with the property of the wave function that it can have a non-vanishing magnitude in regions where the particle potential energy is higher than its energy. For example, the true wave function $\chi_0(z)$ of an inversion layer electron in the ground subband extends into SiO_2 (decaying exponentially with the distance from Si). This means, in particular, that no matter how thick the oxide is, the wave function has a finite magnitude on the metal side, where there may be electronic states at the same energy. These states are not orthogonal to $\chi_0(z)$ and so there is a finite probability that an inversion layer electron will escape to the metal. Except for the very thinnest (a few nanometers) oxides, such tunneling processes are quite unimportant in the operation of MOS devices and one is usually justified in making the approximation of an infinite barrier height (replacing the real difference in the conduction-band energies between Si and SiO_2 of approximately 3.2 eV), which forces the wave function to be exactly zero inside SiO_2. It should be noted, however, that at high gate biases the escape probability becomes appreciable for processes in which electrons first tunnel into the oxide conduction band (Fowler-Nordheim tunneling). Tunneling is much more important in heterostructure barrier systems. This is partly owing to the fact that the typical barrier heights in those systems are an order of magnitude lower than in the Si/SiO_2 structure and partly to the typically lower effective electron masses.

Tunneling in semiconductors is actually a common phenomenon. Ohmic contacts to heavily doped material are possible precisely because of this effect, Fig. 28a. Tunneling is responsible for the breakdown of a reverse-biased Schottky diode, Fig. 28b. A p-n junction of heavily doped semiconductors under a strong reverse bias, Fig. 28c, conducts current via Zener tunneling.[†] A p^+n junction in small range of forward biases conducts because of the tunneling of electrons into empty states in the valence band of a degenerately doped p^+ semiconductor. These processes and related devices, such as the backward diode and the Esaki tunnel diode, are discussed in detail in Ref. 1, and we shall not consider them further.

[†]The avalanche breakdown due to impact ionization is often a more important process for the breakdown of reverse-biased p-n junctions. The operation of most commercially available Zener diodes actually does not depend on the tunneling.

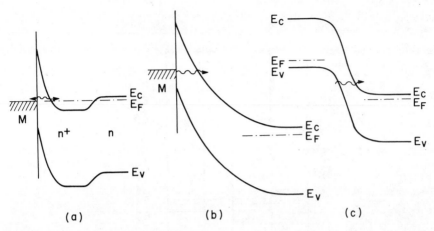

Fig. 28 Tunneling breakdown phenomena in semiconductors. (a) Ohmic contact between a degenerately doped semiconductor and a metal at equilibrium. (b) Tunneling
breakdown of a Schottky diode. Under sufficiently high reverse bias electrons in the
metal are within tunneling reach of states in the semiconductor conduction band. (c)
Zener tunneling. The breakdown occurs at the metallurgical junction of p and n
materials—where the electric field is highest.

In the remaining sections we shall discuss several topics related to the heterostructure barrier tunneling. Modern techniques of crystal growth permit the
implementation of nearly ideal realizations of tunneling structures, familiar to
most readers from the elementary quantum-mechanics courses. Because some of
these structures are rapidly becoming technologically important, it is worthwhile
to review the key concepts from a "device building block" perspective.

2.7.1 Coupled quantum wells

Consider an electron in a double-well potential, Fig. 29. If the separation
between the wells were very large, or the wall separating them were very high,
then the well index (1—left well; 2—right well) would be a good quantum number, and for a symmetric potential the ground-state level E_0 and the excited-state
levels E_n would be doubly degenerate.‡ For a finite "strength" barrier this
degeneracy is lifted and the levels E_n split into doublets. We shall consider only
the case when the barrier transparency is not too great, so that the doublet splitting is much smaller than the interlevel separation $(E_{n+1} - E_n)$.

Whatever the shape of the double-well potential, so long as it is symmetric,
the exact eigenstates are the symmetric Φ^S and antisymmetric Φ^A states with

---

‡Degeneracy means that for a given energy index n any linear combination of the single-well states
$\chi_n^{(1)}$ and $\chi_n^{(2)}$ would be stationary. This degeneracy is in addition to the spin degeneracy, which is
due to the time-reversal symmetry an is always present in single-electron problems in the absence of
a magnetic field.

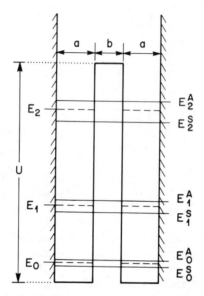

Fig. 29 Electron in a double-well potential. The single-well level E_0, which would be doubly degenerate if the wells were far apart, splits into E_S and E_A in the presence of a tunnel coupling.

respect to the reflection in the plane of symmetry.† From general principles of quantum mechanics we can also assert that the symmetric state will be lower in energy. Indeed any antisymmetric wave function must vanish at the symmetry plane; such a function cannot therefore be the ground state—which must have no nodes.[31] Moreover, in every doublet the Φ_n^A state has one extra node in comparison with the Φ_n^S state and hence it must be higher in energy.

The actual form of the wave functions corresponding to states Φ^S and Φ^A and the respective energy eigenvalues E^S and E^A for a general double-well potential can be found only numerically. If the potential is sufficiently smooth, one can obtain a simple solution quasi-classically.[31] For a rectangular double-well, the exact solution[32] is straightforward, although somewhat cumbersome mathematically. A convenient approach is provided by the widely used approximation, which in the solid-state and molecular physics is called LCAO (linear combination of atomic orbitals); the standard example of a double-well problem is the hydrogen molecular ion H_2^+ (one electron shared by two protons). If the barrier is sufficiently high and wide, so that the doublet splitting $E^S - E^A$ is much smaller than the subband separation in a single well (in the absence of the

---

†To readers familiar with applications of the group theory in quantum mechanics, this statement is perfectly obvious. Since, for a symmetric potential, the Hamiltonian commutes with the reflection operator $\hat{R}$, the stationary states can always be chosen as eigenstates of $\hat{R}$, which are $\hat{R}\,\Phi^S = \Phi^S$ and $\hat{R}\,\Phi^A = -\,\Phi^A$.

other well), then the LCAO functions can be obtained by taking a linear combination of single-well functions corresponding to one subband only:

$$\Phi_n^A = 2^{-\frac{1}{2}} [\chi_n^{(1)} - \chi_n^{(2)}] ; \tag{97a}$$

$$\Phi_n^S = 2^{-\frac{1}{2}} [\chi_n^{(1)} + \chi_n^{(2)}] . \tag{97b}$$

We have considered (Problem 6) the case of an electron in a square-well potential with one wall of a finite height (Fig. 18d). For that problem, the eigenfunctions are of the form

$$\chi_n^{(1)}(z) = \frac{1}{\sqrt{N_n}} \begin{cases} \sin(q_n z) & \text{if } 0 \leq z \leq a \\ \sin(q_n a)\, e^{-\varkappa_n (z - a)} & \text{if } z \geq a \end{cases} \tag{98a}$$

and

$$\chi_n^{(2)}(z) = \chi_n^{(1)}(2a + b - z) , \tag{98b}$$

where $\hbar^2 q_n^2/2m = E_n$, $\hbar^2 \varkappa_n^2/2m = U - E_n$, and $N_n = (a + \varkappa^{-1})/2$. Even though the functions $\chi_n^{1,2}$ are normalized to unity, the Φ's in Eqs. 97 are not because $\chi_n^{(1)}$ and $\chi_n^{(2)}$ are not orthogonal. It is a simple matter to evaluate the expectation value of the Hamiltonian $H = p^2/2m + U(x)$ in the states of Eqs. 97 to first order in the small parameter $e^{-\varkappa_n b}$:

$$E_n^S = \frac{E_n + \Delta_n}{1 + \gamma_n} , \qquad E_n^A = \frac{E_n - \Delta_n}{1 - \gamma_n} , \tag{99}$$

where

$$\gamma_n = \int \chi_n^{(1)} \chi_n^{(2)} \, dz , \quad \text{the "overlap integral" ;} \tag{100a}$$

$$\Delta_n = \int \chi_n^{(1)} H \chi_n^{(2)} \, dz , \quad \text{the "resonance integral".} \tag{100b}$$

If the calculated doublet splitting turns out to be comparable to (or even larger than) the single-well interlevel separation, then one can surely discard the calculation, because the assumption underlying Eqs. 97 is no longer valid.

Exercise 13: Find the normalization factors for the states in Eqs. 97 in terms of the overlap integral. Evaluate γ_n and Δ_n with the functions of Eq. 21 and show that the doublet splitting is given by

$$EAn - ESn = E_n e^{-\varkappa_n(a + b)} \frac{12\hbar}{(a + \varkappa_n^{-1})\sqrt{2mU}} .$$

Find the fallacy in the following "argument": Antisymmetric function, which vanishes in the plane of symmetry, is smaller inside the barrier than the symmetric function. Since it "samples" the high barrier to a lesser extent it should have lower energy.

In the presence of a tunnel coupling, the single-well states $\chi_n^{(1)}$ and $\chi_n^{(2)}$ are not stationary. If, immediately upon the excitation, electrons are prepared in state $\chi_n^{(2)}$, then the subsequent evolution of this state in time is given by

$$\Phi_n(t) = e^{E_n t/i\hbar} \left[\chi_n^{(2)} \cos\left(\frac{\omega_n t}{2}\right) - \chi_n^{(1)} \sin\left(\frac{\omega_n t}{2}\right) \right], \qquad (101)$$

where $i = \sqrt{-1}$, and where $\hbar\omega_n \equiv E_n^A - E_n^S$ is the tunnel splitting between the stationary states of the two-well system. Equation 101 describes an oscillatory electronic motion between the two wells.†

Exercise 14: Derive Eq. 101 by considering the evolution of state $\chi_n^{(2)}$ in accordance with the time-dependent Schrödinger equation—within the approximation of Eqs. 97, that is, neglecting intersubband transitions.

The question of how to "prepare" an electron in state $\chi_n^{(2)}$—to observe the subsequent oscillatory motion—may not be totally impractical, although the experiment is by no means easy. Consider an idealized structure, Fig. 2.30, containing two quantum wells separated by a heterostructure barrier; the wells have identical ground-state levels ($E_1 = E_2 \equiv E_0$) in the conduction band, *but different in the valence band*. This allows the selective preparation of an initial electron state by an ultra-short interband photo-excitation.‡ In a coupled QW system electrons will oscillate between the two wells, giving rise to an oscillating luminescence signal with a period directly related to the tunneling time.[33] In the absence of scattering, the luminescence signals at frequencies ν_1 and ν_2 will oscillate 180° out of phase according to Eq. 101, their intensities being proportional to $\sin^2(\omega t/2)$ and $\cos^2(\omega t/2)$, respectively.

Let us discuss, in the instance of the proposed luminescence oscillation experiment, the question of phase coherence of electronic waves in the presence of

---

†Note that the period T of this oscillatory motion can be *substantially shorter* than the time of a seemingly similar process: tunneling decay of electron from a metastable state inside a quantum well bounded by a barrier of identical transparency. Indeed the latter time τ is inversely proportional to the transmission coefficient of the barrier, and scales as $\exp(2 x_n b)$, whereas $T \equiv 2\pi/\omega \propto \exp(x_n b)$. Compared to a coherent oscillation, the decay into a continuum is inhibited by the rapid phase degradation of states in a continuous spectrum.

‡Photoexcitation of a non-stationary state is a common phenomenon. Duration of the excitation must be shorter than the period T of the oscillatory motion described by Eq. 101. Immediately after the short interaction with an electron, the photon field also will not be in a stationary state—the number of quanta in the interacting mode being no longer sharply determined.

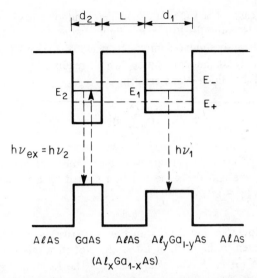

Fig. 30 Schematic illustration of a luminescence oscillation experiment. One of the wells represents a pure GaAs layer, the other is made of an $Al_yGa_{1-y}As$ alloy with a small fraction of aluminum $y \ll x$. The resonance between single-well levels E_1 and E_2 can be "fine-tuned" by an electric field. A non-equilibrium population of holes, necessary for the radiative recombination in the quantum wells, can be maintained by an auxiliary pumping of interband transitions in the cladding layer on the side of the n contact. (After Ref. 33).

scattering.[34] Suppose there is only *elastic* scattering. In this case, electrons will never settle in their stationary states; each electron, however, will oscillate at a slightly different frequency—determined by the local configuration of elastic scatterers in each oscillation channel. An equivalent way of describing this situation is to add a time-dependent phase $\phi_j(t)$ in the argument of the trigonometric functions in Eq. 101—different for each channel j. The relative phases of different electrons will therefore randomize, and upon the time τ_ϕ no luminescence oscillation will be observed. One can even assume that the elastic processes involved do conserve the phase memory, which means only that the functions $\phi_j(t)$ are perfectly deterministic. In principle, this would leave us with the possibility of doing a "time-reversal" trick analogous to the well-known spin-echo experiments in nuclear magnetic resonance. If at $t = t_1$ the momenta of all electrons in the double-well system could be reversed simultaneously, then at $t = 2t_1$ the luminescence oscillations would *re-emerge* to last for another period, $\sim \tau_\phi$. Needless to say, at this time nobody knows how to do such an experiment.

2.7.2 Superlattices

Description of the double-well problem presented above in the weak-coupling limit (doublet splitting small compared to the subband separation) naturally extends to the case of a superlattice in the limit of narrow bands—where the same approach is naturally called the tight-binding approximation.† The two problems are of quite similar nature. In place of the reflection symmetry one now has periodicity of the superlattice, which requires that the exact eigenfunctions must be of the Bloch-Wannier form

$$\Phi^{(k)}(z) = \sum_{j=1}^{N} e^{ikjd} \chi^{(j)}(z) , \tag{102}$$

where d is the superlattice period, N is the number of periods and the set of functions $\chi^{(j)}$ is one and the same function shifted by a whole number of periods, $\chi^{(j)}(z) \equiv \chi(z - jd)$.

Exercise 15: Show that any Bloch function [$\exp(ikz) u_k(z)$ where $u_k(z)$ is periodic] can be written in the form of Eq. 102. Note that this is true in general for a 3D crystal and for *each band individually*. The functions $\chi_n(z - jd)$, called the Wannier functions, are orthogonal if they are centered on different sites j and belong to different bands n.

The Bloch wave vector k, which can take a quasi-continuous spectrum of values‡ replaces the two-valued symmetry label $(S$ or $A)$. In the narrow-band limit we can take for $\chi(z)$ the eigenfunctions $\chi_n(z)$ corresponding to a single subband of an isolated symmetric well.§ (Since here we must include the overlap with both left and right nearest neighbors, the isolated well should be conveniently taken as symmetric, Fig. 18b.) Then we can easily show that the energies $E_n^{(0)}$ of an isolated well in the superlattice split into a band with an energy dispersion relation

$$E_n(k) = \tilde{E}_n + 2\Delta_n \cos(kd) , \tag{103}$$

where Δ_n is the resonance integral, as in Eq. 100b, and $\tilde{E}_n$ is a slightly lowered energy [compared to $E_n^{(0)}$ there is an additional binding energy due to the other wells]. Each band thus has a width of $4\Delta_n$ and contains exactly N states ($2N$—including spin degeneracy).

---

†The tight-binding approximation presented in every solid-state course is used to describe 3D crystals with narrow bands. It is rarely a quantitative approximation for real solids—rather an intuition-guiding model. For semiconductor superlattices, on the other hand, it is often quite a reasonable description.

‡If one imposes periodic boundary conditions, i.e., identifies the $(N + 1)$st well with the 1st, then the allowed values of k are $2\pi \mathscr{L}/Nd$, $\mathscr{L} = 0, \pm 1, \pm 2, \cdots$.

§It is at this point that the generality is lost and the tight-binding approximation sets in. The immediate price is the slight non-orthogonality of functions centered at different sites. The tight-binding approximation, when applicable, results in an enormous simplification of the problem.

Exercise 16: Derive Eq. 103 for the tight-binding band structure. What is the value of the effective mass m near the band eges? How does m scale with d for a fixed value of the barrier height? What is the group velocity of a band electron and how it depends on k?

What happens if one applies an external field to the superlattice? It has been known since the early days of the quantum theory of solids that an electron placed in a lattice field of period a and an additional uniform electric field $\mathcal{E}$ will perform a purely oscillatory motion in the direction of the field with a characteristic frequency $\omega = q\mathcal{E}a/\hbar$. This follows from the fact that the crystal momentum of such an electron increases linearly with time, while both its energy and velocity are periodic function of the crystal momentum. The distance electron travels during one cycle is of the order of $I/q\mathcal{E}$ (where I is the width of the allowed band in which the electron is moving), and the turning points can be interpreted as Bragg reflections of the electron by the periodic lattice. In the language of quantum mechanics the electron energy band in the presence of an electric field splits into a set of discrete levels (the Wannier-Stark ladder) equidistantly separated by energy intervals $q\mathcal{E}a$, each of these levels corresponding to a wave function centered on a particular site and extending in the direction of $\mathcal{E}$ over approximately $I/qa\mathcal{E}$ lattice sites.[35] These assertions are elementary consequences of the fact that a translation by one lattice period changes the Hamiltonian of the system by the constant amount $q\mathcal{E}a$.

It is therefore possible, in principle, for a dc electric field to induce an alternating current in the solid. In real solids this effect (sometimes called the Bloch oscillations) is not observed because collisions usually return an electron to the bottom of the band well before it completes one period (even for the strongest available fields $\mathcal{E}$ one has $\omega\tau \ll 1$, where τ is the scattering time). As a possible remedy to this difficulty, Keldysh[36] suggested the "artificial creation of periodic fields" with periods $d \gg a$, which would split the conduction band into several minibands. He also discussed the possibility of using high-frequency ultrasonic waves to establish a "moving" superlattice.[37]

In an ideal superlattice consisting of a large number of equally spaced identical quantum wells, one can expect a resonant (miniband) transmission, and possibly a negative differential mobility due to the Bragg reflections, if the applied field is such that the potential difference, acquired by an electron over many periods of the superlattice, is less than the width of the lowest miniband I. These effects, particularly the Bragg reflections, are extremely difficult to observe because of scattering and Zener tunneling between electron minibands, as well as domenisation of the electric field in the superlattice. The necessary condition for the realization of the miniband conduction effects is

$$\frac{\hbar}{\tau} < qd\mathcal{E} \ll I,$$

where τ is the electron mean-free-path time. In the opposite limit of a strong electric field, $qd\mathcal{E} > I$, an electron belonging to the band I is localized within one well. In this limit an enhanced electron current will flow at sharply defined values of the external field, $qd\mathcal{E}_j = E_j - E_0$ $(j = 1, 2, \cdots)$, when the ground state in the nth well is degenerate with the first or second excited state in the $(n + 1)$st well, as illustrated in Fig. 31a. Under such conditions, the current is due to electron tunneling between the adjacent wells with a subsequent de-excitation in the $(n + 1)$st well. In other words, electron propagation through the entire superlattice involves a sequential rather than resonant tunneling. Experimental difficulties in studying this phenomenon are usually associated with the non-uniformity of the electric field across the superlattice and the instabilities generated by negative differential conductivity.

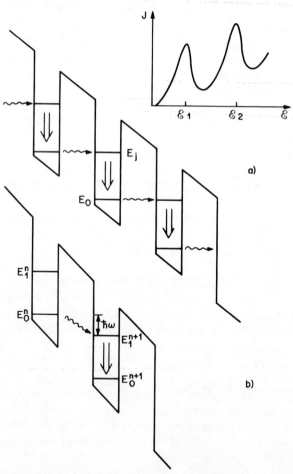

Fig. 31 Illustration of the sequential-tunneling effects in a superlattice. (a) Conductivity peaks, associated with the resonance between the excited and the ground states of adjacent wells. (b) Photon-assisted sequential tunneling process tunable by the applied electric field.

In the regime when the electric field $\mathcal{E}$ exceeds the peak value $\mathcal{E}_j$ (i.e., in the negative differential resistance (NDR) portion of the current-field characteristic), there is a theoretical possibility[38] of the amplification of electromagnetic waves, brought about by the sequential electron tunneling into adjacent wells with a simultaneous emission of photons, Fig. 31b. The maximum gain is predicted to occur at the frequency $\hbar\omega = qd\,(\mathcal{E} - \mathcal{E}_j)$. This exciting effect—a laser action tunable by an applied electric field—has never been observed experimentally. The fact that this device is supposed to operate in the NDR regime presents a fundamental difficulty, since at high currents the uniform field condition cannot be maintained because of the NDR-driven space-charge instability, resulting in domenisation of the field. It is not clear whether this difficulty can be circumvented even in principle.

It should be noted that if one takes a superlattice comprising N identical barriers sandwiched between doped contacts, then most of the applied voltage will fall on the cladding barriers, because the effective resistance associated with the tunneling step in or out of the continuum is typically much higher than that associated with the coherent tunneling between two adjacent quantum wells in resonance (see footnote† on p. 110). To achieve the desired impedance matching, one should design the cladding barriers to be narrower (approximately by a factor of 2) than the internal barriers.[39]

2.7.3 Resonant Tunneling

The miniband conduction in a superlattice consisting of a large number of barriers can be regarded as a generalization of resonant-tunneling (RT) transmission through a double-barrier quantum-well (DBQW) structure. Historically, RT in DBQW structures had been proposed and discussed as an electron-wave phenomenon analogous to the resonant transmission of light through a Fabry-Perot étalon. Consider an electron at energy E incident on a one-dimensional DBQW structure (Fig. 32). When E matches one of the energy levels E_i in the QW, then the amplitude of the electron de Broglie waves in the QW builds up due to multiple scattering, and the waves leaking in both directions cancel the reflected waves and enhance the transmitted ones. Near the resonance one has

$$T(E) \approx \frac{4\,T_1 T_2}{(T_1 + T_2)^2}\,\frac{\gamma^2}{(E - E_i)^2 + \gamma^2}\,, \tag{104}$$

where T_1 and T_2 are the transmission coefficients of the two barriers at the energy $E = E_i$ and $\gamma \equiv \hbar/\tau$ is the lifetime width of the resonant state [quasi-classically, $\gamma \approx E_i(T_1 + T_2)$]. In the absence of scattering, a system of two identical barriers ($T_1 = T_2$) is completely transparent for electrons entering at resonant energies, and for different barriers the peak transmission is proportional to the ratio T_{min}/T_{max}, where T_{min} and T_{max} are, respectively, the smallest and the largest of the coefficients T_1 and T_2. The total transmission coefficient, plotted against the incident energy has a number of sharp peaks, as shown in Fig. 32.

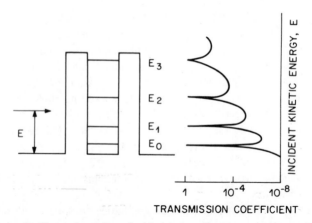

TRANSMISSION COEFFICIENT

Fig. 32 Schematic illustration of a double-barrier electron resonator. The intensity transmission coefficient plotted against the incident kinetic energy in the direction normal to the resonator layers has a number of sharp peaks. In the absence of scattering, a symmetric resonator is completely transparent for electrons entering at the resonant energies (the Fabry-Perot effect).

The modern era in the field of resonant tunneling began after Esaki and Tsu first proposed the idea of semiconductor superlattices and discussed their transport properties.[40] They predicted a negative conductivity associated with electron transfer into negative-mass regions of the mini-zone and the possibility of inducing Bloch oscillations by applying an electric field normal to the superlattice. They have also theoretically considered the transport properties of a *finite* superlattice and showed the possibility of obtaining a negative differential resistance in DBQW structures. These predictions were followed by the first observation of RT in GaAs/$Al_{0.3}Ga_{0.7}As$ heterojunction diodes[41] and GaAs/AlAs heterojunction superlattices.[42] Since the early reports, substantial progress has been achieved in the material quality of heterojunction barrier structures grown by MBE and MOCVD (metal-organic CVD) techniques. The material quality of DBQW diode structures has steadily improved to the point that a pronounced NDR can now be observed at room temperature. Typical current-voltage characteristics of a quantum-well diode are shown in Fig. 33.

In order to understand these characteristics it is not necessary to invoke a resonant Fabry-Perot effect.[43, 44] Physically, the NDR arises from the electron confinement in the quantum-well resonator and the conservation, in the process of tunneling, of the electron momentum component parallel to the plane of the barriers. The emitter electrons form a 3-dimensional Fermi gas; at low enough temperatures they occupy all available states—up to the Fermi level E_F. To a good approximation, the energy of an electron state is proportional to the square of its momentum **p**, and so in the "momentum space" the occupied electron states of the emitter are represented by a sphere, illustrated in Fig. 34. Each point (p_x, p_y, p_z) inside the Fermi sphere corresponds to an occupied state whose

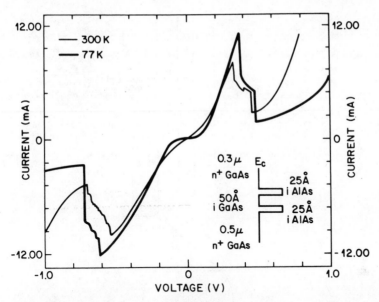

Fig. 33 The *I-V* characteristics of a symmetric DBQW diode that contains an undoped 50-Å thick GaAs QW clad by two undoped 25-Å thick AlAs barrier layers and two n-doped GaAs layers. Diode area is $\approx 2.8 \times 10^{-7} \text{cm}^2$, corresponding to a peak current density of 30 kA/cm² at 300 K. (After Ref. 47).

energy, $(p_x^2 + p_y^2 + p_z^2)/2m$, is proportional to the square of the point's distance from the center of the sphere. States of the quantum well are "confined," which implies that part of the electron energy that corresponds to the motion in the z direction (perpendicular to the plane of the barriers) can take on only discrete values, E_n $(n = 0, 1, 2, \cdots)$. This does not mean that the electron-energy spectrum in the well is discrete, because there is also free motion in the other two directions, that contributes the energy $(p_x^2 + p_y^2)/2m$. For each E_n the allowed energies of quantum-well states span a continuous subband (these subbands overlap partially). In the process of tunneling, an electron conserves both its energy and the components p_x and p_y of its momentum. Consider the tunneling into a particular subband, say E_0. If the energy E_0 is above E_F, then no emitter electron has enough energy, and current is possible only to the extent that electrons get help from the lattice vibrations. If $E_0 < E_F$, then there is a group of emitter electrons that can tunnel into the well: these are shown by the shaded disk in Fig. 34. The tunneling electrons span a range of energies, which increases with an applied bias as the shaded disk moves toward the equatorial plane of the Fermi sphere. At this point the number of tunneling electrons, and hence the current, is highest (the highest number of tunneling electrons per unit area equals $mE_F/\pi \hbar^2$).

Once in the well, electrons tunnel out into the collector in a second step, which may be completely uncorrelated with the first. Further increase of the

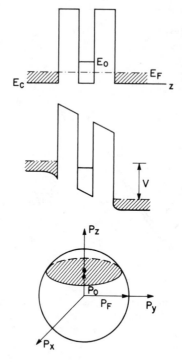

Fig. 34 Mechanism of operation of a double-barrier resonant-tunneling diode. The top figure shows the electron energy diagram in equilibrium. The figure in the middle displays this diagram for an applied bias V, when the energy of certain electrons in the emitter matches unoccupied levels of the lowest subband E_0 in the quantum well. The bottom figure illustrates the Fermi sphere of occupied states for the 3D electron gas in the emitter—the locus of points in the momentum space with $p_x^2 + p_y^2 + p_z^2 \leq p_F^2 \equiv 2mE_F$. Conservation of the lateral momentum in tunneling requires that only those emitter electrons whose momenta lie on a disk $p_z = p_0$ (shaded) can tunnel into the quantum well. Here $p_0^2/2m$ is the energy separation between E_0 and the bottom edge E_C of the conduction band in the emitter. In an ideal double-barrier diode at zero temperature, the resonant tunneling occurs over a voltage range during which the shaded disk moves down from the pole to the equatorial plane of the emitter Fermi sphere. At higher V (when $p_0^2 < 0$) resonant electrons no longer exist, which results in a sharp drop in the current. (After Ref. 43).

bias, when the subband-edge energy E_0 moves below the conduction-band bottom E_C in the emitter, leads to the situation that all electrons in the emitter have higher energies than the quantum-well states corresponding to the same values of the lateral momentum (p_x, p_y), and, therefore, the tunneling current will cease abruptly.† The current will resume rising at still higher bias when the next sub-

---

†This statement is true for low enough temperatures and neglecting the electron interaction with lattice vibrations, imperfections of the structure, and other electrons; in real samples at finite temperatures these factors lead to a non-vanishing "valley" current, as seen in Fig. 33.

band edge F ~~~ing the Fermi sphere, first near the north pole, then sinking tow~~~ ~~~torial plane. Once E_1 moves below E_C, there will be another dr~~~ ~~~. Depending on the number of subbands in the quantum well, on~~~ several peaks in the resonant-tunneling current. Similar argumer~~~ ~~~se, apply to systems of lower dimension, for example, to tunneling ~~~ "quantum wire."[45]

A ~~~ffect should be observable in various *single-barrier* structures in whic~~~ ~~~ng occurs into a two-dimensional system of states. Indeed, according to the ~~~scribed model, in DBQW structures the removal of electrons from the QW occurs via sequential tunneling through the second barrier, but there are other means of electron removal, for example, *recombination*—if the QW is in a *p*-type material.[46] The NDR effect has also been observed in a unipolar single-barrier structure.[47] In these experiments the emitter was separated by a thin tunneling barrier from a QW that was confined, on the other side, by a thin but impenetrable (for tunneling) barrier, Fig. 35. The drain contact to the QW, located outside the emitter area, was electrically connected to a conducting layer underneath. Application of a negative bias to the emitter results in the tunneling of electrons into the QW and their subsequent lateral drift toward the drain contact.

The sequential-tunneling mechanism of NDR is experimentally distinguishable from the Fabry-Perot model. However, it is not clear whether or not such a distinction can be made on the basis of current-voltage characteristics alone. In three-dimensional DBQW diodes, the NDR arises solely as a consequence of the

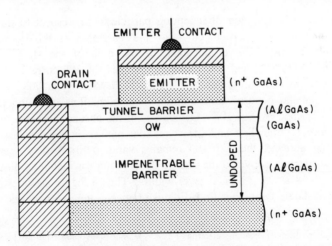

Fig. 35 Single-barrier structure exhibiting the NDR due to tunneling into a QW. Drain contact, concentric with the emitter electrode and located outside the emitter area, is electrically connected to a conducting layer underneath. Application of a drain-emitter voltage results in a nearly vertical electric field lines under the emitter, which allows one to control the potential difference between the emitter and the QW. Of course, this control is much less effective than it would be if the second barrier were nearly as thin as the tunnel barrier. (After Ref. 24).

dimensional confinement of states in a QW and the conservation of energy and lateral momentum in tunneling. This statement being true for both the coherent and the sequential pictures, it should be noted that in the sequential picture the NDR is associated only with the first tunneling step and the device designer is free to describe the NDR by a circuit element in series with an ordinary resistance corresponding to the second tunneling barrier. In contrast, in the coherent picture the NDR is an overall property of the DBQW system. Historically, this motivated a design strategy intended to optimize the Fabry-Perot resonator conditions. In particular, proposals were made for asymmetric barriers that would attain equality of the transmission coefficients $T_1 = T_2$ only with applied bias under the device operating conditions.

Such proposals were largely put to rest after a number of authors[48-50] argued that the predictions of both models for static I-V characteristics were practically indistinguishable. The essence of this argument is that the RT in DBQW diodes is normally observed under large bias ($\gg \gamma$), and that in order to calculate the current one must average over the energy distribution of incoming electrons, which is also typically large: $E_F \gg \gamma$. For the purpose of this averaging, the Lorentzian factor in Eq. 104 reduces to $\pi\gamma\,\delta(E - E_i)$ and the factor γ cancels $(T_1 + T_2)$ in the denominator, leading to $T(E) \approx E_i T_{\min}\delta(E - E_i)$. The calculated current density is then given by

$$J \equiv q \int dE\, n(E)\, v(E)\, T(E) \tag{105}$$

$$\approx qn(E_i)\sqrt{2E_i/m}\; T_{\min}(E_i) \,.$$

Here $n(E)$ is the distribution of incoming particles with respect to the kinetic energy of their motion perpendicular to the barriers; in three-dimensional DBQW structures $n(E_i)$ is proportional to the area of the shaded disk in Fig. 34. Since it is $T_{\min}$ and not T that determines the RT current, it is clear that the coherent mechanism is not sensitive to the barrier asymmetry.†

In the "sequential picture" under similar approximations, the tunneling current is also described by Eq. 105. Of course, in the absence of scattering there is no "incoherent" tunneling and the sequential picture is meaningless. However, it can be shown[50] that Eq. 105 remains valid to first order, even in the presence of scattering, provided the energy distribution for incoming electrons is broader than the scattering-limited level width.‡ Thus a legitimate question may be asked: why should we bother to distinguish between the two pictures?

---

†The situation would be quite different if one were able to study the RT characteristics with quasi-monoenergetic distribution of incoming electrons ($E_F < \gamma$), or if one could design samples in which the resonance would occur at low applied biases—such that $q(E_{F_1} - E_{F_2}) < \gamma$. In the latter case the coherent resistance would be again sensitive to the total transmission coefficient, as given by the Landauer formula, $R^{-1} = (q^2/\hbar)\, T(E_F)$ [such a situation was recently considered[51] by Büttiker]. However, none of the DBQW structures studied to date conforms to these specifications.

‡Nevertheless, it has been argued[52] that scattering reduces the peak-to-valley ratio of the NDR curves in DBQW diodes in a non-equivalent way for the coherent and the incoherent mechanisms; the degradation effect is larger for the coherent process. Thus, a systematic study of the static IV curves with a controlled variation in the strength of scattering may, in principle, distinguish between the two mechanisms.

There are at least two reasons for doing so. *First*, the sequential-tunneling approach provides a natural framework for discussing a new class of three-terminal RT devices, especially those which rely essentially on the NDR property of tunneling into a QW without an attendant (coherent or incoherent) second tunneling step. It can be argued that the most important future applications of RT are associated with multi-terminal devices, because of their potential for enhanced functionality "per terminal" in integrated circuits. *Second*, the question of coherent versus incoherent electron transport transcends in importance the mere analysis of static *I-V* characteristics in DBQW diodes. Distinction between the two processes depends on the relative value of the phase relaxation τ_ϕ time and the tunneling time τ_0. In resonant tunneling, neither of these two quantities is presently free from ambiguities. In the next section, we shall discuss the processes that lead to phase relaxation; in particular, it will be shown that these are not inelastic scattering processes only.

2.7.4 Effect of Electron Scattering on Resonant Tunneling

Consider a single-electron incident on a DBQW structure, assumed for simplicity to have $T_1 = T_2$. The reflected-wave amplitude represents the sum of amplitudes of all quantum-mechanical paths corresponding to multiple reflections from the barriers. At resonance the phases of different amplitudes combine to cancel the net reflected wave. If some of the paths contain an external interaction vertex, which changes the wave function phase by a random amount of order π, the reflected wave will not be canceled. It is unimportant whether the phase-randomizing interaction is inelastic or not.

For example, in a one-dimensional case one can think of a magnetic impurity that flips the electron spin without changing the energy; clearly, partial waves of opposite spin do not cancel each other. For a three-dimensional DBQW structure an elastic scattering event may change the direction of electron momentum in the xy plane (the plane of the barriers). Although the factor describing the electron wave function in the tunneling direction has not changed, the overall phase has, and no cancellation is possible. In a Gedanken experiment measuring the single-electron transmission coefficient, the relevant phase relaxation time τ_ϕ is at least as short as the momentum relaxation time τ_m in the QW, as determined by mobility measurements.†

It should be clearly understood, however, that the phase *memory* is not necessarily lost in an elastic-scattering event, so that another such event can restore the single-electron phase. The interference of scattered waves leads to quantum corrections to the metallic conductivity measured in experiments on a mesoscopic scale. In such experiments, the relevant τ_ϕ^{irr} is determined by the irreversible phase degradation brought about by the electron interaction with an equilibrium reservoir of scatterers. Although this time is sometimes loosely

---

†In fact, one can even have $\tau_\phi \ll \tau_m$, since τ_m gets no contribution from the electron-electron scattering, which is just as important as the impurity and the phonon scattering for altering single-electron phases.

thought of as the inelastic scattering time, strictly speaking, this is not so: an irreversible phase degradation can be also produced by an interaction with a degenerate level of the reservoir. In a double-slit experiment, no interference is possible if one of the interfering paths involves interactions with an external system—leaving that system in a state orthogonal to its initial state. It is thus clear that the question of what is the relevant phase relaxation time can be decided only with respect to a specific experiment. As discussed above, static *I-V* curves of a DBQW diode are rather insensitive in this respect.[†]

With a model estimate for τ_ϕ, one can determine whether the RT is dominated by coherent or incoherent processes by comparing τ_ϕ with the "tunneling time" τ_0, which should be understood as the lifetime of the resonant state limited by its decay due to tunneling.[‡] Quasi-classically, the ratio τ_0/τ_ϕ corresponds to the average number of bounces an electron makes inside the QW relative to the average number of phase-randomizing events it is expected to face while bouncing back and forth.

Quantitatively, the effect of scattering on the characteristics of DBQW diodes is rather poorly understood at this time. As will be discussed in the next section, these characteristics can be strongly affected even by elastic processes that mix the longitudinal and the transverse components of the electron wave function.

2.7.5   Barrier Penetration by a 2DEG in the Presence of Scattering

Theoretically, we are dealing with planar tunneling structures in which the barriers exist for the motion in one direction (z), while in the other two directions the system represents an approximation of translational invariance, which in practice is never perfect. Because of this unavoidable imperfection, the longitudinal (z) component of the wave function gets mixed with the ancillary degrees of freedom (those corresponding to the electron motion in xy plane).

Recently, this problem was discussed with exceptional clarity.[54] Consider the wave function of electrons confined to a QW, bounded by an infinite barrier on one side and a finite-height barrier $V(z)$ on the other, Fig. 36. In the absence of electron-electron interaction and inhomogeneities, the free motion in the xy plane is completely separable from the quantized longitudinal motion. Conse-

---

[†]Even if one could design a DBQW structure in which the energy distribution of incoming electrons were arbitrarily narrow, one would still observe only an inhomogeneous average of the transmission coefficient T over the device area. Only for *mesoscopic* devices with transverse dimensions comparable to the phase coherence length $(D\tau_\phi^{irr})^{1/2}$ the static *IV* curves can be expected to exhibit interference effects associated with the phase memory retention over the time τ_ϕ^{irr}. All presently studied DBQW diodes are macroscopic and these effects are washed out.

[‡]Ultimately, τ_0 is the time that limits the oscillation frequency of DBQW diodes.[53, 44]

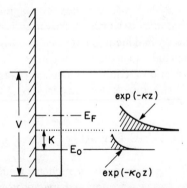

Fig. 36 Penetration of a quantum-well wave function into the confining barrier In the absence of scattering, the wave function factorizes, as in Eq. 83, and all the states belonging to the same subband E_0 decay asymptotically as $\propto \exp(-\varkappa_0 z)$. In the presence of scattering the asymptotic decay is $\propto \exp(-\varkappa z)$ and depends on the in-plane kinetic energy. Here $\hbar^2 \varkappa_0^2 / 2m \equiv V - E_0$ and $\hbar^2 \varkappa^2 / 2m \equiv V - (E_0 + K)$. (After Ref. 54).

quently, the wave function decays into the bulk with the characteristic exponential

$$\psi \propto \exp\left[-\hbar^{-1} \int^z \{2m\,[V(z) - E_0]\}^{1/2}\, dz\right], \qquad (106)$$

where E_0 is the energy of the subband bottom in the QW. The tunneling exponent in Eq. 106 is independent of the kinetic energy $K = \mathbf{p}^2 / 2m$ of the electron motion in xy plane. The situation is qualitatively different in the presence of scattering, which mixes different degrees of freedom. However weak the scattering processes, the asymptotic decay law for the electron density is described by a wave function that would result if the carriers had tunneled in the one-dimensional potential $V(z)$ but with the *total* energy $E \equiv E_0 + K$:

$$|\psi|^2 \propto \exp\left[-2\hbar^{-1} \int^z \{2m\,[V(z) - E]\}^{1/2}\, dz\right]. \qquad (107)$$

This statement has been proven quite generally.[54] At a sufficiently large distance from the QW the decay rate of Eq. 107 is strictly valid; transition to the no-scattering limit, Eq. 106, is described by a pre-exponential factor, which depends on the specific scattering mechanism and which has been evaluated[54] for several model examples, including a short-range model for electron-electron scattering and a model of scattering by inhomogeneities of the structure.

Although these considerations have not been applied to the case of tunneling in DBQW structures, similar effects are likely to have an important role there too. In particular, one can expect a strong effect on the peak-to-valley current ratio. Another consequence of the mixing of longitudinal and transverse motions by elastic scattering that is worth investigating is the dependence of the lifetime of a resonant state on its kinetic energy, $\tau_0 = \tau_0(K)$.

It is reasonable to conclude that any quantitative discussion of the RT in DBQW diodes must be based on a concrete model of scattering processes appropriate for the experimental structure under consideration.

2.8 SUMMARY AND FUTURE TRENDS

We have discussed several basic elements of modern semiconductor devices. It is worth noting that, just as they were ten and twenty years ago, almost all the commercially important devices today are built around the following key elements: *p-n* junctions, metal-semiconductor junctions, and the interface between Si and its thermally grown oxide. Their importance notwithstanding, these elements have been given scant attention in the present chapter for the simple reason that they are covered in detail in the literature. The selected topics deal mostly with the building blocks that engage the imagination of a device physicist and are occasionally used in exploratory devices today. The common feature of these building blocks, their sharply defined layered structure, resolved on a nanometer scale, brings about a need for emphasizing certain aspects of the carrier transport, which are largely peripheral to the operation of the existing commercial semiconductor devices. Understanding of these novel elements can only supplement a good knowledge of the classic device elements.

We have also left out of our consideration all opto-electronic elements, a major field that is currently under explosive development. It is generally believed that lightwave will eventually replace electrical current as the carrier of information signals—both in computer and communications applications, which are the main drivers of innovation in semiconductor devices. Still, the "dark" age of electronics is far from over, and in this chapter we chose to concentrate on the building blocks of modern exploratory transistors, which operate without emitting, absorbing, or transforming light.

The evolution of semiconductor electronics has always been intimately connected with advances in material science and technology. The first revolution in electronics, which replaced vacuum tubes with transistors, was based upon doped semiconductors and relied on newly discovered methods of growing pure crystals. Prior to the 1950s, semiconductors could not be properly termed "doped"—they were dirty. Today, semiconductors routinely used in devices are cleaner (in terms of the concentration of undesired foreign particles) than the vacuum of vacuum tubes.

Subsequent evolution of transistor electronics has been associated with the progress in two areas: (1) miniaturization of device design rules, brought about by advances in the lithographic resolution and doping by ion implantation, and (2) development of techniques for layered-crystal growth and selective doping, culminating in such technologies as MBE and MOCVD, that are capable of monolayer resolution of doping and chemical composition. Of these two areas, the first has definitely had a greater impact in the commercial arena, whereas the second has been mainly setting the stage for the exploration of device physics.

These roles may well be reversed in the future. Development of new and exotic lithographic techniques with a nanometer resolution will be setting the stage for the exploration of various physical effects in mesoscopic devices, while epitaxially grown devices (especially heterojunction transistors integrated with optoelectronic elements) will be gaining commercial ground. When (and whether) this role reversal will take place will be determined perhaps as much by economic as by technical factors. It is anticipated that the lateral miniaturization progress may face diminishing returns when the speeds of integrated circuits and the device packing densities will be limited primarily by the delays and power dissipation in the interconnection rather than individual transistors.[57] Further progress may then require circuit operation at cryogenic temperatures and/or heavy reliance on optical interconnections.[58] Implementation of the latter within the context of silicon VLSI requires hybrid-material systems with heteroepitaxial islands of foreign crystals grown on Si substrates.[59]

All these anticipated developments are likely to be heavily dependent on the progress of material science and techniques for epitaxial growth of semiconductor layers. However muddy our crystal ball may be regarding the future trends in microelectronics, one trend appears to be clear: the device designer of tomorrow will be thinking in terms of multilayer structures defined on an atomic scale. This chapter may serve as an introduction to the electronic transport processes in such structures.

PROBLEMS

1. Electric field in a Mott diode. Find the solution of Eq. 14 appropriate for the intrinsic layer in a Mott diode (Fig. 1) at equilibrium. Assume that the length of the i layer is large compared to the Debye length in the doped contact layer, $\mathcal{L} \gg 1$.

2. Integration of the Poisson equation in a partially depleted layer at equilibrium: beyond the depletion approximation.

    a) Derive Eq. 18 by analogy with elementary mechanics: in Eq. 13b regard ξ as time and ϕ as the position of a particle of mass β. Equation 13b becomes analogous to the second law of Newton with the right-hand side expressing a position-dependent force on the particle.

    b) Derive Eq. 18 by integrating Eq. 13b over ϕ (cf. Ref. 1 p. 367).

3. Thermionic theory of planar-doped barriers.

   a) In the "capacitor" model, which neglects the diffusion of mobile charge into the i layer and assumes that the entire positive charge of "uncovered" donors in the doped contacts is concentrated in an infinitesimally thin region at the boundaries with the i layer, sketch the charge, field, and potential profiles in equilibrium for a silicon PDB diode with $L_1 = 500\,\text{Å}$, $L_2 = 1\,500\,\text{Å}$, $\Sigma/q = 10^{12}\,\text{cm}^{-2}$.

   b) Sketch the same quantities for two applied biases: $V = \pm\,0.5\,\text{V}$ and calculate the uphill barrier heights.

   c) Sketch the approximate behavior of the quasi-Fermi level for both polarities of the applied bias as in part (b). In both contact layers assume the doping $N_D = 10^{18}\,\text{cm}^{-3}$.

   d) Write an expression for the carrier concentration on top of the barrier, assuming that it is given by an equilibrium Boltzmann relation (with respect to the emitter). Calculate the mean carrier velocity on top of the barrier, assuming that no carriers return from the collector side (use Maxwellian distribution in velocities). Find the total carrier flux across the barrier and reduce it to the Richardson form. Evaluate the current for $V = \pm\,0.5\,\text{V}$ at room temperature.

   e) Under the assumptions of part (d) calculate *separately* the drift and the diffusion currents at several points on the uphill slope (say, $0.25L_i$ and $0.75L_i$). You can take any reasonable value for the D and μ coefficients, but make sure that the Einstein relation is not violated. If the calculated currents turn out to be different, find your mistake. Compare these currents with the total current calculated in part (d) and comment on the applicability of the thermionic theory.

4. Transition from weak inversion to strong inversion in an MOS channel. Whenever there is an exponential growth of a physical quantity, something comes about to limit and eventually replace the exponent by a power law. Consider the charge density σ in an MOS channel. In the subthreshold regime it is exponentially related to the gate voltage, $\sigma \propto \exp\left(\beta V_G/n\right)$, where $n > 1$ is an "ideality" factor depending on the relative thicknesses of the oxide and the depletion layer. As the channel charge grows, however, the exponential dependence is replaced by a linear law corresponding to strong inversion. This transition can be described by an equation of the form

$$\sigma = \sigma_0 \exp\left[\frac{\beta(V_G - \alpha\sigma)}{n}\right],$$

similar to that leading to a limitation (Section 2.4.2) of the thermionic charge injection over a triangular barrier. Derive this equation and determine α.

5. SCL current in a film with edge contacts. Consider a thin film with the geometry of contacts illustrated in Fig. 12a. There are no extended electrodes and the solution almost exactly parallels that presented for the bulk diode.

a) Write the 2D charge density $\rho(x, y)$ in the form analogous to Eq. 52.

b) Show that for a filament source, the Green function of Eq. 64 is given by

$$G(x - xP, y - yP) = -\frac{1}{2\pi} \ln [(x - xP)^2 + (y - yP)^2]^{1/2} ,$$

and calculate the x-component of the electric field in the film.

c) In the approximation of constant mobility, derive the integral equation for $n_s(x)$ and bring it into a dimensionless form similar to Eq. 71:

$$\frac{v_s(\xi)}{1 - \xi} \int_0^1 \frac{1 - \xi'}{\xi - \xi'} v_s(\xi')\, d\xi' = 1 .$$

d) An approximate solution of this equation can be represented in the form

$$v_s(\xi) = a\xi^\alpha (1 - \xi)^{1/2}$$

with $a \approx 0.5$ and $\alpha \approx -0.36$. Derive the current-voltage characteristic and show that it is of the form of Eq. 51 with $\zeta_2^a \approx 0.57$.

6. Electron in a square well of finite width on one side (Fig. 18d).

a) Show that bound states correspond to energy eigenvalues E_n determined by the equation

$$\sqrt{2ma^2 E_n/\hbar^2} = (n + 1)\pi - \arcsin\sqrt{E_n/U} , \qquad n = 0, 1, 2, \cdots$$

and their wavefunctions are of the form

$$\chi(z) \propto \begin{cases} \sin[\sqrt{E_n/U}\,(z/a)] & \text{if } 0 \le z \le a \\ \sin[\sqrt{E_n/U}\,] e^{-\varkappa(z - a)} & \text{if } z \ge a \end{cases}$$

where $\varkappa \equiv \sqrt{2m(U - E_n)/\hbar^2}$. Find the normalization factor.

b) Show that the condition for the existence of at least one bound state E_0 is $U \ge U_0 = \pi^2\hbar^2/8ma^2$. As the well deepens, what are the conditions for the emergence of higher-lying bound states?

7. Electron in a triangular potential well (Fig. 18f).

a) First consider the problem of an electron in a uniform electric field $\mathcal{E}$. This corresponds to an unbounded motion and one can expect the spectrum to be continuous. Write the Schrödinger eigenvalue equation in the momentum representation and solve it. Does it have a solution for any value of the energy, E? Transform the eigenfunction, corresponding to some value of E, to the coordinate representation, and express it in the form similar to Eq. 80a (with E in place of E_n), using an integral representation of the Airy function:

$$\text{Ai}(x) = \pi^{-1/2} \int_0^\infty \cos\left[xt + \frac{t^3}{3}\right] dt .$$

b) The triangular well is confined at $z = 0$ by an infinite (perfectly reflecting) wall. The energy level quantization arises from the boundary condition on the wave function $\chi(0) = 0$. Determine the three lowest energy levels and express them in units of $\hbar^2/ma^2$, where $a = (\hbar^2/2mq\mathcal{E})^{1/3}$.

c) Derive the quasi-classical quantization rule,

$$\int_0^{z_n} [E_n - U(z)]^{1/2} \, dz = \pi\hbar(n + \tfrac{3}{4}),$$

where z_n is the classical turning point at energy E_n, for a smooth potential $U(z)$ bounded by an infinite wall on one side ($U = \infty$ for $z \leq 0$). One way of doing it is to use the Bohr-Sommerfeld quantization rule for a symmetric potential $\tilde{U}(z) = U(|z|)$, and note that odd wave functions and energy levels for $\tilde{U}$ have the following correspondence to the wave functions and energy levels for U (prove it):

$$E_n = \tilde{E}_{2n+1} , \qquad \psi_n \sim \tilde{\psi}_{2n+1} .$$

d) Using the above quantization rule, derive an expression for the quasi-classical energy levels E_n in the triangular well. Evaluate the three lowest energy levels and compare them with the exact values from part (b).

e) Consider the width of the triangular-well wave functions, characterized by the expectation values of the operators z and z^2 in the states (80a). Show that

$$<z>_n = \frac{2E_n}{3q\mathcal{E}} = \frac{2}{3}\alpha_{n+1}a ;$$

$$<z^2>_n = \frac{6}{5}[<z>_n]^2 .$$

Make a comparison with the extent of the classical motion defined by the turning points at the energies E_n.

8. Variational function for the lowest subband.[27] Consider a 2DEG at a semiconductor/insulator interface. Make the following simplifying assumptions:

- only one subband is occupied,
- the background doping charge can be neglected,
- exchange energy can be ignored,
- the hypothetical dielectric has the same permittivity as the semiconductor—i.e. there is no image force.

a) Show that the potential energy of a test electron in the Coulomb field of all the subband electrons can be written in the form

$$V = q\mathcal{E} \left[z + \int_0^z (z' - z)\, \chi^2(z')\, dz' \right].$$

b) Assume a trial wave function of the form

$$\chi(z) \sim z\, e^{-bz/2}, \qquad z \geq 0$$

and $\chi(z) = 0$ for $z < 0$. Find the normalization factor and calculate $<z>$.

c) Determine the parameter b variationally, i.e., by minimizing the expectation value in the state $\chi(z)$ of the Hamiltonian $H = p^2/2m + V(z)$ describing a test particle in the field of the subband electrons—all assumed to be in the same state.

9. Density of states in a "relativistic" 2DEG.

a) Derive the density of states per subband in a 2-dimensional, relativistic, free-electron gas characterized by an energy dispersion relation of the form

$$\frac{E^2}{c^2} = (\mathbf{p}^2 + mc^2).$$

In the limit $\mathbf{p}^2/2m \ll mc^2$ express the result as a first-order correction to Eq. 87b.

b) How many electrons (per cm^2) would one need for the relativistic correction to be non-negligible (assume the free-electron mass)? What would be the discontinuity in electric field on crossing a 2DEG of such density (assume the permittivity of vacuum)?

10. Two-dimensional electron gas with an elliptical Fermi surface. Consider a 2DEG with the "elliptical" dispersion relation:

$$E_{n,\, k_x,\, k_y} = E_n + \frac{\hbar^2 k_x^2}{2m_{xx}} + \frac{\hbar^2 k_y^2}{2m_{yy}}.$$

Show that the density of states and the Fermi level can be described by the same expressions as in the case of the isotropic spectrum but with an effective "density-of-states" mass $m \equiv \sqrt{m_{xx}m_{yy}}$. Sketch the Fermi surface.

11. Quantum size effect in polycrystalline silicon grains. There are reasons to believe that in n-type polySi conduction-band electrons are strongly reflected by the grain boundaries (cf. Section 7.2.4). One can therefore view a small grain as a three-dimensional potential well. Because of the quantum size-effect the continuous spectrum of a free conduction-band electron becomes discrete and the density of states in the conduction band at low energies is depressed. This can have a significant effect on the threshold voltage of MOS capacitors with polySi gates.[55]

   a) Model a polySi grain as a cube of side $a \sim 100$ Å filled with a degenerate free-electron gas and show that the electron energy levels are given by

   $$E_{n_1, n_2, n_3} \approx 4 \text{ [meV]} \times [5(n_1^2 + n_2^2) + n_3^2],$$

   where the quantum numbers n_j, $(j = 1, 2, 3)$ assume the values $n_j = 1, 2, \cdots$.

   b) Calculate and plot the number $N(E)$ of states in the grain with energy less than E (including the twofold spin and the sixfold valley degeneracy). Show that under 100 mV there is room for only 72 electrons. Compare with the "classical" number of states N_{bulk} in bulk silicon per grain volume,

   $$N_{\text{bulk}}(E) \equiv (100 \text{ Å})^3 \int_{E_C}^{E} D(E') \, dE'.$$

   c) Plot the number of electrons in the volume of a grain versus the Fermi level at $T = 300$ K both for the bulk and the quantum-confined cases. What is the shift in the Fermi level due to the confinement for the carrier concentration of $2 \times 10^{20} \text{ cm}^{-3}$?

12. Quantum capacitance of a 2DEG. Let σ_1, σ_2, and σ_Q be the charge densities, respectively, on electrodes 1 and 2 and in the quantum well, Fig. 26a.

   a) Show that the energies E_i of the electric field in regions 1 and 2 are given by

   $$E_i = \frac{2\pi d_i \sigma_i^2}{\epsilon_i}, \qquad i = 1, 2, \ldots.$$

   If the total energy E_{tot} were just the sum of E_i, $i = 1, 2, \ldots$, then it would be minimized by placing the entire charge on the middle plate.

b) Calculate the Fermi degeneracy energy E_Q, i.e., the total kinetic energy of electrons in the 2DEG (at zero temperature). Minimizing the total energy $E_{tot} = E_1 + E_2 + E_Q$, show that the ratio of charges on the bottom and middle electrodes equals

$$\frac{\sigma_2}{\sigma_Q} = \frac{\hbar^2 \epsilon_2}{4mg_\nu d_2 q^2} \equiv \frac{C_2}{C_Q},$$

thus proving the equivalent circuit in Fig. 26b.

13. δ-function potential models (Fig. P1). The potential $V_1(x) = \alpha \delta (x - x_0)$ (Fig. P1a) can be regarded as a limiting case of a rectangular potential of height U and width a when $U \to \infty$, $a \to 0$, and $2aU \to \alpha$. The derivative of the wave function $\psi'(x)$ must be discontinuous at x_0, so that (prove it) $\psi'(x_0^+) - \psi'(x_0^-) = (2m\alpha/\hbar^2) \psi (x_0)$.

a) Show that the transmission and the reflection amplitudes for a particle at energy $E = \hbar^2 k^2/2m$ incident on the barrier $V_1(x)$ are given by

$$t(k) = \frac{ik\hbar^2}{ik\hbar^2 - m\alpha} \;\; ; \;\; r(k) = \frac{m\alpha}{ik\hbar^2 - m\alpha}.$$

Note that the phases of t and r differ by $\pi/2$. This is a general property for any symmetric potential (can you prove it?). Evaluate the transmission and the reflection coefficients ($T \equiv |t|^2$ and $R \equiv |r|^2$). Check that $T + R = 1$.

b) Find the wave functions corresponding to the symmetric $|n, S>$ and asymmetric $|n, A>$ eigenstates for a "double well" potential $V_2(x)$ (Fig. P1b). Assuming that $\alpha \gg \hbar^2/ma$ show that the energy eigenvalues for low lying states are given by

$$E_n^A = \frac{\hbar^2 \pi^2 n^2}{2ma^2}, \quad n = 1, 2, \cdots ;$$

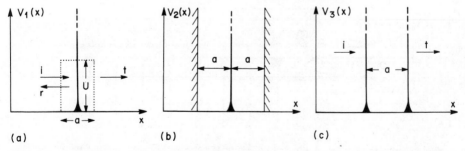

(a)   (b)   (c)

Fig. P1 δ-function models: (a) δ-function potential as a limiting case of a rectangular barrier, (b) model of a double well, and (c) double-barrier transmission.

$$E_n^S \approx \frac{\hbar^2 \pi^2 n^2}{2ma^2} \left[1 - \frac{2\hbar^2}{ma\alpha} \right], \quad n = 1, 2, \cdots \ll \alpha ma/\hbar^2.$$

c) Find the resonant energies at which particles incident on the double-barrier potential $V_3(x)$ (Fig. P1c) are perfectly transmitted ($T = 1$).

d) Show that if $\alpha < 0$, then the potential $V_1(x)$ results in one symmetric bound state of energy $E_0 = -\hbar^2 \varkappa_0^2/2m$, where $\varkappa_0 = \alpha m/\hbar^2$.

14. Kronig-Penney model of a one-dimensional crystal, Fig. P2.

a) Consider a periodic potential in the form of an infinite array of δ-function peaks,

$$V(x) = \sum_j \alpha \delta(x - ja), \quad \alpha < 0.$$

Show that the electron energy dispersion relation $E(k)$ in this potential is implicitly given by the relation

$$\cos(ka) = \cos(\sqrt{\epsilon}) - \frac{\alpha am}{\hbar^2} \frac{\sin(\sqrt{\epsilon})}{\sqrt{\epsilon}} \equiv F(\epsilon),$$

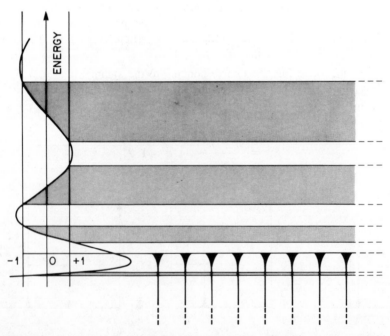

Fig. P2 Electron bands in the Kronig-Penney model. The illustrated example corresponds to $\alpha am/\hbar^2 = -8$. The right-hand side $F(\epsilon)$ of the Kronig-Penney dispersion relation $\cos(ka) = F(\epsilon)$ is plotted against the dimensionless energy ϵ. Allowed bands (shaded) correspond to $|F(\epsilon)| \leq 1$.

where $\epsilon \equiv (2ma^2/\hbar^2) E$ is a dimensionless energy (you may consult Ashcroft and Mermin,[56] p. 146–149). Calculate the effective electronic mass in the allowed bands in the vicinity of the band edges.

b) Assuming that $|\alpha a m/\hbar^2| >> 1$, calculate the tight-binding bandwidth for the lowest band. Compare with the exact solution of the Kronig-Penney model.

REFERENCES

1. S. M. Sze, *Physics of Semiconductor Devices*, 2nd ed., Wiley-Interscience, New York (1981).

2. W. Shockley, *Electrons and Holes in Semiconductors*, Van Nostrand, Princeton (1950).

3. D. M. Caughey and R. E. Thomas, "Carrier Mobilities in Silicon, Empirically Related to Doping and Field," *Proc. IEEE* **55**, 2192 (1967).

4. H. W. Thim "Computer Study of Bulk GaAs Devices with Random One-Dimensional Doping Fluctuations," *J. Appl. Phys.* **39**, 3897 (1968).

5. K. Seeger, *Semiconductor Physics*, 2nd ed., Springer-Verlag, Berlin/Heidelberg (1982).

6. J. G. Ruch, "Electron Dynamics in Short Channel Field-Effect Transistors," *IEEE Trans. Electron Dev.* **ED-19**, 652 (1972).

7. F. Venturi, R. K. Smith, E. C. Sangiorgi, M. R. Pinto, and B. Riccò, "A General Purpose Device Simulator Coupling Poisson and Monte Carlo Transport with Applications to Deep Submicron MOSFET's," *IEEE Trans. Computer-Aided Des.* **CAD-8**, 360 (1989).

8. M. Shur, *GaAs Devices and Circuits*, Plenum, New York (1987).

9. R. F. Kazarinov and S. Luryi, "Charge Injection over Triangualr Barriers in Unipolar Semiconductor Structures," *Appl. Phys. Lett.* **38**, 810 (1981).

10. A. A. Grinberg and S. Luryi, "Space-Charge-limited Current and Capacitance in Double-Junction Diodes," *J. Appl. Phys.* **61**, 1181 (1987).

11. N. F. Mott and R. W. Gurney, *Electronic Processes in Ionic Crystals*, Oxford University Press, Oxford (1948).

12. R. J. Malik, T. R. AuCoin, R. L. Ross, K. Board, C. E. C. Wood, and L. F. Eastman, "Planar-doped Barriers in GaAs by Molecular Beam Epitaxy," *Electron. Lett.* **16**, 836 (1980).

13. J. M. Shannon, "A Majority-Carrier Camel Diode," *Appl. Phys. Lett.* **35**, 63 (1979).

14. R. F. Kazarinov and S. Luryi, "Majority Carrier Transistor Based on Voltage-Controlled Thermionic Emission," *Appl. Phys. A* **38**, 151 (1982).

15. E. H. Rhoderick, "Comments on the Conduction Mechanism in Schottky Diodes," *J. Phys. D* **4**, 1920 (1972).

16. S. E.-D. Habib and K. Board, "Theory of Triangular-Barrier Bulk Unipolar Diodes Including Minority-Carrier Effects," *IEEE Trans. Electron Dev.* **ED-30**, 90 (1983).

17. J. A. Geurst, "Theory of Space-Charge-Limited Currents in Thin Semiconductor Layers," *Phys. Stat. Sol.* **15**, 107 (1966).

18. A. A. Grinberg, S. Luryi, M. R. Pinto, and N. L. Schryer, "Space-Charge-Limited Current in a Film," *IEEE Trans. Electron Dev.* **36**, 1162 (1989).

19. E. H. Rhoderick, *Metal-Semiconductor Contacts*, Clarendon, Oxford (1980).

20. A. C. Gossard, R. F. Kazarinov, S. Luryi, and W. Wiegmann, "Electric Properties of Unipolar GaAs Structures with Ultrathin Triangular Barriers," *Appl. Phys. Lett.* **40**, 832 (1982).

21. E. F. Schubert, J. B. Stark, B. Ulrich, and J. E. Cunningham, "Spatial Localization of Impurities in δ-doped GaAs," *Appl. Phys. Lett.* **52**, 1508 (1988).

22. E. F. Schubert, A. Fisher, and K. Ploog, "The Delta-Doped Field-Effect Transistor (δFET)," *IEEE Trans. Electron Dev.* **ED-33**, 625 (1986).

23. E. F. Schubert, J. E. Cunningham, and W. T. Tsang "Electron-Mobility Enhancement and Electron Concentration Enhancement in δ-doped n-GaAs at $T = 300$ K," *Solid State Commun.* **63**, 591 (1987).

24. G. H. Döhler and K. Ploog, "Doping (n-i-p-i) Superlattices," in *Synthetic Modulated Structures*, L. L. Chang and B. C. Giessen, Ed., Academic, Orlando (1985).

25. E. F. Schubert, B. Ulrich, T. D. Harris, and J. E. Cunningham, "Multi-Subband Photoluminescence in Sawtooth Doping Superlattices," *Phys. Rev.* **B** (1989).

26. T. Ando, A. B. Fowler, and F. Stern, "Electronic Properties of Two-Dimensional Systems," *Rev. Mod. Phys.* **54**, 437-672 (1982).

27. F. F. Fang and W. E. Howard, "Negative Field-Effect Mobility on (100) Si Surfaces," *Phys. Rev. Lett.* **16** 797 (1966).

28. F. Stern, "Self-Consistent Results for n-type Si Inversion Layers," *Phys. Rev.* **B 5**, 4891 (1972).

29. A. Kastalsky and A. A. Grinberg, "Novel High-Speed Transistor Based on Charge Emission from a Quantum Well," *Appl. Phys. Lett.* **52**, 904 (1988).

30. S. Luryi, "Quantum Capacitance Devices," *Appl. Phys. Lett.* **52**, 501 (1988a).

31. L. D. Landau and E. M. Lifshitz, *Quantum Mechanics: Non-Relativistic Theory*, 3rd Ed., Pergamon, London (1977).

32. E. O. Kane, "Basic Tunneling Concepts," in *Tunneling Phenomena in Solids*, E. Burstein and S. Lundquist, Eds., Plenum, New York (1969)

33. S. Luryi, "Possibility of Direct Observation of the Time Evolution in Heterostructure Barrier Tunneling," *Solid State Comm.* **65**, 787 (1988).

34. S. Luryi, "Coherent versus Incoherent Resonant Tunneling and Implications for Fast Devices," *Superlatt. Microstr.* **5**, 375 (1989).

35. R. F. Kazarinov and R. A. Suris, "Electric and Electromagnetic Properties of Semiconductors with a Superlattice," *Fiz. Tekh. Poluprovodn.* **6**, 148 (*Sov. Phys. Semicond.* **5** 120) (1972).

36. L. V. Keldysh, *Zh. Eksp. Teor. Fiz.* **43**, 661 [*Sov. Phys. JETP* **16**, 471 (1963)] (1962).

37. L. V. Keldysh, "Effect of Ultrasound on the Electron Spectrum of a Crystal," *Fiz. Tverd. Tela (Leningrad)* **4**, 2265 (1962) [*Sov. Phys. Solid State* **4**, 1658 (1963)].

38. R. F. Kazarinov and R. A. Suris, "Possibility of the Amplification of Electromagnetic Waves in a Semiconductor with a Superlattice," *Fiz. Tekh. Poluprovodn.* **5**, 797] (*Sov. Phys. Semicond.* **5**, 707) (1971).

39. L. V. Iogansen, "Errors in Papers on Resonant Electron Tunneling in Finite Superlattices," *Pis'ma Zh. Tekh. Fiz.* **13**, 1143 (*Sov. Tech. Phys. Lett.* **13**, 478) (1987).

40. L. Esaki and R. Tsu, "Superlattice and Negative Differential Conductivity in Semiconductors," *IBM J. Res. Dev.* **14**, 61 (1970).

41. L. L. Chang, L. Esaki, and R. Tsu, "Resonant Tunneling in Semiconductor Double Barriers," *Appl. Phys. Lett.* **24**, 593 (1974).

42. L. Esaki and L. L. Chang, "New Transport Phenomenon in a Semiconductor "Superlattice"," *Phys. Rev. Lett.* **33**, 495 (1974).

43. S. Luryi, "Frequency Limit of Double-Barrier Resonant-Tunneling Oscillators," *Appl. Phys. Lett.* **47**, 490 (1985).

44. S. Luryi, "Hot-Electron Injection and Resonant-Tunneling Heterojunction Devices," in *Heterojunction Band Discontinuities: Physics and Device Applications*, F. Capasso and G. Margaritondo, Eds., Elsevier Science, Amsterdam, Chap. 12, pp. 489–564 (1987).

45. S. Luryi and F. Capasso, "Resonant Tunneling of Two-Dimensional Electrons through a Quantum Wire: A Negative Transconductance Device," *Appl. Phys. Lett.* **47**, 1347 (1983) [erratum: **48**, 1693 (1986)].

46. E. A. Rezek, N. Holonyak, Jr., B. A. Vojak, and H. Schichijo, "Tunnel Injection into the Confined-Particle States of $In_{1-x}Ga_xP_{1-z}As_z$ in InP," *Appl. Phys. Lett.* **31**, 703 (1977).

47. H. Morkoç, J. Chen, U. K. Reddy, T. Henderson, and S. Luryi, "Observation of a Negative Differential Resistance Due to Tunneling Through a Single Barrier into a Quantum Well," *Appl. Phys. Lett.* **49**, 70 (1986).

48. M. C. Payne, "Transfer Hamiltonian Description of Resonant Tunneling," *J. Phys. C* **19**, 1145 (1986).

49. M. Jonson and A. Grincwajg, "Effect of Inelastic Scattering on Resonant and Sequential Tunneling in Double Barrier Structures," *Appl. Phys. Lett.* **51**, 1729 (1987).

50. T. Weil and B. Vinter, "Equivalence between Resonant Tunneling and Sequential Tunneling in Double-Barrier Diodes," *Appl. Phys. Lett.* **50**, 1281 (1987).

51. M. Büttiker, "Coherent and Sequential Tunneling in Series Barriers," *IBM J. Res. Devel.* **32**, 63 (1988).

52. R. Gupta and B. K. Ridley, "The Effect of Level Broadening on the Tunneling of Electrons through Semiconductor Double-Barrier Quantum-Well Structures," *J. Appl. Phys.* **64**, 3089 (1988).

53. B. Riccò and M. Ya. Azbel, "Physics of Resonant Tunneling. The One-Dimensional Double-Barrier Case," *Phys. Rev. B* **29**, 1970 (1984).

54. S. V. Meshkov, "Tunneling of Electron from a Two-Dimensional Channel into the Bulk," *Zh. Eksp. Teor. Fiz.* **91**, 2252 [*Sov. Phys. JETP* **64**, 1337] (1986).

55. N. Lifshitz, S. Luryi, and T. T. Sheng, "Influence of the Grain Structure on the Fermi Level in Polycrystalline Silicon: A Quantum Size Effect?" *Appl. Phys. Lett.* **51**, 1824 (1987).

56. N. W. Ashcroft and N. D. Mermin, *Solid State Physics*, Holt, Rinehart and Winston, Philadelphia (1976).

57. R. W. Keyes, "Fundamental Limits in Digital Information Processing," *Proc. IEEE* **69**, 267 (1981).

58. J. W. Goodman, F. J. Leonberger, S.-Y. Kung, and R. A. Athale, "Optical Interconnections for VLSI Systems," *Proc. IEEE* **72**, 850 (1984).

59. S. Luryi and S. M. Sze, "Possible Device Applications of Silicon Molecular Beam Epitaxy," in *Silicon Molecular Beam Epitaxy*, Vol. 1, E. Kasper and J. C. Bean, Eds., (CRC Uniscience, Boca Raton, FL, 1988), Chap. 8.

II FIELD-EFFECT AND POTENTIAL-EFFECT DEVICES

3 The Submicron MOSFET

J. R. Brews
AT&T Bell Laboratories
Murray Hill, New Jersey

3.1 INTRODUCTION

Although design of the MOSFET (metal-oxide-silicon field-effect transistor) began with the intrinsic device, later demands for smaller size required reduction of device parasitics, like junction capacitance and series resistance. Today, because of the use of huge numbers of devices, chip requirements control the circuit and fabrication environment for MOSFET optimization. After outlining this evolution, this chapter introduces the growing list of design trade-offs and presents a simplified algorithm for MOSFET size reduction. We begin with some recent changes in technology, and their changing demands upon the MOSFET.

In nMOS technology (consisting of n-channel enhancement devices with depletion loads) placing more devices on a chip demanded not only smaller devices but closer packing. Pushing devices together introduced trade-offs between devices and their isolation from one another. For example, isolation of devices using field oxides affected threshold voltages, resulting in various width effects upon threshold. Large chips also led to heat removal problems: nMOS power dissipation limited the number of devices that could be housed in an IC (integrated circuit) package, beginning another technology shift.

To reduce power dissipation, nMOS gave way to CMOS (complementary MOS) technology in which both n-channel and p-channel devices are constructed simultaneously on the same substrate. This transition was assisted by other factors related to chip or circuit improvements.[1] For example, the use of complementary devices improved noise margins and allowed use of the full-voltage swing from zero to the supply voltage. Cost advantages of nMOS eroded as miniaturization increased the complexity of processing. Circuit techniques that

High-Speed Semiconductor Devices, Edited by S.M. Sze. ISBN 0-471-62307-5
© 1990 John Wiley & Sons, Inc.

favored *n*MOS over CMOS at large dimensions were less useful at small dimensions. For example, bootstrapping (generation of large-voltage swings to speed circuit response) resulted in high fields and reliability problems, and dynamic charge storage on parasitic capacitances suffered from the poor scalability and fabricational control of parasitics. Whatever the motivations, the switch to CMOS made isolation more demanding. CMOS also involved simultaneous optimization of two different types of device, which meant compromises in processing wherever one device benefited at the expense of the other. It now appears that further miniaturization requires thin epitaxial substrates, separate tubs (tubs are individually doped islands in the substrate) with retrograde *n*- or *p*-doping profiles,† and separate gates (e.g., n^+-gates for *n*MOS and p^+-gates for *p*MOS devices). This complex context for device design is a result of overall goals, beyond the maximum speed or the minimum size of individual MOSFETs.

More recently still, the progressive convergence of CMOS and bipolar processing has reduced the difference in cost between BiCMOS (bipolar devices plus CMOS) and CMOS. Introduction of BiCMOS improves the performance of a given generation of CMOS, allowing amortization of its cost of development over a longer commercial lifetime. Apart from economics, inclusion of bipolar devices diversifies options for trading power, area, and performance at a circuit level, diversifies function building to combine analog and digital circuits, and also allows any mix of bipolar and CMOS circuitry across the chip.[2] The first applications of BiCMOS have used bipolar transistors for driving large capacitive loads, and for sense amplifiers. A developing use is to allow the selective lowering of signal levels to reduce noise.

In the future, bipolar options might moderate interconnection demands in VLSI (very large scale integration). As MOSFETs shrink and chips incorporate more circuits, instead of consolidating functions in blocks with heavy traffic between blocks, cheap replication is used to implement a function where needed. A trivial example is on-site building of an inverter to generate the complement of a clock signal locally, rather than transmitting this signal using a line. This replication cuts communication costs: traffic, wiring space, and noise. For this purpose, the more flexible and cheaper the building of local functions the better, so the bipolar flexibility of BiCMOS might allow a more varied replication-interconnection strategy, and a more distributed architecture.

Evidently, evolving technology introduces processing complexity and problems of compatibility. Device parameters (such as junction depths and channel lengths) couple with circuit parameters (such as voltage swings, noise levels, and delays) and with fabricational parameters (such as implant and oxidation conditions). Fabricational variations affect circuit design, and circuit demands place priority on certain process steps. A robust, clear framework is required to sequence and segment the design, and to identify difficulties that require judgment or additional experimentation.[3,4] How accurate a simulation is needed?

---

†A profile increasing with depth to allow shallow tubs with tolerable values for *I-R* drops, vertical punchthrough voltages, and parasitic capacitances.

Where should it be introduced? Where is experimental input required? For instance, evaluation of hot-electron degradation might be needed only in a late phase of design, or might not be useful until some processes have been decided, for example, processes affecting the drain geometry, or might require a combination of modeling and experiment. The same framework that segments the design also regulates feedback and iteration. Poor parameterization, inappropriate accuracy, or inopportune iteration wastes resources and wastes time.

During the construction of this framework, simplified models of the various trade-offs are very useful. Not only do they help in choosing the parameters to be exchanged between segments, they help to evaluate the efficiency and convergence properties of alternative structuring. These simple models are important for interrelating processing limitations, circuit requirements, and device implications, so decisions are understandable in all these aspects. Also, the simplified models can provide a first attempt at optimization. This inital step is useful because process models must be fitted to experiment, due to inadequate understanding, and circuit models must be fitted to device simulations, due to limited computer capacity. Repetition of this calibration is avoidable if a simplified approach can identify the parameter ranges that bracket the final design.

The simplified treatment of trade-offs requires formulas summarizing basic MOSFET limitations, many of which are empirical and incomplete. For example, electrostatic scaling[5,6] or subthreshold scaling[7] introduces key variables, but ignores relations between them caused by fabrication and circuit constraints. A more complete approach classifies constraints with an ordering dependent on the product under design, for example, memory, gate array, and so on.[8-13] Instead, this chapter simply traces three basic variables of miniaturization: junction depth, oxide thickness, and depletion width, and finds how various constraints couple these parameters.

Interest here is in future developments, so the emphasis is not always on present problems and solutions, but upon problems to be overcome at even higher packing densities and lower voltages. Because of space restrictions, detailed discussion of basic MOSFET terminology and operation is referred to texts[14,15] and review articles.[16] Besides this background, the reader is referred to fundamental discussion of device[17] and circuit limitations.[17-19] For greater depth and more references on special topics, see references 3, 4, and 20. Unless otherwise stated, n-channel devices are discussed.

3.1.1 An Empirical Formula

Among the constraints upon MOSFET miniaturization, historically those due to device physics have received the most attention. One approach to physical constraints is constant-field[5] or electrostatic scaling.[6] This approach chooses a successful large device (e.g., a device from the previous generation) and maintains the same fields inside the small device by correlated reduction of dimensions and voltages. Drawbacks arise where physical mechanisms do not scale with field, for example, where current is due to diffusion, not drift, as in the subthreshold region of operation, and where higher fields are forced built-in junction potentials become larger in comparison to smaller applied voltages.

For the present discussion, miniaturization is introduced using an empirical formula for the minimum size of a MOSFET.[7] This formula is a constraint that keeps subthreshold behavior insensitive to drain bias. As such, it is only one of many constraints, but it appears to be a *limiting* constraint. That is, it must be observed by a successful, smallest device; it is a required relation between the basic variables governing the device. To this constraint others are added, imposing still further relations between these variables. The empirical formula states that the MOSFET channel length has to be larger than L_{min}, where

$$L_{min} = A \left[r_j d \left(W_S + W_D \right)^2 \right]^{1/3} , \qquad (1)$$

where r_j is the junction depth, d is the oxide thickness, and W_S, W_D are the depletion widths of source and drain in a one-dimensional, abrupt junction approximation for uniform doping, namely,[14-16]

$$W_{S,D} = \sqrt{2} \, L_D (\beta \psi_{S,D} - 1)^{1/2} , \qquad (2)$$

with L_D the Debye length for uniform doping, N,

$$L_D = \left(\frac{\epsilon_s}{\beta q N} \right)^{1/2} , \qquad (3)$$

$\beta = q/(kT)$ the inverse thermal voltage, and $\psi_{S,D}$ the potential of the source (drain) relative to the substrate, including the built-in junction potential. With r_j, W_S, W_D in microns, d in angstroms, the fitting parameter is $A = 0.41 \, \text{Å}^{-1/3}$. According to Eq. 1, miniaturization requires shallower junctions (r_j), thinner oxides (d), lower voltages or heavier doping ($W_S + W_D$). Our discussion of the MOSFET is organized around these variables and what sets their lower limits. Equation 1 is illustrated in Fig. 1.

To a limited degree one can understand Eq. 1 intuitively. For instance, it makes sense for a symmetrical MOSFET that Eq. 1 should be unchanged if source and drain are interchanged: indeed, Eq. 1 is unchanged if W_S is exchanged with W_D. It also is reasonable that the minimum size should increase if the depletion widths increase, because such increase enlarges the two-dimensional regions controlled by the source and drain at the expense of the region controlled by the gate, the long-channel MOSFET region. Similarly, increased junction depth increases the source and drain control, so we expect L_{min} to increase with r_j. Finally, as the oxide thickness is increased the screening effect of the gate is reduced, and field lines emanating from the source and drain extend further into the channel region, and fewer of them terminate on the gate. Consequently, two-dimensional source and drain control increases with d. These simple ideas support Eq. 1, but it remains empirical because no derivation exists. A possible replacement for Eq. 1 is examined in Problems 1 and 2.

Equation 1 has limitations. For example, as r_j or d tends to zero, the real L_{min} tends to some limiting value, not to zero as Eq. 1 suggests. The devices used to determine Eq. 1 were too large to establish this limit. Also, Eq. 1 ignores factors like the actual doping profiles of source, drain, and channel.

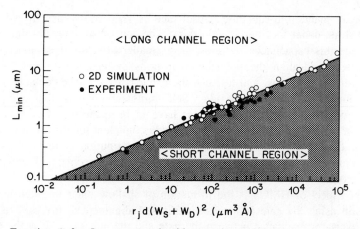

Fig. 1 Equation 1 for L_{min} compared with experimental measurements and computer simulations. The criterion for L_{min} is that for $L = L_{min}$ a 10% variation in drain current occurs as V_{DD} varies from 0.5 to 1 V in the subthreshold regime of gate biases. The units of the abscissa reflect the choice of units: d in Å, the other variables in μm.

Nonetheless, Eq. 1 is useful in keeping surface punchthrough or "drain-induced barrier-lowering" under control, and serves to introduce the variables basic to the organization of this chapter.

3.1.2 The High-Performance MOSFET

To begin with some complications of a real MOSFET, the HMOS (high-performance MOS) structure is shown in Fig. 2. Although proposed long ago, this structure incorporates important aspects of modern MOSFETs. In particular, the substrate doping is not uniform. The region between the source and drain is more heavily doped than the region below the source and drain. The heavily doped surface allows a short source-to-drain spacing. Thus, in Eq. 1 it is the doping between source and drain that must be used. With many MOSFET designs this surface doping is nearly uniform out to the junction depth. For some designs, particularly for low-temperature operation, the doping peaks below the channel region, to reduce dopant-ion scattering in the channel, among other reasons. In such cases, Eq. 1 is a cruder approximation.

The lightly doped epitaxial layer (labeled "p EPI" in Fig. 2) reduces the junction-to-substrate capacitance that is charged and discharged during device operation, speeding response. A lightly doped substrate at the depletion layer edge below the channel also is an advantage in circuits where the source-to-substrate bias varies, shifting the depletion layer edge. An example where such variation occurs is a series chain of MOSFETs, where the drain of one device is coupled to the source of the next, as in the input logic block of a "NAND" gate, or in a chain of transmission gates. For such cases, placement of the depletion layer edge in a more lightly doped region means that any voltage modulation of the source causes less variation in depletion layer charge, and less threshold voltage variation.

As noted, junction-to-substrate capacitance is reduced by use of a lightly doped substrate under the source and drain. The contact structure in Fig. 2 also helps reduce this capacitance by allowing a smaller area for these junctions.[21] The aim is to self-align the polysilicon portions of the source-drain contacts directly over the junctions, and to place the metal portion of the contacts remote from the junctions, over field oxide. Self-alignment minimizes junction area because fewer mask-alignment tolerances have to be included. The remaining metal areas, which can be large and can include alignment tolerances, are placed over thick field oxide where their capacitance contribution is reduced. Also, metal spiking through the polysilicon does not cause problems because the underlying oxide layer is insulating.

Another aspect of HMOS is the formation of shallow junction extensions by implantation using the gate as a mask. This self-alignment to the gate reduces parasitic overlap capacitance between the gate and the junctions. It is the depth of this shallow junction extension that is used in Eq. 1. The entire length of the junction is not made shallow to avoid junction series resistance and leakage problems. Leakage current can arise easily from process damage of shallow junctions, for example, during contact formation. These topics are discussed later.

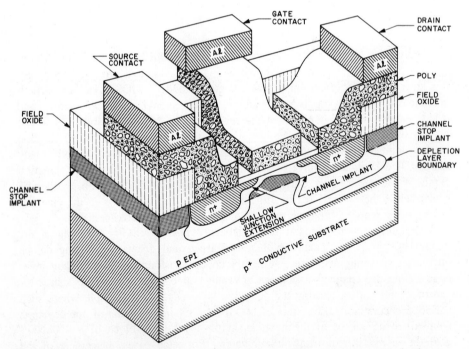

Fig. 2 The HMOS (high-performance MOS) structure. The contact geometry is chosen to place most of the contact area above field oxide so that the junction areas can be minimized, to reduce parasitic capacitance.

Also shown in Fig. 2 is a ballooning of the drain depletion region in the lightly doped substrate. That is, there is a possibility of bulk (subsurface) punchthrough of the drain to the source. To prevent subsurface punchthrough current, the substrate doping cannot be made as low as one would like in trying to reduce junction capacitance and threshold variation under source-to-substrate bias. The substrate doping must be greater than some minimum value, sometimes set crudely as,[22]

$$N_{SUB} \geq \frac{N_{CH}}{10} , \qquad (4)$$

where N_{SUB} is the substrate doping and N_{CH} is the doping in the region between source and drain. Equation 4 sets a lower bound on the junction-to-substrate capacitance.

A final aspect of Fig. 2 worth noting is the conductive substrate employed to prevent coupling through shared voltage drops in the common ground plane, to lower RC charging times, and to prevent latch-up. This substrate is discussed later when isolation is considered.

It is apparent that many structural innovations have occurred that transcend simple scaling ideas. Nonetheless, evaluation of a proposed structure, if not the inspiration for its creation, is based in part upon the limitations on the basic parameters appearing in Eq. 1 now to be discussed one by one.

3.2   JUNCTION DEPTH

As junction depths are reduced, adverse effects arise. First, the cross section of the source and drain junction shrinks, which can increase the parasitic series resistance of the MOSFET, degrading its performance. This series resistance of the source and drain is a large part of the total device resistance for a short device, unless contacts are self-aligned. Second, reduced junction depth increases junction curvature, raising the electric field in the drain region. This increased field is undesirable because it creates energetic carriers that degrade the oxide and because it can limit the maximum voltage that can be used. Third, shallow junctions are prone to damage during fabrication. This damage promotes leakage current, an anathema of CMOS design. Now consider these various limitations on junction depth.

3.2.1   Series Resistance

Series resistance can be divided into four components, as shown in Fig. 3a, namely, contact resistance, R_{CO}, originating in the region between the ohmic contact and the laminar flow region in the heavily doped junction, sheet resistance, R_{SH}, from the laminar flow region, spreading resistance, R_{SP}, corresponding to the current crowding region where the laminar flow pattern is squeezed into a surface accumulation layer formed by the gate at the junction edge next to

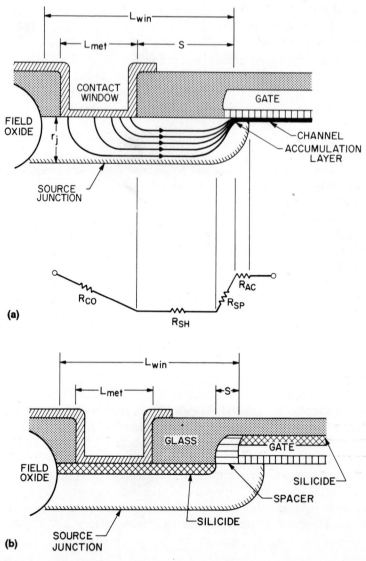

Fig. 3   (a) The parasitic series resistance components of the MOSFET.   (b) A salicide contact, differing from (a) by the silicided extension of the metal window. This extension is separated from the silicided gate by a sidewall spacer of thickness S.   (After Refs. 23, 24 © 1986, 1987 IEEE).

the channel, and accumulation layer resistance, R_{AC}, bridging the spreading resistance region and the actual channel of the device.[23-25]

Series resistance is greatly reduced by using self-alignment that can provide large-area contacts. One approach to self-alignment is discussed in Section 3.2.3. A more common approach is the self-aligned silicide (salicide, for short) contact, shown in Fig. 3b and constructed as follows. Beginning with an opening in the field oxide, a polysilicon gate is formed at a distance L_{win} from the field oxide, freestanding between an exposed source and drain. A spacer (usually an oxide) of thickness S is formed on the sidewalls of the gate. Then a thin metal layer is deposited everywhere and sintered, forming a silicide wherever metal overlies silicon, particularly over the source and the drain and on top of the gate. This sintering self-aligns the silicide: the silicide layer covers the junction right up to the sidewall spacer, and any unreacted metal on the spacer itself is removed with a selective etch that preferentially attacks metal. Alignment of the silicide to the gate is governed by the control of this spacer, which is much tighter than control of mask alignment. Thus, self-alignment can bring the silicide very close to the gate without risking a short-circuit. With the silicide in place, the gate-to-field-oxide window is filled with insulator (called "glass" in Fig. 3b), and a contact hole (of length L_{met}) is positioned and opened to allow metal access to the silicide. The final structure of Fig. 3b resembles that in Fig. 3a, but with a major difference: although the contact metal is fairly remote from the gate, just as in Fig. 3a, this metal is joined to the silicide layer that extends under the insulator from the metal window to within a distance S from the edge of the gate. This silicide extension of the contact increases the length of the contact to $L_{con} \approx L_{win} - S$, increasing the effective contact area. This increased L_{con} is achieved without any increase in L_{win} and, therefore, without any increase in the junction area and junction-to-substrate capacitance. The use of a silicide is important primarily as a means to self-alignment, and is otherwise incidental.

With the silicide layer so close to the gate, there is no laminar sheet resistance region, and the series resistance becomes $R_{CO} + R_{SP} + R_{AC}$. The resistances R_{SP}, R_{AC} depend upon the precise spatial dependence of the junction profile in the surface region under the gate. The current-flow patterns itself to minimize the total resistance,[23-25] and this minimization occurs approximately when $R_{SP} \approx R_{AC}$. An approximate form for R_{SP} is[23,24]

$$R_{SP} \approx \frac{2}{\pi} \frac{\rho}{W} \left\{ \Delta \left(\frac{\delta \rho_{ch}}{\rho} \right) + \ln \left(\frac{r_j}{2r_{ch}} \right) \right\}, \tag{5}$$

where r_{ch} is the surface channel depth, ρ is the bulk junction resistivity, $\delta \rho_{ch}$ is the difference between the junction resistivity at the end of the accumulation layer and that of the bulk junction, and W is the channel width. The second term represents the resistance for abrupt junctions (see Problem 3), and Δ is the additional resistance introduced by the spatial dependence of the junction profile beneath the gate,

$$\Delta \approx E_1 \left(\frac{2r_{ch}}{\pi \lambda_J} \right) - E_1 \left(\frac{2r_j}{\pi \lambda_J} \right) \tag{6}$$

where $E_1(x)$ is a form of the exponential integral,

$$E_1(x) \equiv \int_x^\infty \frac{dt}{t} \exp(-t) ,$$

and the decay length, λ_J, of the profile usually satisfies[23] $\lambda_J \le 200$ Å, so $r_j/\lambda_J \gg 1$. The second term in Eq. 6 is

$$E_1 \left(\frac{2r_j}{\pi\lambda_J} \right) \approx \frac{\exp[-2r_j/(\pi\lambda_J)]}{2r_j/(\pi\lambda_J)} ,$$

which is negligible compared to the first term in Eq. 6 because $r_{ch} < r_j$. As a result, Δ is independent of r_j so R_{SP} depends on junction depth only through the logarithmic term. That is,

$$R_{SP} + R_{AC} \approx 2R_{SP} \approx \frac{4}{\pi} \frac{\rho}{W} \left[\Delta \left(\frac{\delta\rho_{ch}}{\rho} \right) + \ln \left(\frac{r_j}{2r_{ch}} \right) \right] , \qquad (7)$$

which is a weak function of r_j, unlike the sheet resistance term which decreases quickly with r_j. It must be borne in mind that the logarithmic term in Eq. 7 applies only in the case where S is a fairly large multiple of r_j, and $r_{ch} \ll r_j$. When $S \approx r_j$, the case of interest for small devices, the contact resistance region and the region governed by R_{SP} merge, so the contribution $R_{CO} + R_{SP}$ is not easily expressed in a simple formula.[25]

The customary transmission line model for contact resistance provides,[23-25]

$$R_{CO} = \frac{1}{W} \sqrt{\frac{\rho\rho_c}{r_j}} \coth \left(L_{con} \sqrt{\frac{\rho}{\rho_c r_j}} \right) , \qquad (8)$$

with L_{con} the conductive length of the contact, for example, $L_{con} = L_{met}$ in Fig. 3a and $L_{con} = L_{win} - S$ in Fig. 3b. Here ρ is the average bulk resistivity of the junction in Ω-cm and ρ_c is the specific contact resistivity between the contact material and the junction in Ω-cm^2. The coth in Eq. 8 is within 10% of its unity asymptotic value for L_{con} such that

$$L_{con} \ge 1.5 \sqrt{\frac{\rho_c r_j}{\rho}} .$$

Using $\rho \sim 3.3 \times 10^{-3}$ Ω-cm (p^+ junction), $\rho_c \sim 10^{-6}$ to 10^{-7} Ω-cm^2, and $r_j \sim 800$ Å, we find this condition becomes $L_{con} \ge 0.73$ μm $- 0.23$ μm, an inequality that certainly is valid for designs of the near future. Thus Eq. 8 in the large L_{con} limit predicts that contact resistance will be near its least value of

$$R_{CO} \sim \frac{1}{W} \sqrt{\frac{\rho\rho_c}{r_j}} ,$$

which indicates a junction depth dependence, and no dependence on L_{con}. This junction depth dependence reflects the assumption of the transmission line model that the junction is a thin resistive sheet, and so does not include a realistic current flow pattern. These assumptions introduce an exaggerated dependence on r_j and underestimate the dependence on L_{con}.

With no adequate formula for R_{CO} and no published simulations to guide us, only some intuitive guidelines remain. First, for $L_{con} >> r_j$ there is probably only a very weak dependence on L_{con}. As a result, L_{con} should be kept small to minimize contact area that adds parasitic capacitance, but L_{con} should be large compared to r_j. Second, if the gate-to-contact separation S and the junction depth r_j are scaled together, the current-flow pattern probably is unaffected (there are questions about the corresponding voltages, which are reduced, but not fully scaled). Thus, for $S / r_j \approx$ constant, we expect only a weak r_j dependence of series resistance. Additional modeling in this area is desirable.

It is likely that the decay length, λ_J, of the junction profile under the gate is not forced by fabrication to increase as r_j is reduced, so Δ need not increase with smaller r_j (except for the lightly doped drain structures discussed later). Also, it appears likely that methods of dopant activation can be made compatible with small r_j, containing any increase in junction resistivity. Granting these assumptions, reduction of junction depth is contingent upon the self-alignment spacing, S. Now two other mechanisms that might limit the shallowness of junctions are discussed, namely, drain breakdown and junction leakage.

3.2.2 Drain Breakdown

Junction depth affects the maximum field in the drain region with two main results. First, shallower junctions lower the drain breakdown voltage, as discussed in this section. Second, shallower junctions increase the heating of hot carriers, exacerbating the hot-electron damage of the oxide and interface near the drain. This last effect is the most serious, but its discussion is delayed to Section 3.4.2.

One reason that drain breakdown in short-channel MOSFETs differs from simple junction avalanche breakdown is its initiation by carriers injected into the high-field drain region from the channel. Heated in the field near the drain, these injected carriers cause impact ionization, creating secondary holes and electrons. Because of this initiation by injected carriers, feedback can occur: the secondary carriers can themselves alter the injection conditions for the primary carriers, for example, via voltage (*I-R*) drops in the substrate that alter the carrier density in the channel (so called "bipolar breakdown"). See Fig. 4.

The gate electrode plays an important part in this breakdown. First, the gate controls the injection of channel carriers that trigger the avalanche. Second, the gate has an important influence over the field pattern near the drain. Thus, the maximum field is a function of gate bias and oxide thickness, as much as junction depth and substrate doping. Carrier heating is caused by the lateral field near the drain, the field component that drives carriers toward the drain. An

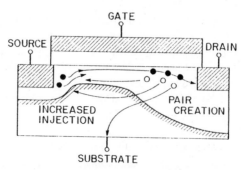

Fig. 4 Injection assisted drain breakdown in the MOSFET. Secondary holes are collected by either the source or the substrate contact, and associated *I-R* drops bias the source to increase electron injection.

often used formula for the lateral field is[22]

$$\mathcal{E}_{\text{lat}} = \frac{V_D - V_G}{[(\epsilon_s/\epsilon_{ox})\, dh]^{1/2}}\,, \tag{9}$$

where h is a length vaguely related to junction depth. For instance, Klaassen[22] suggests for r_j in the vicinity of 0.25 μm, $h \approx 0.5$ μm. Hu and his co-workers[11,26] variously suggested $h \sim 0.016 r_j/d^{1/3}$ (for $d > 150$ Å) and $h \sim 10^{-4} r_j^{2/3}\ L^{2/5}/d^{3/4}$, L = channel length (all lengths in cm), for $d < 150$ Å. Evidently this empirical work needs further development.

Usually hot-carrier degradation (Section 3.4.2) places a more severe constraint on the maximum lateral field than does drain breakdown. Drain breakdown is estimated to occur for fields of the order[22] $\mathcal{E}_{\text{lat}} \sim 0.6$ MV/cm while, as estimated later, hot-carrier degradation is excessive for $\mathcal{E}_{\text{lat}} \sim 0.2$ MV/cm. Thus, drain breakdown is not a major limitation of MOSFET design at small dimensions, and further details are relegated to reference 27. Another type of drain breakdown can occur if the gate creates a field-induced junction by depleting or inverting the surface of the source or drain. See Section 3.3.2.

3.2.3 Junction Leakage

Shallow junctions are prone to damage during fabrication. If this damage is present inside or near the junction depletion layer, leakage current results. The fabrication of shallow p^+/n junctions is particularly difficult because boron channels during implantation and diffuses rapidly during annealing. To avoid channeling, often the silicon is bombarded to make it amorphous, creating damage that then must be annealed after implantation. Trade-offs result between junction leakage, depth of the amorphous layer, amount of annealing, and junction depth.[28] At low anneal temperatures (600 to 700°C) residual damage below the amorphized layer causes very high leakage. At higher temperatures (800 to 1000°C) these defects disappear and only dislocation loops are present. Then leakage depends on their position with respect to the junction depth. Thus, the depth of the amorphous layer is chosen to be just sufficient to stop channeling, and the depletion layer is placed just deep enough to keep dislocation loops outside the depletion layer. If these dislocation loops are remote from the depletion layer, junctions with leakage currents of about[28] 0.7 nA/cm^2 with larger perimeter leakage-current densities of about 10 nA/cm^2 are obtainable. The amorphous layer needs to reside about 300 Å or more above the junction edge.[28]

Control of junction damage requires attention to all aspects of contact formation, not just the placement of the junction dopants. To lower parasitic source and drain resistance, Section 3.2.1 indicated the importance of self-alignment of junction contacts to the gate. Besides possible capacitance trade-offs as electrodes are brought close together, this self-alignment can introduce trade-offs with leakage. For instance, leakage can be increased by the salicide process[29−31] if the heavily doped silicon consumed during the silicide formation brings the silicide-silicon interface too close to the depletion layer of the junction. In shallow n^+/p junctions, experiment indicates that a separation of 80 to 100 nm between the silicide-silicon and the metallurgical p-n junction suffices to avoid this problem.[29] Separation layers and silicon consumption during silicide formation add to junction depth.

A disadvantage of some salicide processes is the combined use of the same silicide both for contacts and for reducing the resistance of interconnections, for example, by using the silicided gate poly for interconnections. Unfortunately, the resistivity of silicided interconnections depends upon the silicide thickness, and this thickness is related to the amount of silicon consumed during the silicide formation. In such cases, salicidation implies an undesirable coupling of junction depth and interconnect resistivity.[29]

One way to self-align the contacts without salicide leakage and without coupling junction depth to silicide resistance is based on a planarization process,[21,32] shown in Fig. 5. Here the gate and sidewall oxide is raised above the field oxide, as shown in Fig. 5a. Following a polysilicon deposition, the poly on top of the high gate protrudes above the poly over the field oxide, Fig. 5b. With addition of a planarization resist, an etch clears the poly off the gate, leaving the poly connection to the junctions intact, Fig. 5c. The resulting self-aligned structure resembles Fig. 2, which, for clarity, does not show the sidewall spacers. A difference in topography between Figs. 2 and 5 is attributable to different assumptions about alignment tolerances. The contact poly can be patterned, Fig. 5d, and a silicide formed on top of the self-aligned polysilicon, rather than immediately over the junctions, eliminating leakage due to silicide formation. The gate also can be silicided. This planarization approach to self-alignment avoids salicide leakage with a small increase in parasitic gate-to-junction capacitance introduced by the parallel, vertical sidewalls of the contacts.

Another advantage of self-alignment by planarization compared to the salicided contact of Fig. 3b is the reduction of parasitic junction-to-substrate capacitances. Unlike the salicide contact of Fig. 3b, the metal contact hole for the structure of Fig. 5 (see Fig. 2) is located over the field oxide, so inaccura-

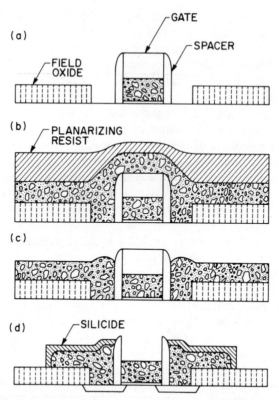

Fig. 5 Several stages in the planarization approach to fabrication of a self-aligned contact. Lateral dimensions are only schematic; relative spacings depend on alignment tolerances. A stacked gate structure is used, consisting of a reactive layer over polysilicon. The reactive layer, which can be metal, doped polysilicon, or doped oxide, reacts with the blanket polysilicon of (b) to enable preferential etching of the blanket polysilicon from the gate area, (c). (After Ref. 21, © 1987 IEEE).

cies in the positioning of this contact hole are not a factor in deciding the junction area. As a result, the junction area for the contact in Fig. 5 can be reduced compared to the salicide contact of Fig. 3b.

3.2.4 Comments

The contact probably should satisfy $L_{con} >> r_j$ to reduce series resistance, but a large L_{con} increases the area contributing to parasitic capacitance. There also might be a series resistance trade-off requiring tighter self-alignment as junction depths are made shallower ($S / r_j \approx$ constant). In summary, two main obstacles in making shallow junctions are (a) achieving self-alignment without introducing serious leakage or capacitance trade-offs and, (b) minimizing junction damage without driving the junction deeper. Already depths of $r_j \approx 500$ Å are feasible, and no fundamental obstacle to further reduction is evident. Aside from feasibility, however, the use of shallow junctions is restricted by hot-electron effects, as discussed in Section 3.4.2.

3.3 OXIDE THICKNESS

Oxide thickness is limited by four basic mechanisms: Fowler-Nordheim leakage, lifetime under time-dependent dielectric breakdown (TDDB), gate-induced drain leakage (GIDL), and oxide tunneling between gate and substrate.

3.3.1 Fowler-Nordheim Leakage and Time-Dependent Dielectric Breakdown

At high enough oxide fields (7 to 8 MV/cm) many electrons tunnel from the electrodes into the conduction band of the oxide, which then conducts a leakage current. See Fig. 6. Because of trapping in the oxide, the field in the oxide

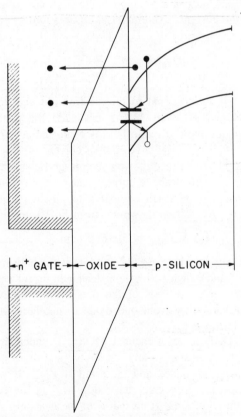

Fig. 6 Fowler-Nordheim leakage in the MOSFET gate oxide. Also shown are some possible interface-trap-assisted leakage paths, with trap levels indicated by short, solid bars.

seldom is uniform but, if it were, the current would follow the Fowler-Nordheim expression[33]:

$$J = A\mathcal{E}_{ox}^2 \exp\left(\frac{-B}{\mathcal{E}_{ox}}\right) \tag{10}$$

where $A = 1.25 \times 10^{-6}$ A/V^2, $B = 233.5$ MV/cm, J is the current density in A/cm^2, and $\mathcal{E}_{ox}$ is the (uniform) oxide field in V/cm. Experimental measurements of A and B are affected by interface and oxide traps so there are uncertainties in their values.

As charge is transported through the oxide a wear-out phenomenon occurs, even at very low currents. Once the total of the charge transported through the oxide accumulates a "charge-to-breakdown" level, Q_{bd}, a short-circuit between the gate and the silicon substrate develops.[34] This charge-to-breakdown is not a well-characterized quantity, and depends on oxide and interfacial quality as well as oxide field, having a value of ≈ 10 C/cm^2 at $\mathcal{E}_{ox} \approx 10$ MV/cm for typical oxides.[35] Many authors suggest a relation to positive charging (hole trapping) somehow related to the current. An empirical observation is[35]

$$Q_{bd} \propto \exp\left(\frac{H}{\mathcal{E}_{ox}}\right) \tag{11}$$

where $\exp(-H/\mathcal{E}_{ox})$ is proportional to the electron-initiated generation rate of positively charged traps, $H \approx 80$ MV / cm and, for a ten-year time-to-breakdown the maximum allowed oxide field is estimated as[36] 7 MV/cm, which according to Eq. 10 allows a Fowler-Nordheim current of 2×10^{-7} A/cm^2, a rather high value. If TDDB in fact allowed such large fields, rather than TDDB it would be Fowler-Nordheim leakage current that set the maximum field. For example, for an oxide leakage small compared to junction leakage, one might specify a current of $\approx 10^{-10}$ A/cm^2, which, from Eq.10, results in a field.

$$\mathcal{E}_{ox} \leq \mathcal{E}_{max} \approx 5.8 \text{ MV / cm} . \tag{12}$$

For the sake of discussion, here Eq. 12 is adopted for both mechanisms, with the result that TDDB and Fowler-Nordheim leakage current set the same limit upon $\mathcal{E}_{max}$. In other words, in view of the present uncertainty in the various parameters and their variability from one oxide to another, we cannot predict which mechanism determines $\mathcal{E}_{max}$.

Let us consider TDDB in an n-channel MOSFET under two different bias conditions: (a) with the gate at V_{DD} and the source and drain grounded and (b) with the gate and source grounded and the drain at V_{DD}. If we assume an n-channel device with an n^+-poly gate, the oxide field in the region where the gate overlaps the drain is different from that over the channel. At zero gate and drain bias the oxide field is zero over the drain, but over the channel the equilibration of the difference in bulk Fermi levels between the n^+-type gate and the p-substrate introduces a field at zero bias of $Q_{s0}/(C_{ox}d)$ where Q_{s0} is the silicon charge density per unit area at zero bias, and C_{ox} is the oxide capacitance per unit area. Thus, as the gate bias is increased, the field in the oxide over the channel always exceeds that in the overlap region, and the wearout condition is worst over the channel. The limiting oxide field, $\mathcal{E}_{max}$, imposes a lower limit on the allowable oxide thickness, which must be larger than d_{min}

$$d_{min} = \frac{V_{DD} - V_{FB} - \psi_s}{\mathcal{E}_{max}}, \tag{13}$$

where V_{DD} is the maximum gate-to-source voltage, V_{FB} is the gate voltage at flatband and, ψ_s is the potential drop across the silicon. See Problem 4.

Now consider the second case of a grounded gate and a drain at V_{DD}. Assuming the substrate also is grounded, the oxide field in the gate-to-drain overlap region now is higher than that in the channel region. A question arises: which of the two bias conditions leads to the larger oxide field? From the argument based on built-in field, it appears that the case with V_{DD} on the gate is more severe. There are two complications to this conclusion, however. The first complication is that sharp edges or corners might arise near the drain or along the gate edge either because of process limitations or errors. Should a field enhancement occur, $\mathcal{E}_{max}$ might be larger in the case with V_{DD} on the drain. The second complication is that there can be oxide currents other than the simple Fowler-Nordheim injection from the grounded gate when the drain is at V_{DD}. These extra currents accelerate wearout and are discussed at the end of the next section on gate-induced drain leakage.

3.3.2 Gate-Induced Drain Leakage

Consider the region where the gate overlaps the n^+-drain, with the gate grounded and the drain at V_{DD}, as shown in Fig. 7. A large field exists in the oxide and, corresponding to Gauss' law, there is a charge Q_s induced in the drain electrode given by

$$\epsilon_{ox} \mathcal{E}_{ox} = Q_s. \tag{14}$$

The charge Q_s is provided by a depletion layer in the drain, which becomes a non-equilibrium, deep-depletion layer for large $\mathcal{E}_{ox}$ because minority carriers that might form an inversion layer are drained laterally to the substrate. The non-equilibrium surface region, which could collect minority carriers except for the lateral draining to the substrate, will be called the "incipient inversion layer" in the drain.

For large enough $\mathcal{E}_{ox}$, the voltage drop in the deep-depletion layer becomes large enough to allow tunneling in the drain via a near-surface trap as shown in Fig. 8. Several trap-assisted events then are possible. As shown in Fig. 8, a trap can emit carriers by either thermal or tunneling emission to either band, although not all choices are available to all the traps. For instance, in regions where the bands do not overlap, at least one of the emission processes must be thermal. For large enough voltages the valence and conduction band overlap,

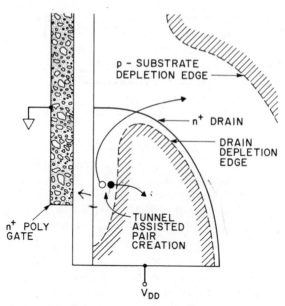

Fig. 7   Gate-induced drain leakage in the MOSFET. A schematic view of the gate-drain overlap region for a grounded gate and the drain biased at V_{DD}. Tunneling created pairs lead to a lateral hole flow in the n^+ drain. This flow prevents formation of an inversion layer inside the drain.

and for large enough fields (a few MV/cm) trap-free, band-to-band tunneling can be significant.[37-44] Temperature independence[37,40] suggests thermal emission plays a minor role. Whatever the mechanism, the minority carriers emitted to the incipient inversion layer are then laterally removed to the substrate, completing a path for a gate-induced drain leakage (GIDL) current.

The assumption that near-surface trap occupancy is decided by emission processes is supported by the following argument. Near the surface the minority-carrier density is lower than in equilibrium because the inversion layer is incipient. The majority-carrier density also is negligible near the surface because of deep depletion. Thus, capture processes are suppressed. For generation-recombination centers when only thermal emission to both bands occurs, the half-occupancy condition applies to a trap with energy level near midgap, maximizing carrier generation. For cases involving tunneling, this critical energy will vary with the tunneling probabilities.

The GIDL current is found to follow the formula

$$J \propto \mathcal{E}^2 \exp\left[\frac{-K}{\mathcal{E}}\right] \qquad (15)$$

where J is the tunneling current/unit volume and $\mathcal{E}$ is the electric field in the depletion layer. The observed value[37-40] of K is $K \sim 19$ to 23 MV/cm except for trap-free samples, which are observed to have a value $K \sim 36$ MV/cm

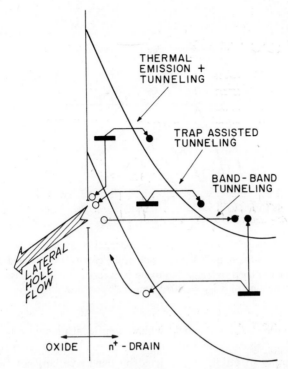

Fig. 8 Various mechanisms that can contribute to gate-induced drain leakage. Vertical transitions are thermal-emission processes. Horizontal transitions are tunneling processes. Holes are prevented from forming an inversion layer by lateral transport to the substrate. This flow is normal to the plane of the diagram, indicated by the slanted arrow, "lateral hole flow," as shown in Fig.7.

appropriate for indirect band-to-band tunneling in silicon.[40] It was shown[40] that junction fabrication using Ge pre-amorphization introduced bulk midgap traps that increase GIDL. It also has been shown that interface traps generated by hot-electron damage[41] or by Fowler-Nordheim tunneling[39] increase GIDL by introducing additional tunneling.

In CMOS circuits this leakage current contributes to standby power, which causes problems with heat dissipation at large device counts. Thus, GIDL must be contained. GIDL can be controlled by increasing the oxide thickness (reducing the field for a given voltage), increasing the doping in the drain (to limit the depletion width and the tunneling volume), or by eliminating traps (assuming voltages and fields low enough that trap-free, band-to-band tunneling is not a factor). Trap elimination requires careful fabrication.

The basis for control of GIDL using drain doping is shown in Fig. 9, which shows the width of the portion of the deep-depletion layer in the drain where significant GIDL occurs. This width is limited by two factors. At low doping GIDL occurs only in that part of the band overlap region where fields satisfy

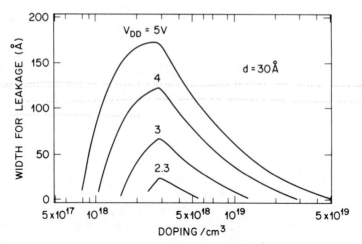

Fig. 9   Width of the portion of the drain deep-depletion layer contributing to GIDL versus doping density in the drain. The left side of the curves is dominated by field, and the right side by band-to-band overlap. An oxide thickness $d = 30$ Å is assumed and a field for significant tunneling of $\mathscr{E}_{GIDL} = 10^6$ V/cm.

$\mathscr{E} \geq \mathscr{E}_{GIDL}$, where $\mathscr{E}_{GIDL}$ is the field where significant tunneling begins. At high doping $\mathscr{E} \geq \mathscr{E}_{GIDL}$ throughout and even outside the band overlap region, and the width contributing to GIDL is restricted by band overlap, not field. The two conditions coincide at a doping, N_{GIDL}, estimated from Gauss's law as

$$N_{GIDL} = \frac{\epsilon_s \mathscr{E}_{GIDL}^2}{2kT(\beta \phi_G - 1)} ,$$

with $\phi_G \approx 1.12$ V, the bandgap voltage. For $\mathscr{E}_{GIDL} \approx 10^6$ V/cm, this expression provides $N_{GIDL} \approx 3 \times 10^{18}/$ cm^3. For the voltages chosen in Fig. 9, the peak in the curves occurs at N_{GIDL}. For larger voltages, which are inappropriate for submicron MOSFETs, the peak can occur for doping below N_{GIDL}. The GIDL current is an integral of a tunneling expression over this region, and has an exponential sensitivity to the width, rather than simple proportionality. Nonetheless, comparison with an evaluation of the tunneling integral[44] shows this width is a good indicator of how doping and oxide thickness affect GIDL.

For the parameters of Fig. 9, the width contributing to GIDL can be greatly reduced by using drain doping levels well above N_{GIDL}. Suppression of GIDL by using low doping levels is undesirable because of increased series resistance. This discussion also favors abrupt junctions, because graded or lightly doped drain structures, like those used to reduce lateral fields, might lead to a significant lateral extension of the drain within which the doping is near N_{GIDL}, causing leakage. Figure 9 also shows that the width of the region causing GIDL rapidly decreases with reduction in voltage, for example, for voltages below $\approx$ 3 V in the case where $d = 30$ Å. As we will see, this voltage is probably above the upper limit of the voltage range acceptable for deep submicron devices, so GIDL will become less significant in the future. See Problem 5.

Assuming GIDL is controlled, consider TDDB in the overlap region. Accompanying GIDL current is the possibility of hot-carrier generation from two sources. First, the lateral minority-carrier current in the incipient inversion layer might become hot enough to allow hot carriers to enter the oxide.[42] Second, the majority carriers present inside the drain after tunneling might become hot enough in the deep-depletion layer to generate secondary carriers that could enter the oxide,[43] as shown in Fig.10 for a pMOS device. In the overlap region, therefore, there is the possibility of oxide current either by Fowler-Nordheim injection or by hot-carrier injection. These currents cause TDDB in the overlap region. Hot-carrier current is more important at larger voltages because more heating can occur. Hence, thinner oxide structures (which necessarily use lower voltages, see Eq. 13) conduct mainly by Fowler-Nordheim currents and, neglecting the acceleration of wearout by tiny hot-hole currents, discussion of TDDB parallels that of Section 3.3.1. Because there is little or no built-in field in the overlap region for n^+-gates over n^+-drains, it is TDDB in the channel region that is more important in deciding oxide thickness.

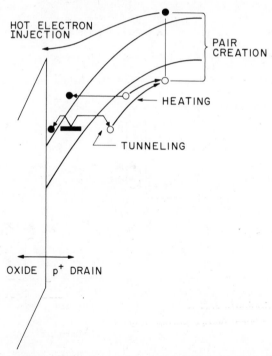

Fig. 10   Possible heating of majority carriers injected into the drain by the tunneling processes causing GIDL. This heating by the normal field results in pair creation, and the minority carrier of the pair can be heated enough in its turn to be injected into the oxide. Lateral heating in the incipient inversion layer also might occur.

3.3.3   Oxide Tunneling

For $d \leq 40$ Å some tunneling through the forbidden gap of the oxide occurs, allowing current to flow directly between the gate and substrate. See Fig. 11. For $d < 30$ Å, an inversion layer cannot form in an MOS capacitor because this tunneling current removes the inversion layer carriers faster than they are supplied by thermal generation[45]. The theory of this tunneling was further developed[46] and compared with experiment on $d = 22$ Å devices.[47] In principle, one would like to set a limit on d using an expression for the oxide tunneling current. A simple expression is not available, however, and the parameters in tunneling expressions are uncertain because of difficulties in characterizing devices. For example, oxide defects, interface traps, and interface morphology affect tunneling. For $d \approx 40$ Å interface traps can appear at the silicon-SiO_2 interface that are characteristic of the metal used for the gate electrode,[48] showing that care in gate fabrication is needed to control oxide tunneling. For boron-doped poly gates, movement of the boron through the oxide does occur, but its influence on oxide tunneling is unknown, as are the effects of silicide formation over poly gates.

Under these circumstances it is expedient simply to set a minimum oxide thickness, d_{min},

$$d_{min} \approx 30 \text{ Å} . \tag{16}$$

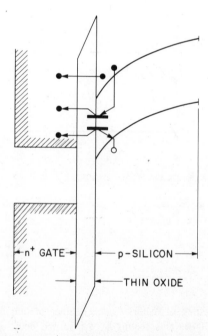

$\leftarrow n^+$ GATE$\rightarrow$ $\leftarrow$ p$-$SILICON $\rightarrow$

THIN OXIDE

Fig. 11   Tunneling through the gate oxide for thin oxides. Shown are both unassisted tunneling and some possible interface-trap assisted tunneling paths, with trap levels indicated by short, solid bars.

This provisional d_{min} can be revised as thin oxides improve and better characterization allows a more confident choice.

3.3.4   Comments

There are three effects that could limit oxide thickness: TDDB lifetime, Fowler-Nordheim leakage current, and oxide tunneling. Another limitation that cannot be evaluated yet is a difficulty in manufacture. TDDB lifetime and Fowler-Nordheim leakage limit the maximum field allowed in the oxide, taken here as $\mathcal{E}_{max} \approx 5.8$ MV/cm. There is too much uncertainty in $\mathcal{E}_{max}$ to allow a decision between TDDB lifetime and Fowler-Nordheim leakage. Improvements in oxide quality might make a clear decision possible in the future. The third effect, tunneling between gate and substrate through the oxide forbidden gap, sets $d_{min} \approx 30$ Å.

For low standby power dissipation, GIDL must be controlled. For small devices, this leakage is best reduced by using heavily doped, abrupt junctions and maintaining low trap densities, rather than using thick oxides. Next we discuss factors limiting the applied voltage and doping, to determine whether Eq. 13 or Eq. 16 for d_{min} is decisive in setting the oxide thickness.

3.4   DEPLETION WIDTHS

To decrease the depletion widths W_S and W_D in Eq. 1 the doping level can be maximized to reduce the Debye length, or the voltage can be minimized. First consider increasing the channel doping.

3.4.1   Doping, Threshold Voltage, and Driving Ability

Although increasing the channel doping can lower the breakdown voltage of the drain, this effect does not limit the maximum doping. The real limitation is the increase in threshold voltage as channel doping increases. As a rough approximation, for long channels and uniform doping, V_T often is taken as[14−16]

$$V_T = V_{FB} + 2\psi_B + \frac{qN}{C_{ox}}\sqrt{2}L_D(2\beta\psi_B - 1)^{1/2} , \qquad (17)$$

where

$$\beta\psi_B = \ln \frac{N}{n_i} , \qquad (18)$$

with n_i the intrinsic carrier density per unit volume and V_{FB} the flatband voltage. As Eq. 17 shows, when N is increased to allow smaller devices, V_T increases. To counter this increase in V_T by reduction of d to increase C_{ox} in Eq. 17 is not an option because d already is set by V_{DD} and $\mathcal{E}_{max}$ in Eq. 13, or by Eq. 16, or by manufacturability. Hence, the maximum doping is set by how closely V_T and V_{DD} can approach each other.

The relation between V_T and V_{DD} might vary with the application. Consider two very common parts in CMOS circuits: transmission gates and inverter stages. Transmission gates are useful in minimizing power and interconnection demands[17] and a number of implementations are discussed by Annaratone.[18] For a transmission gate in a clocked CMOS circuit (see Fig. 12), V_T is chosen high enough to avoid leakage current when the gate is grounded. The allowable leakage is determined by how long a time the gate must hold a voltage on a capacitive node. Such a gate represents an RC delay in a circuit, and to minimize the resistance contribution, V_{DD} should exceed V_T sufficiently to avoid excessive MOSFET resistance. Analysis suggests an adequate V_{DD} is[9]

$$V_{DD} \approx 4V_T . \tag{19}$$

According to Eq. 19, for a transmission gate the minimum V_{DD} is set by V_T, which in turn is set by off-current leakage requirements.

Another major CMOS component is the inverter shown in Fig. 13a. The calculated quasi-static transfer characteristics of the inverter (i.e., the V_{out} versus V_{in} relation for slowly varying V_{in}) are shown in Fig. 13b for several threshold voltages. As with the transmission gate, the closer V_T approaches V_{DD}, the slower the circuit. For example, consider the case where $V_T = V_{DD}/2$, corresponding to the sharpest transfer characteristic in Fig. 13b. As the input voltage increases from logic 0 to logic 1, switching begins only at $V_{in} = V_{DD}/2$. The p-channel device then switches off, while the n-channel device has only barely switched on. As a result, for a transient ramp input voltage, there is very little current to discharge the output node and, even for very small output node capacitances, V_{out} will drop only slowly as V_{in} continues to increase. For reasonable

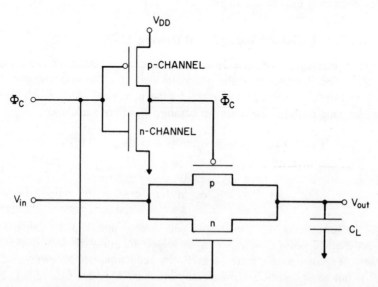

Fig. 12 A CMOS transmission gate driven by a clock signal, Φ_c, and its complement, $\overline{\Phi}_c$, generated locally by an inverter.

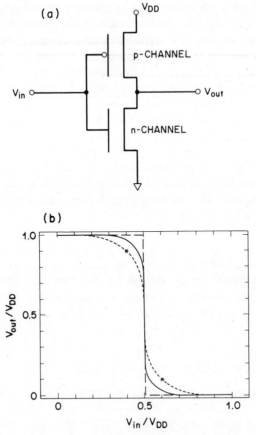

Fig. 13 (a) CMOS inverter and (b) long-channel CMOS inverter transfer characteristics for three threshold-to-supply voltage ratios, V_T / V_{DD}; — — — = 0.5, —————— = 0.25, - - - - - - - - - = 0.1. Asterisks indicate the unity-gain points for the 0.1 ratio.

speed, the supply voltage should be chosen to exceed threshold by some margin, and we adopt Eq. 19 for the inverter also.

Another concern with the inverter is noise immunity. A measure of the noise immunity of the inverter is the location of the two unity-gain points (slope = -1) on the inverter curve. On Fig. 13b the unity-gain points for $V_T = 0.1 V_{DD}$ are shown as asterisks. For input voltages between the unity-gain points, small-signal noise is amplified rather than reduced on passage through the inverter. Input voltage levels that intrude upon this region are likely to result in output errors, so the narrower this region the better. In Fig. 13b, as threshold drops the curves are degraded in this respect, but not enough to become a factor in a zeroth-order design. As threshold varies over the entire range $0 \leq V_T \leq V_{DD} / 2$, the admissible 0 input voltage range varies from $3/8 \, V_{DD}$ to $1/2 \, V_{DD}$ and, because of leakage and speed restrictions on V_T, only a portion of this entire range for V_T is really available. Therefore, we accept the noise margin provided by the solid-line characteristic in Fig.13b corresponding to Eq.19.

Besides leakage and noise, power consumption places another requirement on minimum V_T. For $V_T = V_{DD}/2$, the p-channel device switches off just as the n-channel device switches on. Consequently, both devices are not simultaneously above threshold, and the small current flow directly from the supply terminal to ground through both devices leads to little power consumption. For values of V_T less than $V_{DD}/2$, however, both devices are simultaneously on during part of the high-low transition of the inverter, leading to larger supply-to-ground current and greater power dissipation for lower V_T. This dissipation, sometimes called short-circuit power dissipation (Ref. 18, p. 55), is additional to that caused by charging and discharging of any capacitive load through the various circuit resistances. It is assumed here that short-circuit power dissipation is not excessive when Eq. 19 applies.

To summarize, the major requirement upon V_T is taken to be the drive limitation upon the closeness of V_T to V_{DD}. To reduce leakage, noise, and power dissipation as much as possible, V_T is set as near to V_{DD} as the drive requirement allows. Equation 19 is adopted for this purpose, thereby coupling the maximum doping to the applied voltage: to make a device small by decrease of the depletion layer widths the doping must increase, in turn making V_T higher, and driving up V_{DD}. The limits on maximum V_{DD} are discussed next.

3.4.2 Drain Voltage and Hot-Electron Damage

The maximum voltage applied to the gate affects TDDB and GIDL. For any gate voltage, however, the oxide thickness and junction doping can be increased to bring these quantities under control. The remaining limitation on voltage stems from V_{DD} on the drain, which creates a large lateral electric field, $\mathcal{E}_{lat}$. This lateral field drives carriers toward the drain, increasing their energy, or "heating" them. This heating causes degradation indirectly, because some of these energetic carriers interact with the oxide and/or the interfacial region, causing threshold shift and transconductance degradation.

This carrier heating occurs near the drain at large drain voltages. In a long-channel model of the MOSFET, the spatial extent of this region and the fields near the drain are largely independent of channel length, but the current entering this region varies inversely with the length of the long-channel region. Hence, the shorter the device, the larger the flux of carriers that are heated [$\propto I_D$], and degradation accelerates for shorter devices. For somewhat smaller devices, short-channel effects intrude. Drain control over the channel affects the current, not only through "drain-induced barrier lowering" in subthreshold, but also in the saturation region. Thus, the incident current increases more rapidly than $(1/L)$ at short channel lengths. For even smaller devices, the heavier doping and shallower junctions increase fields in the drain region, because voltages are not scaled to maintain constant fields. As a result velocity saturation occurs, limiting the current entering the heating region. This effect also is reflected in I_D, which continues to represent the flux available for heating.

These considerations are included in an empirical formula for the device life-time[26,49],

$$\tau = BW \frac{(I_{SUB}/I_D)^{-m}}{I_D} , \tag{20}$$

where I_D is the drain current and I_{SUB} is the substrate current given quasi-empirically by[50]

$$I_{SUB} = \gamma I_D \left(\frac{\mathscr{E}_{lat}}{\mathscr{E}_i}\right) \exp\left(-\frac{\mathscr{E}_i}{\mathscr{E}_{lat}}\right) , \tag{21}$$

with

$$\gamma \approx 3 \times 10^{10} \mathrm{cm}^{-2} , \quad \mathscr{E}_i \approx 2 \times 10^6 \mathrm{\ V/cm} , \tag{22}$$

and m a function of oxide field that varies from author to author, B a constant that depends on the injection conditions, the oxide quality and the criterion chosen for lifetime (e.g., a common choice is the time for the drain current at a given gate and drain bias to shift 10% due to damage). The lateral field $\mathscr{E}_{lat}$ in Eq. 21 can be estimated using a fitted formula,[26,50]

$$\mathscr{E}_{lat} = \frac{V_{DD} - V_{SAT}}{L_{SAT}} , \tag{23}$$

with the saturation value of drain bias,

$$V_{SAT} = \frac{(V_{GS} - V_T)L\mathscr{E}_{SAT}}{V_{GS} - V_T + L\mathscr{E}_{SAT}} , \tag{24}$$

and $\mathscr{E}_{SAT}$ the field at which velocity saturation occurs, $\mathscr{E}_{SAT} \approx 5 \times 10^4$ V/cm for electrons. The length L_{SAT} is found empirically to be[50]

$$L_{SAT} \approx 1.7 \times 10^{-2} d^{1/8} r_j^{1/3} L^{1/5}, \quad (d < 150 \text{ Å} , L < 0.5 \text{ } \mu m) , \tag{25a}$$

$$\approx 0.22 d^{1/3} r_j^{1/3}, \quad\quad\quad\quad (d > 150 \text{ Å}) , \tag{25b}$$

where d, r_j, and L are in centimeters. For thick oxides, a square-root dependence on r_j also has been suggested.[20,26] Hot-carrier degradation is noticeable only for fields $\mathscr{E}_{lat}$ high enough to cause velocity saturation. In saturation, the current is

$$I_D \approx C_{ox} W v_{SAT} \frac{(V_{GS} - V_T)^2}{V_{GS} - V_T + \mathscr{E}_{SAT} L} \tag{26}$$

where v_{SAT} is the saturation velocity of the carriers, $v_{SAT} \sim 10^7$ cm/s for electrons. Using Eq. 26 in Eq. 20 for the lifetime,

$$\frac{v_{SAT}\tau}{B} = \left(\frac{I_D}{I_{SUB}}\right)^m \frac{1}{C_{ox}(V_{GS} - V_T)} \left[1 + \frac{\mathscr{E}_{SAT} L}{V_{GS} - V_T}\right] , \tag{27}$$

in which the right-hand side is a known function of $\mathcal{E}_{lat}$ and the left side allows us to impose the constraint of constant lifetime.

Intuitively, Eq. 20 or 27 has a rough interpretation taking I_{SUB}/I_D as a measure of the fraction of the carriers that are hot, and $(I_{SUB}/I_D)^m$ as the fraction that are hot enough to cause damage. (More accurately, I_{SUB} reflects the number hot enough to cause impact ionization.) As discussed, the extra factor in $(1/I_D)$ is a measure of the incident flux. Equation 21 for (I_{SUB}/I_D) is a function primarily of $\mathcal{E}_{lat}$ and, as such, it averages over aspects of the heating region such as its spatial extent and the spatial variation of its field, and also averages the distribution of the hot carriers in energy and in space.

Equation 27 shows that lifetime can be regarded as a function of $\mathcal{E}_{max}$ through m and the factor $C_{ox}(V_{GS} - V_T)$ and of $\mathcal{E}_{lat}$ (through the factor in I_D/I_{SUB}), with the result that specification of lifetime determines a maximum allowable $\mathcal{E}_{lat}$ ($\mathcal{E}_{max}$ already has been specified by the discussion of Section 3.3). As an estimate we adopt the value,[12]

$$\mathcal{E}_{lat} \approx 2 \times 10^5 \text{ V / cm} . \tag{28}$$

Specifying $\mathcal{E}_{lat}$ places an upper bound upon V_{DD} through Eq. 23. This constraint upon V_{DD} varies with channel length through V_{SAT} of Eq. 24. Using $V_{DD} \approx 4 V_T$ there are then two relations between channel length and V_T, namely, Eq. 1 and Eq. 27. Therefore, their combination determines the minimum length for given input parameters which, for discussion here, have the following values: (a) minimum $r_j \approx 500$ Å, (b) maximum oxide field, $\mathcal{E}_{max} \approx 5.8 \times 10^6$ V/cm, and (c) maximum lateral field, $\mathcal{E}_{lat} \approx 2 \times 10^5$ V/cm.

3.4.3   Minimum Sized Devices

Combining the constraints to find the predicted minimum L is not altogether straightforward. One way is to combine the conditions relating to $\mathcal{E}_{max}$ first. When the oxide field is $\mathcal{E}_{max}$, the potential drop ψ_s in the silicon in a one-dimensional approximation is given by Gauss's law[14-16] as

$$\epsilon_{ox} \mathcal{E}_{max} = qN\sqrt{2} L_D \left[e^{\beta(\psi_s - 2\psi_B)} + \beta\psi_s - 1 \right]^{1/2} \tag{29}$$

which can be rearranged as

$$\beta\psi_s = 2\beta\psi_B + \ln\left[\frac{1}{2}\left(\frac{\epsilon_{ox}}{\epsilon_s} \right)^2 (\beta\mathcal{E}_{max}L_D)^2 + 1 - \beta\psi_s \right] \tag{30}$$

Equation 30 can be solved quickly by iteration using $\psi_s \approx 2\psi_B$ as the first guess. Assuming an n^+-poly gate, the threshold is estimated as

$$V_T = V_{FB} + 2\psi_B + \sqrt{2}\left(\frac{\epsilon_s}{\epsilon_{ox}} \right)\left(\frac{d}{L_D} \right)\frac{kT}{q}(2\beta\psi_B - 1)^{1/2} \tag{31}$$

and, neglecting interfacial fixed and trapped charge,

$$V_{FB} \approx -\frac{\phi_G}{2} - \psi_B . \qquad (32)$$

with ϕ_G the bandgap voltage, $\phi_G \approx 1.12$ V at 300 K.

The required oxide thickness now is found using the requirement that at a gate bias of $V_{DD} = 4V_T$ the field in the oxide should be $\mathscr{E}_{max}$, that is, from Eq. 13,

$$\mathscr{E}_{max} = \frac{4V_T - V_{FB} - \psi_s}{d} . \qquad (33)$$

Substituting for V_T, V_{FB}, and ψ_s and collecting the terms involving d one finally obtains

$$\frac{d}{L_D} = \beta \frac{-3\phi_G/2 + 5\psi_B - \psi_s}{\beta \mathscr{E}_{max} L_D - 4\sqrt{2} \left[\frac{\epsilon_s}{\epsilon_{ox}} \right] (2\beta\psi_B - 1)^{1/2}} , \qquad (34)$$

which determines d as a function of doping density, N, for the given $\mathscr{E}_{max}$. Thus, for any N we can find d, assuring the requirements that $V_{DD} = 4V_T$ and that the oxide field is $\mathscr{E}_{max}$. With this N and d, L can be evaluated using Eq. 1, and $\mathscr{E}_{lat}$ found from Eq. 23. See Problem 10. A sample of the results is in Table 1.

If the basis for Table 1 is correct, the hot-electron constraint is less severe for thinner oxides, mostly because thin-oxide devices employ lower V_{DD}. None of the devices satisfy the criterion of Eq. 28, $\mathscr{E}_{lat} \leq 2 \times 10^5$ V/cm so, for comparison, L in the last column is computed using $\mathscr{E}_{lat} = 2 \times 10^5$ V/cm and Eqs. 23 to 25. From the table, it is clear that for $\mathscr{E}_{max} = 5.8$ MV/cm the criterion of

TABLE 1 Minimum Sized Devices with Maximum Oxide Field[a]

N (10^{18}/cm^3)	d (Å)	V_{DD} (V)	$\mathscr{E}_{lat}^{(b)}$ (MV/cm)	L^b (μm)	L^c (μm)
1.29	30	1.8	0.33	0.091	0.29
1.23	35	2.1	0.38	0.099	0.41
1.18	40	2.4	0.42	0.107	0.55
1.11	50	3.0	0.51	0.122	0.90
1.01	80	4.7	0.73	0.164	2.60

[a] N chosen as independent variable, d from Eq. 34, V_T from Eq. 31, $\mathscr{E}_{max} = 5.8$ MV/cm, $r_j = 0.05\,\mu m$, $V_{DD} = 4\,V_T$.

[b] L from Eq. 1, $\mathscr{E}_{lat}$ from Eq. 1, Eqs. 23-25.

[c] L from Eqs. 23 to 25 using $\mathscr{E}_{lat} = 2 \times 10^5$ V/cm.

Eq. 1 based on subthreshold drain bias independence allows shorter L than does the hot-electron criterion with $\mathcal{E}_{lat} = 0.2$ MV/cm.

Table 1 is constructed at the limit of the largest oxide field possible. Because this approach allows the largest V_T possible, and hence the largest doping levels, the devices in Table 1 are short. Because $(V_{DD} - V_T)$ is large, driving ability should be good, and because V_T is large, there is a large margin for V_T variations due to process or circuit noise. There are disadvantages with this approach, however. First, large voltages imply large power dissipation and more heat. Second, large voltages imply isolation difficulties between devices, and with larger spacing the overall chip optimization might suffer. Third, for the thickest oxides in Table 1 voltages are so high that junction breakdown intervenes. Fourth, all the devices in Table 1 have values of lateral field in excess of 0.2 MV/cm. In view of the exponential dependence of lifetime from Eqs. 20 and 21 upon lateral field, none of the devices in Table 1 look promising from the viewpoint of hot-electron degradation. That is, to use the devices in Table 1 requires improvements in oxide resistance to hot-electron damage,[51-53] or special structures like lightly doped drain junctions or buried drains.[3,20] Such structures are used at large dimensions, but good control is difficult at small dimensions.

One way to improve lifetime is to lower voltages. In Table 2 the devices are resized by choosing d, combining Eq. 1 with Eqs. 24 to 26, and letting $\mathcal{E}_{ox}$ take whatever value is necessary. The resulting devices satisfy all the constraints, and as Table 2 shows, none reaches an oxide field as large as $\mathcal{E}_{max} = 5.8$ MV/cm, the limit of Eq. 12. The voltages are lower than those in Table 1, with corresponding benefits to power dissipation, isolation and, of course, acceptable

TABLE 2 Minimum Sized Devices with Maximum Lateral Field[a]

L (μm)	d (Å)	N (10^{17}/cm^3)	V_T (V)	$\mathcal{E}_{ox}$ (MV/cm)
(i) $r_j = 0.05$ μm				
0.10	30	7.8	0.31	4.0
0.13	40	5.5	0.35	3.4
0.15	50	4.1	0.37	2.9
0.22	80	2.3	0.45	2.3
(ii) $r_j = 0.08$ μm				
0.12	30	9.5	0.36	4.7
0.14	40	6.6	0.40	3.9
0.17	50	5.1	0.44	3.4
0.25	80	2.9	0.52	2.6

[a] $\mathcal{E}_{lat} = 0.2$ MV/cm, $V_{DD} = 4V_T$, L satisfies L_{min} from Eq. 1 and hot-electron constraints.

lifetime. Because the voltages are as large as possible given $\mathcal{E}_{lat} \approx 0.2$ MV/cm, V_T and the doping levels also are as large as possible subject to this constraint, but not as large as those in Table 1. As a result, the devices of Table 2 are longer than those in Table 1. They also have poorer driving ability and less margin for process or circuit noise in V_T.

Table 2 also shows that r_j is not a critical parameter in achieving a short device. For example, at $r_j = 0.05$ μm an $L = 0.13$ μm at $V_T = 0.35$ V is possible, while at $r_j = 0.08$ μm an $L = 0.12$ μm at $V_T = 0.36$ V is possible. However, the first example requires $d = 40$ Å and the second $d = 30$ Å. Hence, there is a trade-off here between shallower junctions or thinner oxides. This trade-off is evident for fixed L_{min} directly from Eq. 1, but Table 2 also incorporates a constant $\mathcal{E}_{lat}$.

This discussion has revolved around the empirical formulas Eq.1 and Eqs. 23 to 27. To get some idea of the accuracy of these formulas, they are compared in Fig. 14 with selected results from a study[11] of devices with $r_j = 0.2$ μm and $V_T = 0.65$ V. From our discussion, V_T should be considered a design variable but, to make a comparison with this study possible, V_T was fixed in Fig. 14. The intersection of the curve labeled "$\mathcal{E}_{lat} \sim 0.20$ MV/cm" with the curve labeled "EQ (1)" is marked in Fig. 14 by a small circle, and it falls inside the shaded region of acceptability.

Let us examine the curves in Fig. 14 more carefully. First, consider the curves for TDDB. Figures 14a and 14b show that TDDB limits minimum oxide thickness and maximum voltage independent of L, and Fig. 14c shows TDDB forces $V_{DD}(max) / d(min) \approx$ constant. These results are consistent with our discussion, which treats TDDB as determining $\mathcal{E}_{max}$, independent of L.

Second, consider the short-channel threshold shift, the curves labeled ΔV_T. Figure 14a shows ΔV_T sets a maximum d that increases with L. Figures 14b and 14c show that very large changes in V_{DD} are needed to affect either the minimum L at a chosen d or the maximum d at a chosen L. This behavior can be compared to Eq. 1, which also is intended to control these short-channel effects. In Fig. 14, the curves marked EQ (1) are computed from Eq. 1 subject to the constraint $V_T = 0.65$ V. This threshold constraint forces the doping to decrease as d increases, approximately according to $d / L_D \approx$ constant (see Eq. 31) and, therefore, $d \propto L^3 / L_D^2 \propto L$ (see Eq. 1). It can be seen that Eq. 1 is qualitatively similar to the ΔV_T curve, but is more cautious, in part because Eq. 1 is based on gate biases well below threshold, while the ΔV_T curve uses a threshold criterion $\Delta V_T = 0.1$ V, where ΔV_T is the departure from long-channel threshold obtained from a plot of threshold voltage versus channel length. See Problem 6.

Third, consider the curves for hot-electron lifetime, τ. Superimposed on Fig. 14 are the curves (labeled with a value of $\mathcal{E}_{lat}$) obtained using Eqs. 23 and 24. These curves for fixed $\mathcal{E}_{lat}$ do not agree well with the curves for constant τ, especially the length dependence. Because the constant-τ curves are based on a lifetime estimate extrapolated from measured shifts in drain current it would appear that either Eqs. 23 to 25 are not very satisfactory for $\mathcal{E}_{lat}$ estimation or

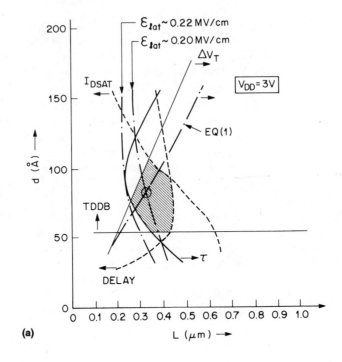

(a)

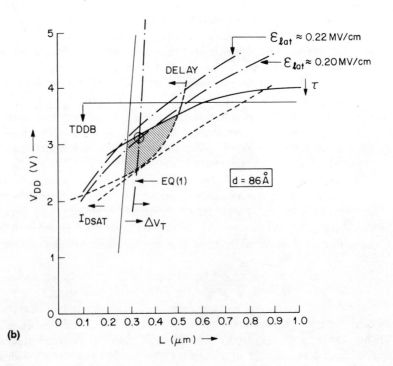

(b)

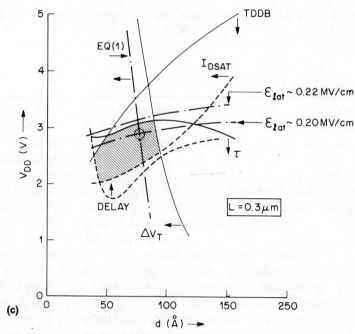

Fig. 14 Comparison of the results from Eq. 1 for L_{min} and Eqs. 23 to 27 for hot-electron effects (labeled $\mathcal{E}_{lat}$) with the results of Jeng et al.[11] (a) Oxide thickness versus channel length for $V_{DD} = 3\,V$. (b) V_{DD} versus channel length for $d = 86\,Å$. (c) V_{DD} vs. oxide thickness for $L = 0.3\,\mu m$. V_T fixed at 0.65 V, $r_j = 0.2\,\mu m$. The various curves of Jeng et al. are based on measurements using the following criteria: (i) $\Delta V_T = 0.1\,V$ on V_T versus L plot; (ii) $\tau = 10$ years under hot-electron degradation using 10% limit on ΔI_{DS}, (iii) TDDB based on 5% loss of devices in 10 years on 5 mm² area. Also shown dashed are curves for ring oscillator delay of 120 ps/stage and drive current in saturation (I_{DSAT}) of 0.5 mA/μm at $V_{GS} = 3V$. (After Ref. 11, © 1988 IEEE).

that the variation of the parameters B and m distort lifetime estimates from those based on simply on $\mathcal{E}_{lat}$. See Problems 7 and 8.

Finally, consider the delay curves shown on Fig 14, computed for a ring oscillator delay of 120 ps/stage. A constant-delay condition is partially approximated in Tables 1 and 2 by the drive constraint of Eq. 9. The delay curves in Fig. 14 are an active constraint only in Fig. 14c, where the delay limits the minimum oxide thickness to perhaps $d > 30\,Å$. This limit might reflect a trade-off of capacitive delay against driving ability, a conjecture supported by Table 2. Table 2 shows that in the range $d \sim 30$ to $40\,Å$ an increase in d is accompanied by a decrease in doping level, keeping threshold nearly fixed. Hence, in a minimum size device, larger oxide thicknesses lead not only to lower oxide capacitances, but to lower junction capacitances. Apparently the driving ability condition of Eq. 9 should be supplemented by capacitance considerations, suggesting the fastest design need not result from the thinnest oxide nor the shortest device in circuits like ring oscillators where device capacitances dominate circuit response times.[3]

3.4.4 Coping with Hot Carriers

The previous section shows that the minimum size of MOSFETs is restricted by hot-carrier degradation of device lifetime. To relieve this limitation involves progress on several fronts. First is increasing the resistance of oxides to hot-carrier damage,[51–53] allowing a larger $\mathcal{E}_{lat}$ than Eq. 28. Second is understanding the causes of degradation, to enable accurate simulation of lifetime for structural optimization and to improve processes. Here simulation and measurements work together: while theory unravels the variety of effects intertwined in experiment, experiment can identify unimportant detail, helping to simplify the physical models. A third front is development of structures less sensitive to degradation, such as structures with modified drain geometries like the lightly doped drain MOSFET that attempts to reduce carrier heating by lowering the lateral field, or the buried-drain structures that attempt to keep hot carriers away from the oxide by subsurface collection.[3] Below we discuss a first step toward better understanding, the use of simulation to strip spatial nonuniformity from experimental measurements. Then the lightly doped-drain structure is described.

Spatial Nonuniformity and Hot-Carrier Damage. Hot-electron damage is located above or adjacent to the drain, and lifetime is complicated by this spatial nonuniformity—a device that focuses hot carriers in a small area behaves differently than one that spreads them over a wider area. The extent of the damaged region is a function of device geometry, of the duration and conditions of stress, and of the spatial distribution of oxide and interfacial defects.

Several recent studies find that damage occurs initially at the position of maximum $\mathcal{E}_{lat}$. In both these studies the peak in $\mathcal{E}_{lat}$ is positioned on the drain side of the field-reversal point so, oddly enough, damage occurs most rapidly in the region where the field component normal to the interface opposes entry of electrons into the oxide. A simulation of electron trapping in n-channel devices[54] found for abrupt junctions ($n^+ = 10^{20}$ cm^{-3}) and a stress condition $V_D > V_G$ that maximum trapping occurs initially at the interface about an oxide thickness from the drain, and moves in toward the drain as trapping continues, initially following the position of maximum $\mathcal{E}_{lat}$. Although slower, trapping also occurs on the source side of the damaged region, which grows slowly toward the source. Measurements and simulations of the density and position of interface traps generated by electrons[55] found interface traps also are created at the position of maximum $\mathcal{E}_{lat}$ for $d \leq 100$ Å. This study of trap generation used a non-abrupt junction profile and, as a result, the maximum $\mathcal{E}_{lat}$ occurred above the drain junction in the gate-drain overlap region for $V_D = 2V_G$. The effects of the damage region upon device characteristics can be qualitatively understood based upon a three-section model.[56,57]

The deconvolution of spatial nonuniformity is only the first step in modeling hot-carrier damage. More ambitious is the development of models for hot-carrier transport and hot-carrier chemistry. The hot carriers that can be injected into the oxide of a MOSFET involve only the high-energy tail of the carrier distribution function, and modeling of this tail involves dynamics different from that for the main body of the distribution. In particular, a Maxwell distribution, with an effective hot-carrier "temperature" adjusted to describe the majority of the carriers, fails to provide the correct high-energy tail to describe oxide injection. Development of transport models is an active area, but its discussion is too complex for this chapter.[3] We now turn to the amelioration of hot-carrier effects by structural modification.

Modified Drain Structures. Hot-carrier degradation can be reduced by reducing the lateral field that heats the carriers. This field can be reduced by changing the doping profile of the drain from abrupt to graded, trading lower fields for increased parasitic series resistance (see Section 3.2.1). In addition, a graded drain structure invites GIDL (Section 3.2.2). Figure 15 is a schematic version of such a structure in which the drain is split into two regions: a conventional heavily doped drain and a lightly doped drain (LDD) extension. The parameters to be optimized for this design are (a) the doping of the lightly doped region, (b) the length of the lightly doped region L_n, and (c) the length of the overlap of the gate over the lightly doped region L_g.

Study shows[58] that the maximum field reduction occurs if the gate extends completely over the lightly doped region. This geometry has good and bad points. The advantages are that this "fully overlapped" geometry has good degradation resistance—not only is the field reduction larger than for partially

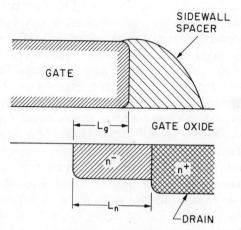

Fig. 15 Schematic cross section of a lightly doped drain structure showing the gate overlap length, L_g, and the length of the lightly-doped portion of the drain, L_n. Also shown is the sidewall spacer, usually but not necessarily an oxide. This spacer serves as a mask for the n^+-drain implant.

overlapped cases, but the spacer oxide is protected from hot carriers. This spacer oxide often is susceptible to degradation because of growth conditions (e.g., deposition or wet oxidation) or later etching or implant damage. The disadvantage of the fully overlapped geometry is that it increases parasitic gate-to-drain capacitance.

Adopting the fully overlapped geometry, now we find the length of the lightly doped region. Analysis shows[58] that field reduction improves with increasing length of the lightly doped region approximately as

$$\mathcal{E} \approx \mathcal{E}_{lat} \left[\frac{1 + \exp(-L_n / L_{sat})}{3 - \exp(-L_n / L_{sat})} \right]$$

with $\mathcal{E}_{lat}$ from Eq. 23, and where L_{sat} is the length of the region where the carriers are velocity saturated, given very approximately by,

$$L_{sat} = \left[\frac{\epsilon_s d \, r_j}{\epsilon_{ox} \eta} \right]^{1/2}$$

with η a fitting parameter ~ 1.2. (Contrast this formula with the denominator in Eq. 9 and with Eq. 25. Empiricism could use some refinement.) Thus, increasing L_n beyond L_{sat} serves only to increase parasitic resistance with no benefit to field reduction. The prediction of a maximum value for a useful L_n is experimentally supported.[59]

Finally, what is the best choice of doping? Simple bounds on doping come from limiting cases—at high doping the "lightly" doped drain becomes simply an extension of the heavily doped drain, and the maximum field results at the tip of the LDD extension, adjacent to the channel. This field is large because the LDD extension is shallow. At the other extreme, a very lightly doped drain hardly changes the structure from a conventional MOSFET with a high-field region at the heavily doped drain. Study shows the lowest lateral field occurs for that doping at which the position of maximum lateral field is poised to switch from one extreme position to the other, from the tip of the drain extension to the edge of the heavily doped region.[58] No intermediary position for the maximum field is possible, at least within a simple model for the fully overlapped geometry.[58] As already noted,[54,55] the position of maximum $\mathcal{E}_{lat}$ is where most hot-electron damage occurs. This damage can charge the interface and deplete the drain junction, increasing parasitic drain resistance. Hence, it is better to have the maximum $\mathcal{E}_{lat}$ at the heavily doped junction, where damage-induced depletion is more difficult, than at the tip of the lightly doped drain. To force this position for maximum $\mathcal{E}_{lat}$, the doping of the drain extension is chosen somewhat below the ideal, "poised" value.

To control the gate overlap over the heavily doped portion of the junction, an inverse-T type of gate is one proposal.[59,60] Figure 16 shows how one exploratory version is made, the gate overlapped device (GOLD) structure.[59] One

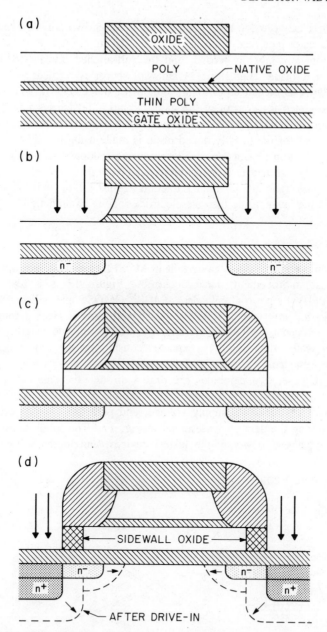

Fig. 16 Fabrication sequence for the gate overlapped LDD structure. (a) Beginning sandwich structure with very thin native-oxide etch-stop layer. (b) Gate etch followed by lightly doped drain implant. (c) Sidewall spacer formation, including etching of bottom poly gate layer. (d) Sidewall oxidation of the bottom poly gate layer to control gate-drain overlap for maximum lateral field reduction and minimum parasitic capacitance. Heavily doped drain implant follows and then a drive-in, activation anneal.

begins with a native oxide layer sandwiched between two poly layers (Fig 16a). On top is a mask patterned to define the gate. Using the native oxide as an etch stop, the top poly layer is etched, and the self-aligned implant for the lightly doped drain is made (Fig 16b). A sidewall spacer is formed, and the lower, thin poly layer is etched to form the bottom of the inverse-T gate (Fig. 16c). Then the key step for control of the gate overlap is made, a sidewall oxidation of the bottom poly layer under the sidewall spacer (Fig. 16d). Finally the self-aligned implant for the heavily doped drain is made using the sidewall spacer as mask, and the drain dopants are driven in during a dopant-ion activation anneal. Control of this structure is considered adequate for channel lengths above 0.3 μm.[59] The trade-offs of field reduction against GIDL and the implications of an insulating layer inside the gate electrode require examination.

3.4.5 Speed

Although chip considerations can result in MOSFETs that are not optimized for speed, great improvements have occurred. Figure 17 is a scatter plot of observed delays/stage for unloaded nMOS, CMOS, and silicon-on-insulator (SOI) ring oscillators versus effective gate length, and it gives a rough idea of the speed attained by MOSFET technology at a given gate length. Figure 17 shows less scatter than might be expected considering the wide range of technologies represented.

To aid the eye, two parabolas are drawn through the data. A parabola in gate length is suggested by analytic delay expressions,[61] with coefficients that depend on oxide thickness, supply voltage, and parasitics, among other factors that vary among the devices represented in Fig. 17. The solid curve is an optimistic view chosen to weight the fastest devices, and forced to zero delay at

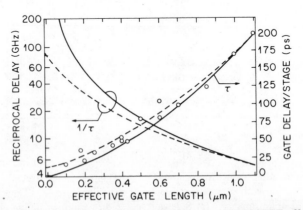

Fig. 17 Ring oscillator delay/stage and reciprocal delay versus MOSFET effective channel length. Data points are reported values, and the two curves are parabolas fitted to the data. The solid curve is an optimistic (fastest) estimate, and the dashed curve is a conservative estimate. Reciprocal delays are about 4 to 5 times the clocking rate of a real circuit.

zero gate length (an exaggeration). The dashed curves are a fit to the data allowing non-zero delay at zero gate length. The dashed curves are thus a moderate expectation for MOSFET performance. Figure 17 also shows the reciprocal delay (with a vertical logarithmic scale) to allow an estimate of circuit speed. Depending on the circuit, this frequency should be divided by some factor (about 4 or 5) to predict a clock rate. Some parameters for the smaller devices used in Fig. 17 are shown in Table 3.

3.4.6 Comments

This section brings together circuit and device constraints. Simplified relations that are supported by computer simulation help in examining the options available. According to the models used, the hot-electron-lifetime constraint will continue to control maximum V_T levels. Thin oxides and shallow junctions are desirable, and minimum oxide thickness is not limited by Fowler-Nordheim leakage or TDDB, but by manufacturability and oxide tunneling. The next section examines constraints introduced by the isolation between MOSFETs and between MOSFETs and bipolar devices.

3.5 ISOLATION AND THE MOSFET

Circuit density is decided by design rules, not simply by MOSFET size. But close packing of devices places demands upon isolation between devices, and blurs the interface between the device and the isolation as the devices are pushed closer together, altering device behavior. This section presents some of these effects without a full assessment of isolation itself.

There are two types of isolation: insulating and *p-n* junction isolation. Insulating isolation, usually an oxide, isolates the active source, drain, and channel regions of a device from those of its neighbor. The objective is to surround

TABLE 3 Some Existing Small MOSFETs

L (μm)	d (Å)	N (10^{17}/cm^3)	r_j (μm)	V_{DD} (V)	V_T (V)
0.10[a]	45		~0.07	1.3	~0.3
0.16[b]	35	5	0.08	2.5	0.5
0.18[c]	70	2–3	0.11	2.5	0.4
0.20[d]	75		0.20	3.0	~0.7

[a]G. A. Sai-Halasz et al., *IEEE Electron Dev. Lett.*, **EDL-9** (12), 633 (1988);
[b]M. Miyake et al., *IEEE Trans Electron Dev.*, **ED-36** (2), 392 (1989);
[c]B. Davari et al., *Tech. Dig. Int. Electron Dev. Meet.*, p. 56 (1988);
[d]J. Chung et al., *IEEE Electron Dev. Lett.*, **EDL-9** (4), 186 (1988).

each device, preferably deeper than the depth of its junctions, by a wall of oxide. To do this, various approaches are available: (a) an oxide is grown everywhere and windows opened; inside the windows the devices are made, either directly on the bare silicon at the bottom of the windows, or upon an epitaxial silicon grown to fill the windows, or (b) the device areas are masked and a local oxidation (LOCOS) of the isolation area performed or (c) trenches are cut around the device areas and then the trench walls are thinly oxidized and the trench filled. In choosing a method, many considerations enter: manufacturability, planarity of the surface, defect generation, and parasitic devices.

The second type of isolation, *p-n* junction isolation, is used in the silicon substrate to prevent current flows from coupling adjacent devices, and to prevent capacitive coupling of the devices which could occur via merging of their depletion layers. The basic strategy is twofold: to prevent depletion layer expansion by dopant-ion placement, and to keep current flows where they are wanted, with low voltage drops, by providing high-conductivity paths where possible. For example, the use of buried n^+- and p^+-layers is discussed at the end of this section.

The demands placed upon isolation increase as one goes from nMOS to CMOS to BiCMOS. In the case of bipolar devices especially, device area is a strong function of isolation technology, and minimizing this area drives improvement of isolation even at the expense of greater complexity. Discussion begins with the narrow-width effect, goes on to parasitic MOSFETs, and concludes with buried layers and bipolar compatibility.

3.5.1   Narrow-Width Effect

Looking at Fig. 2, the MOSFET cross section in the width direction resembles Fig. 18a. This figure shows the gate electrode extends across the thin gate oxide region over the channel, and also over the thick field-oxide region on both sides of the channel. The field oxide extends all over the chip surface, except for windows where the active devices are made, and it is made thick so voltages cannot induce conducting channels in the silicon underneath, which cause unwanted leakage paths between devices. To further inhibit these undesired channels, a channel-stop implant is placed under the field oxide to make the threshold voltage of the isolation region very high. The cross section of Fig. 18a also could appear as in Fig. 18b or 18c depending upon the fabrication method.[62-65] For example, LOCOS (local oxidation of silicon) can produce a structure somewhat like Fig. 18b, and trench isolation one like Fig. 18c.

Toward the sides of the channel in Fig. 18a, the field lines spread apart and some spill outside the thin-oxide region, reducing gate control over the channel edges, and leading to a high-threshold channel-edge region. As widths reduce, this high-threshold portion becomes a significant fraction of the total device width. This rise in threshold voltage with decreasing width was the first width effect to be observed, and is commonly called the "narrow-width" effect although many effects of narrow width exist.[63] The channel-stop implant modi-

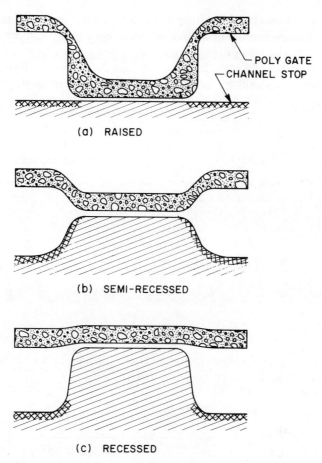

POLY GATE
CHANNEL STOP

(a) RAISED

(b) SEMI-RECESSED

(c) RECESSED

Fig. 18 Various width cross sections of a MOSFET using different field-oxide geometries. (a) Raised field oxide, obtained by etching an existing oxide. (b) Semi-recessed field oxide obtained by an idealized local oxidation without "bird's beak." (c) Recessed field oxide, found in various trench isolation methods.

fies this effect, introducing a parasitic gate-to-channel-stop capacitance. This parasitic capacitance exists for all three geometries, and its size depends upon isolation geometry and the channel-stop distribution under the gate edges. Regardless of the cross section, the possibility exists for the channel-stop dopant to move laterally into the MOSFET by diffusion during process steps that follow the channel-stop implant. If this movement occurs, the threshold at the edges of the MOSFET is raised, and then the edges conduct less than the center, again causing the MOSFET to appear narrower than its geometrical width.

In Fig. 18c the gate field is focused by the edge geometry of the channel, intensifying gate control of the edge region. As a result, these edges turn on earlier and more rapidly than the center of the channel, the opposite effect to

Fig. 18a. This lowering of edge threshold, called the "inverse narrow-width" effect, introduces off-current leakage and humps in the current versus gate bias curve.[62,65] The turn-off behavior of the recessed structure is delayed by edge conduction (see Fig. 19), but substrate bias dependence is less for the recessed structure, and sensitivity to width variation is much reduced. Figure 20 shows

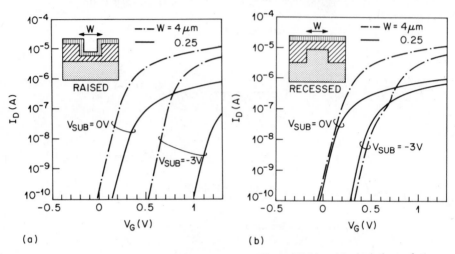

Fig. 19 Turn-off behavior of MOSFETs with different field oxide isolation schemes. Device parameters are $L = 2\,\mu m$, $d = 200\,Å$, $N = 2 \times 10^{16}/cm^3$ $V_D = 0.05\,V$. (After Ref. 64, © 1988 IEEE).

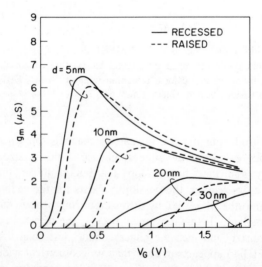

Fig. 20 Transconductance of short and narrow devices for two field-oxide geometries. Device parameters are $W = 0.25\,\mu m$, $L = 0.3\,\mu m$, $r_j = 0.15\,\mu m$, $N = 2 \times 10^{17}/cm^3$, $V_D = 0.05\,V$ and zero substrate bias. (After Ref. 64, © 1988 IEEE).

the advantage of a recessed isolation for the transconductance of a short and narrow device from three-dimensional simulation.[64] For oxide thicknesses at or
below 50 Å the difference in transconductance between the raised and recessed
structures is secondary to other considerations, such as zero-bias leakage current
and hot-electron stress resistance.

Another type of width effect is a high density of defects near the gate edges.
Due to processing difficulties, such as stresses introduced during field oxidation,
the defect density near the gate edges can be high enough to affect device behavior. Hot-electron damage near the gate edges can differ from that in the center
of the channel,[66] and defect generation related to LOCOS isolation occurs along
the gate edges.[67,68] At the source and drain ends of the channel, additional
defects can be introduced by the junction processing.[40]

At this point it is clear that the isolation geometry must be designed with
attention to the devices. Qualitatively, thinner gate oxides and heavier doping
for the active device reduce the width effects.[64] The width effects just discussed
are edge effects—their size depends only indirectly on device spacing. In contrast, the role of the parasitic MOSFETs now to be discussed increases with
closer packing.

3.5.2 Parasitic MOSFETs

Stray current paths can occur in several ways, so a check of isolation involves a
number of structures. To begin, consider adjacent MOSFETs in the same p-tub
or n-tub with a common gate conductor that can induce a channel under the field
oxide, coupling the two devices with a parasitic MOSFET. See Fig. 21.

Consider this parasitic device as the packing density is increased, making its
source-drain separation, L, smaller. Although Eq. 1 is not quantitative for non-
planar geometries, qualitatively Eq. 1 suggests the threshold for forming a parasitic channel is independent of L down to some minimum L_{min}. For $L < L_{min}$
threshold lowering occurs, and Eq. 1 suggests several remedies. For example,
L_{min} can be reduced simply by reducing oxide thickness, and to maintain threshold voltages for thinner field oxides, N can be increased so d/L_D = constant.
(See Eq. 31 for V_T.) This strategy results in packing varying according to
$L_{min} \propto (dL_D^2)^{1/3} \propto N^{-1/2}$. While controlling parasitic channels, this approach
also could reduce fringing fields that cause cross-talk, because lines are brought
closer to ground planes. A disadvantage is greater line capacitance, leading to
larger drive requirements for devices. A second strategy is to leave the field-
oxide thickness the same, but simply increase the doping to counteract threshold
lowering. In this case, packing varies according to $L_{min} \propto (dL_D^2)^{1/3} \propto N^{-1/3}$, so
that to obtain the same packing without oxide-thinning requires a heavier
channel-stop doping. Larger channel-stop doping is undesirable for several reasons: (a) it increases the narrow-width effect due to lateral dopant diffusion
from the channel-stop into the active device area, (b) it increases parasitic sidewall capacitance between the source and drain and the channel-stop region and,
(c) it lowers the junction breakdown voltages.

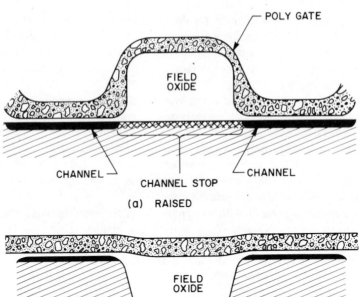

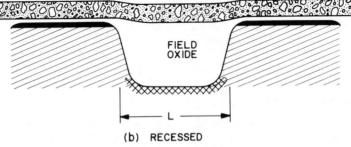

Fig. 21   A possible parasitic MOSFET allowing leakage between two adjacent channels. (a) Raised field oxide. (b) Recessed field oxide. The "channel length" of the parasitic MOSFET is L, the channel separation of the active devices.

Doping and oxide thickness changes are not the only ways to reduce this particular parasitic. A recessed field oxide is better than a raised field oxide.[63,69] The reason is that the bottom corners (edges in three dimensions) of the field-oxide wall in Fig. 21b are coupled only weakly to the gate, leading to a high threshold for inversion in the corners. Until the corners turn on, isolation is maintained. This corner effect has a strong dependence on the radius of curvature.

Other parasitic MOSFETs are made up of an n-channel device sharing a gate with a p-channel device. Figure 22a shows a CMOS inverter. When $V_{in} = V_{DD}$, if the isolation is inadequate the p-well can invert under the field oxide, joining the n-channel to the n-tub, losing isolation. Figure 22b shows that current then could flow from the n-channel source to the n-tub acting as drain. Applying Eq. 1 qualitatively to the parasitic MOSFET of Fig. 22b, the drain depth, r_j, is the tub depth, suggesting better isolation for shallower n-tubs and heavier p-tub doping.[70] This leakage mechanism is less important than punchthrough of the n-tub to the n^+-source or drain, and is corrected the same way.[71] For our discussion, the impact of these remedies on the active MOSFET

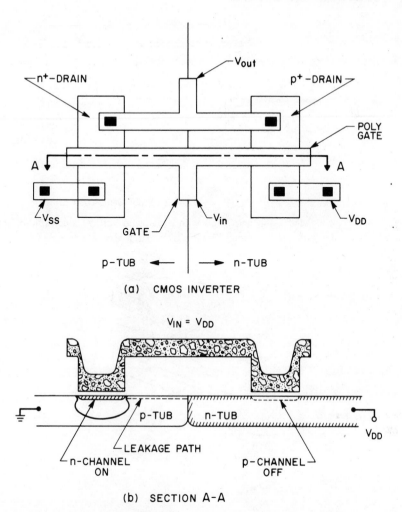

Fig. 22 A possible parasitic MOSFET straddling a p-tub n-tub boundary under an inverter. (a) Inverter layout showing position of Section A-A. (b) Cross-section through A-A showing a possible leakage path. The n-tub serves as the drain of the parasitic MOSFET, biased at V_{DD}.

is clear: increased junction capacitance and source-to-substrate bias dependence of threshold voltage.

To reduce the parasitic of Fig. 22 and also to prevent dopant compensation at the n-tub to p-tub interface, a trench can be introduced between the tubs as shown in Fig. 23. For example, a ¼ μm recessed isolation technology that allows 4 μm deep oxide walls has been developed.[72] A vertical parasitic MOSFET can result, however, with the positively biased n-tub acting as both a gate for the sidewall channel and as a drain to collect the resulting current. This parasitic MOSFET is suppressed using a channel-stop implant at the bottom of

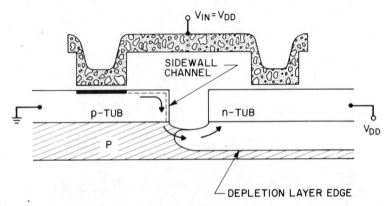

Fig. 23   A trench isolated *p*-tub and *n*-tub with a parasitic vertical sidewall MOSFET (*n*-tub as gate) and a parasitic horizontal MOSFET (V_{in} strap as gate).

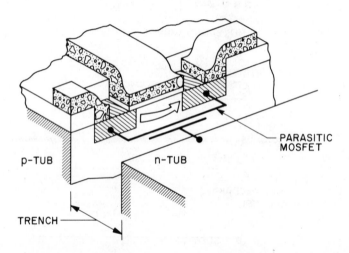

Fig. 24   A parasitic sidewall MOSFET with *n*-tub as the gate and the trench serving as gate oxide.

the trench. To take maximum advantage of the trench, however, the active devices should be right at the tub edge. As shown in Fig. 24, sidewall inversion then introduces leakage paths that can short-circuit single devices. Even if the devices are moved back from the edge somewhat, the trench corner parasitic can connect devices to the sidewall. An example is shown in Fig. 25 where two inverters in series have a stray dc leakage path when V_{in} = logic 1.

To remedy this problem several steps have been taken. First, a good-quality thermal SiO_2 layer is used inside the trench, to minimize interface fixed and trapped charge that tends to lower the inversion threshold of the *p*-tub. Second, a sidewall channel-stop has been introduced on the *p*-sidewall by direct

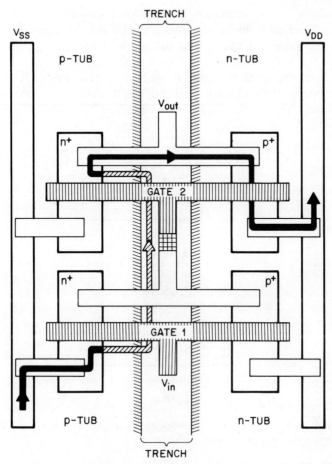

Fig. 25 A dc leakage path through two inverter stages in series. The solid black path is a good conducting route and the shaded path is a parasitic route. The parasitic consists of a field oxide corner under gate 1 (cf. Fig. 23), a sidewall path along the trench (cf. Fig. 24), and a corner path again under the V_{out} strap.

implant[65,73] or by a polysilicon diffusion source.[74] These doping techniques have been used for $\approx$ 0.8 μm wide trenches. For 0.5 μm widths, control of the sidewall doping and geometry is difficult.[65] Whatever the methods, sidewall doping affects the active devices through increased junction capacitance and source-substrate bias dependence. Control of sidewall parasitic MOSFETs is a major obstacle to miniaturizing isolation.[72]

3.5.3 Substrate Isolation and Bipolar Compatibility

As noted earlier, a conductive substrate helps avoid undesired feedback such as *I-R* biasing of junctions (latch-up and parasitic bipolar breakdown) and coupling between circuits via resistive voltage drops induced by ground currents. The simple use of n^+- or p^+-substrates is precluded, however, if bipolar compatibility is desired. Figure 26 shows some typical bipolar structures. A common feature is the buried n^+-layer: in the case of the *n-p-n* device[75] it is a collector, and for the side-contact *p-n-p* it is a base.[76] If the bipolar device were placed on a p^+-substrate, the buried n^+-layer would have an unacceptable parasitic capacitance to ground. Consequently, a lightly-doped substrate is necessary. The combined BiCMOS structure then looks like Fig. 27, with buried conductive

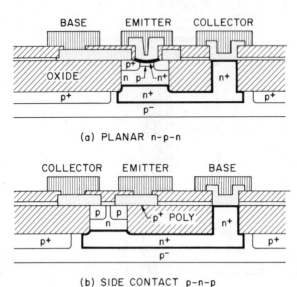

Fig. 26 Non-aligned bipolar structures showing the need for a buried n^+ contact, and thus also for a lightly-doped substrate.

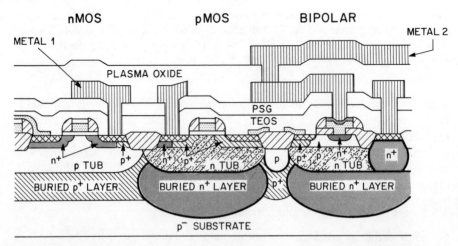

Fig. 27 A BiCMOS structure using semi-recessed field oxide and *p-n* junction substrate isolation. Alternatively, trench isolation could be used. (After Refs. 77, 78 © 1987 1988 IEEE).

layers that are placed selectively under the corresponding devices, making the use of a common p^+- or n^+-substrate impossible, and requiring top-side buried-layer contacts.[77,78]

For use in BiCMOS, a typical gate is shown in Fig. 28. During operation, MOSFETs charge and discharge the bases of the *n-p-n* bipolar devices. Once activated, the BiCMOS gate supplies a larger current to an output load than can a simple CMOS gate. Thus, for capacitive loads demanding much more charge than the base charge needed to activate the bipolar devices, BiCMOS gates are faster than simple CMOS gates. Delay versus load capacitance curves, however, always show a crossover in favor of CMOS over BiCMOS for small capacitive loads, with a crossover capacitance that depends on the constraints of the comparison, such as power, area, or processing trade-offs.

Some of the bipolar–MOSFET trade-offs are seen in a typical 2-μm BiCMOS design.[78] Figure 29 shows how the MOSFET junction capacitances depend on the epitaxial layer thickness. The smaller the thickness, the more the dopant from the buried layer encroaches upward into the active device area, raising the capacitances. Figure 30 shows that MOSFET threshold voltages also are affected for epitaxial layer thicknesses below 1.3 μm.

Turning to the bipolar device, Fig. 31 shows cutoff frequency, f_T, and small-signal gain, h_{FE}, as a function of current level. This figure shows that large collector current (heavy injection) degrades the bipolar device, lowering f_T and h_{FE}.[78,79] In BiCMOS, high currents are needed, introducing design trade-offs.[4] For example, the impact of heavy injection might be reduced (a) by increasing the *n*-tub doping, which increases the parasitic base-collector and *p*MOS junction capacitances, (b) by increasing the emitter area, which increases

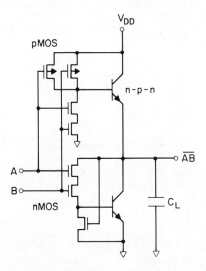

Fig. 28 A typical BiCMOS NAND gate. In addition to the MOSFETs normally found in a CMOS gate, bipolar transistors are provided. Additional MOSFETs speed movement of the bipolar base charge. (After Ref. 78 © 1987 IEEE).

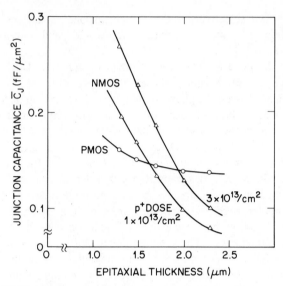

Fig. 29 MOSFET junction capacitance versus epitaxial layer thickness. Two buried layer p^+-implant doses are indicated (After Ref. 78 © 1987 IEEE).

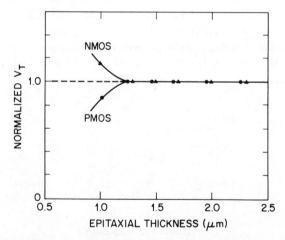

Fig. 30 MOSFET threshold voltages versus epitaxial layer thickness. To avoid influence of the buried layers on threshold, the thickness must exceed $1.3\,\mu m$. (After Ref. 78 © 1987 IEEE).

the base charge required and restricts usefulness of BiCMOS gates to larger values of load capacitance, or (c) by thinning the epitaxial layer, as discussed next.

Figure 32 shows the dependence of f_T and the knee current (i.e., the current at which h_{FE} is half its maximum) on epitaxial layer thickness. Because of increased collector resistance and heavy injection, f_T drops with increasing epi-

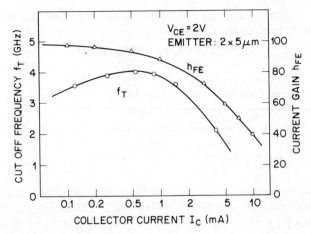

Fig. 31 Cutoff frequency and gain versus current level for the bipolar device. (After Ref. 78 © 1987 IEEE).

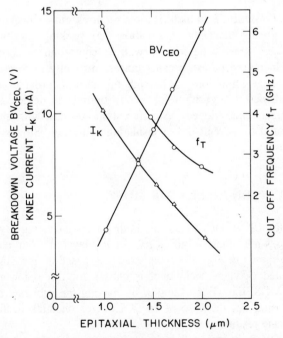

Fig. 32 Cutoff frequency and knee current versus epitaxial layer thickness. These parameters both are degraded as the thickness increases. Also shown is the collector-emitter breakdown voltage that improves as the epitaxial layer thickness increases, and sets the minimum acceptable value at $1.5\,\mu m$ to obtain the $BV_{CEO} = 10\,V$ needed for this particular application. (After Ref. 78 © 1987 IEEE).

taxial thickness despite the reduced base-collector capacitance. To obtain the collector-emitter breakdown voltage $BV_{CEO} = 10$ V demanded in this design example, Fig. 32 shows the epitaxial layer thickness must be chosen no smaller than 1.5 μm. This choice in turn affects isolation, as the epitaxial thickness influences parasitic leakage. To obtain the required BV_{CEO} with a thinner epitaxial layer, the base design of the bipolar device could be changed, by making the base thicker or more heavily doped. Both choices adversely affect the bipolar transistor, either lowering its frequency response or its driving ability and, as already mentioned for Fig. 30, an epitaxial layer thickness below 1.3 μm adversely affects the MOSFETs unless the entire process is changed to reduce dopant up-diffusion.

This discussion provides examples of how epitaxial layer thickness affects both the bipolar transistor and the MOSFET in a BiCMOS process. In addition, of course, many process steps affect both devices, and those processes that impact the base of the bipolar device become critical, including contact formation and thermal steps subsequent to base formation. These trade-offs still are under examination.[4]

3.5.4   Comments

Full advantage of MOSFET miniaturization requires closer packing of active devices, placing greater demands on isolation. As packing density reaches its limits, it introduces parasitic devices as well as compromises upon the active devices through increased parasitic capacitances, difficulty in threshold control and, possibly, voltage limitations. Although this chapter provides examples of these trade-offs, a design algorithm for jointly optimizing device and isolation requirements remains undeveloped. A different way to deal with these isolation problems is to build the circuit on an insulating substrate. Silicon-on-insulator construction is discussed next.

3.6   THE SILICON-ON-INSULATOR MOSFET

Silicon-on-insulator (SOI) MOSFETs are built on an insulating layer, so substrate isolation does not required buried p^+- or n^+-layers.[80] An example of an SOI MOSFET is shown in Fig. 33. This structure, based on the SIMOX (*Sepa*ration by *IM*planted *OX*ygen) technique, is chosen for discussion as a general case of the SOI MOSFET. For example, by making the substrate a good conductor and increasing d_b, the case of an SOI device on thick insulator with a grounded back plane results.

Even the SIMOX technique still is in evolution, and its success depends on further improvements in the material.[81] Efforts to improve the material include (a) annealing at temperatures near the melting point of silicon ($T_m = 1412$ °C) to remove oxygen precipitates and to form a sharp Si-SiO$_2$ boundary for the buried SiO$_2$ layer,[82] (b) reducing threading dislocations,[83] important if bipolar struc-

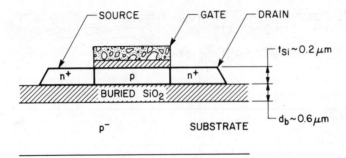

Fig. 33 A MOSFET built on a SIMOX substrate.

tures are to be made because the usual density of dislocations ($\sim 10^9$ cm^{-2}) is far too high for base electrodes and, (c) gettering of metal defects to reduce leakage currents[84] from the usual 10^{-3} A/cm^2 to 10^{-5} A/cm^2, at the expense of a poor quality buried interface. Leakage currents of 17 μA/cm^2 and mobilities of 580 cm^2/V-s for electrons, 205 cm^2/V-s for holes with $N \approx 2 \times 10^{16}$/cm^3 are reported.[85] In short, SIMOX has materials problems, but progress is being made. Already 16K SRAMs with 1.25 μm design rules have been manufactured.[81]

3.6.1 Floating Substrates

Because the structure of Fig. 33 does not have a substrate contact, there is no direct access to the majority carriers in the body of the device. As a result, when the external contacts deplete the substrate, the majority carriers cannot leave the device except by recombination or diffusion into the junctions. A comparison of an SOI MOSFET and a bulk MOSFET is shown in Fig. 34.[86] In this simulation holes are majority carriers, and hole storage in the SOI device is evident, quite different from the hole distribution in the bulk MOSFET. Because recombination or generation is not included,[86] as the gate and drain voltages increase, holes communicate with the outside world only by diffusion into the source. Figure 34 also shows the reduced field in the SOI device between the source and drain. The buried oxide layer spreads the potential drop with some assistance from the 10^{16}/cm^3 hole density on the bottom interface.

Also, when the drain bias is large enough to cause pair creation, the generated majority carriers accumulate until a balance is reached between generation and recombination, and the device body can become charged due to excess carriers. The substrate "floats" to a potential dependent upon the number of stored majority carriers; this potential is determined by internal charge storage and current and is not controlled directly by external bias. The floating substrate causes kinks in the *I-V* characteristics of SOI MOSFETs when holes generated by impact ionization near the drain diffuse toward the source, causing increased electron injection.[87] A balance is reached when the increased recombination and

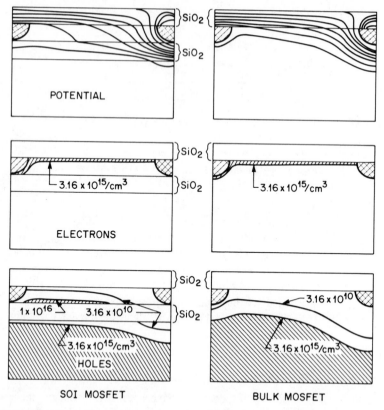

Fig. 34 Comparison of an SOI with a bulk MOSFET showing (a) potential contours, (b) electron concentration contours, (c) hole concentration contours. Device parameters are $d = 500\,\text{Å}$, $t_{Si} = 0.25\,\mu\text{m}$, $L = 2\,\mu\text{m}$, $V_{DS} = 5\,\text{V}$, $V_{GS} = 3\,\text{V}$. (After Ref. 86 © 1984 IEEE).

out-diffusion accompanying hole storage balances the generation near the drain.[87] If a balance cannot be reached, channel-assisted breakdown occurs.[88] This kink effect is sensitive to the lifetime of majority carriers. At long lifetimes ($\approx$ 1 to 100 μs) simulation shows[89] kinks in the *I-V* curves are pronounced while at short lifetimes ($\leq$ 0.1 ns) they are absent.

The kink effect is difficult to model numerically, because it depends on a delicate balance of the hole reservoir. As a practical matter, the kink effect can cause current overshoot and provides poor characteristics for analog applications.[90] It is desirable to avoid the kink effect, and one way to suppress it is to use poor lifetime material. Better devices result from a different remedy supported by modeling, the use of fully depleted SOI devices as discussed next.

3.6.2 Fully Depleted Devices

For thick silicon films, an SOI MOSFET differs from a bulk MOSFET through better isolation, and lower parasitic capacitance, because the bottom of the source and drain are on an insulating layer. These advantages are achieved at the cost of a floating substrate and materials problems. For thin silicon films, however, the SOI MOSFET is fully depleted, and the device behavior depends on both the bottom and the top interface. This two-sided behavior lowers the fields inside the device, tending to reduce hot-electron effects, short-channel effects, and to increase the driving ability of the device. The idealized effect of a top and a bottom gate on short-channel behavior shows that because both gates can terminate field lines from the drain, the screening of the drain is increased when two gates are used. This screening allows a lighter substrate doping compared to a conventional device for the same short-channel effect on threshold.[91] The fully depleted SOI device also has no floating substrate region, and a reduced dependence on majority-carrier storage.

Absence of a floating region in a depleted structure is shown in Fig. 35, which compares the potential between source and drain for the fully and partially depleted cases. There is a smaller potential well to trap holes, the holes are pushed toward the source where exit or recombination is more likely, and the potential near the drain is less steep, meaning much less pair creation in the fully depleted structure. (The lower field near the drain also means less hot-electron damage to the oxide.) Because of these differences, the kink effect is not a problem for fully depleted structures.[90]

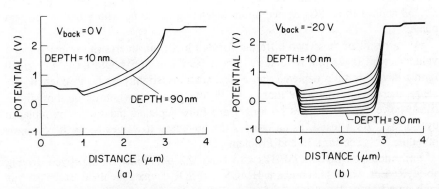

Fig. 35 Potential versus distance from source-to-drain for (a) a fully depleted and (b) a partially depleted SOI MOSFET. Device parameters are $d = 250 \text{Å}$, $t_{Si} = 0.1 \, \mu\text{m}$, buried oxide thickness $= 0.35 \, \mu\text{m}$, $N = 8 \times 10^{16}/\text{cm}^3$ (p-type). Contours are shown for two depths: 100 and 900 Å from the surface. (After Ref. 90 © 1988 IEEE).

Subthreshold slope also can improve in a depleted structure. Let S denote the gate swing needed to reduce subthreshold current one decade,[14-16]

$$S \approx \left[\frac{kT}{q} \ln 10 \right] \frac{dV_G}{d\psi_s} , \tag{35}$$

where ψ_s is the potential drop between the silicon surface and ground and V_G is the gate-to-substrate voltage. The derivative in Eq. 35 can be expressed as (ignoring interface traps),[14-16]

$$\frac{dV_G}{d\psi_s} = 1 + \frac{C_s}{C_{ox}} , \tag{36}$$

where C_s is the capacitance between the silicon surface and ground, and C_{ox} is the gate oxide capacitance, both per unit area. Although in a bulk device

$$\frac{1}{C_s} = \frac{W}{\epsilon_s} \tag{37}$$

with W the width of the surface depletion layer, in a fully depleted SOI device the analog of this depletion layer is comprised of three regions: the depleted silicon film of thickness t_{Si}, the buried oxide layer of thickness d_b, and the depletion layer in the supporting (grounded) silicon substrate of thickness W_{SUB}, that is, for SOI devices in full depletion

$$\frac{1}{C_s} = \frac{W_{SUB}}{\epsilon_s} + \frac{d_b}{\epsilon_{ox}} + \frac{t_{Si}}{\epsilon_s} . \tag{38}$$

In short, the substrate capacitance is the series combination of the capacitances of the three layers. If C_s from Eq. 38 is less than C_s from Eq. 37, the SOI device has a lower S than the bulk device for the same C_{ox}. For $W_{SUB} = 0$, Eq. 38 applies to an SOI device on an insulating substrate with a bottom ground plane.

As a result of these two different forms for C_s, an interesting transition in S values occurs if we plot S versus t_{Si} for a family of SOI MOSFETs. For small t_{Si} the device is fully depleted and S depends only slightly on t_{Si} through W_{SUB}. As t_{Si} increases to the point that $t_{Si} = W$ (the depletion width in the thin silicon film), however, the devices no longer are fully depleted and Eq. 38 is replaced by Eq. 37. As a result, as t_{Si} crosses the boundary to values $t_{Si} > W$, S jumps in value.[80,90] See Fig. 36 and Problem 11.

Fully depleted SOI MOSFETs also offer the possibility of improved driving ability.[92] For example, compare HMOS and SOI devices identical except one is built on a lightly doped substrate, the other on a buried oxide layer. For discussion, assume both have the same threshold, and assume that in both devices the depletion width at threshold satisfies $W = t_{Si}$, with t_{Si} the thickness of the surface doped layer in the HMOS and the thickness of the silicon film for the SOI devices. Then for gate biases above threshold and low drain biases, the two devices provide comparable currents.

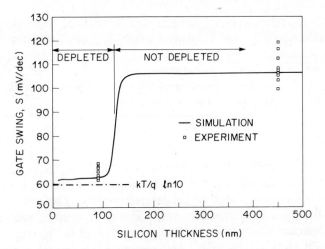

Fig. 36 The transition in subthreshold gate swing, S, needed to reduce subthreshold current by one decade versus thickness of the silicon film t_{Si}, in an SOI device. The transition occurs as t_{Si} exceeds W, the depletion-layer width in the silicon film. Device parameters are $d = 250\,\text{Å}$, $D_{it} = 10^{11}/\text{cm}^2$ (front interface), $D_{it} = 2 \times 10^{11}/\text{cm}^2$ (back interface), $N = 8 \times 10^{16}/\text{cm}^3$, $d_b = 1\,\mu\text{m}$. The data points are for two sets of devices with $t_{Si} = 0.09$ and $0.45\,\mu\text{m}$. (After Ref. 80 © 1987 IEEE).

As drain voltage is increased, however, the channel I-R drop forces the surface potential to increase as one approaches the drain, implying a reduction in oxide field $(V_G - \psi_s)/d$ and hence a reduction in total charge. For the bulk device this reduction in field reduces the inversion layer carrier density not only because the total charge (inversion layer plus bulk) is reduced but also because the increased band bending actually increases the bulk charge by an amount

$$\delta Q = C_s \, \delta\psi_s \tag{39}$$

with C_s from Eq. 37. As a result the inversion layer charge changes by an amount

$$q\delta N_{inv} = -(C_{ox} + C_s)\delta\psi_s \tag{40}$$

which is the sum of the reductions due to the variations in the oxide field and the bulk charge. This reduction in channel strength is well known as the cause for the current-voltage curves of the long-channel HMOS device to bend and eventually to saturate at large V_D.[14-16]

For the SOI device, an increase $\delta\psi_s$ also increases the field in the region beneath the inversion layer, but the increased bulk charge is now in the grounded silicon layer below the buried oxide. Thus Eqs. 39 and 40 still apply, but C_s is given by Eq. 38. Hence, the effect of the I-R drop upon inversion layer charge is smaller in the SOI device if C_s is smaller. Because the HMOS structure has a bulk punchthrough limitation upon the minimum substrate doping, such as Eq. 4, while the SOI device has an oxide layer and a support (grounded) silicon layer with no punchthrough restriction, it is possible to design the SOI structure to have a smaller C_s than the HMOS structure. Experimental improvements in saturation current of 25 to 30% result for long-channel devices.[92] Some improvement in transconductance also is expected for fully depleted SOI devices because velocity saturation is less pronounced due to the reduction of lateral field evidenced in Fig. 35a.

3.6.3   Comments

SOI devices have a number of advantages over bulk MOSFETs, at least in the regime of 1 μm channel lenghts. As devices continue to shrink, these advantages may be less evident if thin silicon layers ($\approx$ 500 Å) become necessary, bringing the front and back interfaces very close to each other, and requiring good-quality interfaces both front and back. Another difficulty with thin SOI structures is BiCMOS compatibility, because buried layers are not possible. Compromise bipolar designs have problems with gain degradation at high current levels.[80] Nonetheless, advantages in circuit speed, packing density, and hot-electron reliability might be decisive. Combination with selective epitaxial growth presages a variety of structures delayed for now by processing.

3.7   SUMMARY AND FUTURE TRENDS

We have explored miniaturization satisfying the constraints of (a) good subthreshold behavior, (b) reasonable reliability, and (c) good driving ability. This discussion was based upon rudimentary formulas summarizing computer simulations. According to this approach, despite thinner oxides and lower voltages, device size will continue to be limited by hot-electron damage.

This analysis needs improvement. First, of course, fitted formulas are imperfect. Not based on mechanism and opaque to intuition, they are known to be valid only over the fitted range, and are unreliable guides to imagination. Equation 25, for the length of the high-field region near the drain, is an example: as devices shrink, Eq. 25 changes form abruptly with reduction in oxide thickness, undermining confidence that it exhibits a correct dependence upon junction depth and channel length, and making extrapolation to smaller devices risky. Another example is Eq. 1, untested below 0.3 μm. Models based on explicit approximations are preferable to identify where and why the models are appropriate. Failing this goal, a substitute is to use models fitted over a wide range of parameters, a range extending well beyond immediate concerns, to allow room to optimize.

Second, there are many unquantified trade-offs, such as those related to isolation. As the scope of optimization widens, treatment of these trade-offs needs development. Otherwise it is not apparent which path to take for isolation, nor which processes to use. Another uncertainty is the status of bipolar devices in

VLSI (very large scale integration). Bipolar devices already have a role for output drivers and some special on-chip applications, but more impact on circuits and chips requires bipolars that can be used anywhere without severe penalties in area or yield. This proliferation of bipolars imposes demands in isolation and processing more stringent than for CMOS.

As complexity increases, opportunity opens for novel approaches. Silicon-on-insulator is an option that modifies not only the fabricational picture and power-speed-isolation trade-offs, but even the intrinsic device trade-offs because the electrical basis of a dual-gate structure without majority-carrier access is different from a standard MOSFET. Another novel approach is low-temperature operation.[93] Here again overall trade-offs decide viability, trade-offs that include refrigeration, repair, and packaging as well as changed device physics, particularly for hot-electron damage. Both of these novel approaches, SOI and low-temperature operation, have bipolar compatibility problems. In SOI the problem is unavailability of buried contacts and high dislocation densities. For low-temperature operation, difficulties with bipolar devices include reduced gain[94] and increased base currents,[95] although work has begun to reduce these problems.[96,97]

Unlike device or small-circuit design, VLSI involves more trade-offs, and they couple chip, circuit, device, and fabrication more closely. Efficient balancing of these trade-offs requires careful development of a strategy, not only to ensure convergence to a good design, but to make sure resources and time are well deployed. As understanding improves models and processes, frequent and sometimes radical changes in this organization are expected. This continuous modification and exploration of alternatives prompts further development and enlargement of simple models like those discussed here.

ACKNOWLEDGMENT

The author is pleased to thank many colleagues for discussions and suggestions: M.-L. Chen, T. -Y. Chiu, W. T. Lynch, C. S. Rafferty, S. M. Sze and, especially, A. G. Lewis and K. K. Ng. He also thanks Margo R. Baker for typing the manuscript many times, and Norman Erdos for editorial assistance.

PROBLEMS

1. In subthreshold, the long-channel part of a MOSFET is the part of the channel where the surface band bending, ψ_s , is constant. Assuming the equipotential $\psi = \psi_s$ parallels the metalurgical junction contours as shown in the diagram P1, the length of the long-channel portion of the channel, L , is approximately

$$L = L_m - (W_S - W_O) - (W_D - W_O) \qquad \text{(P1)}$$

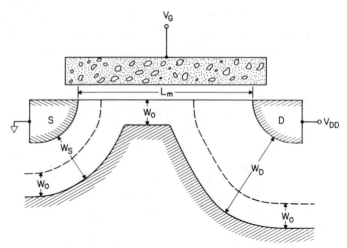

Diagram P1

with L_m the separation of the metallurgical source and drain junctions, and where W_S is the width of the depletion layer under the source, W_D is the depletion layer under the drain, and W_O is the distance from the contour $\psi = \psi_s$ to the depletion-layer edge which, in a one-dimensional uniform doping approximation, is $W_O = \sqrt{2}\,L_D(\beta\psi_s - 1)^{1/2}$. From Eq. P1, the smallest device occurs when $L = 0$, indicating that the sum $(W_S + W_D - 2W_O)$ helps decide the size of the smallest device. This argument suggests that Eq. 1 of the text might be replaced by

$$L_{\min}' = A' \left[r_j d\,(W_S + W_D - 2W_O)^2 \right]^{1/3}. \tag{P2}$$

Equation P2 leads to smaller $L_{\min}$ as gate bias increases, because larger gate biases increase W_O as ψ_s increases. Compare this replacement for $L_{\min}$ with Eq. 1 for the following cases. Assume an n^+ gate, source and drain all short-circuited to the p-substrate with substrate doping varied from $10^{15}/\text{cm}^3$ to $10^{18}/\text{cm}^3$ and for temperatures 290 and 77 K, and two oxide thicknesses, $d = 100$ and 1000 Å. Assume $r_j = 0.15\ \mu\text{m}$, and adopt the temperature dependencies provided in Problem 4. Show that for $A' = 0.478$ Å$^{-1/3}$ Eq. 1 of the text and Eq. P2 agree for the case $r_j d(W_S + W_D)^2 = 10\ \mu\text{m}^3 - \text{Å}$ and $d = 1000$ Å. Explain why different curves result for the two oxide thicknesses. Is the proposed replacement for $L_{\min}$ in reasonable agreement with the data of Fig. 1?

2. Compare the replacement Eq. P2 of Problem 1 with Eq. 1 of the text for a fixed inversion-layer carrier density, $N_{\text{inv}} = 10^8/\text{cm}^2$. Compare the formulas for $d = 1000$ Å and $T = 77$ and 290 K. Use $r_j = 0.15\ \mu\text{m}$, as for Problem 1. Take the temperature dependencies from Problem 4 and neglect dopant freezeout. Show that the two formulas agree at $r_j d(W_S + W_D)^2 =$

$10 \ \mu m^3 - \text{Å}$ and $T = 290$ K if $A' = 1.30 \ \text{Å}^{-1/3}$ Contrast the implications of the two predictions for a program to operate integrated circuits at liquid nitrogen temperature. Explain whether the comparison at fixed N_{inv} is more reasonable than the comparison of Problem 1, where gate, source, and drain were short-circuited to the substrate.

3. A conformal mapping from H. Kober, *Dictionary of Conformal Transformations*, 2nd edition, Dover Publications, N.Y., 1957, p. 161, namely:

$$x + jy = \frac{r_j}{\pi} \cosh^{-1} \left[\frac{2w - s^2 - 1}{s^2 - 1} \right] \qquad \text{(P3)}$$

$$- \frac{r_{ch}}{\pi} \cosh^{-1} \left[\frac{(s^2 + 1)w - 2s^2}{(s^2 - 1)w} \right]$$

maps the $z = x + jy$ plane onto the $w = u + jv$ plane, as shown in diagram P2, where $s = r_j / r_{ch} \geq 1$, $r_j = $ junction depth, $r_{ch} = $ channel depth, $j = \sqrt{-1}$. Using this conformal mapping, find the resistance of

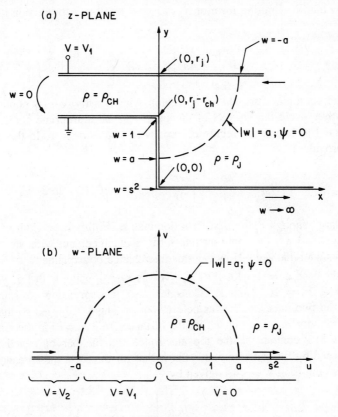

(a) z - PLANE

(b) w - PLANE

Diagram P2

this step in cross section. Compared to the resistance of a bar of cross section Wr_j and resistivity ρ_J in series with another bar of cross section Wr_{ch} and resistivity ρ_{CH}, show that the spreading resistance between the junction of depth r_j and the channel of depth r_{ch} is given by

$$R_{SP} = \left(\frac{\rho_J}{\pi W} \right) \ln \left[\left(\frac{s+1}{s-1} \right)^{1/s} \frac{s^2 - 1}{4a} \right] \qquad \text{(P4)}$$

$$- \left(\frac{\rho_{CH}}{\pi W} \right) \ln \left[\left(\frac{s-1}{s+1} \right)^{s} \frac{4s^2}{(s^2 - 1)a} \right]$$

with ρ_J the resistivity in Ω-cm of the junction, ρ_{CH} that of the channel, and W the channel width. In Eq. P4 the channel resistivity is taken to apply inside the region extending from $x = -\infty$ up to a boundary given by an equipotential of quasi-Fermi level located at $|w| = a$, and the junction resistivity applies from this equipotential to $x = +\infty$.

Show that the limiting form of R_{SP} as $s \rightarrow \infty$ (the deep-junction limit) is

$$R_{SP} = \frac{\rho_J}{\pi W} \ln \left(\frac{s^2}{4a} \right) + \frac{\rho_{CH}}{\pi W} \left[2 - \ln \left(\frac{4}{a} \right) \right]. \qquad \text{(P5)}$$

In the common case where the region with resistivity ρ_{CH} continues only slightly outside the channel of thickness r_{ch}, an appropriate choice of equipotential to bound the region of resistivity ρ_{CH} is $a = 1$. In this case, Eq. P5 provides

$$R_{SP} = \frac{2\rho_J}{\pi W} \ln \left(\frac{r_j}{2r_{ch}} \right) + \frac{2\rho_{CH}}{\pi W} (1 - \ln 2).$$

For large enough r_j the first term dominates. In this same limit of $s \rightarrow \infty$, but for $a = s^2$, the same leading term is obtained, but with the two resistivities interchanged. How can this interchange be explained?

For the case $a = 1$, plot WR_{SP}/ρ_J versus r_j/r_{CH} using both Eq. P4 and P5 for $\rho_{CH}/\rho_J = 10^{-4}$ and 1. Consider also the spreading resistance compared to two bars in series as before, but of lengths adjusted so that the bar of resistivity ρ_{CH} is extended by the distance from $x = 0$ to the endpoint of the $a = 1$ contour on the top electrode, and the bar of resistivity ρ_J is shortened by the same amount. Show that for this case the spreading resistance in the limit $s \rightarrow \infty$ is given by

$$R_{SP} = \frac{2\rho_J}{\pi W} \ln \left(\frac{r_j}{2r_{ch}} \right) + \frac{2\rho_{CH}}{\pi W} \left[1 - \sqrt{2} + \frac{1}{2} \ln \left(\frac{3 + 2\sqrt{2}}{4} \right) \right]$$

Why does one spreading resistance increase with decreasing ρ_{CH} while the other decreases? What are the major influences upon spreading resistance? What do these calculations suggest about the gate-bias dependence of spreading resistance?

4. For a MOSFET with an n^+ source-drain and an n^+-poly gate over a p-type substrate, calculate the minimum oxide thickness, d versus gate-source voltage, based upon a maximum oxide field of 7 MV/cm. Take temperatures of 290 and 400 K, and a doping level of $N = 10^{17}/cm^3$. Assume the Fermi level in the gate and source lies at the band edge. Compute the intrinsic carrier density neglecting the temperature dependence of the effective mass using (Ref. 14, p. 19)

$$n_i = 1.45 \times 10^{10}(T/300)^{3/2}\exp\left[\frac{\phi_G(300K)}{2V_{th}(300K)} - \frac{\phi_G(T)}{2V_{th}(T)}\right]/cm^3$$

and assume the bandgap is given by $E_G(T)/q = \phi_G(T) = 1.17 - \alpha T^2/(\beta + T)$, with $\alpha = 4.73 \times 10^{-4}$ and $\beta = 636$ (Ref. 14, Fig. 8, p. 15). Here the thermal voltage $V_{th}(T) \equiv kT/q$. Explain the reasons for the results qualitatively, and suggest some practical implications.

5. Consider minimizing gate-induced drain leakage by adjusting the doping of the drain. Assume an n^+-gate over an n-type drain of doping N/cm^3. Suppose the applied drain bias is fixed at V_{DD}. Find an expression for the thickness of the region available to cause GIDL as a function of drain doping. Assume there is a minimum field above which significant GIDL occurs once band overlap arises, namely $\mathcal{E}_{GIDL}$. Plot a curve showing the spatial extent of the region contributing to GIDL for $d = 30$ and 500 Å as a function of doping level in the drain for $V_{DD} = 5$ and 3 V. Use $5 \times 10^{17}/cm^3 \le N \le 5 \times 10^{19}/cm^3$ and assume $\mathcal{E}_{GIDL} = 10^6$ V/cm.

6. Write a FORTRAN 77 program to compute the curves labeled "EQ(1)" in Fig. 14a.

7. Write a FORTRAN 77 program to compute the curves labeled $\mathcal{E}_{lat}$ in Fig. 14a.

8. Compare curves of d versus L for constant $\mathcal{E}_{lat}$ as shown in Fig. 14a with curves of d versus L for constant lifetime τ from Eq. 27. Assume $m = 4$ and that B is a constant. Lump all the constant terms together as a fitting parameter called "time," for convenience. Assume the parameters of Jeng et al.[11] from the caption to Fig. 14, and adopt $V_{GS} = V_{DD} = 3$ V. Do devices with the same $\mathcal{E}_{lat}$ have approximately the same lifetimes?

9. Write a FORTRAN 77 program to compute the inverter transfer characteristics in Fig. 13 for arbitrary V_T/V_{DD}.

10. Derive Eq. 34 of the text.

11. Compute some curves for subthreshold slope similar to the one shown in Fig. 36. Assume interface charge is given by $Q_{it}^f = -C_{it}^f \psi_{it}^f$ at the front surface, where ψ_s^f is the potential at the front surface, and assume a similar expression for the back surface charge. Assume the device has an n-type substrate sandwiched between an n^+-poly gate and a degenerately doped n^+ ground. Suppose the work functions of the poly gate and the grounding substrate are the same. We suppose that source and drain are grounded as well. Use the depletion approximation for depletion widths, for example, $W_f = \sqrt{2} \, L_D \, (\beta\psi_s^f - 1)^{1/2}$.

a) Treat four choices of gate-to-grounded-substrate bias, $V_G = 0.25, 0.5, 1.0,$ and 1.65 V. Adopt the following parameter values: $d = 250$ Å, $d_b = 1 \, \mu m$, $N = 8 \times 10^{16}/cm^3$, $D_{it} = 0$, both front and back.

b) Using the same thicknesses as in (a), for the case of a gate bias of 1.65 V, compute curves for the following combinations of front and back interface trap densities/eV-cm^2: $(0,0)$, $(0,10^{12})$, $(10^{12},0)$, and $(10^{12},10^{12})$. Assume the silicon thickness is in the range $0 \le t_{Si} \le 0.2 \, \mu m$.

REFERENCES

1. J. Y. Chen, "CMOS—The Emerging VLSI Technology," *IEEE Circuits Dev. Mag.*, **2**(2), 16 (1986).

2. M. Kubo, I. Masuda, K. Miyata, and K. Ogiue, "Perspective on BiCMOS VLSIs," *IEEE J. Solid-State Circuits*, **23**(1), 5 (1988).

3. J. R. Brews, K. K. Ng, and R. K. Watts, "The Submicrometer Silicon MOSFET," in *Submicron Integrated Circuits*, R. K. Watts, Ed., Wiley, New York, Chap. 1, 1989, pp. 9–86.

4. A. R. Alvarez, Ed., *BiCMOS Technology and Applications*, Kluwer Academic, Norwell, MA, 1989.

5. R. H. Dennard, F. H. Gaensslen, H.-N. Yu, V. L. Rideout, E. Bassous, and A. R. LeBlanc, "Design of Ion-Implanted MOSFETs with Very Small Physical Dimensions," *IEEE J. Solid-State Circuits*, **SC-9**(5), 256 (1974).

6. G. Baccarani, M. R. Wordeman, and R. H. Dennard, "Generalized Scaling Theory and its Application to a ¼ Micrometer MOSFET Design," *IEEE Trans. Electron Dev.*, **ED-31**(4), 452 (1984).

7. J. R. Brews, W. Fichtner, E. H. Nicollian, and S. M. Sze, "Generalized Guide to MOSFET Miniaturization," *IEEE Electron Dev. Lett.*, **EDL-1**(1), 2 (1980).

8. E. Takeda, G. A. C. Jones, and H. Ahmed, "Constraints on the Application of 0.5 μm MOSFETs to ULSI Systems," *IEEE Trans. Electron Dev.*, **ED-32**(2), 322 (1985).

9. J. R. Pfiester, J. D. Shott, and J. D. Meindl, "Performance Limits of CMOS ULSI," *IEEE J. Solid-State Circuits,* **SC-20**(1), 253 (1985).

10. K. Tanaka and M. Fukuma, "Design Methodology for Deep Submicron CMOS," *Tech. Dig. Int. Electron Dev. Meet.,* p. 628 (1987).

11. M.-C. Jeng, J. Chung, J. E. Moon, G. May, P. K. Ko, and C. Hu, "Design Guidelines for Deep-Submicrometer MOSFETs," *Tech. Dig. Int. Electron Dev. Meet.,* p. 386 (1988).

12. W.-H. Lee, T. Osakama, K. Asada, and T. Sugano, "Design Methodology and Size Limitations of Submicrometer MOSFETs for DRAM Applications," *IEEE Trans. Electron Dev.,* **ED-35**(11), 1876 (1988).

13. C. G. Sodini, S. S. Wong, and P. K. Ko, "A Framework to Evaluate Technology and Device Design Enhancements for MOS Integrated Circuits," *IEEE J. Solid-State Circuits,* **SC-24**(1), 118 (1989).

14. S. M. Sze, *Physics of Semiconductor Devices,* 2nd ed., Wiley, New York 1981, Chap. 8.

15. Y. P. Tsividis, *Operation and Modeling of the MOS Transistor,* McGraw-Hill, New York, 1987.

16. J. R. Brews, "Physics of the MOS Transistor," in *Silicon Integrated Circuits, Applied Solid State Science,* Suppl. 2, Part A, D. Kahng, Ed., Academic, New York, 1981, Chap. 1.

17. C. Mead and L. Conway, *Introduction to VLSI Systems,* Addison-Wesley, Reading, MA, 1980.

18. M. Annaratone, *Digital CMOS Circuit Design,* Kluwer Academic, Boston, 1986.

19. N. Weste and K. Eshraghian, *Principles of CMOS VLSI Design,* Addison-Wesley, Reading, MA, 1985.

20. N. G. Einspruch and G. Sh. Gildenblat, Eds., *Advanced MOS Device Physics,* Vol. 18 in *VLSI Electronics Microstructure Science,* Academic, New York, 1989

21. W. T. Lynch, "Self-Aligned Contact Schemes for Source-Drain in Submicron Devices," *Tech. Dig Int. Electron Dev. Meet.,* p. 354 (1987); W. T. Lynch, P. D. Foo, R. Liu, J. Lebowitz, K. J. Orlowsky, G. E. Georgiou, and S. J. Hillenius, "UPMOS—a New Approach to Submicron VLSI," *Solid State Devices* , Proc. of the 17th European Solid State Device Research Conf., ESSDERC'87, 1988, p. 25.

22. F. M. Klaassen, "Design and Performance of Micron-Size Devices," *Solid-State Electronics,* **21**(3), 565 (1978).

23. K. K. Ng and W. T. Lynch, "Analysis of the Gate-Voltage Dependent Series Resistance of MOSFETs," *IEEE Trans. Electron Dev.,* **ED-33**(7), 965 (1986).

24. K. K. Ng and W. T. Lynch, "The Impact of Intrinsic Series Resistance on MOSFET Scaling," *IEEE Trans. Electron Dev.,* **ED-34**(3), 503 (1987).

25. J. M. Pimbley, "Two Dimensional Current Flow in the MOSFET Source-Drain," *IEEE Trans. Electron Dev.*, **ED-33**(7), 986 (1986).

26. C. Hu, S. C. Tam, F.-C. Hsu, P. K. Ko, T.-Y. Chan, K. W. Terrill, "Hot-Electron-Induced MOSFET Degradation—Model, Monitor, and Improvement," *IEEE Trans. Electron Dev.*, **ED-32** (2), 375 (1985).

27. F.-C. Hsu, P. K. Ko, S. Tam, C. Hu, and R. S. Muller, "An Analytical Breakdown Model for Short-Channel MOSFETs," *IEEE Trans. Electron Dev.*, **ED-29**(11), 1735 (1982).

28. E. Landi and S. Solmi, "Electrical Characterization of p^+/n Shallow Junctions Obtained by Boron Implantation into Preamorphized Silicon," *Solid-State Electron.*, **29** (11), 1181 (1986).

29. L. Van den Hove, R. Wolters, K. Maex, R. F. Keersmaecker, and G. J. DeClerck, "A Self-Aligned $CoSi_2$ Interconnection and Contact Technology for VLSI Applications," *IEEE Trans. Electron Dev.*, **ED-34**(3), 554 (1987).

30. R. Liu, D. S. Williams, and W. T. Lynch, "Mechanisms for Process-Induced Leakage in Shallow Silicided Junctions," *Tech. Dig. Int. Electron Dev. Meet.*, p. 58 (1986).

31. D. L. Kwong, Y. H. Ku, S. K. Lee, E. Louis, N. S. Alvi, and P. Chu, "Silicided Shallow Junction Formation by Ion-Implantation of Impurity Ions into Silicide Layers and Subsequent Drive-in," *J. Appl. Phys.*, **61**(11), 5084 (1987).

32. T. -Y. Chiu, G. M. Chin, M. Y. Lau, R. C. Hanson, M. D. Morris, K. F. Lee, A. M. Voschenkov, R. G. Swartz, V. D. Archer, and S. N. Finegan, "A High-Speed Super Self-Aligned Bipolar-CMOS Technology," *Tech. Dig. Int. Electron Dev. Meet.*, p. 24 (1987).

33. Y. Nissan-Cohen, J. Shappir, and D. Frohman-Bentchkowsky, "Measurement of Fowler-Nordheim Tunneling Currents in MOS Structures under Charge Trapping Conditions," *Solid-State Electron.*, **28**(7), 717 (1985).

34. D. R. Wolters and J. J. van der Schoot, "Dielectric Breakdown in MOS Devices. Parts 1-3," *Philips J. Res.*, **40**, 115, 137, 164 (1985).

35. I. C. Chen and C. Hu, "Accelerated Testing of Time-Dependent Breakdown of SiO_2," *IEEE Electron Dev. Lett.*, **EDL-8**(4), 140 (1987).

36. R. Moazzami, J. Lee, I.-C. Chen, and C. Hu, "Projecting the Minimum Acceptable Oxide Thickness for Time-Dependent Dielectric Breakdown," *Tech. Dig. Int. Electron Dev. Meet.*, p. 710 (1988).

37. T. Y. Chan, J. Chen, P. K. Ko, and C.Hu, "The Impact of GIDL on MOSFET Scaling," *Tech. Dig. Int. Electron Dev. Meet.*, p. 718 (1987).

38. C. Chang and J. Lien, "Corner Field Induced Drain Leakage in Thin Oxide MOSFETs," *Tech. Dig. Int. Electron Dev. Meet.*, p. 714 (1987).

39. I. C. Chen, C. W. Teng, D. J. Coleman, and A. Nishimura, "Interface-Trap Enhanced Gate-Induced Leakage Current in MOSFET," *IEEE Electron Dev. Lett.*, **EDL-10** (5), 216 (1989).

40. D. -S. Wen, S. H. Goodwin-Johansson, and C. M. Osburn, "Tunneling Leakage in Ge-Preamorphized Shallow Junctions," *IEEE Trans. Electron Dev.*, **ED-35**(7), 1107 (1988).

41. C. Duvvury, D. J. Redwine, and H. J. Stiegler, "Leakage Current Degradation in *n*-MOSFETs Due to Hot-Electron Stress," *IEEE Electron Dev. Lett.*, **EDL-9**(11), 579 (1988).

42. J. Chen, T. Y. Chan, P. K. Ko, and C. Hu, "Gate Current in Off-State MOSFET," *IEEE Electron Dev. Lett.*, **EDL-10**(5), 203 (1989).

43. I. -C. Chen, D. J. Coleman, and C. W. Teng, "Gate Current Injection Initiated by Electron Band-to-Band Tunneling in MOS Device," *IEEE Electron Dev. Lett.*, **EDL-10** (7), 297 (1989).

44. R. Shirota, T. Endoh, M. Momodomi, R. Nakagama, S. Inoue, R. Kirisawa, and F. Masuoka, "An Accurate Model of Sub-Breakdown Due to Band-to-Band Tunneling and Its Application," *Tech. Dig. Int. Electron Dev. Meet.*, p. 26 (1988).

45. V. A. K. Temple, M. A. Green and J. Shewchun "Equilibruim-to-Nonequilibrium Transition in MOS (Surface Oxide) Tunnel Diodes," *J. Appl. Phys.*, **45**(11), 4934 (1974).

46. J. G. Simmons, F. L. Hsueh, and L. Faraone, "Two-Carrier Conduction in MOS Tunnel Oxides—Theory," *Solid-State Electron.*, **27**(12), 1131 (1984).

47. F. L. Hsueh, L. Faraone, and J. G. Simmons, "Two-Carrier Conduction in MOS Tunnel Oxide—Experimental Results," *Solid-State Electron.*, **27**(6), 499 (1984).

48. S. Kar and W. E. Dahlke, "Interface States in MOS Structures with 20 − 40 Å Thick SiO$_2$ Films on Nondegenerate Si," *Solid-State Electron.*, **15**, 221 (1972).

49. T. Y. Chan, C. L. Chiang, and H. Gaw, "New Insight into Hot-Electron-Induced Degradation of *n*-MOSFETs," *Tech, Dig. Int. Electron Dev. Meet.*, p. 196 (1988).

50. J. Chung, M.-C. Jeng, G. May, P. K. Ko, and C. Hu, "Hot-Electron Currents in Deep-Submicrometer MOSFETs," *Tech. Dig. Int. Electron Dev. Meet.*, p. 200 (1988).

51. R. Harata, Y. Ohji, Y. Nishioka, I. Yoshida, K. Mukai, and T. Sugano, "Improvement of Hardness of MOS Capacitors to Electron-Beam Irradiation and Hot-Electron Injection by Ultra-Dry Oxidation of Silicon," *IEEE Electron Dev. Lett.*, **EDL-10**(1), 27 (1989).

52. P. J. Wright, N. Kasai, S. Inoue, and K. C. Saraswat, "Hot-Electron Immunity of SiO$_2$ Dielectrics with Fluorine Incorporation," *IEEE Electron Dev. Lett.*, **EDL-10**(8) 347 (1989).

53. T. Hori and H. Iwasaki, "Improved Hot-Carrier Immunity in Submicron MOSFETs with Reoxidized Nitrided Oxides Prepared with Rapid Thermal Processing," *IEEE Electron Dev. Lett.*, **EDL-10**(2), 64 (1989).

54. P. Roblin, A. Samman, and S. Bibyk, "Simulation of Hot-Electron Trapping and Aging in *n*-MOSFETs," *IEEE Trans. Electron Dev.*, **ED-35**(12), 2229 (1988).

55. M. G. Ancona, N. S. Saks, and D. McCarthy, "Lateral Distribution of Hot-Carrier Interface Traps in MOSFETs," *IEEE Trans. Electron Dev.*, **ED-35**(12), 2221 (1988).

56. H. Haddara and S. Cristoloveanu, "Two-Dimensional Modeling of Locally Damaged Short-Channel MOSFETs Operating in the Linear Region," *IEEE Trans. Electron Dev.*, **ED-34**(2), 378 (1987).

57. C. Nguyen-Duc, S. Cristoloveanu, and G. Ghibaudo, "A Three-Piece Model of Channel-Length Modulation in Submicrometer MOSFETs," *Solid-State Electron.*, **31**(6), 1057 (1988).

58. K. Mayaram, J. C. Lee, and C. Hu, "A Model for the Electric Field in Lightly-Doped Drain Structures," *IEEE Trans. Electron Dev.*, **ED-34**(7), 1509 (1987).

59. R. Izawa, T. Kure, and E. Takada, "Impact of the Gate-Overlapped Device (GOLD) for Deep Submicrometer VLSI," *IEEE Trans. Electron Dev.*, **ED-35**(12) 2088 (1988).

60. T.-Y. Huang, W. W. Yao, R. A. Martin, A. G. Lewis, M. Koyanagi, and J. Y. Chen, "A New LDD Transistor with Inverse-T Gate Structure," *IEEE Electron Dev. Lett.*, **EDL-8**(4), 151 (1987).

61. D. C. Mayer and W. E. Perkins, "Analysis of the Switching Speed of a Submicrometer-Gate CMOS/SOS Inverter," *IEEE Trans. Electron Dev.*, **ED-28**(7), 886 (1981).

62. K. K. -L. Hsueh, J. J. Sanchez, T. A. Demassa, and L. A. Akers, "Inverse-Narrow Width Effects and Small-Geometry MOSFET Threshold Voltage Model," *IEEE Trans. Electron Dev.*, **ED-35**(3), 325 (1988).

63. M. Sugino, L. A. Akers, and J. M. Ford, "Optimum *p*-Channel Isolation Structure for CMOS," *IEEE Trans. Electron Dev.*, **ED-31**(12), 1823 (1984).

64. N. Shigyo, S. Fukuda, T. Wada, K. Hieda, T. Hamamoto, H. Watanabe, K. Sunouchi, and H. Tango, "Three Dimensional Analysis of Subthreshold Swing and Transconductance for Fully Recessed Oxide (Trench) Isolated ¼ μm-Width MOSFETs," *IEEE Trans. Electron Dev.*, **ED-35** (7), 945 (1988).

65. K. Ohe, S. Odanaka, K. Moriyama, T., Hori, and G. Fuse, "Narrow-Width Effects of Shallow Trench-Isolated CMOS with n^+-Polysilicon Gate," *IEEE Trans. Electron Dev.*, **ED-36**(6), 1110 (1989).

66. M. Bourcerie, B. S. Doyle, J.-C. Marchetaux, A. Boudou, and H. Mingam, "Hot-Carrier Stressing Damage in Wide and Narrow LDD NMOS Transistors," *IEEE Electron Dev. Lett.*, **EDL-10**(3), 132 (1989).

67. J. Hui, P. V. Voorde, and J. Moll, "Scaling Limitations of Submicron Local Oxidation Technology," *Tech. Dig. Int. Electron Dev. Meet.*, p. 392, (1985).

68. J. C. Marchetaux, B. S. Doyle, and A. Boudou, "Interface Traps under LOCOS Bird's Beak Region," *Solid-State Electron.*, **30**(7), 745 (1987).

69. S. H. Goodwin and J. D. Plummer, "Electrical Performance and Physics of Isolation Structures for VLSI," *IEEE Trans. Electron Dev.*, **ED-31**(7), 861 (1984).

70. A. G. Lewis, J. Y. Chen, R. A. Martin, and T.-Y. Huang, "Device Isolation in High-Density LOCOS Isolated CMOS," *IEEE Trans. Electron Dev.*, **ED-34**(6), 1337 (1987).

71. J. Y. Chen and D. E. Snyder, "Modeling Device Isolation in High Density CMOS," *IEEE Electron Dev. Lett.*, **EDL-7**(2), 64 (1986).

72. N. Kasai, N. Endo, A. Ishitani, and H. Kitajima, "¼ μm CMOS Isolation Using Selective Epitaxy," *IEEE Trans. Electron Dev.*, **ED-34**(6), 1331 (1987).

73. G. Fuse, M. Fukumoto, A. Shinohara, S. Odanaka, M. Sasago, and T. Ohzone, "A New Isolation Method with Boron-Implanted Sidewalls for Controlling Narrow-Width Effects," *IEEE Trans. Electron Dev.*, **ED-34**(2), 356 (1987).

74. B. Davari, C. Koburger, T. Furukawa, Y. Taur, W. Noble, A. Megdanis, J. Warnock, and J. Mauer, "A Variable-Size Shallow Trench Isolation (STI) Technology with Diffused Sidewall Doping for Submicron CMOS," *Tech. Dig. Int. Electron Dev. Meet.*, p. 92 (1988).

75. K. Kikuchi, S. Kameyama, M. Kajiyama, M. Nishio, and T. Komeda, "A High-Speed Bipolar LSI Process Using Self-Aligned Double Diffusion Polysilicon Technology," *Tech. Dig. Int. Electron Dev. Meet.*, p. 420 (1986).

76. K. Nakazato, T. Nakamura, and M. Kato, "A 3 GHz Lateral *p-n-p* Transistor," *Tech. Dig. Int. Electron Dev. Meet.*, p. 416, (1986).

77. R. A. Chapman, D. A. Bell, R. H. Eklund, R. H. Havemann, M. G. Harward, and R. A. Haken, "Submicron BiCMOS Well Design for Optimum Circuit Performance," *Tech. Dig. Int. Electron Dev. Meet.*, p. 756 (1988).

78. T. Ikeda, A. Watanabe, Y. Nishio, I. Masuda, N. Tamba, M. Odaka, and K. Ogiue, "High-Speed BiCMOS Technology with a Buried Twin Well Structure," *IEEE Trans. Electron Dev.*, **ED-34**(6), 1304 (1987).

79. H. Stübing and H.-M. Rein, "A Compact, Physical, Large-Signal Model for High-Speed Bipolar Transistors at High Current Densities—Part 1: One-Dimensional Model," *IEEE Trans. Electron Dev.*, **ED-34**(8), 1741 (1987).

80. J. P. Colinge, "Thin-Film SOI Devices: A Perspective," *Microelectronic Eng.*, **8**, 127 (1988)

81. C.-E. Daniel Chen and P. Chatterjee, "Silicon-on-Insulator: Why, How, and When," in *Deposition and Growth: Limits for Microelectronics*, American Institute of Physics Conference Proceedings, No. 167, G. W. Rubloff, Ed., American Institute of Physics, New York, 1988, p. 310.

82. G. K. Celler, "SOI Structures by Ion-Implantation and Annealing in a Temperature Gradient," *Solid State Dev.*, p. 583 (1988).

83. D. Hill, P. Fraundorf, and G. Fraundorf, "The Reduction of Dislocations in Oxygen Implanted Silicon-on-Insulator Layers by Sequential Implantation and Annealing," *J. Appl. Phys.*, **63**(10), 4993 (1988).

84. B.-Y. Mao, R. Sundaresan, C.-E. D. Chen, M. Matloubian, and G. Pollack, "The Characteristics of CMOS Devices in Oxygen Implanted Silicon-on-Insulator Structures," *IEEE Trans. Electron Dev.*, **ED-35**(5), 629 (1988).

85. J. R. Davis, K. Reeson, and P. L. F. Hemment, "Small Geometry SOI/CMOS Devices on SIMOX Substrates," *Solid State Dev.*, p. 595 (1988).

86. K. Yamaguchi, "Mathematical Modeling of Semiconductor-on-Insulator (SOI) Device Operation," *IEEE Trans. Electron Dev.*, **ED-31**(7), 977 (1984).

87. K. Kato, T. Wada, and K. Taniguchi, "Analysis of Kink Characteristics in Silicon-on-Insulator MOSFETs Using Two-Carrier Modeling," *IEEE Trans. Electron Dev.*, **ED-32**(2), 458 (1985).

88. K. K. Young and J. A. Burns, "Avalanche-Induced Drain-Source Breakdown in Silicon-on-Insulator n-MOSFETs," *IEEE Trans. Electron Dev.*, **ED-35**(4), 426 (1988).

89. S. P. Edwards, K. J. Yallup, and K. M. DeMeyer, "Two Dimensional Numerical Analysis of the Floating Region in SOI MOSFETs," *IEEE Trans. Electron Dev.*, **ED-35**(7), 1012 (1988).

90. J.-P. Colinge, "Reduction of Kink Effect in Thin-Film SOI MOSFETs," *IEEE Electron Dev. Lett.*, **EDL-9**(2), 97 (1988).

91. T. Sekigawa and Y. Hayashi, "Calculated Threshold Voltage Characteristics of an XMOS Transistor Having an Additional Bottom Gate," *Solid-State Electron.*, **27**(8/9), 827 (1984).

92. J. C. Sturm, K. Tokunaga, and J.-P. Colinge, "Increased Drain Saturation Current in Ultra-Thin Silicon-on-Insulator (SOI) MOS Transistors," *IEEE Electron Dev. Lett.*, **EDL-9**(9), 460 (1988).

93. G. Sh. Gildenblat, "Low Temperature CMOS," in *Advanced MOS Device Physics*, N. G. Einspruch and G. Sh. Gildenblat, Eds., Academic, New York, 1989, Chapter 5, pp. 191-236.

94. D. D. Tang, "Scaling the Silicon Bipolar Transistor," in *Submicron Integrated Circuits*, R. K. Watts, Ed., Wiley, New York, 1989. Chap. 2, esp. p. 100.

95. J. C. S. Woo, J. D. Plummer, and J. M. C. Stork, "Non-Ideal Base Current in Bipolar Transistors at Low Temperatures," *IEEE Trans. Electron Dev.*, **ED-34**(1), 130 (1987).

96. J. C. S. Woo and J. D. Plummer, "Optimization of Silicon Bipolar Transistors for High Current Gain at Low Temperatures," *IEEE Trans. Electron Dev.*, **ED-35**(8), 1311 (1988).

97. J. M. C. Stork, D. L. Harame, B. J. Meyerson, and T. N. Nguyen, "Base Profile Design for High Performance Optimization of Bipolar Transistors at Liquid Nitrogen Temperature," *IEEE Trans. Electron Dev.*, **ED-36**(8),1503 (1989).

4 Homogeneous Field-Effect Transistors

M. A. Hollis and R. A. Murphy

Lincoln Laboratory
Massachusetts Institute of Technology
Lexington, Massachusetts

4.1 INTRODUCTION

In recent years the field of high-speed transistors has greatly expanded to include devices made of sophisticated heterostructures. However, the most prevalent and well-developed high-speed transistors remain those in which carriers flow through a homogeneous semiconductor layer and are controlled by the influence of the voltage placed on a control electrode, usually referred to as a gate or base electrode. The high-speed properties of these devices are achieved through careful materials selection, careful control of the critical dimensions of the device, and minimization of parasitics. Although homogeneous field-effect transistors (FETs) do not take advantage of the special properties of heterojunctions, many offer greater simplicity and ease of fabrication because they do not depend as critically upon the precise fabrication of thin layers and sharp interfaces. FETs are unipolar or majority-carrier devices and do not exhibit the minority-carrier effects that can hamper the speed and thermal stability of bipolar transistors. Homogeneous FETs offer high performance at low cost and have established a proven track record of utility and reliability in important systems applications. Their characteristics and performance are the subjects of this chapter.

This chapter is written from the point of view of the device technologist. The qualitative features that provide the high-speed performance of FETs are discussed, and basic models that describe the device operation are given. The chapter is intended to place homogeneous FETs in context, both with respect to

High-Speed Semiconductor Devices, Edited by S.M. Sze. ISBN 0-471-62307-5
© 1990 John Wiley & Sons, Inc.

one another and to other types of devices, and to provide the reader with a conceptual basis to design new devices. Many of the general principles given in this chapter apply to heterojunction FETs and metal-oxide-semiconductor FETs (MOSFETs) as well.

4.1.1 Nature of Control Mechanism

The operation of a homogeneous field-effect transistor is discussed in terms similar to the charge-control concepts of Refs. 1 and 2. A simple diagram illustrating a field-effect device is shown in Fig. 1a. Each gate of the device induces a depletion region in the adjacent semiconductor channel. In this generalized three-terminal device, the current due to carrier flow from the source to the drain electrode is modulated by changes in the depletion width caused by changes in the gate voltage. The gate voltage in such a FET effectively modulates the cross-sectional area of the channel, in contrast to the MOSFETs of the previous chapter and the heterojunction FETs of Chapter 5 where the gate modulates the density of flowing charge in a quasi-two-dimensional channel.

Examining this control process in more detail, a change in the gate voltage is caused by the addition of charge ΔQ_G to the gate electrode. If the channel through which the carriers flow maintains space-charge neutrality, then the charge ΔQ_G added to the gate causes the depletion region boundary to shift so that the total space charge within the depletion region changes by $-\Delta Q_G$. This process, illustrated in Fig. 1b, causes the total number of carriers in the channel to change by $\Delta Q/q$, where q is the magnitude of the electronic charge and

$$\Delta Q = \Delta Q_G \tag{1}$$

for n-type channels and

$$\Delta Q = -\Delta Q_G \tag{1a}$$

for p-type channels. ΔQ is the increment to the total flowing charge Q in the channel, and is produced by the addition of ΔQ_G to the gate. The efficacy of this process can be described by the rate at which the current changes with respect to gate charge, namely

$$\frac{1}{\tau} = \left. \frac{\partial I_{DS}}{\partial Q_G} \right|_{V_{DS}}, \tag{2}$$

where I_{DS} is the current that flows from drain to source and V_{DS} is the drain-source voltage. I_{DS} is positive for n-channel FETs and negative for p-channel FETs. Combining Eqs. 1 and 2,

$$\Delta I_{DS} = \frac{\Delta Q_G}{\tau} = \frac{\Delta Q}{\tau} \tag{3}$$

for n-channel devices and

$$\Delta I_{DS} = \frac{\Delta Q_G}{\tau} = -\frac{\Delta Q}{\tau} \tag{3a}$$

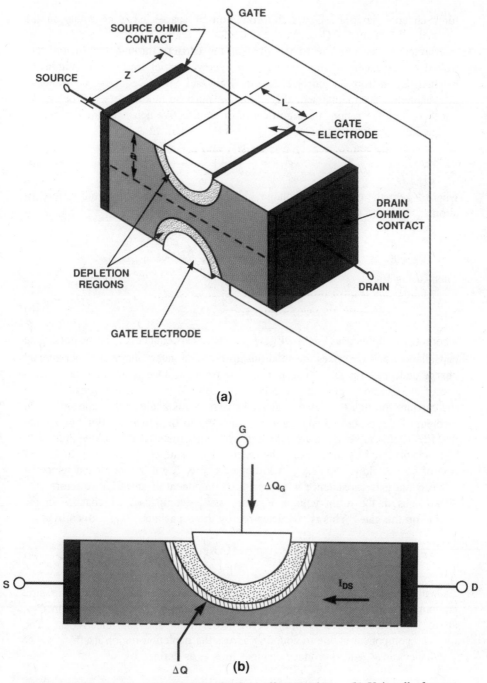

Fig. 1 (a) Simple generalized model of a field-effect transistor. (b) Unit cell of generalized FET in (a), illustrating the response of the depletion region boundary to ΔQ_G.

for *p*-channel devices. Under the condition of space-charge neutrality in the channel, the quantity ΔQ exists at the depletion region boundary, and τ can be interpreted as the average transit time of the carriers through the region controlled by the gate. In Fig. 1b, τ is the average carrier transit time within the hatched region from its source end to its drain end. If space-charge neutrality in the channel is not maintained, this argument must be modified (see Problem 2), but the quantity τ retains the significance of an effective delay for the current.

The intrinsic transconductance g_{mi} of a FET is defined as

$$g_{mi} = \left. \frac{\partial I_{DS}}{\partial V_{GS}} \right|_{V_{DS}} \tag{4}$$

where V_{GS} is the gate-source voltage. Using the chain rule for partial differentiation,

$$g_{mi} = \left. \frac{\partial I_{DS}}{\partial V_{GS}} \right|_{V_{DS}} = \left. \frac{\partial I_{DS}}{\partial Q_G} \right|_{V_{DS}} \left. \frac{\partial Q_G}{\partial V_{GS}} \right|_{V_{DS}} \tag{5}$$

Then, using Eq. 2,

$$g_{mi} = \frac{C_G}{\tau} , \tag{6}$$

where C_G is defined as $\left. (\partial Q_G / \partial V_{GS}) \right|_{V_{DS}}$. This capacitance C_G is the total gate capacitance, and describes the relationship between gate voltage and incremental charge added to the gate. The intrinsic transconductance is an important measure of a transistor's gain and depends inversely on the effective transit time of the carriers through the device. It is clear from this analysis that one important element of the design of any transistor must be to minimize τ. For homogeneous FETs, this is achieved by selecting a channel material that allows high carrier velocity, and by minimizing the extent of the gate electrode along the direction of carrier flow. Since g_{mi} also depends directly on the gate capacitance, it is clear that gate capacitance is an intrinsic component of any FET structure.

Changes in the drain voltage V_{DS} are also accompanied by changes in the charge on the gate. This is characterized by the capacitance C_{DG}, given by

$$C_{DG} = - \left. \frac{\partial Q_G}{\partial V_{DS}} \right|_{V_{GS}} \tag{7}$$

As with all other capacitances in the FET, this capacitance contains a parasitic component as well as an intrinsic component. The parasitic component in C_{DG} arises from electrostatic coupling between the drain and gate electrodes. The intrinsic component is due to coupling across the depletion region itself.

Changes in V_{DS} also produce changes in I_{DS}, as defined by

$$G_o = \left. \frac{\partial I_{DS}}{\partial V_{DS}} \right|_{V_{GS}} \tag{8}$$

where G_o is the drain-to-source conductance. Assuming that the carrier velocity

v is constant in the control region of the transistor, it can be shown (see Problem 3) that

$$G_o = \frac{C_o v}{L_{\text{eff}}},$$
(9)

where L_{eff} is the effective control region length for the transistor, respectively, and

$$C_o = \frac{\partial Q}{\partial V_{\text{DS}}} \bigg|_{V_{\text{GS}}},$$
(10)

where C_o is the output capacitance.

4.1.2   Equivalent Circuit and High-Speed Figures of Merit

In addition to the circuit elements discussed above, which are intrinsic to the device, two important resistive parasitic elements are inevitable. These are the gate resistance R_G, which represents the resistance of the gate electrode(s), and the source resistance R_S, which represents the series resistance of the source ohmic contact and the semiconductor material between the gate and source electrodes. A simple equivalent-circuit model of a homogeneous FET is shown in Fig. 2. The input conductance G_π is a measure of the leakage of the gate input diode, and is usually negligible for FETs for V_{GS} values below the forward turn-on voltage of the diode. C_{GS} in Fig. 2 is defined as

$$C_{\text{GS}} = C_G - C_{\text{DG}}.$$
(11)

Aside from the parasitic elements R_G, R_S, and G_π, this equivalent circuit is derived from the intrinsic control mechanisms of the transistor. In practice, real

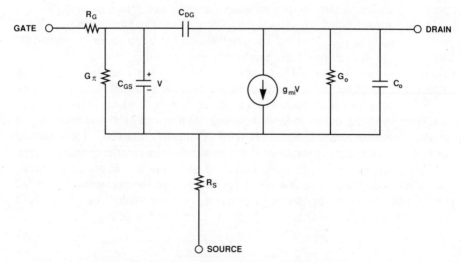

Fig. 2   General equivalent-circuit model of a FET.

FETs must be modeled by a more complicated equivalent circuit having additional parasitics—a more complete equivalent circuit is presented in Section 4.2.4. In addition, the equivalent circuit of Fig. 2 may not properly model the gate-to-drain feedthrough and drain-to-gate feedback seen in real devices (see Problem 4).

The characterization of high-frequency transistors is an important and sometimes difficult task. The difficulty primarily stems from the fact that the performance of a device in a circuit can be as dependent on the circuit as the device. As a result, high-frequency transistors are often characterized by gains, deduced from network-analyzer measurements, that describe their capabilities in a manner that is reasonably independent of the circuit in which they are placed. One example of such a gain is h_{21}, which is the forward current gain with the output short-circuited. The frequency at which this gain extrapolates to one is the unity short-circuit current-gain frequency f_T. As is discussed throughout this chapter, f_T is an important figure of merit for both analog and digital applications. It can be shown (see Problem 4) that the intrinsic f_T of a FET is given by

$$f_T = \frac{g_{mi}}{2\pi(C_{GS} + C_{DG})} . \tag{12}$$

Or, using Eqs. 6 and 11 in Eq. 12,

$$f_T = \frac{1}{2\pi\tau} . \tag{13}$$

Equation 13 shows that f_T is a measure of a quantity that is fundamental to the transistor, the delay time of the current through the control region.

The maximum available gain (MAG) is the power gain of a two-port network when both the input and the output are conjugately impedance-matched[3] and is the highest gain that can be achieved by lossless tuning on the input and output ports. However, simultaneous conjugate matching in a transistor amplifier can only be achieved when the transistor is unconditionally stable, meaning that it will not oscillate regardless of its terminations. If a transistor is potentially unstable, one description of the amplifier's gain capability that is often used is the maximum stable gain (MSG), which is the highest power gain that can be achieved if the input and/or output ports are resistively loaded to achieve stability.[4] Another measure is the unilateral power gain (U), which is the gain that would exist if the output-to-input feedback in the transistor was canceled by another feedback network, with both ports conjugately matched. The unilateral gain is always defined, regardless of the stability of the amplifier, and is invariant under lossless reciprocal embedding.[4,5] This means that U provides a measure of a device's capability that is independent of both the measuring circuit and the manner in which the device is connected in the circuit. Assuming G_π is negligible, an approximate expression for U for the circuit of Fig. 2 is

$$U \approx \frac{1}{4} \frac{(f_T/f)^2}{G_o(R_G + A^2 R_S) + g_{mi} B R_G} , \tag{14}$$

where

$$A \equiv \frac{C_{GS}}{C_{GS} + C_{DG}}$$

and

$$B \equiv \frac{C_{DG}}{C_{GS} + C_{DG}} .$$

Equation 14 includes only first-order corrections for R_G and R_S.

The maximum frequency of oscillation, f_{max}, is the frequency at which U becomes unity. MAG also equals unity at the same frequency.[3] The maximum frequency of oscillation is an important figure of merit because it indicates the highest frequency at which a device can amplify the power of a signal. Setting U of Eq. 14 equal to 1 yields an expression for f_{max} given by

$$f_{max} \approx \frac{f_T}{2[G_o(R_G + R_S) + 2\pi f_T C_{DG} R_G]^{1/2}} . \tag{15}$$

The f_{max} of a device may be either above or below its f_T; the exact relationship of these figures of merit depends on the specific values of the terms that govern them. If $f_{max} > f_T$, the power gain obtainable at frequencies between f_{max} and f_T is due to voltage gain resulting from an output/input impedance ratio greater than unity. The power gain at frequencies below f_T in this case derives from both current gain and voltage gain. For the case where $f_{max} < f_T$, power gain above f_{max} is not possible due to excessive voltage attenuation from input to output.

The various quantities and figures of merit discussed in this section have been derived under the assumption of small-signal, linear operation. There are numerous other figures of merit that are more specific to individual applications, both small-signal and large-signal, and these are discussed in the appropriate sections of the rest of the chapter.

4.2   THE METAL-SEMICONDUCTOR FIELD-EFFECT TRANSISTOR (MESFET)

4.2.1   Background

As discussed in Chapter 3, Si MOSFETs can be fabricated that have very low ($< 10^{10}$ cm^{-2}) trap density at the Si/SiO$_2$ interface. This low trap density allows the gate, which is separated from the channel by tens of nanometers or more of SiO$_2$, to induce and control the inversion layer in the channel. Many III-V compounds have electron mobilities and peak non-equilibrium electron velocities that are superior to those of Si. Attempts to make MOSFETs with these compounds have been stymied over the years by the lack of a stable native oxide and by the fact that many of these compounds such as GaAs display a large interface trap density at the boundary with other materials. One way around these problems was shown in 1966 when the first MESFET was produced by evaporating an Al Schottky-barrier gate directly onto an *n*-type GaAs channel.[6] A generic MESFET structure is shown in Fig. 3. In a GaAs MESFET, the interface traps that are formed at the metal/GaAs boundary merely help create the Schottky-barrier depletion layer in the channel and do not hamper the effective modulation of the depletion layer thickness with gate bias.

The *n*-channel GaAs MESFET presently dominates the world transistor markets for many microwave and high-speed digital applications. It is comparatively easy and inexpensive to make, and offers enhanced performance over Si devices because of the superior electronic properties of GaAs. The Schottky-barrier heights ϕ_{Bn} of metals on *n*-GaAs fall within a narrow range from 0.72 to 0.90 V and are insensitive to the common pre-deposition surface treatments.[7,8] Si MESFETs have been demonstrated, but are more difficult to make than Si MOSFETs due to the great care required to prevent native oxide formation at the metal/Si interface. In addition, Si MESFETs are relatively unpopular because they are generally outperformed by Si MOSFETs and bipolar transistors. MESFETs made of other III-V compounds such as InP and InGaAs have been studied, and most suffer from low Schottky-barrier heights that allow thermionic gate leakage current and/or from interface-trap (barrier height) con-

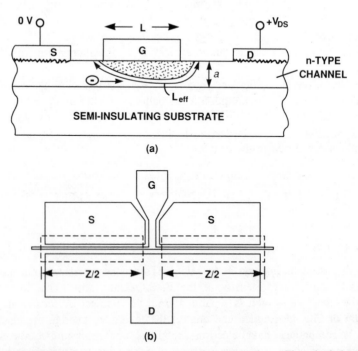

Fig. 3 Basic MESFET structure. (a) Cross section of active region. The stippled area is the gate depletion region. (b) Top view of a small microwave MESFET. The active areas lie within the dashed boundaries.

trol problems. Solutions to these problems are presented by the *p-n* junction FET concept of Section 4.3 and the heterojunction FET technologies of Chapter 5.

4.2.2 Basic Structure

A cross section through the active area of a basic *n*-channel MESFET is shown in Fig. 3a. The drain is normally biased positive with respect to the source; as a result, the gate depletion region is asymmetrically shifted toward the drain. The dimensions *a* and *L* are known as the channel thickness and the physical gate length, respectively. From the discussion in Section 4.1.1, the effective gate length L_{eff} is defined as the total length of the curvilinear depletion region boundary in the channel of the device. As discussed in the next section, the achievement of maximum f_T demands that the device be designed such that L_{eff} is prevented from being significantly longer than the physical gate length *L*.

Figure 3b shows the overall top view of a small discrete microwave MESFET. The source, gate, and drain pads all extend outside the active areas to provide enough pad area for wirebonding into a circuit. The active areas can be defined by either mesa etching or ion implantation—the general fabrication of MESFETs is discussed further in Section 4.2.7. The dimension Z/2 for each respective active area is commonly referred to as the gate-finger width. The total gate width for the device of Fig. 3b is therefore Z. An important rule of thumb for microwave-FET design is that the distance from the gate feed point to the end of each gate finger (Z/2) be less than $\lambda/16$, where λ is the wavelength along the gate at the frequency of operation.[9,10] This ensures that the ends of the device will not be operating significantly out of phase with the portion near the feed point. Considering the gate/channel structure as a microstrip-like transmission line for the purpose of estimating λ, upper limits on Z/2 range from about 5 mm at 1 GHz to 50 μm at 100 GHz.

For high-power applications, many unit cells, each having a gate-finger width less than $\lambda/16$, are placed close together and fed in parallel to produce a power-FET structure. None of the boundaries of this configuration may exceed the $\lambda/16$ criterion, however. In addition to these size constraints, there are also impedance and thermal constraints on power-FET design. With common microwave techniques, it is difficult to match typical transmission line impedances to FET input impedances that are below a few ohms. Because power-FET impedance varies inversely with the number of cells connected in parallel, the impedance-matching restriction may limit the number of cells that can be used in a given power FET. The spacing between cells in a power FET may also have to be restricted to some minimum value to avoid excessive temperatures in each cell, thereby restricting the total number of cells as well. These size, impedance, and thermal restrictions place a limit on the amount of output power that may be obtained from one device at a given frequency; realization of higher power requires that the outputs of several power FETs be combined using circuit techniques.

MESFETs used in digital circuits are typically much smaller than those used in microwave applications, having gate widths of the order of 10 μm to minimize the power consumed per logic gate. The digital devices do not have the large bonding pads but are connected to each other by relatively narrow metal interconnects.

4.2.3   DC Characteristics and Basic Design

The typical *I-V* characteristics of an *n*-channel depletion mode (normally on) MESFET are shown in Fig. 4. Drain current flows for gate-source voltages V_{GS} that are less negative than a threshold voltage V_T. For a given channel of uniform doping N_D and dielectric permittivity ϵ_s, the Schottky-barrier depletion equation in Ref. 7 may be used to define a pinch-off voltage V_P, given by

$$V_P = \frac{qN_D a^2}{2\epsilon_s} \, , \tag{16}$$

which represents the net potential required to deplete a channel of thickness a under the gate. The threshold voltage is then defined as

$$V_T = V_{bi} - V_P \, , \tag{17}$$

where V_{bi} is the Schottky-barrier built-in voltage. Unlike depletion mode MESFETs, enhancement mode (normally off) MESFETs are designed so that the channel is fully pinched off at zero gate bias. Enhancement mode, *n*-

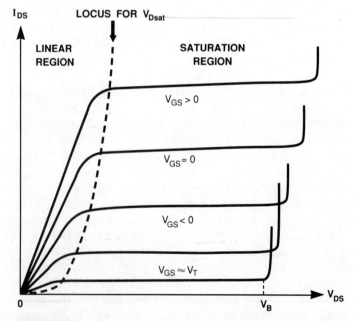

Fig. 4 Typical *I-V* characteristics of an *n*-channel depletion mode MESFET.

channel MESFETs have characteristics that are similar in shape to those of Fig. 4, but have threshold voltages that are usually slightly above 0 V. The upper limit for V_{GS} for both types of devices is reached when the gate begins to draw significant current, which for n-GaAs MESFETs is around $+0.7$ V. The channels of depletion mode MESFETs are thicker and/or doped more heavily than those of enhancement mode MESFETs, therefore providing more drain current at maximum V_{GS}. The voltage V_B in Fig. 4 defines the point at which avalanche breakdown occurs between the gate and drain and represents an upper operating limit for drain-source and drain-gate voltage. The same general discussion applies to p-channel FETs, given appropriate changes of sign for voltages and currents. The circuit symbols for MESFETs and JFETs are the same and are given in Fig. 5.

The dashed parabola in Fig. 4 is the locus of saturation voltage. The saturation voltage V_{Dsat} is the drain-source voltage at which the drain current saturates, for a given V_{GS}. The region to the left of this parabola is known as the linear region because, for a given gate voltage, the drain current is linear with V_{DS} until saturation effects become dominant as V_{Dsat} is approached. The region between the parabola and avalanche breakdown is known as the saturation region. The saturated drain current for $V_{GS} = 0$ is often referred to as I_{DSS}.

A simple but accurate model for n-GaAs MESFET operation in the saturation region is derived with the aid of Fig. 6. This model, known as the saturated-velocity model,[11,12] is valid for devices with gate lengths between roughly 0.5 and 2 μm. For gate lengths below 0.5 μm, the electron transport may be in the

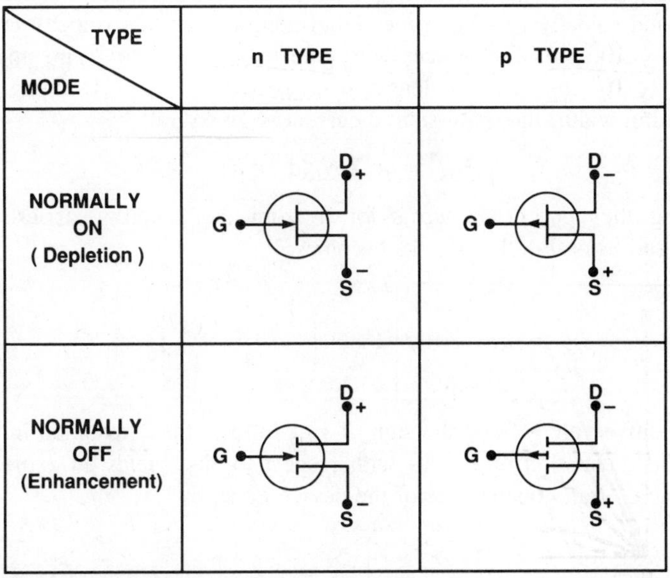

TYPE / MODE	n TYPE	p TYPE
NORMALLY ON (Depletion)		
NORMALLY OFF (Enhancement)		

Fig. 5 Symbols for n-channel and p-channel depletion mode and enhancement mode MESFETs and JFETs. (After Ref. 11)

near-ballistic regime to be discussed shortly. Devices having gate lengths much above 2 μm may not exhibit electron velocity saturation in the channel, but these devices are relatively slow and are outside the scope of this book. In the saturation region, the electric field in the channel under the gate depletion region is significantly above the critical field for the onset of intervalley transfer, which is about 3.5 kV/cm for GaAs. The electrons traveling from source to drain therefore scatter from the high-group-velocity, lower-energy Γ valley to the low-group-velocity, higher-energy L valleys. The average electron velocity in the channel effectively saturates and is approximately constant for electric fields above about 10 kV/cm. This electron velocity saturation produces the saturation of drain current noted in the transistor characteristics. Drain-current saturation occurs by this same mechanism in FETs made of other materials such as InGaAs, which, like GaAs, have high low-field electron mobility. FETs made of materials such as Si, SiC, and diamond, which have low low-field mobility, also demonstrate drain-current saturation, but it derives mainly from the geometric constriction of the channel as described by the model in Section 6.2 of Ref. 11.

The velocity saturation in the channel is responsible for the formation of the space-charge dipole domain shown in Fig. 6a. Since the electron velocity is saturated, current continuity requires an accumulation of electrons in the most constricted portion of the channel and a depletion of electrons just beyond the constriction toward the drain. The formation of the domain is discussed in greater detail in Ref. 11 and its respective references.

Under typical operating conditions in the saturation region, it is an accurate approximation to assume that electrons travel under the gate at a constant effective saturated velocity v_s. Experiments have shown that this velocity is typically about 1.2×10^7 cm/s for devices with gate lengths of 1 to 2 μm, and rises to about 1.5×10^7 cm/s as gate length decreases to 0.5 μm. Defining h as the gate depletion width, the drain-source current is given by[11,12]

$$I_{DS} = q v_s Z N_D (a - h) . \tag{18}$$

Substituting the appropriate terms for h from the Schottky-barrier depletion equation, and using Eq. 16, Eq. 18 becomes

$$I_{DS} = q a v_s Z N_D \left[1 - \left(\frac{V_{bi} - V_{GS}}{V_P} \right)^{1/2} \right] \tag{19}$$

V_{bi} is a positive voltage, and the sign of V_{GS} follows the convention indicated in Fig. 4. Differentiation of Eq. 19 with respect to V_{GS} yields an expression for the intrinsic dc transconductance of the device given by

$$g_{mi} = \frac{v_s \epsilon_s Z}{a} \left(\frac{V_P}{V_{bi} - V_{GS}} \right)^{1/2} = v_s Z \left[\frac{q N_D \epsilon_s}{2(V_{bi} - V_{GS})} \right]^{1/2} \tag{20}$$

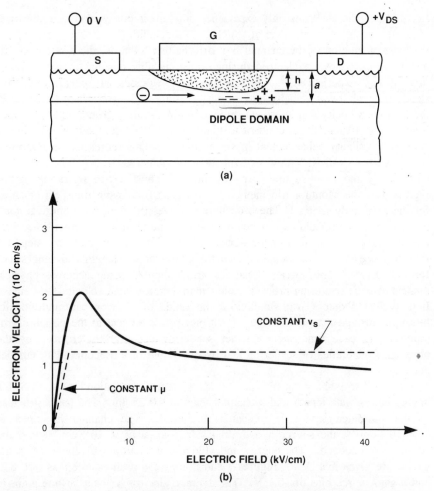

Fig. 6 (a) A dipole domain forms in the channel due to velocity saturation. The gate depletion width h is constant over the portion of the channel in which the carrier velocity is saturated. (b) Solid line is the steady-state velocity-field curve of GaAs, valid for samples of length greater than roughly 0.5 μm. Intervalley transfer occurs at fields above 3.5 kV/cm. The dashed line indicates a common piecewise approximation.

As discussed in Section 4.2.4, the extrinsic transconductance seen at the terminals of a real device is limited by parasitic source resistance, and is always less than that given by Eq. 20. More sophisticated analytic[12–14] and heuristic[15–17] models have been developed that give a more detailed treatment of FET behavior.

For III-V FETs having gate lengths significantly less than 0.5 μm, the electron transit through the high-field region under the gate may be so fast that significant intervalley transfer may not occur before the electrons reach the drain.

In this case, the electrons only experience phonon, impurity, and carrier-carrier scattering in the channel, none of which is catastrophic. Under these conditions, the electron motion in the channel is nearly ballistic, that is, the average kinetic energy acquired by an electron during transit is almost equal to the potential drop it has experienced.[18] Average electron velocities of several times the steady-state saturated velocity may occur. In some cases (see Chapters 2 and 7) the average velocity can approach the maximum Γ-valley group velocity, which for GaAs is 10^8 cm/s. In contrast to many of the III-V compounds, Si exhibits very little velocity enhancement in short devices. This represents an important advantage for small III-V devices over their Si counterparts.

Probably the most accurate way to model a short device operating in this regime uses the Monte Carlo method to solve the Boltzmann transport equation for the carrier dynamics.[19] The procedure involves the iterative numerical computation of potential and carrier behavior at each point of a two-dimensional grid model of the cross section of the device. The point spacings and time steps for such a process are much smaller than the characteristic length and time scales for the device. The carrier dynamics are computed using accurate physical models of each scattering process, rather than a steady-state velocity-field curve. In a typical Monte Carlo simulation, thousands of electrons are individually followed through a model device. Each electron is subject to the possibility of scattering at each time step; whether scattering occurs is determined by the physical probabilities of scattering and by the output of a pseudorandom number generator.

Figures 7 to 9 show the results of a Monte Carlo simulation[20] of a MESFET having both a gate length and a channel length of 0.25 μm. The basic structure of the simulated device is indicated in Fig. 7. The channel is doped at 7×10^{16} cm^{-3}, the source and drain n^+ regions at 7×10^{17} cm^{-3}, and the substrate is undoped. The energy and velocity distributions of electrons in the device are given in Fig. 8. Figures 7 and 8 are for a drain-source bias of 0.8 V and a gate-source bias of –0.2 V. The average electron velocity at one point in the channel in Fig. 8 reaches 7×10^7 cm/s, which is about a factor of six higher than the saturated velocity typically assumed for longer devices. Since f_T scales inversely with the transit time for carriers through the control region, such short devices operating in the near-ballistic regime are capable of f_T values far in excess of 100 GHz. Figure 9 shows the computed I-V characteristics of the simulated device assuming a gate width of 20 μm. The transconductance per unit gate width is 643 mS/mm at $V_{DS} = 0.8$ V and $V_{GS} = -0.05$ to -0.2 V.

Several features seen in the curves of Fig. 9 arise from what are known as short-channel effects. First, the drain-current saturation for the higher (least negative) gate biases is weaker than is typically seen in devices having longer channels. This is due to the fact that in short devices many of the electrons whose energies rise above the threshold for transfer to the L valleys may not have time to transfer before reaching the drain. In such cases the average electron velocity in the channel is a stronger function of the drain-source electric field than in longer channels where the electron population in the L valleys

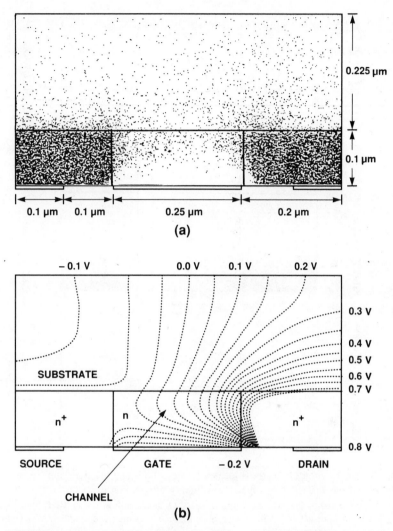

Fig. 7 MESFET structure for Monte Carlo simulation. (a) Spatial distribution of electrons frozen at one instant in time. Note the significant penetration of electrons into the substrate. (b) Distribution of equipotential lines. Note the smooth variation of potential along the conducting channel, and the absence of dipole domain formation. (After Ref. 20)

reaches equilibrium and the average carrier velocity saturates for electric fields above 10 kV/cm. Second, no drain-current saturation is seen at all for the more negative gate voltages near pinchoff. This is due to the fact that electrons can flow from the source to the drain through the substrate. This parasitic leakage path is not well controlled by the gate, and is exacerbated when the source and drain are placed close together. Substrate leakage also causes a negative shift in the threshold voltage V_T. The leakage can be suppressed by placing a

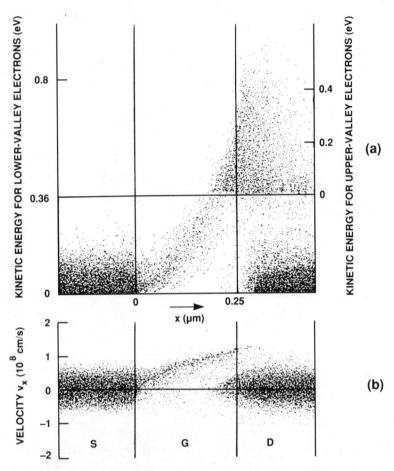

Fig. 8   Distribution of electrons located between 0.09 and 0.1 μm from the plane of the contacts of the MESFET of Fig. 7. (a) Energy distribution for both the Γ valley and the L valleys. Most of the electrons that transfer to the L valleys do so after entering the n^+ drain region, and a few of these scatter or diffuse back into the channel. (b) Distribution of lateral component of velocity. Note the high-velocity stream of electrons that constitutes the drain current. The average lateral electron velocity reaches 7×10^7 cm/s at 0.16 μm from the source. (After Ref. 20)

complementary-doped or heterostructure buffer layer between the active layer and the substrate. The leakage effects are also minimized by some of the following design practices.

A few simple design rules are followed almost universally in successful MESFET design. In general, the channel material selected should allow the highest average carrier velocity under the expected conditions of operation, and the physical gate length L should be the smallest practicable given the lithography techniques available. These conditions assure that f_T, f_{max}, and the overall

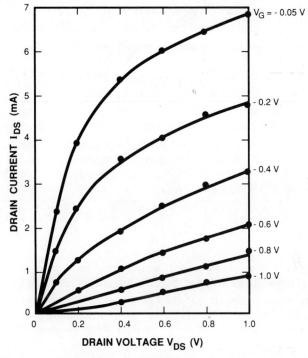

Fig. 9   Simulated characteristics of the MESFET of Fig. 7.   (After Ref. 20)

performance of the device are maximized.   The selection of material and gate length may also be subject to cost, availability, and/or manufacturing constraints—the performance may have to be traded off against these additional constraints in many cases.   Once L has been determined, the channel depth a is determined using the rule that the ratio L/a should be around 4 to 5 for an optimal trade-off between f_T, parasitics, and short-channel effects.[11,12] Too large an L/a ratio reduces current capacity unnecessarily, and too small an L/a ratio results in high G_o and enhanced short-channel effects.   In addition, for small L/a ratios, the effective gate length L_{eff} may be significantly larger than L over much of the gate bias range, thereby unnecessarily reducing the f_T in this bias range.

Once L and a range for a have been chosen, the requirements of the circuit application can be used to determine the channel doping level N_D.   For digital devices, V_T is matched to the specific logic level of the circuit by using Eqs. 16 and 17 to compute an appropriate value of N_D.   For analog applications, designing the right values of breakdown voltage, drain current, output I-V swing, transconductance, and/or others may be more important.   Many of the relevant quantities can be worked out using Eqs. 16 to 20.   For channel doping-thickness products $N_D a$ less than 2.3×10^{12} cm^{-2}, the drain-gate breakdown voltage

V_{BDG} for GaAs FETs is given approximately by[11,12,21]

$$V_{BDG} \approx \frac{5 \times 10^{13}}{N_D a} . \tag{21}$$

This relation takes into account the distribution of the electric field in FET channels that meet the doping-thickness criterion, and assumes that the avalanching occurs in the gate depletion region on the drain side. For $N_D a > 2.6 \times 10^{12}$ cm^{-2}, the drain-gate breakdown voltage is equal to that for a Schottky-barrier diode on bulk material. Figure 10 shows the breakdown voltage as a function of impurity concentration for one-sided abrupt p-n junctions and Schottky-barrier diodes on selected bulk materials. The breakdown voltage for this case is also given by[22]

$$V_{BDG} \approx 60 \left[\frac{E_g}{1.1} \right]^{3/2} \left[\frac{10^{16}}{N_D} \right]^{3/4} , \tag{22}$$

which is a universal relation for semiconductors where E_g is the room-temperature bandgap in eV and N_D is in units of cm^{-3}. Due to the two-dimensional nature of the electric field in the regime of Eq. 21, breakdown voltages for that case can be as much as three times higher than those for the regime of Fig. 10 and Eq. 22.[11,12,21] Power FET designers often use this to advantage, as discussed in Section 4.2.5. The relation between drain-gate breakdown voltage and the voltage V_B shown in Fig. 4 is

$$V_B = V_{BDG} + V_T . \tag{23}$$

More detailed discussions of FET design are given in the following sections. Many of the equations used both here and in the literature are approximations, and serve only as guidelines for FET performance. More accurate calculations of FET behavior can be obtained with two-dimensional numerical simulations. In general, the realization of high-performance FETs is an iterative cut-and-try process. The fine tuning of critical performance parameters may require a few rounds of fabrication in which relevant design details are carefully adjusted.

4.2.4   Small-Signal Operation

Figure 11a shows a detailed small-signal equivalent circuit for a MESFET with most of its parasitics. Figure 11b depicts a cross section of the MESFET indicating the origin of each of the elements of the equivalent circuit. Under typical bias conditions, C_{GS} is several times larger than the other capacitances in the circuit. R_S and R_D each have a component due to ohmic contact resistance and a component due to resistance in the channel itself. R_i represents the resistance of the low-field region under the source end of the gate where the velocity is not saturated, and contributes along with R_S and R_G to the input RC charging time. G_o is a function of the velocity saturation and dipole-domain formation, and was defined by Eq. 8. C_{DC} is the effective capacitance of the dipole domain. It has

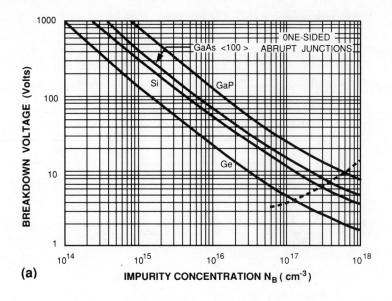

(a)

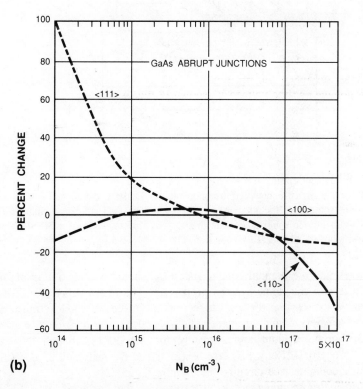

(b)

Fig. 10 (a) Avalanche breakdown voltage versus impurity density for one-sided abrupt *p-n* junctions and Schottky-barrier diodes on Ge, Si, <100>-oriented GaAs, and GaP. The dashed line indicates the maximum doping beyond which tunneling dominates the voltage breakdown characteristics. (b) Crystal-orientation dependence of avalanche breakdown voltage for GaAs normalized to the curve in (a) above. (After Ref. 22)

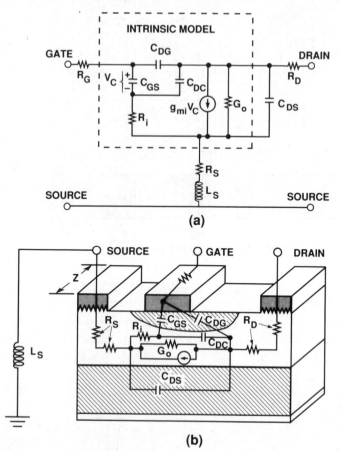

Fig. 11 (a) Equivalent circuit of a MESFET. The relation $g_{mi} = g_{mi}(dc) e^{-j\omega\tau}$ is commonly used for the high-frequency transconductance. Leakage conductances G_π, G_{DG}, and/or G_{DC} are sometimes included in parallel with their respective capacitances C_{GS}, C_{DG}, and C_{DC} to provide a more accurate model. (b) Cross section of a MESFET indicating the origins of the elements. (After Ref. 11)

been suggested that the domain conductance G_{DC} should be negative in order to model the effect of the domain more properly.[23] The phase delay $e^{-j\omega\tau}$ in the high-frequency expression for g_{mi} is commonly used to model transit-time effects of the carriers under the gate. The general expressions for f_T and f_{max} given in Section 4.1.2 hold approximately for MESFETs, though the exact small-signal properties of any FET are best obtained by analyzing its equivalent circuit using a computerized circuit-simulation program.

The extrinsic transconductance g_m measured at the terminals of a FET is reduced from the intrinsic value by the effect of R_S, and is given by

$$g_m = \frac{g_{mi}}{1 + g_{mi}R_S}. \tag{24}$$

In addition, in real FETs there is always some inductance in the connection between the source contact pad and the true ground of the circuit. This source inductance L_S can reduce the high-frequency stability of the device and cause a faster rolloff of the power gain with increasing frequency. Good FET design seeks to minimize all parasitics as much as possible; the minimization of R_S and L_S is especially important for most applications.

In the practical design and scaling of FETs, the small-signal power gain A_P of a stable FET well below f_{max} is often assumed to roll off as $1/(f^2)$ or –6 dB/octave. From the discussion of Section 4.1, A_P at a given frequency obeys the rough rule of thumb that

$$A_P \sim (f_{max})^2 \sim (f_T)^2 \sim \left[\frac{1}{L}\right]^2, \qquad (25)$$

where the " $\sim$ " symbol denotes "goes as" or "varies as." From this argument emerges another simple scaling rule that keeps A_P roughly constant as a FET design is scaled in frequency:

$$L \sim \frac{1}{f}. \qquad (26)$$

For a given value of A_P, FETs that operate at high frequencies require proportionately shorter gate lengths than their low-frequency counterparts.

The amount of noise that an amplifier adds to the amplified signal is often just as important as its gain. The noise spectrum for a typical GaAs MESFET is shown in Fig. 12. The noise properties of FETs used for low-noise, small-signal amplification have been described by an empirical model.[12,24] The broad-

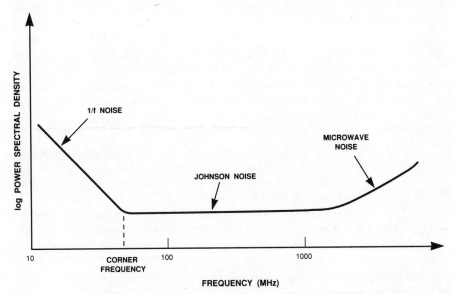

Fig. 12 Typical noise spectrum for a GaAs MESFET.

spectrum noise generated within FETs at microwave frequencies has two basic sources. The first is thermal or Johnson noise in parasitic resistances such as R_S and R_G in Fig. 11. The Johnson noise that occurs in the input circuit is amplified by the FET and appears at the output. The second source is thermally induced statistical fluctuation in the local carrier density in the channel under the gate. This channel noise appears at the output as well. The noise figure (NF) of a device is defined as the signal-to-noise ratio at the device input divided by the signal-to-noise ratio at the device output.[25] This definition assumes the input signal is associated with a noise source at 290 K. Expressions for device noise figure in dB in the microwave range are given by[12,24]

$$
\mathrm{NF} = 10 \log \left[1 + \frac{T_n}{290} \right] \approx 10 \log \left[1 + 2\pi f K_f C_{\mathrm{GS}} \left(\frac{R_G + R_S}{g_{\mathrm{mi}}} \right)^{1/2} \right]. \quad (27)
$$

The parameter T_n is known as the equivalent noise temperature and is in units of degrees Kelvin. The term K_f is an empirical parameter that can be used to fit experimental data to Eq. 27. The value of K_f for a given device is a function of factors such as geometry and channel material quality. The quantity inside the square brackets in Eq. 27 is sometimes referred to as $F_{\min}$, and the quantity $2\pi f K_f C_{\mathrm{GS}}[(R_G + R_S)/g_{\mathrm{mi}}]^{1/2}$ is known as the noise measure. Note that the noise measure can also be expressed as $(f/f_T)K_f[(R_G + R_S)g_{\mathrm{mi}}]^{1/2}$.

An associated gain G_a is often quoted together with noise figure and is the gain of the device under the conditions in which the noise figure was measured. This gain is important because the system noise figure of a cascaded series of amplifiers is dominated by the first amplifier if its gain is sufficient. The system noise figure is optimized (reduced) by using a first-stage amplifier having high gain and a minimal noise figure.[25]

Examination of the above expressions for noise measure reveals that the noise figure is minimized by reducing the parasitic resistances and increasing the f_T of the device. Reducing the parasitic resistances reduces their Johnson noise. Increasing the f_T improves the gain of the device, thereby allowing it to amplify the desired signal to a higher level above the channel noise at the output. Consistent with these criteria, low-noise devices have high values of f_T and small gate-finger widths for minimal R_G.

At low frequencies the noise in FETs is dominated by a type of noise whose power spectral density varies inversely with frequency. The magnitude of this so-called $1/f$ noise drops below the thermal noise floor at a corner frequency that is typically between 100 kHz and 100 MHz. In FETs the $1/f$ noise is due primarily to carriers hopping in and out of deep levels in the substrate, buffer layer, and/or active layer and in and out of surface traps on channel surfaces. While not a problem in amplifiers that operate above the corner frequency, $1/f$ noise is a problem in oscillators and mixers because it is up-converted by the intrinsic non-linearities of these devices and appears as a skirt about the carrier or intermediate frequency (IF). $1/f$ noise is minimized by controlling the deep-level density and by passivation of active-layer surfaces.

4.2.5 Large-Signal Analog Operation

The most common large-signal analog application for MESFETs is in power amplifiers. Power-amplifier design usually seeks to maximize power output or overall efficiency or both. The power output and efficiency of a FET amplifier are illustrated with the aid of Fig. 13. The maximum power output is obtained in Class A operation, and is given by

$$P_m = \frac{I_m(V_B - V_{Dsat})}{8}.$$ (28)

Here it has been assumed that the drain current swings from its maximum value I_m to zero. The drain efficiency η_D is defined as the ratio of rf power output to dc power input and is given by[12]

$$\eta_D = \frac{\eta_{max}(V_B - V_{Dsat})}{V_B + V_{Dsat}}.$$ (29)

The factor η_{max} is a function of the dependence of g_m on V_{GS} and of the class of operation.[26] Assuming Class A operation and a constant $g_m(V_{GS})$, η_{max} is 50%. The second assumption may be fairly realistic for real FETs, because the source resistance has a linearizing effect on g_m through Eq. 24. The power-added effi-

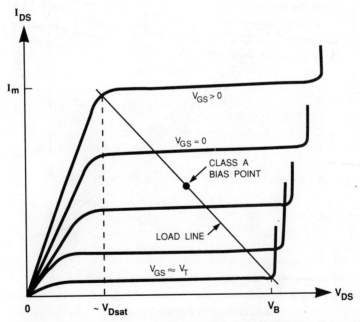

Fig. 13 *I-V* characteristics of an *n*-channel depletion mode MESFET showing the load line for maximum power output. I_m is the maximum value of drain current that is reached before the gate Schottky-barrier diode conducts significantly.

ciency η_{PA} is defined as

$$\eta_{PA} = \frac{P_{out}(rf) - P_{in}(rf)}{P_{in}(dc)} = \eta_D \left[1 - \frac{1}{G} \right] , \tag{30}$$

where G is the large-signal power gain. The power-added efficiency is concerned with the net boost given to a signal that passes through an amplifying stage, and is an important parameter in the design of systems where dc power is limited or overall heat dissipation is of concern.

Examination of Eqs. 28 to 30 reveals that the power output and the efficiency of a FET amplifier are improved by reducing V_{Dsat}. Reducing R_S and R_D of Fig. 11 will minimize the contribution of extrinsic parasitics to V_{Dsat}. The use of semiconductors having a high low-field mobility will reduce the intrinsic component of V_{Dsat}.

The scaling rules for power and efficiency in FETs are more complicated than it seems from Eqs. 28 to 30. Inspection of these equations implies that a

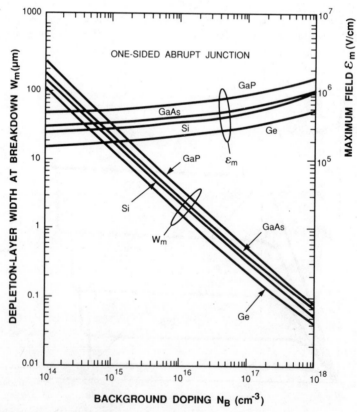

Fig. 14 Depletion layer width and maximum electric field at breakdown for selected semiconductors. (After Ref. 22)

straightforward way to raise the power and efficiency would be to increase V_B. There are trade-offs, however, involving V_B, I_m, f_T, and G.

The first of these trade-offs can be derived in the following way. From Eq. 23,

$$V_B = V_{BDG} + V_T \sim V_{BDG} \qquad (31)$$

to first order, since V_{BDG} is usually at least a few times larger than V_T. From avalanche breakdown theory,[22] V_{BDG} is proportional to the maximum electric field $\mathscr{E}_m$ at breakdown. $\mathscr{E}_m$ is a fundamental material property that is related to the bandgap of the semiconductor, and is shown in Fig. 14 for a few selected semiconductors. The drain-source electric field $\mathscr{E}_B$ in the channel of a FET at drain-gate breakdown is proportional to V_B/L_{eff}. $\mathscr{E}_B$ is also proportional to $\mathscr{E}_m$, since V_B and V_{BDG} are related. The product of f_T and V_B can then be written as[27]

$$V_B f_T \approx V_B \left(\frac{v_a}{2\pi L_{\text{eff}}} \right) \sim \frac{\mathscr{E}_B v_a}{2\pi} \sim \frac{\mathscr{E}_m v_a}{2\pi}, \qquad (32)$$

where v_a is the average carrier velocity under the gate. This relation is valid for scaling a FET of a given geometry, assuming V_T remains fixed and that the $N_D a$ product does not shift enough to alter the breakdown regime from that of Eq. 21 to that of Eq. 22, or vice versa. Equation 32 represents a trade-off between f_T and V_B that is governed only by the semiconductor-related quantities $\mathscr{E}_m$ and v_a.

A trade-off also exists between V_B and I_m. Assuming that $V_B \gg V_{D\text{sat}}$ and that $V_{BDG} \gg |V_T|$, the maximum rf output power that can be obtained from a FET is

$$P_{\max} = \frac{I_m V_B}{8} \approx \frac{I_m V_{BDG}}{8}. \qquad (33)$$

Substituting from Eqs. 19 and 21,

$$P_{\max} = \frac{I_m V_B}{8} \approx \frac{(q v_a Z)(5 \times 10^{13})}{8} = K_1 \mathscr{E}_m v_a Z, \qquad (34)$$

where K_1 is a constant and the dependence of V_{BDG} on $\mathscr{E}_m$ has been indicated explicitly. To first order, the maximum power and the $I_m V_B$ product vary only with the material-dependent quantities $\mathscr{E}_m$ and v_a and with the FET gate width, Z.

The quantity $K_1 \mathscr{E}_m v_a$ is defined as the maximum power density $P_{d,\max}$, and is commonly expressed in units of W/mm of gate width. $P_{d,\max}$ is constant with frequency below the f_T of a FET[28] and decreases as $1/f$ at frequencies above f_T.[27] Here it is assumed that the small-signal concept of f_T is approximately valid for the large-signal case as well. The $1/f$ dependence above f_T results from the fact that the average carrier cannot transit the high-field region of a device in a time that is less than $1/2\pi f_T$. Below f_T, the carrier transits the full high-field region and acquires an energy from the field that is equal to the total potential drop across the device. Above f_T, the carrier only transits part way in one cycle, and only acquires part of the total energy. A related mechanism produces a $1/f$ dependence in the power-frequency performance of IMPATT diodes (see Chapter 9).

The computed value of $P_{d,\max}$ for GaAs power MESFETs below f_T is very close to that observed experimentally. Substituting a saturated drift velocity v_s of 1.2×10^7 cm/s into Eq. 34 yields a value of 1.2 W/mm for $P_{d,\max}$. Most reported laboratory results fall just below this value, with the highest reported being 1.4 W/mm.[29]

An important rule that generally must be followed in scaling power FETs with frequency comes from the fact that circuit losses and bandwidth criteria set a practical lower limit to the impedance that can be matched with microwave circuitry. Since C_{GS} dominates the input impedance in well-designed FETs, this places an upper limit on the value of C_{GS} that can be used at a specific frequency. This restriction on C_{GS} translates into an upper limit on gate width Z and the maximum power $P_{\max}$ that can be obtained at a given frequency. A simple rule that is valid below f_T is therefore given by

$$P_{\max} \sim Z \sim \frac{1}{f} . \tag{35}$$

This relation assumes that V_T is kept constant, and that L and a scale as $1/f$ so that the input capacitance per unit gate width is constant. Note that Eq. 35 enforces the same $1/f$ dependence as the $\lambda/16$ rule discussed in Section 4.2.2. At frequencies above f_T, $P_{\max}$ varies as $1/(f^2)$ since $P_{d,\max}$ also has a $1/f$ dependence in this range. Device or circuit parasitics can cause the frequency dependence of $P_{\max}$ to be different than these theoretical considerations predict. The maximum power output of GaAs MESFETs is discussed further in Section 4.2.7.

In the design and scaling of power FETs, the large-signal gain G is generally assumed to behave the same way as the small-signal power gain. As a result, the gate length L for power FETs is usually selected and scaled with frequency in the manner discussed in Sections 4.2.3 and 4.2.4. The large-signal gain of a FET in a power amplifier is usually much less than MAG, MSG, or U, however, because the load line and class of operation are chosen to maximize the output power and/or efficiency at the expense of G. The single-stage, large-signal gain for power amplifiers usually must be at least 4 dB but rarely exceeds 10 dB. In many cases, constraints imposed on I_m and V_B in scaling a power FET design to a different frequency will change the optimal load resistance for maximum power output, possibly reducing G to an unacceptable value. This can often be avoided if V_T is allowed to vary as power FETs are scaled in frequency. (Problems 6 and 7 illustrate the application of scaling rules to power FETs.)

Successful power FET design requires a careful analysis of the circuit requirements and their interrelationships with the trade-offs and scaling rules above. Various permutations of the scaling principles can be derived for different sets of conditions, for example, a flexible V_T versus a fixed one. Many power FETs are designed with $N_D a < 2.3 \times 10^{12}$ cm^{-2} so that the enhanced gate-drain breakdown voltage given by Eq. 21 is obtained. For GaAs power FETs, this gives $I_m < 400$ mA/mm and $I_{DSS} \approx 350$ mA/mm.[21,30] Power FETs that must operate above about 15 GHz may not be able to take advantage of the enhanced V_B regime, however, since the f_T and gain may be degraded too severely by the $V_B f_T$ trade-off.[29] GaAs power FETs almost always use a recessed gate structure such as that discussed in Section 4.2.7. Heat sinking and thermal management are important considerations for power FET design—general discussions of these topics are given in Refs. 12, 28, and 31. Power FETs that operate near or above their f_T may draw a significant amount of rf gate current, necessitating the use of a low-resistance gate-finger geometry to minimize the voltage drop and power dissipation in each finger.

Equations 32 and 34 are useful not only for scaling FET designs but also for evaluating the suitability of different materials as power FET channels. $P_{d,\,max}$ and the $V_B f_T$ product are governed by the fundamental, material-related quantities $\mathscr{E}_m$ and v_a. For a given gate length, the materials that have the highest promise for power operation have the highest $\mathscr{E}_m v_a$ product. Recall that v_a can be a strong function of the gate length, especially for some III-V compounds that exhibit high non-equilibrium carrier velocities for $L < 0.5\ \mu$m and lower velocities in longer devices. $\mathscr{E}_m$ is a very weak function of doping, and is usually assumed to be a constant for a given material in such comparisons. Consideration of the thermal conductivity of channel and substrate materials may also be necessary in overall material comparisons. A material with a higher thermal conductivity can operate at higher power levels for a given channel temperature than one of lower conductivity. These topics are discussed further in Section 4.5.

The non-linearities present in FETs are undesirable for some types of large-signal analog operation. As one example, the non-linearity of $g_{mi}(V_{GS})$ causes both harmonic generation and intermodulation distortion. It has been shown that a linear $g_{mi}(V_{GS})$ characteristic can be obtained with vertical doping profiles in which the channel doping density $N_D a$ is concentrated in a plane at the bottom of the channel rather than spread uniformly throughout.[11,12,32] As another example of non-linearity, the intrinsic capacitances and resistances of a FET vary substantially with bias, a fact that can affect the power output.[30] These non-linear properties prevent a linear RLC load from conjugately matching the FET output impedance at most points along the load line. The drain current and drain-source voltage thus deviate from their nominal 180° phase relationship in large-signal operation, converting the load line into an ellipse.

4.2.6 Digital Operation

The universal goal of device and circuit design for digital operation is to obtain the highest possible switching speed for the lowest overall static and dynamic power dissipation. To this end, this section closely couples the discussion of device design with an understanding of the requirements of digital circuits, and follows much of the material in Refs. 33 and 34.

As with the analog FET development discussed above, MESFET- and JFET-based logic circuits have primarily employed n-channel GaAs devices. A truly complementary logic technology with practically no static power dissipation such as Si CMOS is desirable for GaAs, but its development has been hindered by the poor hole transport and the lack of a MOS technology in GaAs. GaAs devices have a number of important advantages for high-speed, low-power logic, however, as will become apparent both in this chapter and subsequent chapters.

The simplest inverters that can be constructed with MESFETs or JFETs are shown in Fig. 15. The inverter is the basic building block of both memory cells and logic circuits. The inverter in Fig. 15a uses a simple resistor load, whereas the one in Fig. 15b uses an active load composed of a depletion mode FET with its gate connected to its source. The characteristics of these inverters for logic operation are shown in Fig. 15c. The resistive load gives the classic resistive load line, whereas the load characteristic for the active load is a mirror image of the drain-source characteristic of a depletion mode FET for $V_{GS} = 0$. An advantage of the active load over the resistive one is that the active load characteristic is saturated over much of the logic swing, thereby providing a rapid output-voltage excursion over the saturated portion for a small input swing. FETs used for active loads are often small to minimize the load current and power dissipation of the stage and to minimize the capacitive loading on the switching transistor. In many cases they will consume less wafer area than a suitable resistive load. Another type of load, called a current limiter, is a hybrid between the resistive and active loads and consists of a FET channel with ohmic contacts at both ends but with no gate. With proper design, carrier velocity in the channel will saturate just as in the active load, but the current limiter has no gate to contribute a loading capacitance to the switching transistor.

The simplest logic family that can be made up of the inverters of Fig. 15 and their logic-circuit permutations is known as direct-coupled FET logic (DCFL). The logic family composed of elements like those of Fig. 15b is often referred to as enhancement/depletion DCFL, or E/D DCFL. In DCFL, the output of a logic stage, which is the drain of the switching FET, is connected directly to the gate(s) of the switching FET(s) of the following stage(s). For this case, the logic operation as described in Fig. 15c is valid only for $V_{DD} < V_{GM}$, where V_{GM} is the forward turn-on voltage of the Schottky-barrier or p-n junction gate diode of the switching transistor in the following stage. If $V_{DD} > V_{GM}$ in DCFL, the high logic level is clamped at $V_{OH} = V_{GM}$ by the turn-on of the gate diode in the following stage. As will be shown shortly, this clamping action improves the effective rise time of the inverter output-voltage waveform. A DCFL stage run in this overdrive mode will dissipate more power in the high logic state than for the case of Fig. 15c, however, since the load must supply the gate current for the following stage while sustaining a voltage of $V_{DD} - V_{GM}$. V_{GM} at room temperature is about 0.8 V or higher for GaAs

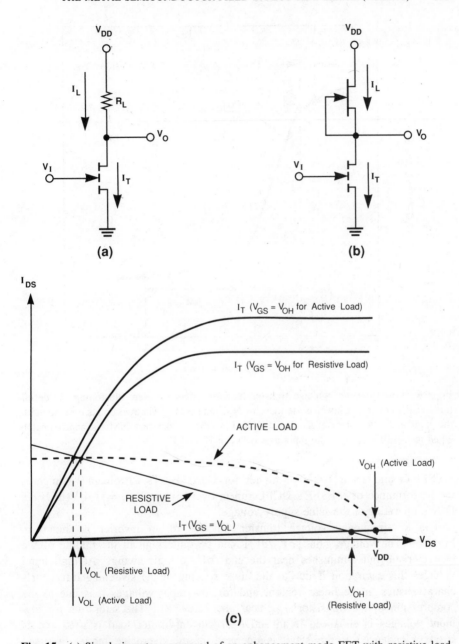

Fig. 15 (a) Simple inverter composed of an enhancement mode FET with resistive load. (b) FET inverter with active load. (c) Characteristics of the inverters in (a) and (b). V_{OH} is the output voltage in its high or logic 1 state. V_{OL} is the output voltage in its low or logic 0 state. Note that the active load offers a larger logic swing than the resistive load when both are designed for a given current at V_{OL}. The quantitative advantage for the active load improves as this current is reduced, since V_{OH} for the resistive load may drop significantly.

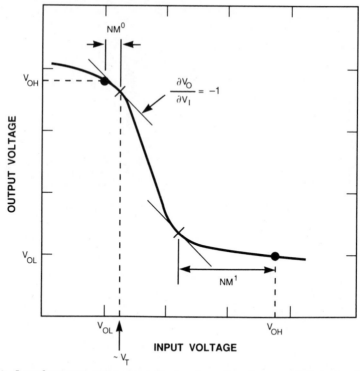

Fig. 16 Low-frequency voltage transfer function of an inverter, illustrating the definitions of the upper and lower noise margins NM^1 and NM^0. The two points correspond to the high and low logic levels depicted in Fig. 15c. Note that NM^0 is approximately equal to the difference on the input axis between V_T and V_{OL}.

MESFETs and about 1.4 V or higher for GaAs JFETs. Problem 9 compares the performance of MESFET DCFL run in overdrive at $V_{DD} = 1.4$ V to JFET DCFL operating at the same supply voltage.

The low-frequency voltage transfer function of an inverter is shown in Fig. 16. The voltage gain $|\partial V_O/\partial V_I|$ can be quite high in the middle of the characteristic but diminishes near the ends of the logic swing. At high input voltages this saturation is due to the close spacing of the switching transistor's characteristics in the linear region, and for low input voltages it is due to the compression of the transistor's g_m near and below V_T. The saturation for low input voltages is enhanced if an active or current-limiter load is used and is strongly enhanced if the stage is operated in the overdrive mode discussed above. The noise margin (NM) for each logic level is defined in Fig. 16 as the voltage difference, measured on the input axis, between the logic level and the nearest point with unity voltage gain. Noise margins have traditionally been a measure of the immunity of a circuit to spurious triggering by noise or crosstalk. Wide noise margins are also important for preventing false triggering or latchup in integrated circuit (IC) technologies that have significant variations in V_T across a wafer or from wafer lot to wafer lot.

The choice of V_T for the enhancement mode switching transistor in DCFL circuits is a trade-off between NM^0 and logic swing. V_T must be high enough to obtain a reasonable NM^0, yet low enough that the logic swing is not significantly hampered. Since they are restricted to small logic swings by the $V_{OH} \leq V_{GM}$ condition, GaAs MESFET switches in DCFL circuits are usually designed for low V_T values of 0.1 to 0.3 V. The small logic swing and NM^0 in these circuits demand the utmost uniformity in V_T from production technology. A common rule for the maximum tolerable size of threshold voltage variations across one wafer and from wafer to wafer is

$$\sigma(V_T) \leq \frac{V_{OH} - V_{OL}}{20}, \qquad (36)$$

where $\sigma(V_T)$ is the standard deviation of V_T. For GaAs MESFET DCFL, $\sigma(V_T)$ should be of the order of 0.025 V, a value that has been sporadically demonstrated in the laboratory[34] but is still approximately a factor of two beyond the capability of present manufacturing technology. In the near future, further manufacturing refinements should enable this degree of threshold control to be achieved. Early difficulty achieving such uniformity led to an interest in GaAs JFET DCFL, where both V_T and the logic swing can be larger, and to the development of circuits that use depletion mode MESFETs and are less sensitive to $\sigma(V_T)$. JFET DCFL is discussed further in Section 4.3, and the depletion mode MESFET circuits are covered shortly.

The dynamic switching performance of an inverter is illustrated with the aid of Fig. 17. Shifting the logic level of the drain of the FET involves charging or discharging all of the capacitors shown in Fig. 17a. The capacitances C_{DG} and C_{GS} in Fig. 17a are the large-signal equivalents of those defined previously for

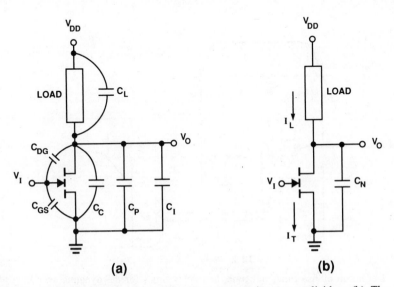

Fig. 17 (a) Basic inverter circuit showing all capacitances explicitly. (b) The capacitances have been lumped into one equivalent node capacitance. (After Ref. 33)

the FET, and C_C represents an effective large-signal drain-source capacitance. C_L is the capacitance of the load, C_P is the capacitance due to interconnect lines and fringing fields, and C_I is the effective input capacitance of the driven gate(s). The effective node capacitance C_N of Fig. 17b is given by

$$C_N = C_C + C_{\mathrm{DG}} \left[1 - \frac{\partial V_I}{\partial V_O} \right] + C_L + C_P + C_I, \qquad (37)$$

where C_I includes the Miller-effect capacitance $C_{\mathrm{DG}}(1 + \partial V_O/\partial V_I)$ and C_{GS} of the subsequent stage(s). Except for C_P and possibly C_L, all of the capacitances of Eq. 37 are functions of bias. Accurate calculations of the switching response of an inverter must include the bias dependence of not only these elements but of other elements of the FET such as its g_m and internal conductances. The heuristic models mentioned in Section 4.2.3 are well suited for these calculations and are easily employed in computer simulations of inverter performance.

Figure 18 illustrates the rise time t_R, the fall time t_F, and the delay times $t_{D,R}$ and $t_{D,F}$ for an inverter. The rise of the inverter output is due to the charging of C_N by the load current I_L, while the fall of the output is due to the discharge of C_N through the FET. Expressions for the characteristic times are given by[33]

$$t_F = 0.8(V_{\mathrm{OH}} - V_{\mathrm{OL}}) < \frac{C_N}{I_T - I_L} >, \qquad (38)$$

$$t_{D,F} = 0.5(V_{\mathrm{OH}} - V_{\mathrm{OL}}) < \frac{C_N}{I_T - I_L} >, \qquad (39)$$

$$t_R = 0.8(V_{\mathrm{OH}} - V_{\mathrm{OL}}) < \frac{C_N}{I_L} >, \qquad (40)$$

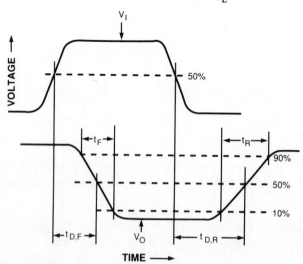

Fig. 18 Definition of rise time, fall time, and delay times for the dynamic switching of an inverter. (After Ref. 33)

and

$$t_{D,R} = 0.5(V_{OH} - V_{OL}) < \frac{C_N}{I_L} > . \qquad (41)$$

The brackets around the right-hand term in each equation signify the average value of that term over the appropriate voltage excursion defined in Fig. 18. The exact value of these bracketed terms is not always easy to derive analytically, due to the bias dependence of C_N, I_T, and I_L and to the reactive effects in the circuit. Figure 19 illustrates the improvement in the rise time t_R due to the clamping action in the overdrive mode. The delay time $t_{D,R}$ is improved by this mechanism as well.

When a voltage step propagates through a digital circuit, roughly half of the involved gates turn on and the remainder turn off. The average propagation delay time per gate, t_D, is approximately the average of $t_{D,F}$ and $t_{D,R}$ above. The maximum clock frequency f_{mcf} at which an inverter can be driven is limited to $f_{mcf} \approx 1/(2 t_D)$. The gate delay t_D is often measured using a ring oscillator circuit, which consists of an odd number N of inverters that are connected sequentially around a continuous loop. Such a circuit will oscillate at the highest frequency f_0 at which the loop gain is unity and the total phase delay around the loop is an odd multiple of π. The value of t_D is derived by measuring f_0

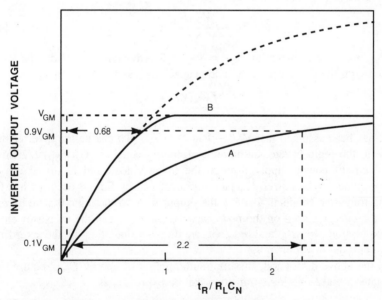

Fig. 19 Output voltage waveforms for an inverter having a load of resistance R_L. Curve A corresponds to the case where $V_{DD} = V_{GM}$, and curve B to the case where $V_{DD} > V_{GM}$. Curve B initially rises faster and is clamped at V_{GM}, thus its normalized rise time is only 0.68 as opposed to 2.2 for curve A. The dashed curve is the locus that B would follow in the absence of clamping. (After Ref. 33)

and using the relation

$$t_D = \frac{1}{2Nf_0} ,$$
(42)

Multiple traveling wavefronts may exist simultaneously in a ring oscillator. To prevent equally spaced wavefronts from yielding an artificially high f_0, N, in addition to being odd, should be a prime number. If N is prime, the period of the measured waveform will always be twice the propagation delay around the entire ring in accordance with Eq. 42.

The value of t_D obtained from a ring oscillator represents the best case for that particular device and circuit design. Most gates in real logic circuits have a fanout of 2 or 3, meaning that they must drive 2 or 3 subsequent gates rather than just one as in a ring oscillator. As fanout increases, t_D degrades because the value of C_N increases. Frequency divider circuits are frequently used to assess t_D for realistic circuit applications, since the gates in these circuits have typical fanouts. As will be discussed in Section 4.2.7, t_D is observed experimentally to improve with increasing V_{DD} in the overdrive mode because of the progressive shortening of $t_{D,R}$ by the clamping action.

The power dissipation of an inverter driven at a frequency f is composed of a static component that is independent of f and a dynamic component that is a function of f. Assuming the switching transistor conducts half the time and that $V_{DD} < V_{GM}$, the static power dissipation is given by[33]

$$P_{st} \approx \frac{V_{DD}I_{LL}}{2} ,$$
(43)

where I_{LL} is the load current at $V_{DS} = V_{OL}$. The dynamic power dissipation is associated with the discharge of the node capacitance and is given by

$$P_{dyn} \approx \frac{fC_N(V_{DD})^2}{2} ,$$
(44)

where it has been assumed that $V_{OH} - V_{OL} \approx V_{DD}$. To charge the inverter output node to the logic 1 state, an amount of energy equal to $[C_N(V_{DD})^2]/2$ flows from the circuit power supply through the inverter load and is stored on the node capacitance. The power dissipation represented by Eq. 44 occurs when the switching transistor conducts, driving the output node to the logic 0 state by dissipating the energy stored on the node capacitance as heat in the transistor channel. (Expressions for P_{st} and P_{dyn} of an inverter operating in the overdrive mode are derived in Problem 9.)

From the above discussion, the minimum amount of energy E_{min} required by a logic gate to make one logic transition can be expressed as

$$E_{min} = t_D P_D \approx \frac{C_N(V_{DD})^2}{2} ,$$
(45)

where P_D is the dynamic power that flows during the transition time given approximately by t_D. The quantity $t_D P_D$ is known as the power-delay or speed-power product. The power-delay product is an important figure of merit for evaluating logic circuits and devices, the goal being to minimize $t_D P_D$. This is especially important for high-density, high-speed integrated circuits that not only need low gate delays but need very low power dissipation per gate to avoid overheating. Equation 45 shows that the only way to decrease the power-delay product is to decrease the node capacitance and/or the supply voltage.

At this point, a general discussion can be made of FET design for digital operation. The first requirement for digital FETs is that they have the highest possible f_T. Even though f_T is a small-signal concept, FETs that have high values of f_T along the logic swing have a high overall transconductance/capacitance ratio in large-signal operation. A high g_m enables a FET to develop a large I_{DS} quickly in order to discharge a capacitance. FET structures with a high g_m/Z need minimal gate width to drive the node capacitance in a logic circuit. Such small FETs with low internal capacitance do not contribute excessively to the node capacitance that the previous logic stage must drive.

As with the analog FETs discussed previously, digital FETs should use the minimum practical gate length L to maximize f_T. The optimization of L/a ratio, R_S, and N_D for digital FETs generally follows the discussion of Sections 4.2.3 and 4.2.4. Substrate leakage should be suppressed so that the FET turns on and off abruptly at $V_{GS} = V_T$ and develops a high transconductance just above V_T. The breakdown voltage V_B for digital FETs need only be slightly larger than V_{DD} or the logic swing, which are typically quite low. Such a low V_B enables the $V_B f_T$ trade-off of Eq. 32 to be used to advantage to optimize f_T. The selection of the gate width for FETs in digital circuits is a careful balance between a FET being large enough to drive its node capacitance, yet small enough that it does not dominate the node capacitance of the previous stage. For the fanouts and parasitic interconnect and load capacitances encountered in typical circuits, Z values of the order of 10 μm are usually optimal. The minimization of interconnect and load capacitance and of wiring delays is at least as important in the design of high-performance digital ICs as the optimization of the switching devices themselves.

GaAs and other III-V devices have a number of advantages over Si devices for low-power, high-speed logic. At the low V_{DD} levels that are desirable for such logic, the III-V devices offer average electron velocities v_a that are several times higher than those offered by Si. For example, a device with $L = 1$ μm biased at $V_{OH} = 0.5$ V has a channel electric field of the order of 5×10^3 V/cm. As shown in Chapter 1, the steady-state electron velocity for GaAs at this field is about 1.8×10^7 cm/s, and for Si it is about 5×10^6 cm/s. This velocity disparity becomes even larger in devices where $L \ll 0.5$ μm because GaAs displays significant near-ballistic transport whereas Si does not. The parameters f_T, g_m, and f_{mcf} are all proportional to v_a, giving GaAs and other III-V compounds a decisive advantage for speed. Si CMOS is superior to all other digital technologies in static power dissipation, but due to the gradual nature of the Si velocity-field curve it must be run at high V_{DD} levels to achieve speeds approaching those of GaAs FET logic of equal gate length. As Eq. 44

predicts, the dynamic power dissipation for CMOS at high clock rates and high V_{DD} is substantial, and dilutes its static power advantage. For high clock rates, the overall power dissipation of GaAs DCFL is less than or comparable to that of all Si technologies, and GaAs DCFL offers higher speed.

The design principles and constraints for digital MESFETs have given rise to a number of alternate circuit approaches beyond simple DCFL. The gate delay of a simple DCFL inverter like those of Fig. 15 degrades in proportion to the number of subsequent gates that it must drive. The circuit of Fig. 20a alleviates this problem by adding a quasi-complementary output buffer to the inverter, enabling the stage to drive a large capacitive load with little speed degradation. The output buffer adds no additional static power dissipation if $V_{DD} < V_{GM}$. As mentioned earlier, MESFET DCFL requires very tight threshold control, which is not easy to achieve in manufacturing. The buffered-FET logic (BFL) approach shown in Fig. 20b overcomes this tight restriction by using depletion mode FETs with negative V_T values. Buffered-FET logic offers relaxed threshold control requirements, larger logic swing, and an output buffer stage for high fanout capability. The price paid for these advantages includes high power dissipation, higher consumption of wafer area per gate, and the need for both dc level-shifting diodes and a second voltage supply. Schottky-diode FET logic (SDFL), shown in Fig. 20c, also uses depletion mode FETs and provides relaxed threshold control requirements with lower power dissipation than BFL. The input diodes in SDFL perform the logical OR function, so that each gate in SDFL is capable of a more complex logic operation than the inverters of BFL or DCFL. As a result, an SDFL circuit consumes less wafer area than the equivalent BFL circuit. A logic approach that enables depletion mode FETs to be used in circuits with a single voltage supply is shown in Fig. 20d. Known as capacitor-coupled FET logic (CCFL), this approach couples the logic signal from one stage to the next through either a capacitor or the capacitance of a reverse-biased diode. CCFL eliminates the need for level-shifting diodes and allows a wide tolerance for variations in V_T. The main drawback to CCFL is that it is ac coupled, and therefore may require periodic refreshes to maintain the logic states. Logic design often involves the combination of elements of these various classic approaches into new circuits that are optimized for a specific task. An example of this is shown in Fig. 20e, where elements of quasi-complementary buffered DCFL, BFL, and CCFL have been combined into one gate.

Of the common FET-based logic families, only DCFL and its simple derivatives can satisfy the power dissipation and area/gate guidelines given by Table 1 for VLSI and ULSI. BFL and SDFL will probably be limited to LSI levels of integration—BFL by its power consumption and area/gate requirements and SDFL by its power consumption. Simple DCFL is often believed to have a disadvantage in speed compared to the more complex approaches, but a carefully controlled study has shown that DCFL offers comparable speed with substantially lower power dissipation.[37] The tight manufacturing tolerances on V_T for MESFET DCFL will become progressively easier to achieve as the industry

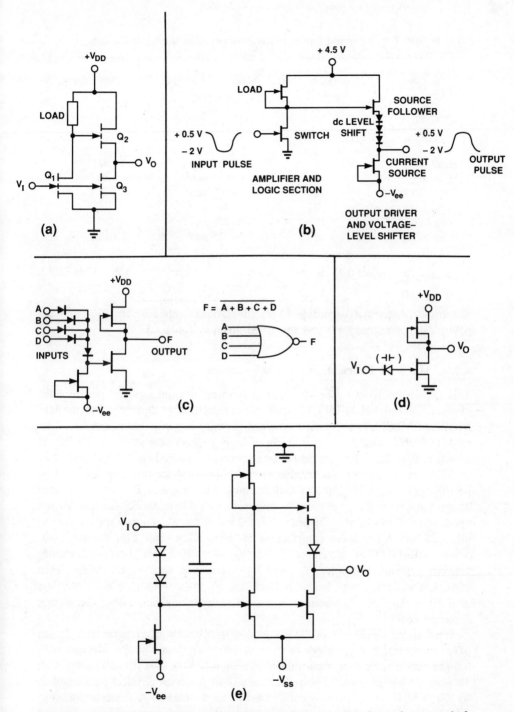

Fig. 20 Representative approaches for FET-based logic. (a) Quasi-complementary buffered DCFL inverter. (b) BFL inverter. (c) SDFL NOR gate. (d) Capacitor-coupled FET logic (CCFL) inverter. (e) Low-power super-buffer CCFL inverter. [(a) After Ref. 33; (b)–(d) After Ref. 34; (e) After Ref. 36]

TABLE 1 Area and Power Dissipation per Gate for Logic Circuits[a]

Integration Scale	Number of Gates per Chip	Maximum Average Area per Gate	Maximum Average Dissipation per Gate (mW)
Small (SSI)	10	2.5 mm^2	500
Medium (MSI)	10^2	0.25 mm^2	50
Large (LSI)	10^3	0.025 mm^2	5
Very Large (VLSI)	10^4	2500 μm^2	0.5
Ultra Large (ULSI)	10^5	250 μm^2	0.05
Giant (GSI)	10^6	25 μm^2	0.005

[a]Based on a chip area of 0.25 cm^2 and a power dissipation per chip of 5 W. (After Ref. 33)

matures. (Problem 10 considers the limitations imposed on V_{DD} by the power/gate guidelines for VLSI and ULSI given in Table 1.)

4.2.7 MESFET Fabrication and Performance

This section briefly describes the typical fabrication sequences and performance of analog and digital MESFETs. A more comprehensive description of the fabrication of MESFETs and their associated circuitry is given throughout Ref. 38.

A typical fabrication sequence for a high-performance analog MESFET is shown in Fig. 21. The gate fabrication process is tailored so that the gate electrode has maximal cross-sectional area with a minimal footprint, which simultaneously provides both low R_G and minimal gate length. The gate is placed nearer the source than the drain to minimize R_S. Those MESFETs that do not operate in the enhanced V_B regime of Eq. 21 may use a deep gate recess to minimize R_S and R_D and the contribution of the surface traps to $1/f$ noise. For power MESFETs in which enhanced V_B is desired, the best compromise between parasitic resistance, I_m, and V_B is achieved with a gate-recess depth equal to the surface-trap depletion depth and $N_D a^*$ between gate and drain less than 2.3×10^{12} cm^{-2}, where a^* is the active layer thickness minus the surface depletion depth.[21]

Small-signal MESFETs of the design of Fig. 21 with gate lengths of 0.25 μm have demonstrated f_{max} values as high as 180 GHz,[39] and similar devices with 0.3-μm gate lengths have shown noise figures as low as 0.73 dB at 12 GHz with 10.5-dB associated gain.[40] Figure 22 is a plot of f_T versus physical gate length L for GaAs MESFETs, and was constructed using the highest f_T values reported to date for experimental devices. The higher values of f_{max} and f_T are often not measured directly, but are derived instead by extrapolating the measured small-signal gain from lower frequencies using a rolloff of –6 dB/octave. It has been

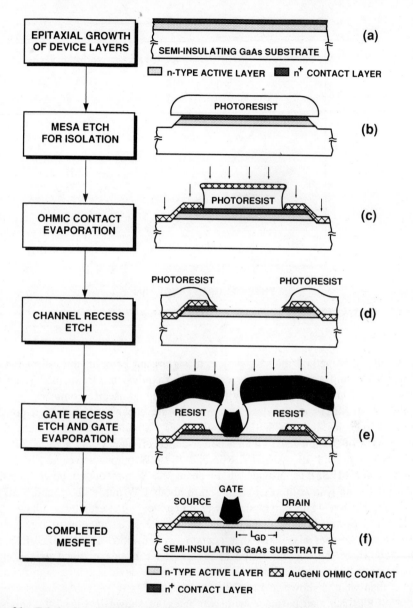

Fig. 21 Fabrication sequence of a high-performance analog GaAs MESFET using the so-called double-recess process. The n^+ contact layer improves the source and drain ohmic contact resistances, and is etched away over the active channel. The gate is offset toward the source to minimize R_S, and is made as thick as practical to minimize R_G. The effect of surface depletion on R_S and R_D is minimized by making the channel layer extra thick and placing the gate in a recess. The length L_{GD} is designed to be greater than the depletion width at gate-drain breakdown.

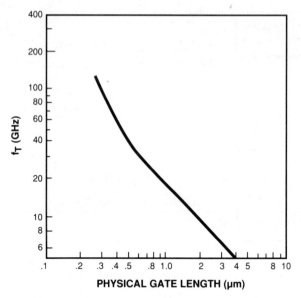

Fig. 22   Plot of f_T versus physical gate length L for GaAs MESFETs.

shown that this method may be subject to significant error in the estimation of f_{max}.[23] To obtain a more accurate estimate, an equivalent circuit model can be constructed from the low-frequency measured data and used to compute f_{max}.

Power MESFETs of the design of Fig. 21 are capable of very high power-added efficiencies and substantial output power. Small-gate-width devices with gate lengths of 0.4 μm have yielded power-added efficiencies as high as 59% in Class AB with 10.4-dB associated gain and 132-mW output power at 10 GHz.[32] GaAs power MESFETs provide up to about 0.5 W/mm output power density with little or no gain compression and up to about 1 W/mm with around 1 dB of gain compression. The maximum power output for commercially available discrete GaAs power MESFETs is plotted in Fig. 23 as a function of frequency. Commercial devices also exist that contain two or more power MESFETs on one chip where the outputs of the MESFETs have been combined on the chip using circuit techniques. The power output of these devices can be significantly higher than what is shown in Fig. 23.

MESFETs can also be made with two gates, as shown in Fig. 24. This so-called dual-gate MESFET structure is roughly equivalent to two FETs connected in cascade and offers high gain as well as additional capabilities such as gain control and mixing, which single-gate FETs cannot easily offer.[12] As a high-gain amplifier, the input is usually applied to G_1, and G_2 is grounded for ac, forming a cascode stage. An amplifier with gain control is realized by connecting G_2 to a variable dc voltage supply. The gain of the stage is adjusted by varying the dc bias on G_2. A mixer is realized by applying one signal to be mixed to G_1 and the other to G_2.

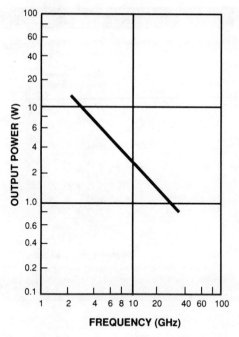

Fig. 23 Approximate maximum output power of discrete commercial GaAs power MESFETs as a function of frequency. The plot does not represent the output power of one MESFET as a function of frequency, but rather depicts the output power that can be obtained at a given frequency from a single commercial MESFET designed to provide maximum power at that frequency.

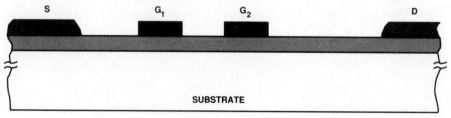

Fig. 24 Cross section of a dual-gate MESFET.

A growing trend in the microwave industry is the movement away from hybrid integrated circuits toward monolithic integrated circuits. Hybrid circuits have traditionally been made by individually attaching discrete active and passive components to circuits fabricated on metallized alumina or quartz substrates. The high cost, marginal repeatability, and electrical parasitics of this technique make it less attractive than a monolithic fabrication process that forms active devices, passive components, and transmission lines on a single semiconductor wafer in one process sequence. The present monolithic microwave-integrated-circuit (MMIC) technology is primarily GaAs-based, taking advantage of the excellent performance of GaAs active devices and the availability of

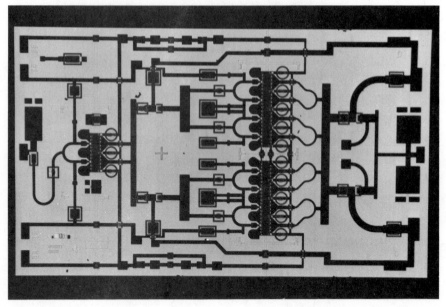

(a)

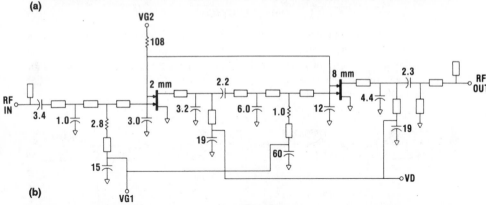

(b)

Fig. 25 (a) Photomicrograph of the top surface of a GaAs 3-W variable-gain amplifier MMIC. The wafer measures 6.6 by 3.8 mm. (b) Circuit schematic of the MMIC, with element values given in picofarads and ohms. Two dual-gate power MESFETs are used: a 2-mm gate-width device for the first stage and an 8-mm gate-width device for the output stage. The 2-mm power MESFET is located to the left of the veritcal column of three small circles in (a). The 8-mm power MESFET is located to the left of the vertical column of ten small circles. (After Ref. 41 © 1988 IEEE).

high-quality, semi-insulating GaAs substrates. The resistivity of these substrates is sufficiently high at 10^8 Ω-cm that low-loss transmission lines can be fabricated on them. A good discussion of MMIC technology is given in Ref. 38.

A representative GaAs MMIC[41] is shown in Fig. 25a. The maximum output power of the circuit is over 3 W from 8.0 to 11.5 GHz with more than 20% power-added efficiency and up to 13 dB of large-signal gain. The gain of the MMIC can be attenuated by up to 20 dB by varying VG2 in Fig. 25b.

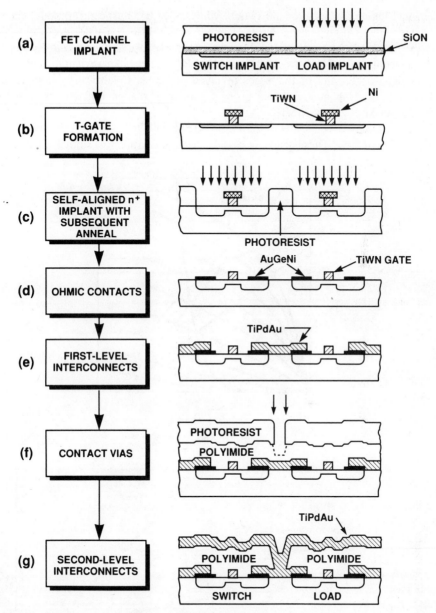

Fig. 26 Fabrication process for MESFET DCFL circuits with active loads. Note that the n^+ source and drain regions are self-aligned to the gate. (After Ref. 42)

Due mainly to its lower cost, ion implantation directly into semi-insulating GaAs substrates is favored over epitaxial techniques for the fabrication of MESFET-based digital integrated circuits. A representative fabrication sequence for MESFET DCFL circuits[42] is shown in Fig. 26. In this process, n^+ source and drain regions are self-aligned to the gate of each FET. A relatively light channel implant is used for the switch, which must be enhancement mode, and a heavier one is used for the depletion mode active load. A gate recess is typically not used for such digital IC fabrication because the uniformity of etch depth has been difficult to control, leading to unacceptable variation in V_T. A very detailed account of the issues involved in GaAs IC fabrication is given in Ref. 43. An implantation-based process similar to that in Fig. 26 can also be used to make low-cost MMICs.[44]

Representative performance of GaAs MESFET DCFL gates[45] is shown in Figs. 27 and 28. These gates were fabricated by a process similar to that in

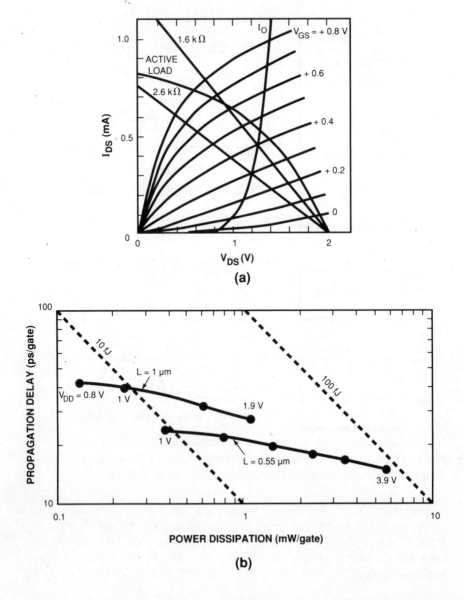

(a)

(b)

Fig. 26. Figures 27b and 28 illustrate the trade-off between t_D and power dissipation in the overdrive mode that was discussed in Section 4.2.6. As V_{DD} is increased, t_D slowly improves, but the power dissipation increases markedly faster. As shown in Fig. 27b, the shortest gate delay observed was 15 ps at a supply voltage of 3.9 V. The lowest power dissipation and power-delay product were 0.04 mW/gate and 2.4 fJ/gate, respectively, and are depicted in Fig. 28 at $V_{DD} = 0.7$ V.

Another self-aligned MESFET fabrication process, called the self-aligned implantation for n^+-layer technology (SAINT) process,[46] is shown in Fig. 29. The gates in the SAINT process are formed after the implants and the subsequent high-temperature implant activation anneal, rather than before as in the process of Fig. 26. Figure 30 shows three variants of the basic SAINT MESFET.[47] The structure in Fig. 30a, known as a buried p-layer SAINT (BP-SAINT) FET, uses an implanted p-layer to minimize short-channel effects in the

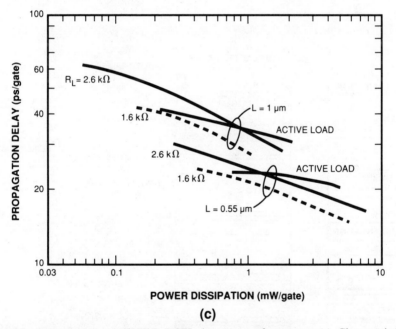

(c)

Fig. 27 Representative MESFET DCFL inverter performance. (a) Characteristics of switching FET with $L = 0.55$ μm plotted together with three load characteristics used for study. Note the presence of short-channel effects in the FET and active load. The curve labeled I_O is the gate-source diode characteristic of the switching FET of the following stage. This curve must be subtracted from each load characteristic to obtain the effective operating locus when the circuit is operated in overdrive. (b) Performance versus V_{DD} for ring oscillators made with switching FETs having gate lengths of 1 and 0.55 μm . A resistive load of 1.6 kΩ was used. Each dashed line represents a constant power-delay product. (c) Performance comparison of inverters with different gate lengths and loads. (After Ref. 45)

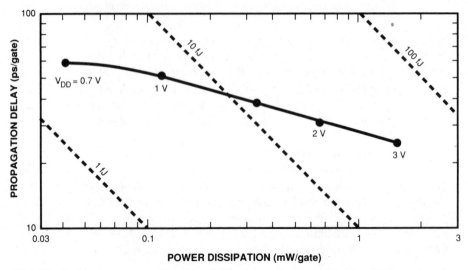

Fig. 28 Performance versus V_{DD} for a DCFL inverter with a switching-FET gate length of 0.55 μm and a resistive load of 2.75 kΩ. (After Ref. 45)

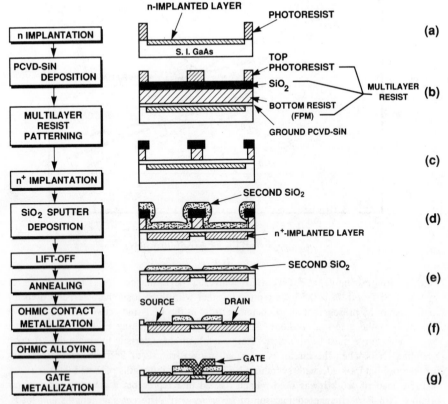

Fig. 29 Basic SAINT process sequence. FPM stands for polytetrafluoropropyl methacrylate. (After Ref. 46)

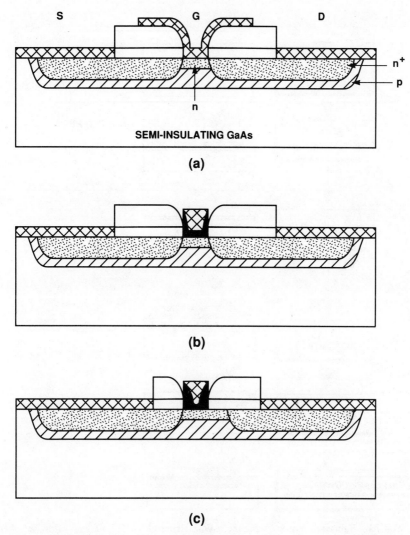

Fig. 30   (a) BP-SAINT FET.   (b) FG-SAINT FET.   (c) Asymmetric FG-SAINT FET. (After Ref. 47)

manner discussed earlier. The structure in Fig. 30b is known as a flat-gate or FG-SAINT FET, and offers reduced parasitic capacitance due to the reduced overlap of the gate metal on the adjacent dielectric. Figure 30c shows an asymmetric FG-SAINT FET in which the channel between gate and drain is extended. This structure offers higher breakdown voltage and reduced C_{DG}, G_o, and short-channel effects compared to the other SAINT FET designs, and is the favored SAINT structure for use in analog MMIC applications. The fabrication process for the asymmetric FG-SAINT FET is shown in Fig. 31.

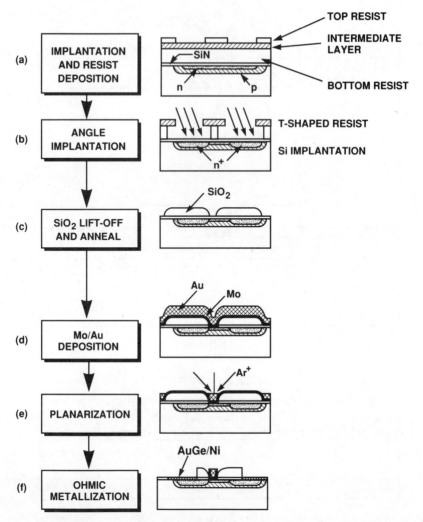

Fig. 31 Fabrication process for the asymmetric FG-SAINT FET. (After Ref. 47)

SAINT MESFETs have been fabricated with gate lengths as small as 0.1 μm.[48] These devices have demonstrated f_T and f_{max} values as high as 90.8 and 78 GHz, respectively. A ring oscillator made of these devices has shown a gate delay as low as 5.9 ps/gate with 31.2 mW/gate power dissipation and 184 fJ/gate power-delay product.

4.3 THE *p-n* JUNCTION FIELD-EFFECT TRANSISTOR (JFET)

JFETs differ from MESFETs in that the depletion region of a *p-n* junction, rather than a Schottky barrier, is used to modulate the channel conduction. All of the MESFET discussion and equations of Section 4.2 hold generally for

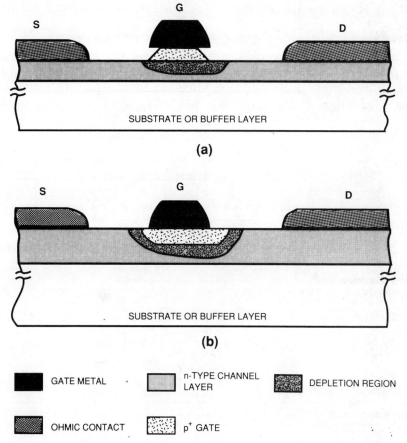

Fig. 32 Basic JFET types. (a) Grown-junction JFET. (b) Diffused- or implanted-gate JFET. The gate metal in both cases makes ohmic contact to the p^+ region. The extent of depletion into the heavily doped p^+ region is negligible.

JFETs as well, the one exception being that V_{bi} for JFETs is determined by the semiconductor bandgap and the doping of the *p-n* junction, rather than by a Schottky-barrier height. An exact expression for V_{bi} for *p-n* junctions is given in Ref. 22; V_{bi} is approximately equal to E_g/q for doping levels of interest in JFETs.

The two basic types of *n*-channel JFETs are shown in Fig. 32. The *p* and *n* regions are interchanged for *p*-channel devices. The semiconductor gate region of a JFET is usually doped very high to minimize R_G. The grown-junction JFET in Fig. 32a is fabricated by epitaxially growing a p^+ layer on top of the *n*-type channel layer and then etching it away everywhere except for the portion that forms the gate. A heterojunction JFET is formed if this p^+ layer is of a different material than the channel. The JFET structure of Fig. 32b is made by either diffusing or implanting acceptors into the channel layer to form the gate.

JFETs have a few minor disadvantages compared to MESFETs. For III-V devices, the gate formation process for JFETs is more difficult than that for MESFETs, where the metal Schottky gate is simply evaporated onto a clean channel surface. (The opposite is true for Si, where the MESFET gate formation is hampered by the native oxide.) It is often more difficult to obtain submicrometer gate lengths in JFETs than in MESFETs. JFETs may also have somewhat higher relative R_G than MESFETs. For JFETs, R_G consists not only of the resistance of the metal gate, as in a MESFET, but also includes the ohmic-

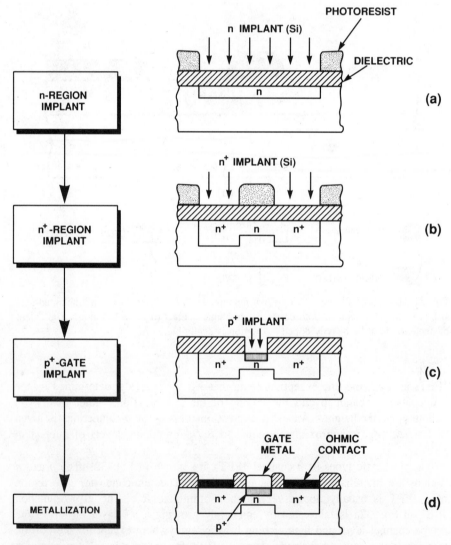

Fig. 33 Fabrication process for enhancement mode JFET DCFL switch. (After Ref. 34)

contact resistance of the gate metal/semiconductor interface and the resistance of the semiconductor gate region itself.

A principal advantage of the JFET concept is that it allows FETs to be fabricated using channel materials that are not suitable for MESFETs, MOSFETs, or metal-insulator-semiconductor FETs (MISFETs). Prime examples of such materials include $In_{0.53}Ga_{0.47}As$ and InP, neither of which has demonstrated sufficient Schottky-barrier height for practical MESFETs or sufficiently stable MOS or MIS structures for reliable MOSFETs or MISFETs. There is a strong need to produce transistors in these materials because they are easily integrated with GaInAsP-based lasers and detectors in monolithic optoelectronic integrated circuits (OEICs) for optical-fiber communications in the 1.3- to 1.6-μm wavelength region. Implanted-gate InP JFETs with f_T values of 9 GHz have been demonstrated in simple OEICs.[49] Diffused-gate $In_{0.53}Ga_{0.47}As$ JFETs with gate lengths of 1.2 μm have demonstrated f_{max} values above 30 GHz for similar applications.[50] As is discussed in Section 4.5, InP is a promising material for power FETs, and InP power JFETs having 1-μm gate lengths have demonstrated an output power density of 1 W/mm at 4.5 GHz with 3-dB associated gain.[51]

As mentioned in Section 4.2.6, an important advantage for JFETs over MESFETs in GaAs DCFL logic is the higher forward turn-on voltage of a *p-n* junction compared to a Schottky barrier. This provides JFET DCFL with larger logic swing, larger NM^0, and higher yields due to less stringent manufacturing tolerances for V_T. Figure 33 shows the fabrication sequence for an implanted-

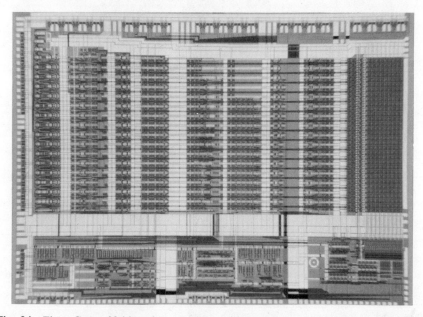

Fig. 34 First GaAs 32-bit microprocessor, fabricated in enhancement mode JFET DCFL. The wafer measures 8.2 by 11.2 mm, and contains 21,606 transistors. (After Ref. 52, © 1988 IEEE).

gate JFET DCFL process. Such a process has been used to produce the first 32-bit GaAs microprocessor,[52] shown in Fig. 34. This microprocessor uses both *n*- and *p*-channel devices as switching FETs, depending on the local circuit requirements. Resistive loads are used and consist of lightly implanted n^- regions with ohmic contacts at each end. The microprocessor has demonstrated an operation rate of 60 million instructions per second (MIPS).

4.4 THE PERMEABLE-BASE TRANSISTOR (PBT)

The permeable-base transistor[53] is a novel transistor design capable of high-performance analog operation at millimeter-wave frequencies. The basic structure of a PBT is shown in Fig. 35. In contrast to the planar FETs previously

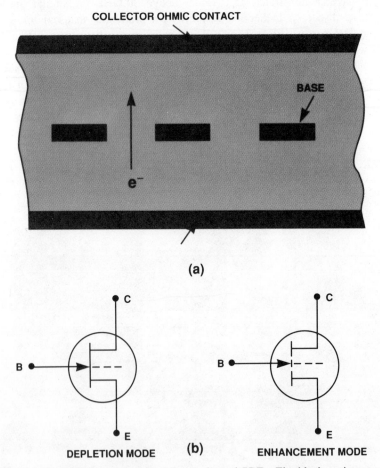

Fig. 35 (a) Conceptual cross section of an *n*-channel PBT. The black regions are metal base grating lines that run normal to the page. (b) Circuit symbols for *n*-channel depletion mode and enhancement mode PBTs.

discussed, the PBT is a "vertical" device in which the current flow is normal to the wafer surface rather than parallel to the surface. The emitter and collector regions are separated by a metal grating that forms the base of the transistor. The grating lines make Schottky-barrier contacts to the semiconductor channels of the device. Voltage applied to the base controls the current flow from the collector to the emitter, much in the manner of a solid-state version of a vacuum triode. The emitter, base, and collector nomenclature is used because the PBT concept grew out of earlier work on metal-base transistors in which a solid metal layer was used for the base rather than a grating. Unlike a bipolar transistor, the PBT is a majority-carrier device and functions essentially as a FET over most of its operating bias range.

Because of the vertical disposition of the PBT, the control-region length is determined by the channel carrier concentration and the thickness of the base metallization, and can be made short for high f_T. Monte Carlo simulations of GaAs PBTs have predicted ultimate f_T values of 200 and 300 GHz for grating periodicities of 0.32 and 0.2 μm, respectively.[54] The vertical geometry provides a number of additional advantages over horizontally disposed FETs. Each vertical channel in the PBT is controlled by two opposing gates, thus the PBT does not suffer from the substrate leakage discussed in Section 4.2. The PBT structure combines many individual active channels in parallel in a very compact area, and as such can control large currents without the gate-voltage-phasing problems mentioned in Section 4.2.2.

High-frequency PBTs have thus far been fabricated in GaAs and Si. The GaAs PBT structure is depicted in Fig. 36, and Fig. 37 illustrates its fabrication

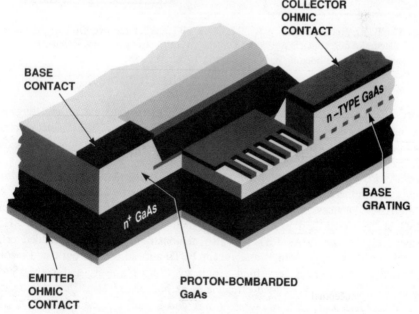

Fig. 36 Cutaway diagram of the GaAs PBT. (After Ref. 57)

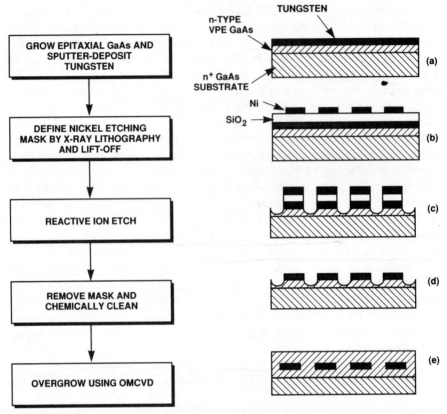

Fig. 37 Basic process sequence for the GaAs PBT. After the overgrowth, the emitter and collector ohmic contacts are deposited, the overall active areas are isolated by proton implantation, and the bonding pads are added. (After Ref. 55)

process. The key element of the process is the encapsulation of the tungsten grating lines within a single crystal of GaAs using overgrowth by metal-organic chemical vapor deposition (MOCVD). The issues involved in this novel technology are reviewed in Ref. 55.

GaAs PBTs having a grating periodicity of 0.32 μm have demonstrated MSG values as high as 21.3 dB at 18 GHz and extrapolated f_{max} values as high as 223 GHz.[55] Devices with 0.24-μm grating periodicities have shown 20.1-dB MSG at 26.5 GHz and extrapolated f_{max} and f_T values of 265 and 38 GHz, respectively.[56] The GaAs PBT has also demonstrated high power-added efficiency due to its high gain,[57] and recent PBTs and amplifiers built at Lincoln Laboratory have shown very high efficiency, output power, and gain. Such PBTs have shown power-added efficiencies as high as 53% in Class AB operation at 21 GHz with 75-mW output power and 8.7-dB associated gain. A 22-

GHz PBT power amplifier has demonstrated an output power of 501 mW in Class AB with 30% power-added efficiency and 5-dB gain. Single-stage PBT amplifiers have demonstrated small-signal gains of 6 to 7.5 dB over the 91- to 94-GHz range.

The transport properties of electrons in Si limit the frequency capabilities of the Si PBT to frequencies below that of its GaAs counterpart. However, the Si PBT provides a practical, high-performance Si microwave device that takes advantage of the advanced fabrication technology and superior thermal conductivity of Si.[58] A schematic drawing of a Si PBT is given in Fig. 38. Since surface-trap-induced depletion can be minimized in Si, a structure that does not involve overgrowth is possible for Si PBTs. Grooves are etched into an n-type Si layer, and platinum is deposited on the tops of the ridges and in the bottoms of the grooves and then annealed to form the respective PtSi emitter and base metallizations. Selective ion implantation is typically used to dope the active regions and to achieve device isolation.

Si PBTs with a grating periodicity of 0.32 μm have shown f_{max} values of 30 GHz and f_T values of 22 GHz.[59] Because of its low concentrations of crystal defects and surface traps, the Si PBT has demonstrated very low $1/f$ noise and excellent performance in low-phase-noise oscillators up to 20 GHz.[59] The noise 100 kHz away from the carrier frequency in these oscillators is typically 17 dB lower than that of an identical oscillator fabricated using a commercial GaAs MESFET.

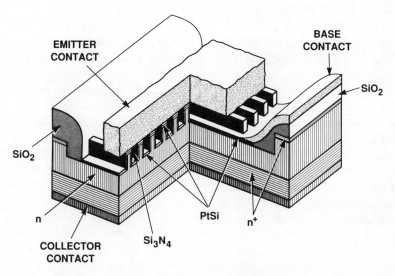

Fig. 38   Cutaway diagram of the Si PBT.   (After Ref. 59)

4.5 SUMMARY AND FUTURE TRENDS

GaAs FETs have been the workhorses for both the microwave and the high-speed digital industries for over a decade, and will no doubt continue in this role for many years to come. GaAs MESFETs and JFETs are easy to fabricate and offer high performance at low cost. The gain and speed of these devices are consistently higher than those of their Si competitors.

As a class, FETs possess a number of advantages over bipolar transistors for both analog and digital applications. The control electrode for FETs is generally made of metal, providing a lower parasitic control-lead resistance than the semiconductor base of a bipolar transistor. Being essentially capacitive, the input impedance of a FET is extremely high at dc and low frequencies—FETs effectively require only a voltage bias on the gate as opposed to bipolar transistors, which require a dc base current as well. The $g_m(V_{GS})$ characteristic of FETs is less nonlinear than the exponential $g_m(V_{BE})$ characteristic of bipolar transistors, which is advantageous for analog applications requiring low distortion. FETs do not suffer from the minority-carrier storage effects that degrade the speed of bipolars. Since the carrier mobility in semiconductors usually decreases as temperature increases, FET channels are inherently stable against thermal runaway. Bipolar transistors unfortunately can exhibit both thermal runaway and second breakdown (see Chapter 6 and Ref. 60).

GaAs MESFETs and JFETs do have a few limitations compared to bipolar transistors and more advanced III-V FETs. As mentioned in earlier sections, both types of FETs can suffer from substrate leakage. The heterojunction FET designs of the next chapter provide greater freedom from this effect, and PBTs and bipolars do not exhibit it at all. The maximum gate periphery that can be used for a planar FET at a given frequency is limited by the phase restriction discussed in Section 4.2.2. Compact, vertically oriented devices like the PBT and the bipolar transistor offer greater freedom from this restriction as well. The threshold voltage of digital FETs is sensitive to both doping and geometry variations in the devices, thus threshold uniformity is not always easy to obtain. On the other hand, V_T for bipolar transistors is set by the bandgap, and is insensitive to process variations.

Advanced devices such as PBTs, heterojunction FETs, and the heterojunction bipolar transistor (HBT) of Chapter 6 have already demonstrated excellent performance even though they have not reached full maturity. Though the advanced devices require more sophisticated fabrication technology, each holds great promise for high-speed applications.

The development sequence for a high-speed device technology begins with the recognition that a new material or device design may have advantageous properties. This is followed by the development of the materials-growth and fabrication technology for discrete active devices, and finally by the fabrication of integrated circuits based on the new material and/or device. The high-yield, cost-effective fabrication of digital ICs, OEICs, and MMICs in a given material requires a stable, mature technology that takes time to develop. For logic applications, Si- and GaAs-based digital ICs are expected to dominate the field for many years to come since they are the only technologies with sufficient maturity. GaAs devices are expected to retain their edge in speed over Si devices, giving GaAs-based SSI, MSI, and LSI circuits a speed advantage over their Si counterparts. The speed advantage of GaAs is essentially negated in VLSI and ULSI circuits, though, because the overall speed of these complex circuits is only a very weak function of the device speed. However, GaAs-based VLSI and ULSI integrated circuits implemented in low-voltage DCFL may still possess an advantage in overall power dissipation. Most OEICs are either GaAs- or InP-based, and have not yet reached the level of maturity of digital ICs. Progress is rapid in this field, however, and low-cost, high-performance monolithic OEICs should become a reality soon. Both Si and GaAs MMICs share the market below 10 GHz, but GaAs-based MMICs presently dominate the market above 10 GHz and should continue to do so.

A whole range of advanced materials are being explored for future-generation high-speed transistors and integrated circuits. As discussed in Ref. 50 and Chapters 5 and 6, the $In_xGa_{1-x}As$ alloys show great promise for many analog and digital applications because of their high low-field electron mobility and high electron velocities in the near-ballistic regime. $In_xGa_{1-x}As$ has a lower breakdown field than GaAs, however, which may limit many $In_xGa_{1-x}As$-based analog devices to small and medium power levels.

Several other materials are receiving increased attention for use in high-power analog FETs. As discussed in Section 4.2.5, three fundamental quantities that are especially relevant for power devices are breakdown field, average carrier velocity, and thermal conductivity. Figure 39 presents a comparison of several semiconductors for power operation that is based on the $V_Bf_T \sim \mathcal{E}_mv_a$ relation of Eq. 32. Table 2 lists the thermal conductivity of selected materials at room temperature. Figure 39 and Eqs. 32 and 34 predict that InP, SiC, and diamond are all superior in potential power density and speed to GaAs, Si, and Ge. InP has about 50% more thermal conductivity and roughly twice the $\mathcal{E}_mv_a$ product of GaAs, giving it a potential power density of about three times that of GaAs. InP power MISFETs have indeed demonstrated $P_{d,max}$ values as high as 4.5 W/mm,[64,65] which is over three times higher than the highest reported for GaAs power FETs.[29] The long-term reliability of InP MISFETs has been less than desired, but should improve with further research. The β phase of SiC has a bandgap of 2.2 eV, a high theoretical saturated electron velocity of 2×10^7 cm/s, a high breakdown field of 5×10^6 V/cm, and a high thermal conductivity.[66] The chemical and thermal stability of SiC make it an excellent candidate for use in devices that operate at high power levels and/or in high-temperature environments. The fabrication of basic SiC MESFETs and JFETs is reported in Ref. 66. Diamond also shows excellent promise for high-power and/or high-temperature applications.[61] It has a theoretical saturated electron velocity of 2.3×10^7 cm/s, a very high breakdown field of 10^7 V/cm, and the highest thermal conductivity of any solid at room temperature. Its high bandgap of 5.5 eV should enable devices to operate above 1000°C, and diamond point-

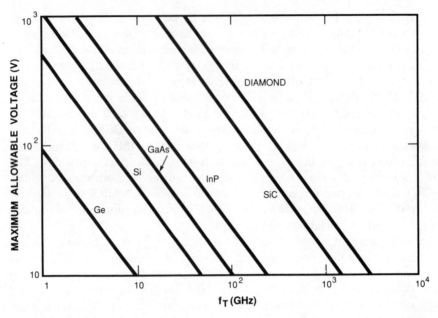

Fig. 39 Plot of the maximum operating voltage for transistors made of selected semi-conductors as a function of estimated f_T. The f_T estimates are based on the steady-state velocity-field curve for each material. (After Ref. 61)

TABLE 2 Thermal Conductivity of Selected Materials at 300 K

Material	Thermal Conductivity (W-cm^{-1}-K^{-1})	Reference Number
C (diamond)	22.5	61
Cu	4.6	61
Ge	0.6	62
Si	1.5	62
GaAs	0.4	62
InP	0.68	63
SiC	3.5	66

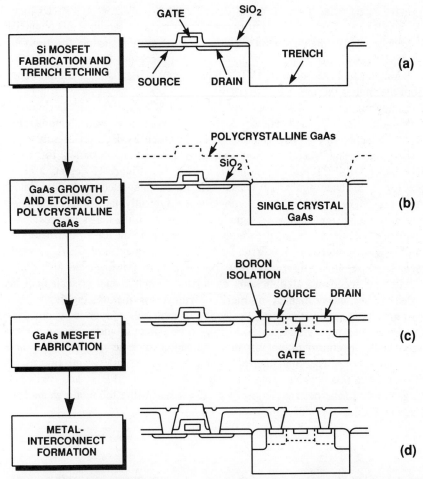

Fig. 40 Fabrication sequence for the integration of GaAs MESFETs with Si MOSFETs on a Si wafer. (After Ref. 70)

contact transistors have demonstrated power gain at 510°C.[67] A non-optimized diamond PBT has been fabricated,[61] and numerical simulations of diamond PBT performance indicate that it is capable of producing ten times the power output of a Si PBT of the same geometry.[68] Diamond transistor technology is still in its infancy, however, and areas such as the epitaxial growth and doping of diamond require further study.

As mentioned in Chapter 1, another burgeoning area of materials development is that of growing GaAs and other III-V compounds on Si substrates. Such a monolithic GaAs/Si (MGS) technology allows the integration of high-data-rate optoelectronic transceivers directly on Si wafers, enabling high-speed, inexpensive optical interconnects to be used between Si-based subsystems. This technology also allows Si integrated circuits to take advantage of the speed of

GaAs devices by using them in the circuit to perform selected high-speed functions, thereby enhancing the overall system performance. Another advantage of MGS technology is that it enables the integration of the analog functions of GaAs MMICs with the computational and signal-processing functions of Si ICs. In addition, GaAs-based power MMICs fabricated on Si substrates should benefit from the higher thermal conductivity of Si.

The integration of Si MOSFETs with GaAs MESFETs, light-emitting diodes, semiconductor lasers, and photoconductive detectors has been demonstrated using MGS technology.[69] The viability of GaAs-based MMICs on Si substrates has also been established. A representative fabrication sequence for a Si CMOS/GaAs MESFET process is shown in Fig. 40. Figure 41 depicts a cross section of a Si wafer showing GaAs MESFETs integrated with Si CMOS devices. In MGS processes, the Si devices are typically fabricated first since they require higher temperatures for their formation than the GaAs devices.

The one fundamental material parameter that is important for all transistors in all high-speed applications is carrier velocity. The quantities f_T, f_{max}, gain, transconductance, noise figure, current, power output, power-added efficiency, and switching speed are all functions of carrier velocity, and as such may be regarded as secondary figures of merit. Critical for maximizing high-speed transistor performance is the careful selection of materials and device geometries that provide the highest carrier velocity. This process must of course be blended with a consideration of the application-specific requirements and figure(s) of merit. The fundamental principles, scaling rules, and figures of merit covered in this chapter apply not only to MESFETs, JFETs, and PBTs but apply generally to the heterojunction FETs discussed in the following chapter as well as many other types of transistors.

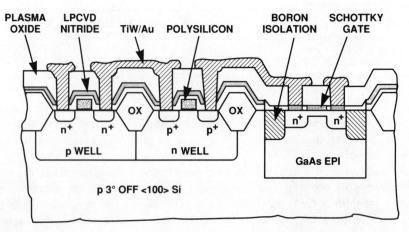

Fig. 41 Cross section of a Si wafer showing GaAs MESFETs integrated with Si CMOS devices. (After Ref. 70)

ACKNOWLEDGMENT

The authors express their sincere thanks to their many colleagues both at Lincoln Laboratory and throughout the high-speed device community for many useful discussions and suggestions over the years on these topics. This work was supported by the Department of the Air Force.

PROBLEMS

1. Charge-control concepts can be applied to two-terminal devices as well as three-terminal devices. Consider the simple parallel-plate device shown in Fig. P1. Both contacts are ohmic contacts to the conducting medium between them. This medium is characterized by its conductivity σ and its permittivity ϵ. The device has a cross-sectional area of A and a spacing between the contacts of d. A voltage V is placed across it and a total current I flows through it.

    a) Show that the charge induced on contact 2, Q_2, is given by

    $$Q_2 = \frac{\epsilon V A}{d} .$$

    b) Show that the current I is given by

    $$I = I_c + \frac{\partial Q_2}{\partial t} ,$$

    where I_c is the conduction current through the device.

    c) If the effective transit time for charge is defined as $1/\tau = \partial I_c / \partial Q_2$, show that $\tau = \epsilon/\sigma$.

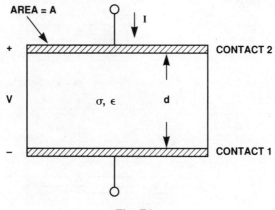

Fig. P1

2. It is instructive to consider charge control in an idealized miniature vacuum triode. Though this device is not a solid-state device, electronic conditions that approximate those within it can be achieved by reducing doping concentrations in permeable-base transistors. In addition, such devices may prove to be very useful in their own right, because they take advantage of the ballistic transport and the high breakdown strength of vacuum. The idealized triode is shown in Fig. P2. Assume that the gate electrode has infinite conductivity, that all electrons pass through it unimpeded, and that no gate current flows. Further, assume that the electron density at the source electrode is given by

$$n = -\alpha \mathcal{E}_0 \, ,$$

where $\mathcal{E}_0$ is the electric field normal to the source electrode at the source electrode. Also, assume that the electrons transit through the structure at a constant z-directed velocity v and that the device has depth w normal to the page. Direct current voltages of V_G and V_D are applied to the gate and drain, respectively.

a) Show that the electronic charge density in the region between the source and gate electrodes is

$$n = \frac{\alpha V_G}{\mathcal{L}_G(1 + \alpha q \mathcal{L}_G/2\epsilon)} \, .$$

b) The transconductance g_m is defined by

$$g_m = \frac{\partial I_D}{\partial V_G} \, .$$

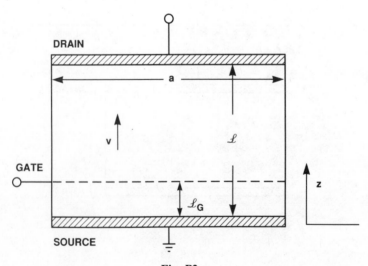

Fig. P2

Using this definition, show that

$$g_m = \frac{aw\alpha qv}{\mathcal{L}_G(1 + \alpha q\,\mathcal{L}_G/2\epsilon)} .$$

c) Show that the charge induced on the gate electrode is given by

$$Q_G = aw \left[\frac{q\,\mathcal{L}\alpha V_G}{2\mathcal{L}_G(1 + \alpha q\,\mathcal{L}_G/2\epsilon)} + \frac{\epsilon V_G}{\mathcal{L}_G} + \frac{\epsilon(V_G - V_D)}{\mathcal{L} - \mathcal{L}_G} \right] .$$

Does the charge induced on the gate electrode completely neutralize the electronic charge flowing through the device? If not, why not?

d) Show that the total gate capacitance is given by

$$C_G = aw \left[\frac{q\alpha\,\mathcal{L}}{2\mathcal{L}_G(1 + \alpha q\,\mathcal{L}_G/2\epsilon)} + \frac{\epsilon}{\mathcal{L}_G} + \frac{\epsilon}{\mathcal{L} - \mathcal{L}_G} \right] .$$

e) Neglect the inter-electrode terms in the formula for the capacitance above, namely, the $\epsilon/\mathcal{L}_G$ and $\epsilon/(\mathcal{L} - \mathcal{L}_G)$ terms. Show that the unity-current-gain frequency $f_T = g_m/(2\pi C_G)$ is given by

$$f_T = \frac{1}{2\pi\tau} ,$$

where $\tau = \mathcal{L}/(2v)$. Note that the expression for τ represents an effective delay time for the drain current and is not what would be expected from simple transit-time arguments. The reason for this is that the flowing space charge induces complementary charge on the source and drain electrodes as well as on the gate electrode.

3. a) By expressing the drain current I_{DS} as a function of the gate-source voltage V_{GS} and the drain voltage V_{DS}, and alternately expressing I_{DS} as a function of the channel charge Q and V_{DS}, show that

$$\left.\frac{\partial I_{DS}}{\partial V_{DS}}\right|_{V_{GS}} = \left.\frac{\partial I_{DS}}{\partial V_{DS}}\right|_{Q_G} + \left.\frac{\partial I_{DS}}{\partial Q_G}\right|_{V_{DS}} \left.\frac{\partial Q_G}{\partial V_{DS}}\right|_{V_{GS}}.$$

b) Assume that the electrons transit through the control region of the device at a constant velocity v_s and that the gate region can be characterized by an effective control length L_{eff}. Use the result of (a) and Eq. 18 of this chapter to show that the output conductance G_o is given by

$$G_o = \frac{C_o v_s}{L_{eff}} ,$$

where

$$C_o = \left.\frac{\partial Q}{\partial V_{DS}}\right|_{V_{GS}} .$$

4. a) Derive an expression for f_T for the following simple equivalent circuit of a FET. Assume that the portion of the input current that is fed forward through C_{DG} is negligible compared to $g_{mi}V$. (This simple lumped-element model does not accurately simulate the nonreciprocal nature of feedback and feedthrough observed in real devices. For example, for the circuit of Fig. P3, Im$[y_{21}]$ = Im$[y_{12}]$, whereas for real FETs, Im$[y_{21}]$ and Im$[y_{12}]$ are generally unequal. More sophisticated equivalent circuits such as the one of Fig. 11 or those of Refs. 71 and 72 are required for proper simulation of these effects. See Refs. 71 and 72 for a further discussion of this general topic.)

b) Does the value of the gate resistance R_G affect f_T? Why?

c) Could the values of R_S, G_o, C_o, and G_π of Fig. 2 affect f_T?

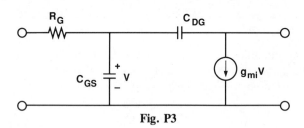

Fig. P3

5. The optimal noise figure of a device usually occurs in the saturation region but at relatively low values of I_{DS} and V_{DS}. Why? How can this fact and the $V_B f_T$ trade-off be used to optimize the noise figure of a FET that is to be designed for use in a low-noise, small-signal amplifier?

6. A power FET having a gate length L_1, a gate width Z_1, a uniform doping density of N_{D1}, and a channel thickness a_1 operates at frequency f_1 giving a maximum of X watts of output power with Y dB of associated large-signal power gain. A new power FET is to be designed to operate an octave higher at frequency f_2 with the same V_T as the original FET. Assume that f_1 and f_2 are below the respective f_T values for each FET, and that the input impedance of the original FET is the smallest that can be properly matched at f_1. Also assume that the average carrier velocity is constant and that C_{GS} dominates the input impedance Z_{in} for each FET.

a) First determine the gate length L_2 for the new FET in terms of L_1 by using Eq. 26, and then find the following for the new FET by scaling the parameters of the original FET: the channel thickness a_2, the f_T, the channel doping N_{D2}, and the gate width Z_2. (Hint: To obtain Z_2, use $C_{GS} = \epsilon_s LZ/h$ and scale h with N_D.)

b) Assume that $V_{Dsat} \approx 0$, that $V_{BDG} \gg |V_T|$, and that $N_D a < 2.3 \times 10^{12}$ cm^{-2} for both FETs. Express the maximum drain-source voltage for the new FET, V_{B2}, in terms of that for the original FET, V_{B1}. Express I_m for the new FET in terms of that for the original FET. On an I_{DS} versus V_{DS} plot similar to Fig. 12, sketch the load line that yields maximum output power for the original FET, labeling its endpoints. Plot the corresponding load line for the new FET on the same plot. What has happened to the optimal load resistance in scaling the device from f_1 to f_2? Express the maximum output power of the new FET in terms of that of the original FET.

c) Express the large-signal power gain G_2 for the new FET in terms of Y. (Hint: Use the relation $G =$ voltage gain $\times$ current gain.)

d) Assume that $V_{B1} = 20$ V, that $V_{Dsat} = 1$ V, that $Y = 7$ dB, and that $\eta_{max} = 50\%$. Compute η_D and η_{PA} for both the original FET and the new FET. Comment on the importance of low V_{Dsat} and high gain.

7. Work Problem 6 again, letting V_T vary by setting V_P for the new FET equal to $V_P/2$ for the original FET. What are the benefits of allowing V_T to vary in the frequency scaling of power FETs?

8. f_{max} is usually considered to be less relevant than f_T as a figure of merit for logic operation, since some parameters such as R_G that are important for f_{max} are not important for the switching operation of a logic gate. Explain, using the circuit diagram in Fig. P4, why the value of R_G does not control the rates of charging and discharging of C_I. Assume that $|Z_L| \gg R_G$.

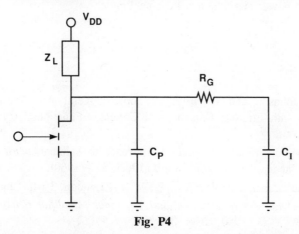

Fig. P4

9. a) Derive an expression for the static power dissipation of a DCFL inverter operating in the overdrive mode with a duty cycle of 50%. Assume that the load current in the logic 1 state is known as I_{LH}.

b) Derive an expression for the dynamic power dissipation of a DCFL inverter in the overdrive mode. Assume that V_{OL} is not negligible.

c) A simple MESFET DCFL inverter with an active load has a node capacitance of 30 fF and is to be run at a clock frequency of 5 GHz with a duty cycle of 50%. Assume that $V_{OL} = 0.1$ V and that

$I_{LL} = 0.5$ mA. Compute the overall power dissipation for this inverter with $V_{DD} = 0.8$ V, assuming that the inverter is not operating in overdrive. Do not neglect V_{OL}.

d) Compute the overall power dissipation for a JFET DCFL inverter operating under the conditions of (c), with the exception that $V_{DD} = 1.4$ V.

e) Compute the overall power dissipation for the MESFET inverter of (c) operating in overdrive with $V_{DD} = 1.4$ V and $V_{GM} = 0.8$ V. Assume that the current through the active load is constant at 0.5 mA for voltages across the load that range from $V_{DD} - V_{GM}$ to $V_{DD} - V_{OL}$. How does the power dissipation of this MESFET inverter in overdrive compare to that of the JFET inverter operating at the same supply voltage? How would the gate delay t_D for a circuit made of these overdriven MESFET inverters qualitatively compare to that of the same circuit made of the JFET inverters?

f) Of the inverter approaches presented in (c), (d), and (e), which one best meets the guidelines for VLSI given in Table 1? Why?

g) Compute t_R and $t_{D,R}$ for the overdriven MESFET inverter of (e).

10. Assuming a clock frequency of 5 GHz and a node capacitance of 30 fF, calculate the upper limit to V_{DD} for both VLSI and ULSI using the power/gate restrictions from Table 1. Assume that I_{LL} will be held very low so that the static power dissipation will be negligible.

REFERENCES

1. E. O. Johnson and A. Rose, "Simple General Analysis of Amplifier Devices with Emitter, Control, and Collector Functions," *Proc. IRE*, **47**, 407 (1959).

2. R. D. Middlebrook, "A Modern Approach to Semiconductor and Vacuum Device Theory," *Proc. IEE Suppl.*, **106B**, 887 (1960).

3. R. Spence, *Linear Active Networks*, Wiley, London, 1970, Chap. 7.

4. W.-K. Chen, *Active Network and Feedback Amplifier Theory*, McGraw-Hill, New York, 1980, Chap. 3.

5. E. S. Kuh and R. A. Rohrer, *Theory of Linear Active Networks*, Holden-Day, San Francisco, 1967, Chap. 5.

6. C. A. Mead, "Schottky Barrier Gate Field Effect Transistor," *Proc. IEEE*, **54**, 307 (1966).

7. S. M. Sze, *Physics of Semiconductor Devices*, 2nd ed., Wiley, New York, 1981, Chap. 5.

8. C. J. Palmstrom and D. V. Morgan, "Metallizations for GaAs Devices and Circuits," in M. J. Howes and D. V. Morgan, Eds., *Gallium Arsenide*, Wiley, Chichester, 1985, Chap. 6.

9. G. C. Taylor and R. L. Camisa, "Prospects for Power from GaAs FETs at Millimeter-Wave Frequencies," *RCA Rev.*, **47**, 509 (1986).

10. P. H. Ladbrooke, "Some Effects of Wave Propagation in the Gate of a Microwave M.E.S.F.E.T.," *Electron. Lett.*, **14**, 21 (1978).

11. S. M. Sze, *Physics of Semiconductor Devices*, 2nd ed., Wiley, New York, 1981, Chap. 6.

12. B. Turner, "GaAs MESFETs," in M. J. Howes and D. V. Morgan, Eds., *Gallium Arsenide*, Wiley, Chichester, 1985, Chap. 10.

13. M. A. Khatibzadeh and R. J. Trew, "A Large-Signal, Analytic Model for the GaAs MESFET," *IEEE Trans. Microwave Theory Tech.*, **MTT-36**, 231 (1988).

14. C.-S. Chang and D.-Y. S. Day, "Analytic Theory for Current-Voltage Characteristics and Field Distribution of GaAs MESFET's," *IEEE Trans. Electron Dev.*, **36**, 269 (1989).

15. Y. Tajima, B. Wrona, and K. Mishima, "GaAs FET Large-Signal Model and Its Application to Circuit Designs," *IEEE Trans. Electron Dev.*, **ED-28**, 171 (1981).

16. W. R. Curtice, "GaAs MESFET Modeling and Nonlinear CAD," *IEEE Trans. Microwave Theory Tech.*, **MTT-36**, 220 (1988).

17. K. Tanaka and Y. Kawakami, "A MESFET Model for the Design of GaAs Digital Integrated Circuits," *Electron. Comm. Japan*, Part 2, **71**, 37 (1988).

18. E. Constant, "Non-Steady-State Carrier Transport in Semiconductors in Perspective with Submicrometer Devices," in L. Reggiani, Ed., *Hot-Electron Transport in Semiconductors*, Topics in Appl. Phys., Vol. 58, Springer-Verlag, Berlin, 1985, Chap. 8.

19. L. Reggiani, "The Monte Carlo Method," in L. Reggiani, Ed., *Hot-Electron Transport in Semiconductors*, Topics in Appl. Phys., Vol. 58, Springer-Verlag, Berlin, 1985, p. 23.

20. Y. Awano, K. Tomizawa, N. Hashizume, and M. Kawashima, "Monte Carlo Particle Simulation of a GaAs Short-Channel MESFET," *Electron. Lett.*, **19**, 20 (1983).

21. S. H. Wemple, W. C. Niehaus, H. M. Cox, J. V. DiLorenzo, and W. O. Schlosser, "Control of Gate-Drain Avalanche in GaAs MESFET's," *IEEE Trans. Electron Dev.*, **ED-27**, 1013 (1980).

22. S. M. Sze, *Physics of Semiconductor Devices*, 2nd ed., Wiley, New York, 1981, Chap. 2.

23. R. J. Trew, "Equivalent Circuits for High Frequency Transistors," Proc. IEEE/Cornell Conf. on Adv. Concepts in High Speed Semicond. Devices and Circuits, Cornell University, Ithaca, NY (1987), p. 199.

24. H. Fukui, "Optimal Noise Figure of Microwave GaAs MESFET's," *IEEE Trans. Electron Dev.*, **ED-26**, 1032 (1979).

25. H. L. Kraus, C. W. Bostian, and F. H. Raab, *Solid State Radio Engineering*, Wiley, New York, 1981, Chap. 2.

26. L. J. Kushner, "Output Performance of Idealized Microwave Power Amplifiers," *Microwave J.*, **32**, 103 (October 1989).

27. E. O. Johnson, "Physical Limitations on Frequency and Power Parameters of Transistors," *RCA Rev.*, **26**, 163 (1965).

28. J. V. DiLorenzo and W. R. Wisseman, "GaAs Power MESFET's: Design, Fabrication, and Performance," *IEEE Trans. Microwave Theory Tech.*, **MTT-27**, 367 (1979).

29. H. M. Macksey and F. H. Doerbeck, "GaAs FETs Having High Output Power Per Unit Gate Width," *IEEE Electron Dev. Lett.*, **EDL-2**, 147 (1981).

30. Y. Crosnier, H. Gerard, and G. Salmer, "Analysis and Understanding of GaAs MESFET Behaviour in Power Amplification," *IEE Proc.*, **134**, 7 (1987).

31. J. V. DiLorenzo and D. D. Khandelwal, Eds., *GaAs FET Principles and Technology*, Artech, Dedham, 1982.

32. R. P. Smith, D. A. Seielstad, P. Ho, P. M. Smith, D. H. Reep, W. Fabian, and J. M. Ballingall, "Impulse-Doped GaAs Power FETs for High Efficiency Operation," *Electron. Lett.*, **24**, 597 (1988).

33. K. Lehovec and R. Zuleeg, "Analysis of GaAs FET's for Integrated Logic," *IEEE Trans. Electron Dev.*, **ED-27**, 1074 (1980).

34. B. M. Welch, R. C. Eden, and F. S. Lee, "GaAs Digital Integrated Circuit Technology," in M. J. Howes and D. V. Morgan, Eds., *Gallium Arsenide*, Wiley, Chichester, 1985, Chap. 13.

35. S. M. Sze, *Physics of Semiconductor Devices*, 2nd ed., Wiley, New York, 1981, Chap. 1.

36. A. Fiedler, J. Chun, and D. Kang, "A 3 ns 1K × 4 Static Self-Timed GaAs RAM," *Tech. Dig.*, IEEE GaAs IC Symp., 1988, pp. 67–70.

37. H. P. Singh, R. A. Sadler, A. E. Geissberger, D. G. Fisher, J. A. Irvine, and G. E. Gorder, "A Comparative Study of GaAs Logic Families Using Universal Shift Registers and Self-Aligned Gate Technology," *Tech. Dig.*, IEEE GaAs IC Symp., 1986, pp. 11–14.

38. M. J. Howes and D. V. Morgan, Eds., *Gallium Arsenide*, Wiley, Chichester, 1985.

39. P. C. Chao, P. M. Smith, K. H. G. Duh, and J. C. M. Hwang, "60-GHz GaAs Low-Noise MESFET's by Molecular-Beam Epitaxy," *IEEE Trans. Electron Dev.*, **ED-33**, 1852 (1986).

40. I. Banerjee, P. W. Chye, and P. E. Gregory, "Unusual C-V Profiles of Si-Implanted (211) GaAs Substrates and Unusually Low-Noise MESFET's Fabricated on Them," *IEEE Electron Dev. Lett.*, **9**, 10 (1988).

41. R. B. Culbertson and D. C. Zimmerman, "A 3-Watt X-Band Monolithic Variable Gain Amplifier," *Tech. Dig.*, IEEE Microwave and Millimeter-Wave Monolithic Circuits Symp., 1988, pp. 121–124.

42. H. P. Singh, R. A. Sadler, J. A. Irvine, and G. E. Gorder, "GaAs Low-Power Integrated Circuits for a High-Speed Digital Signal Processor," *IEEE Trans. Electron Dev.*, **36**, 240 (1989).

43. C. G. Kirkpatrick, "Making GaAs Integrated Circuits," *Proc. IEEE*, **76**, 792 (1988).

44. A. E. Geissberger, I. J. Bahl, E. L. Griffin, and R. A. Sadler, "A New Refractory Self-Aligned Gate Technology for GaAs Microwave Power FET's and MMIC's," *IEEE Trans. Electron Dev.*, **35**, 615 (1988).

45. R. A. Sadler and L. F. Eastman, "A Performance Study of GaAs Direct-Coupled FET Logic by Self-Aligned Ion Implantation," Proc. IEEE/Cornell Conf. on High-Speed Semicond. Devices and Circuits, Cornell University, Ithaca, NY (1983), p. 267.

46. K. Yamasaki, K. Asai, and K. Kuramada, "GaAs LSI-Directed MESFET's with Self-Aligned Implantation for n^+-Layer Technology (SAINT)," *IEEE Trans. Electron Dev.*, **ED-29**, 1772 (1982).

47. T. Enoki, K. Yamasaki, K. Osafune, and K. Ohwada, "0.3-μm Advanced SAINT FET's Having Asymmetric n^+-Layers for Ultra-High-Frequency GaAs MMIC's," *IEEE Trans. Electron Dev.*, **ED-35**, 18 (1988).

48. Y. Yamane, T. Enoki, S. Sugitani, and M. Hirayama, "5.9 ps/Gate Operation with 0.1 μm Gate-Length GaAs MESFET's," *Tech. Dig.*, IEEE Intl. Elect. Dev. Mtg., 1988, pp. 894–896.

49. S. J. Kim, G. Guth, G. P. Vella-Coleiro, C. W. Seabury, W. A. Sponsler, and B. J. Rhoades, "Monolithic Integration of InGaAs *p-i-n* Photodetector with Fully Ion-Implanted InP JFET Amplifier," *IEEE Electron Dev. Lett.*, **9**, 447 (1988).

50. R. Schmitt and K. Heime, "InGaAs Junction FETs with Frequency Limit (MAG = 1) Above 30 GHz," *Electron. Lett.*, **21**, 449 (1985).

51. J. B. Boos, S. C. Binari, G. Kelner, P. E. Thompson, T. H. Weng, N. A. Papanicolaou, and R. L. Henry, "Planar Fully Ion-Implanted InP Power Junction FET's," *IEEE Electron Dev. Lett.*, **EDL-5**, 273 (1984).

52. D. L. Harrington, G. L. Troeger, W. C. Gee, J. A. Bolen, C. H. Vogelsang, T. P. Nicalek, C. M. Lowe, Y. K. Roh, K. Q. Nguyen, J. F. Fay, and J. Reeder, "A GaAs 32-Bit RISC Microprocessor," *Tech. Dig.*, IEEE GaAs IC Symp., 1988, pp. 87–90.

53. C. O. Bozler and G. D. Alley, "Fabrication and Numerical Simulation of the Permeable Base Transistor," *IEEE Trans. Electron Dev.*, **ED-27**, 1128 (1980).

54. Y. Awano, K. Tomizawa, and N. Hashizume, "Electrical Performances of GaAs Permeable Base Ballistic Electron Transistors," Proc. 11th Intl. Symp. GaAs and Related Compounds, Biarritz, *Inst. Phys. Conf. Ser.*, No. 74, p. 623 (1984).

55. M. A. Hollis, K. B. Nichols, R. A. Murphy, and C. O. Bozler, "Advances in the Technology for the Permeable Base Transistor," *Proc. Advanced*

Processing of Semicond. Devices, Vol. 797, SPIE Bay Point Symp. on Advances in Semicond. and Semicond. Structures, Bay Point, FL, 1987, p. 335.

56. K. B. Nichols, R. H. Mathews, M. A. Hollis, C. O. Bozler, A. Vera, and R. A. Murphy, "GaAs Permeable Base Transistors Fabricated with 240-nm-Periodicity Tungsten Base Gratings," *IEEE Trans. Electron Dev.*, **35**, 2446 (1988).

57. K. B. Nichols, M. A. Hollis, C. O. Bozler, M. A. Quddus, L. J. Kushner, R. Mathews, A. Vera, S. Rabe, R. A. Murphy, and D. L. Olsson, "High Power-Added Efficiency Measured at 1.3 and 20 GHz Using a GaAs Permeable Base Transistor," Proc. IEEE/Cornell Conf. on Adv. Concepts in High Speed Semicond. Dev. and Circuits, Cornell University, Ithaca, NY (1987), p. 307.

58. D. D. Rathman, B. A. Vojak, D. C. Flanders, and N. P. Economou, "Silicon permeable base transistors," Ext. Abs. 16th Conf. on Solid State Devices and Materials, Kobe, Japan (1984), p. 305.

59. D. D. Rathman and W. K. Niblack, "Silicon Permeable Base Transistors for Low-Phase-Noise Oscillator Applications Up to 20 GHz," *Digest*, IEEE MTT-S Intl. Microwave Symp., Vol. I, IEEE, New York, 1988, pp. 537–540.

60. S. M. Sze, *Physics of Semiconductor Devices*, 2nd ed., Wiley, New York, 1981, Chap. 3.

61. M. W. Geis, N. N. Efremow, and D. D. Rathman, "Summary Abstract: Device applications of diamonds," *J. Vac. Sci. Technol. A*, **6**, 1953 (1988).

62. S. M. Sze, *Physics of Semiconductor Devices*, 2nd ed., Wiley, New York, 1981, Appendix H.

63. S. Adachi, "Lattice thermal resistivity of III–V compound alloys," *J. Appl. Phys.*, **54**, 1844 (1983).

64. M. Armand, D. V. Bui, J. Chevrier, and N. T. Linh, "High Power Microwave Amplification with InP MISFET's," Proc. IEEE/Cornell Conf. on High-Speed Semicond. Devices and Circuits, Cornell University, Ithaca, NY (1983), p. 218.

65. L. Messick, D. A. Collins, R. Nguyen, A. R. Clawson, and G. E. McWilliams, "High-Power High-Efficiency Stable Indium Phosphide MISFET's," *Tech. Dig.*, IEEE Intl. Elect. Dev. Mtg., 1986, pp. 767–770.

66. G. Kelner, S. Binari, K. Sleger, and H. Kong, "β-SiC MESFET's and Buried-Gate JFET's," *IEEE Electron Dev. Lett.*, **EDL-8**, 428 (1987).

67. M. W. Geis, D. D. Rathman, D. J. Ehrlich, R. A. Murphy, and W. T. Lindley, "High-Temperature Point-Contact Transistors and Schottky Diodes Formed on Synthetic Boron-Doped Diamond," *IEEE Electron Dev. Lett.*, **EDL-8**, 341 (1987).

68. M. W. Geis, D. D. Rathman, J. J. Zayhowski, D. Smythe, D. K. Smith, and G. A. Ditmer, "Homoepitaxial Semiconducting Diamond," in G. H.

Johnson, A. R. Badzian, and M. W. Geis, Eds., *Diamond and Diamond-Like Materials Synthesis*, Materials Research Society, Pittsburgh, PA, 1988, p. 115.

69. G. W. Turner, H. K. Choi, J. P. Mattia, C.-L. Chen, S. J. Eglash, and B.-Y. Tsaur, "Monolithic GaAs/Si Integration," *Proc. Mat. Res. Soc. Symp.*, **116**, 179 (1988).

70. H. Shichijo, R. J. Matyi, and A. H. Taddiken, "Monolithic Process for Co-Integration of GaAs and Silicon Circuits," *Tech. Dig.*, IEEE Intl. Elect. Dev. Mtg., 1988, pp. 778–781.

71. W. Fischer, "Equivalent Circuit and Gain of MOS Field Effect Transistors," *Solid-State Electron.*, **9**, 71 (1966).

72. R. H. Dawson, "Equivalent Circuit of the Schottky-Barrier Field-Effect Transistor at Microwave Frequencies," *IEEE Trans. Microwave Theory Tech.*, **MTT-23**, 499 (1975).

5 Heterostructure Field-Effect Transistors

S. J. Pearton and N. J. Shah

AT&T Bell Laboratories
Murray Hill, New Jersey

5.1 INTRODUCTION

The heterostructure field-effect transistor (HFET) has made an important contribution to the field of ultra-high-speed microwave and digital electronics. HFETs demonstrate extremely high performance, with noise figures of 1.3 dB at 60 GHz and a gain of 9.5 dB[1] for discrete microwave HFETs, while digital circuits with propagation delays of 10 ps/gate[2] and static random-access memory (SRAM) access times of 0.5 ns have been reported.[3]

The initial concept of the accumulation of charge at a heterojunction interface and its potential for devices was introduced in the late 1960s.[4] The development in the 1970s of extremely high precision epitaxial-growth techniques has now made the fabrication of suitably high-quality heterostructures in the III-V compound semiconductors a reality. Starting in the 1980s, the application of these techniques to heterojunction field-effect transistors led to the achievement of exceptional performance from both digital and microwave circuits.

The majority of the work on HFETs has been on n-channel devices in the AlGaAs/GaAs materials system. In this type of device, a large-bandgap doped material (AlGaAs) is grown heteroepitaxially on an undoped lower-bandgap material (GaAs). Such a structure is often referred to as a "modulation-doped" heterostructure and it was first demonstrated in 1978.[5] The physics of heterojunction structures is such that the charge transport within an electronic device can be optimized by tailoring the layer thicknesses and doping levels.

In the literature, devices based on the modulation doping principle are referred to by various acronyms, notably selectively doped heterostructure transistor (SDHT), high electron mobility transistor (HEMT), two-dimensional elec-

High-Speed Semiconductor Devices, Edited by S.M. Sze. ISBN 0-471-62307-5

tron gas FET (TEGFET), and modulation doped FET (MODFET). The more general family to which the devices belong shall be referred to as heterostructure FETs or HFETs. While the n^+-AlGaAs/GaAs materials system is the most developed for HFETs, there is now considerable interest in strained-layer InGaAs-channel HFETs on GaAs, and n^+-InAlAs/InGaAs heterostructures on InP, which offer better carrier transport in the channel. These devices will also be covered in this chapter.

In Section 5.2 we present the basic structure of an HFET and a discussion of its mode of operation. This includes the fundamental properties of heterostructures in semiconductors, the nature of bandgap discontinuities, and the calculation of charge control and threshold voltage in HFETs. In Section 5.3, we deal with the transport of carriers in HFETs under the bias conditions in a real device and their influence on device characteristics. Here we also describe the effects of scaling gate length of the HFETs to submicron dimensions on device characteristics. The materials technology and the fabrication of HFETs are covered in Section 5.4. A survey of HFETs other than the conventional modulation-doped n^+-AlGaAs/GaAs HFET is conducted in Section 5.5, where we consider alternative structures and materials systems. The device and circuit performance of HFETs is presented in Section 5.6 to illustrate the progress in the materials, fabrication, and device technologies, which have resulted in ultra-high-speed circuits. We summarize the chapter in Section 5.7, with a discussion of the future directions for HFET technology.

5.2   DEVICE PRINCIPLES: BAND STRUCTURE AND DEVICE BEHAVIOR

In this section we outline some of the basic characteristics of semiconductor heterojunctions and their application to HFET devices. We begin with a description of the HFET structure, to show the reader what the device consists of, after which we move onto a discussion of heterostructures. Finally, we discuss calculation of charge control within the HFET, which leads to some trade-offs in the design of the HFET structure.

5.2.1   The Structure of the Conventional n^+-AlGaAs/GaAs HFET

The cross section of a conventional HFET fabricated by a recess-gate technology is illustrated in Fig. 1. The source and drain contacts and the gate metallization are entirely analogous to those in either Si-MOS or GaAs-MESFET devices. The epitaxial layer structure of the material from which this device was fabricated is shown in Fig. 2. The device is grown on a semi-insulating GaAs substrate to achieve isolation between individual devices in an integrated circuit. The typical resistivity of semi-insulating GaAs is 10^8 Ω-cm. The layers grown on top of this substrate are a semi-insulating or lightly p-type GaAs "buffer" layer, an undoped AlGaAs layer (the "spacer layer"), n^+-AlGaAs (the

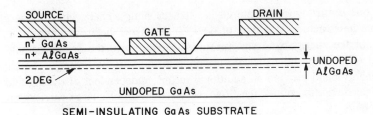

Fig. 1 Schematic of a conventional recess-gate HFET.

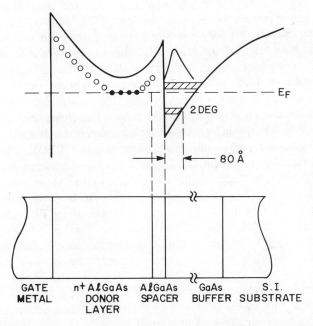

Fig. 2 Epitaxial layer structure and conduction-band diagram for an HFET under positive gate bias.

"donor layer") and an n^+-GaAs "cap" layer. The role of each of these layers will become apparent in this section. The thicknesses of the layers and their doping have a direct influence on the device properties of the HFET. Typical gate lengths vary from 1.0 to < 0.1 μm according to speed, application, and yield requirements. Further details of the fabrication process are covered in Section 5.4.

5.2.2 Heterojunctions in Semiconductors

In conventional semiconductor devices only one type of semiconductor is used throughout, and rectification of current flow is achieved by creating a junction within the structure. This type of device is termed a homostructure. If more than one semiconductor material is used, causing a change in the energy bands within the structure, this type of device is termed a heterostructure. The ability to tailor the energy-band structure adds flexibility to the design of new devices based on doping and material variations in the various layers. The changes in the energy band provide an additional means, independent of doping and applied external fields, to control the flow and distribution of charge carriers throughout these devices.

When two semiconductor materials with different bandgaps are joined together to form a heterojunction, discontinuities in both the conduction- and valence-band edges occur at the heterointerface. For the heterojunction FET, the wide-bandgap material, for example, AlGaAs, is doped n-type with donors. The added charges bend the band edges and create a triangular potential well in the conduction-band edge of the lower bandgap material, for example, GaAs. Electrons accumulate in this well and form a sheet of charge analogous to the inversion channel in an SiO_2/Si metal-oxide semiconductor (MOS) structure. The thickness of this channel is typically only 100 Å, which is much smaller than the de Broglie wavelength of the electrons. Therefore, the electrons are quantized in a two-dimensional system at the heterointerface, and so the channel of the HFET is called a two-dimensional electron gas (2DEG). The physical separation of the electrons from the donors (impurities) reduces the impurity scattering and, therefore, enhances the mobility as well as the effective velocity of the electrons under the influence of an electric field. This sheet of high-mobility electrons can be used as the active channel of an FET and can be modulated by the field effect from a gate electrode. This, in brief, is the basis of the operation of HFET devices for high-speed applications. Heterostructures in the Ge_xSi_{1-x} materials system have also been fabricated into modulation-doped FETs.

It is noteworthy that heterostructures in III-V and group I-V semiconductors also play a large role in the operation of both optical and electronic devices. Transport across heterostructures is used in III-V
tors, for example, to suppress hole current into the emitter and to create hot electron injection of electrons into the base of an n-p-n device. Resonant tunneling and quantum-confined optical and electronic devices owe their properties to the control and quality of heterostructures.[6]

Quality of Heterostructures. There are a number of requirements for producing a high-quality heterojunction system. First, the two components should, in principle, have the same lattice constant. The growth of lattice-mismatched layers on top of each other, in general, leads to the formation of misfit dislocations at the heterointerface between the two materials. Carrier transport through such a disordered region is usually far from ideal, leading to poor device characteristics. In heterojunction devices such as lasers these non-radiative dislocations may cause rapid degradation of the laser output and considerably reduced operational lifetimes. The exception to the lattice-matching requirement occurs when one of the layers in the heterostructure is so thin (usually ≤ 100 Å) that its lattice is actually distorted by strain from the underlying material, and is constrained to have the same lattice constant as this material. We will discuss these strained-layer or pseudomorphic systems in greater detail in Section 5.5.

The second major requirement for a good heterostructure system basically relates to the ability to grow such a system reproducibly. This entails having two materials whose growth conditions, especially temperature, are compatible with each other; the ability to dope the materials controllably, and the availability of appropriate substrates on which to grow the heterostructure. In some cases it also is useful to be able to grade the composition of the materials near the heterointerface to improve the transport properties in that region. In the choice of semiconductor materials it is also most appropriate if the individual lattice components are isoelectronic with each other, in order to avoid uncontrolled cross-doping at the heterointerface. In other words, the best quality heterostructures are obtained with III-V semiconductor layers grown on other III-V layers. In these systems interdiffusion at the interface does not cause doping in either layer. This is not the case with, for example, the growth of III-V layers on group *I-V* substrates or epitaxial layers (such as GaAs on Si).

A final requirement for high-quality heterostructures is that there be predictable values for the conduction- and valence-band offsets at the heterointerface. This is important in tailoring the properties of the heterostructure for different transport properties. These band offsets are calculated from an electron affinity rule. The most widely used and best understood heterostructure is that of $GaAs/Al_xGa_{1-x}As$. The value of the composition x, may be varied from 0 to 1, and the two materials are lattice matched over this entire range. Another important system is that of $In_{0.52}Al_{0.48}As-In_{0.53}Ga_{0.47}As$, which is lattice matched to InP at these particular compositions.

Conduction- and Valence-Band Diagrams. When two dissimilar semiconductors are grown on top of each other, the fact that they have different bandgaps may cause energy offsets between the two conduction bands and between the two valence bands. These discontinuities in the band edges act as potential steps and may be used in conjunction with external applied potentials to control the flow of current within the heterostructure. In attempting to calculate these band offsets the simplest approach is to assume that the potential energy at the heterointerface can be described by a linear superposition of overlapping atomic potentials from both sides of the interface.[7] In practice there usually is considerable difficulty in accurately describing these potentials because of charge and potential readjustment at the interface.

The most widely used method for calculating conduction- and valence-band discontinuities is the electron-affinity rule.[8] This is shown schematically in Fig. 3, with all energies referenced to the vacuum level. The conduction-band offset, or discontinuity, is given by

$$\Delta E_c = \chi_1 - \chi_2 , \qquad (1)$$

where χ_1 and χ_2 are the electron affinities, or energies required to promote an

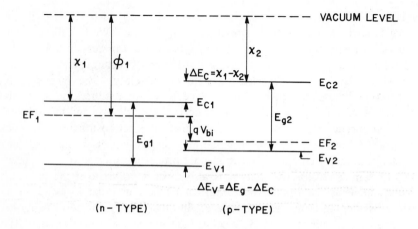

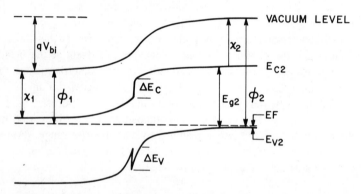

Fig. 3   Band diagrams of an *n*-type and a *p*-type semiconductor (top), and the abrupt heterojunction formed from them (bottom). The discontinuities ΔE_c and ΔE_v are calculated from the electron-affinity rule.

electron from the bottom of the conduction band into vacuum, of the respective materials in the heterostructure. These energies differ from the work functions of the materials, which are defined as the energy separation between the Fermi level and vacuum. Similarly, the valence-band offset is given by

$$\Delta E_v = \Delta E_g - \Delta E_c . \tag{2}$$

Here, ΔE_g is the difference in the bandgaps between the two materials.

The Fermi level is constant throughout the heterostructure. If we designate this energy by E_F, then the electron (*n*) and hole (*p*) densities are given by

$$n = N_c \exp\left[-\frac{E_c - E_F}{kT}\right] \tag{3}$$

and

$$p = N_v \exp\left[-\frac{(E_F - E_v)}{kT}\right],$$ (4)

where N_c is the density of states in the conduction band for each component material and N_v is the density of states in the valence band. At an infinite distance from the heterointerface χ_1 and χ_2 and their respective work functions ϕ_1 and ϕ_2 are unchanged from their bulk values. Also, the individual bandgaps are constant on either side of the heterointerface and the vacuum level is continuous.

The general procedure for constructing an energy-band diagram is first to determine the conduction- and valence-band offsets from the electron-affinity rule, and the second, to determine the built-in potential (V_{bi}) at the heterointerface, where

$$V_{bi} = \phi_2 - \phi_1 = V_n + V_p$$ (5)

for a heterojunction consisting of n- and p-type materials. V_p and V_n are defined below. In the depletion approximation where mobile carriers are neglected and there is no interface charge,

$$N_A W_p = N_D W_n,$$ (6)

where $W_{p,n}$ are the depletion depths on each side of the interface. Then

$$\frac{V_n}{V_p} = \frac{N_D W_n^2 \epsilon_p}{N_A W_p^2 \epsilon_n}$$ (7)

$$= \frac{N_A \epsilon_p}{N_D \epsilon_n}$$ (8)

where ϵ_n and ϵ_p are the dielectric permittivities of the n- and p-type materials. This yields

$$V_n = \frac{N_A \epsilon_p V_{bi}}{N_A \epsilon_p + N_D \epsilon_n}$$ (9)

and

$$V_p = \frac{N_D \epsilon_n V_{bi}}{N_A \epsilon_p + N_D \epsilon_n}.$$ (10)

The depletion depths W_n and W_p are given by

$$W_n = \left[\frac{2N_A \epsilon_n \epsilon_p V_{bi}}{qN_D(\epsilon_n N_D + \epsilon_p N_A)}\right]^{1/2}$$ (11)

and

$$W_p = \left[\frac{2N_D \epsilon_n \epsilon_p V_{bi}}{qN_A(\epsilon_n N_D + \epsilon_p N_A)}\right]^{1/2}$$ (12)

In the case of an applied bias V, the parameter V_{bi} is replaced by $(V_{bi} - V)$. A plot of the carrier concentration with distance from the heterojunction can then be constructed from these relations. The effect of including minority carriers on the lightly doped side of the junction makes little contribution to V_n and V_p but leads to a higher value of electric field strength at the interface compared to that calculated using the depletion approximation.

Experimentally, conduction- and valence-band offsets can be measured by a number of methods, including X-ray photoelectron spectroscopy (XPS), capacitance-voltage and current-voltage measurements, or by the internal photo-emission technique. The most accurate method is XPS, which measures the energy distribution of Auger electrons emitted from core levels in the valence band. This gives a direct measurement of ΔE_v, and ΔE_c is then obtained from $\Delta E_g - \Delta E_v$.

One of the problems with AlGaAs of this composition range is the presence of deep-level states, commonly called DX centers, which are believed to be caused by the band structure of the AlGaAs pushing shallow donor levels down into the gap from their original position near the conduction-band edge. Figure 4 is a plot of the energies of the Γ-, L-, and X-valleys in the GaAs-AlGaAs materials system, as well as the relative position of the DX center trap. These DX centers give rise to two unusual phenomena:

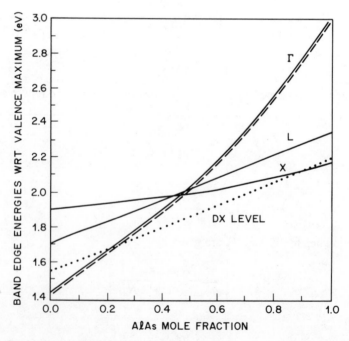

Fig. 4 Conduction-band diagram for the GaAs-AlGaAs material system showing the relative position of DX centers (dotted lines) and the shallow donor impurities (dashed line).

1. Collapse of the drain I-V characteristics in FET structures based on selectively doped heterostructures. This is manifested by a bias-dependent distortion of the I-V characteristics when the HFET is cooled to cryogenic temperatures in the absence of light. A practical temperature for the testing and the potential operation of HFETs is 77 K, the temperature of liquid nitrogen. Therefore, much of the data on this topic consists of comparisons of room temperature and 77 K measurements.

2. Persistent photoconductivity, in which illumination of the heterostructure at reduced temperatures causes an increase in the 2DEG density because of population of other subbands in the quantum well at the heterointerface. When the illumination is removed, however, the increased 2DEG density remains for extended periods (up to days at reduced temperatures).

The effects of these DX centers can be reduced by spatially separating the Si dopants from the Al in the AlGaAs.[9] In this case, the AlGaAs is replaced by an AlAs-GaAs superlattice in which only the GaAs is doped with Si. The band structure of such a superlattice is similar to that of the modulation-doped heterostructure, and thus it can be used to fabricate devices of comparable performance to conventional heterostructure FETs.

5.2.3 Aluminum Mole Fraction

The ratios of the band edge discontinuities (ΔE_c and ΔE_v) to the total energy bandgap difference (ΔE_g) are among the most important parameters in determining the electrical properties of a heterojunction. For $Al_xGa_{1-x}As$ with an aluminum mole fraction less than approximately 0.45, the alloy has a direct bandgap and the energy bandgap difference of a n^+-AlGaAs/GaAs heterojunction at 300 K is given by[10]

$$\Delta E_g \text{ (eV)} = 1.247x \quad ; \quad x \leq 0.45. \tag{13}$$

The ratio of the valence-band-edge discontinuity ΔE_v to the direct Γ bandgap difference has been studied extensively. The majority of the measured ratios lie in the range 0.33 to 0.41 and are independent of the aluminum mole fraction. An empirical relation for the valence-band-edge discontinuity for $x \leq 0.45$ is given by[11]

$$\Delta E_v = 0.45x \tag{14}$$

and, therefore, the conduction-band-edge discontinuity is

$$\Delta E_c = 0.797x . \tag{15}$$

For aluminum mole fractions larger than 0.45, the X-valley drops below the Γ-valley and the conduction-band-edge discontinuity decreases with increasing aluminum mole fraction. The dependence of conduction-band-edge discontinuities on aluminum mole fraction is shown in Fig. 5.[12]

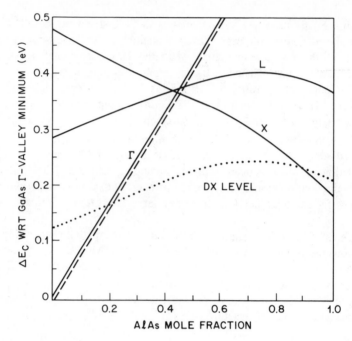

Fig. 5 Dependence of conduction band edge discontinuities in $Al_xGa_{1-x}As$ as a function of AlAs mole fraction.

The significance of the conduction-band-edge discontinuity is that this parameter determines the extent of confinement of charge in the potential well at the heterointerface. The confinement is directly related to the efficiency of the control of the charge in the channel by the field due to the gate. In addition, a deeper well (and a larger bandgap discontinuity) reduces the emission of hot electrons into the donor layer. As discussed later, the strained-layer n^+-AlGaAs/InGaAs HFET derives many of its superior characteristics from the larger ΔE_c. The dependence of ΔE_c for the InGaAs/GaAs heterojunction is roughly

$$\Delta E_c = -0.8y , \qquad (16)$$

where y is the InAs mole fraction. Therefore, a device that has a $x = 0.3$ AlGaAs donor layer and GaAs channel has a similar ΔE_c to a device with a $x = 0.2$ AlGaAs donor layer and $y = 0.1$ InGaAs channel.

Table 1 is a collection of values of physical parameters relevant to the design of HFETs for III-V semiconductors.

5.2.4 Charge Control and Device Characteristics

This section deals with the control of charge in the HFET structure under the influence of the potential from the gate. This is a one-dimensional approach,

TABLE 1 The Principal Physical Parameters of Selected III-V Materials

Material	m^*/m_0	μ_0 (cm^2/V-s)	v_{sat} (10^{7}cm/s)	$\Delta E(\Gamma - L)$ (eV)	E_g (eV)
GaAs	0.063	4600	1.8	0.31	1.42
Al$_{0.3}$Ga$_{0.7}$ As	0.092	1000	—	0.10	1.65
In$_{0.15}$Ga$_{0.85}$ As	0.062	5800	2.0	0.43	0.55
InP	0.08	2800	2.4	0.69	1.35
In$_{0.53}$Ga$_{0.47}$ As	0.032	7800	2.1	0.55	0.78
InAs	0.022	16000	3.5	0.87	0.35

where no account is taken of the field between source and drain or the transport of the carriers across the channel. This more complicated situation is discussed in Section 5.3.

Charge Control in the HFET. Although the basic operation principles of the n^+-AlGaAs/GaAs resembles those of the Si MOSFET, in many aspects the details of material parameters and device physics are different. In particular, the requirement for a more exact determination of the charge in the channel is necessary, due to the much smaller effective mass of electrons in GaAs.[13] However, the final results are similar to those of MOSFETs. The charge control in the HFET is defined as the behavior of charge in the device as a function of an externally applied field from the gate electrode. The influence of source-drain bias must also take into account the transport of this charge in the device.

The sheet charge n_s to applied voltage V_g relationship is given by[14]

$$n_s = \frac{\epsilon}{q\,(d_d + d_i + \Delta d)}\,(V_g - V_{th})\,,\qquad(17)$$

where d_d is the AlGaAs donor-layer thickness and d_i is the AlGaAs spacer-layer thickness. The threshold voltage is then given by

$$V_{th} = \phi_b - \Delta E_c - \frac{qN_d d_d^2}{2\epsilon} + \Delta E_{FO}\,.\qquad(18)$$

For the AlGaAs-GaAs material system, the following material-dependent parameters are used to calculate the threshold voltage: $\Delta d \sim 80$ Å and $\Delta E_{FO} \sim 0$ and 25 mV at 300 and 77 K, respectively. Δd and ΔE_{FO} are fitting parameters used in the linear approximation of Fermi potential-2DEG charge-density relationship. Using this threshold voltage formula, the threshold voltage of any arbitrary HFET with any doping distribution can be determined explicitly. This equation can also be used to derive the "sensitivity" of a threshold voltage to the variation in either the thickness or doping concentration of a particular structure.[14] The threshold voltage of the HFET is a critical parameter in determining its use in depletion (D) and enhancement-depletion (E/D) mode circuits. Much of the literature on device fabrication revolves around the concept of minimizing the sensitivity of threshold voltage to materials and process variations in order to enhance the yield of digital integrated circuits (ICs).

The detailed spatial distribution of charge due to the field effect of the gate on the conducting channel is determined by the doping, composition, and thickness of the layers between the gate and the channel, as well as by the background doping in the channel itself. An approximate solution of the charge-control problem can be solved using the Poisson equation invoking Fermi-Dirac statistics. However, a more accurate solution is afforded by the self-consistent one-dimensional quantum-mechanical solution of the coupled Poisson and Schrödinger equations. This yields the structure of the conduction- and valence-band edges as well as the charge distribution throughout the structure as a function of the gate bias.

In this exact formulation, the Poisson equation

$$\frac{d^2V}{dz^2} - \frac{q\rho_f}{\epsilon} = 0 \tag{19}$$

and the Schrödinger equation

$$\frac{-h^2}{2\pi^2 m_e^*} \frac{d^2\zeta_n}{dz^2} + V\zeta_n = E_n\zeta_n \tag{20}$$

are solved iteratively in a self-consistent manner. The quantity V in these equations is the depth (z)-dependent potential energy representing the conduction band. It is related to the electrostatic potential by the relationship

$$V = q\phi - \chi_e , \tag{21}$$

where q is the magnitude of the electric charge, ϕ is the electrostatic potential and χ_e is the electron affinity in the Poisson equation, ρ_f is the net free charge density, given by

$$\rho_f = qN_D - qn_e - qN_A , \tag{22}$$

where N_D is the ionized dopant (donor) density (reduced by the density of electrons bound to donor sites) and n_e is the free-electron concentration, and N_A is the background doping in the p^- GaAs buffer layer. The material constant ϵ is the dielectric permittivity. The other quantities are Planck's constant, h, effective electron mass, m^*, the nth eigenfunction or eigenstate (ζ_n), and the nth eigenenergy, E_n.

In general, these equations are solved at discrete points in the one-dimensional device and then iterated to arrive at a self-consistent solution. Given a charge distribution, ρ_f, the Poisson equation is solved to determine the conduction-band profile as a function of depth, $V(z)$. This $V(z)$ is solved with the Schrödinger equation for the eigenenergies and eigenmodes, E_n and ζ_n. The

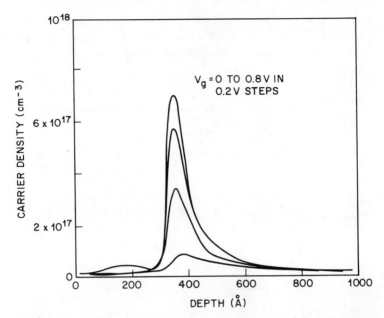

Fig. 6 Plot of carrier density as a function of depth for a GaAs-AlGaAs HFET, for different applied gate voltages. The lowest curve corresponds to 0 V gate bias (V_g), and each higher curve is for 0.2, 0.4, 0.6 and 0.8 V respectively. The latter two give the same result.

free electron concentration is then derived by the relationship

$$
n_e = \frac{4\pi k_b T m_e^*}{h^2} \sum_n \zeta_n^2 \ln \left[1 + \exp \frac{E_F - E_n}{kT} \right]. \tag{23}
$$

This gives a new value of n_e, the electron charge density at a point, and this is then fed back into the calculation of the Poisson equation to get $V(z)$; k is the Boltzmann constant, and E_F is the Fermi energy.

This method was used to produce the plots in Fig. 6 for a conventional HFET for a range of gate-bias voltages. The carrier density and the conduction-band edge are plotted as a function of depth and gate voltage. The curves show the accumulation of charge at the AlGaAs/GaAs heterointerface, which is the two-dimensional quantum-confined electron gas of thickness ~ 100 Å discussed previously, and which constitutes the channel of the HFET. However, at a large enough forward-bias voltage on the gate, a substantial number of free carriers are also accumulated in the AlGaAs layer. Figure 7 is a plot of sheet carrier density as a function of gate-to-channel voltage. The free carriers in AlGaAs are called N_{sc}, whereas carriers bound at a DX center are called N_{SB}. This charge N_{sc} is referred to as the parasitic "parallel AlGaAs MESFET" in the HFET. Because the transport of electrons in AlGaAs is inferior to that in GaAs, the result of this charge is to reduce the effectiveness of

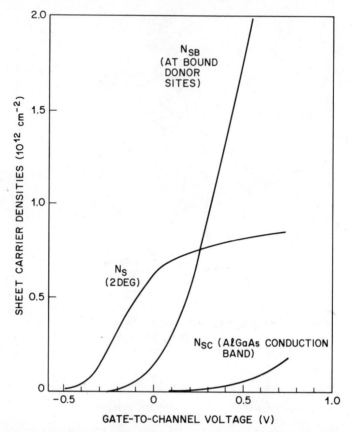

Fig. 7   Sheet carrier density in a GaAs-AlGaAs HFET structure as a function of gate-to-channel voltage. N_S represents the carrier density in the 2DEG, N_{SC} the free carriers in the AlGaAs donor layer, and N_{SB} the carriers bound at DX centers.

transport in the FET and to contribute an excess capacitance, which becomes a large parasitic for the device. This leads us to the importance of charge control and the concept of "modulation efficiency" in HFETs.

Modulation Efficiency.   A nomograph of the doping density and thickness of the AlGaAs layers required to achieve the maximum 2DEG concentration in a conventional n^+-AlGaAs/GaAs HFET structure is illustrated in Fig. 8.[15] The chart shows that if the n^+-AlGaAs region is too thick or is doped above the design limit, the AlGaAs layer remains undepleted. The "minimal shunting zone" is the region where there is no free carrier occupancy in the n^+-AlGaAs. This is where the HFET should operate to obtain the maximum benefit from the confinement of the channel in the 2DEG. For too low a doping or too small a doped AlGaAs thickness, the 2DEG concentration falls below the maximum value (typically 1×10^{12} cm^{-2}) observed in this device.

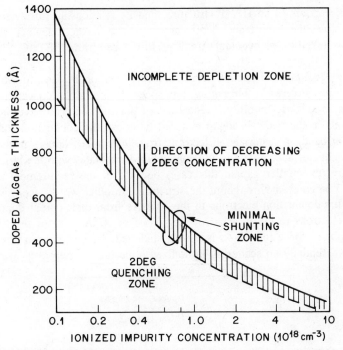

Fig. 8 Nomograph relating doping density and thickness of doped AlGaAs layer to depletion of this layer. (After Ref. 15)

A quantifiable method of determining the effectiveness of modulation of the charge in the 2DEG has been proposed.[16] The degradation of device performance under bias conditions was determined to be due to the inefficient modulation of the charge in the HFET, as a result of thermal electrons in the AlGaAs, and a gradient of the carriers in the channel. In particular, the de-confinement of charge and its dependence on the heterojunction barrier, ΔE_c, showed that the larger the value of ΔE_c, the better the overall device performance.

5.3 DEVICE PRINCIPLES: CARRIER TRANSPORT IN THE HFET

This section deals with HFET operation when the device is under the normal source-drain bias conditions used in a real circuit. The first issue is the mobility of the carriers and the effective velocity of the carriers in the channel. The trade-offs, the aspect ratio (channel length : gate-channel spacing) of the device with submicron gates, are addressed.

5.3.1 Transport in Heterostructure Devices

The transport of carriers in the HFET is a critical aspect in understanding the mode of operation of the device. This section deals with the enhancement of mobility of carriers in an HFET structure, and its relationship to device operation. We stress the importance of the carrier velocity in these devices, and how it is closely linked to the eventual circuit performance of an HFET IC.

Mobility in Modulation-Doped Heterostructure.   One of the keys to the usefulness of heterostructures in fabricating high-speed devices is the concept of selective doping to enhance mobility.[17] The basis of this method is to have the donor energy levels in the wider-bandgap material lie above the conduction-band edge of the lower-bandgap materials. Electrons from the ionized donors in the wide-gap semiconductor accumulate in the conduction-band states with the larger electron affinity (or smaller gap in this case) because of the requirement that the Fermi level be constant throughout the structure. In other words, there is a separation of the conduction electrons in the AlGaAs from their parent donor ions, and these electrons reside in a narrow region of the GaAs side of the AlGaAs-GaAs interface. Since this GaAs is usually undoped, the electrons in this region move with essentially no scattering by ionized impurities. Therefore, at reduced

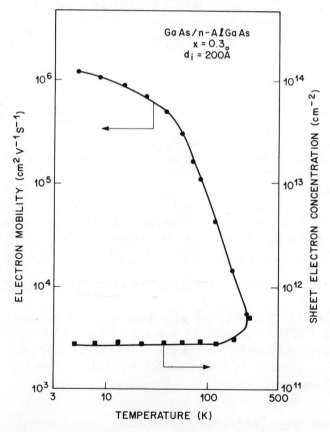

Fig. 9   Temperature dependence of 2DEG electron mobility in a GaAs-AlGaAs HFET structure with a 200 Å AlGaAs spacer layer.   (After Ref. 18)

temperatures, where this scattering mechanism is dominant, the electron mobility is enhanced significantly over the usual value in epitaxial GaAs of equivalent doping density. An example of the magnitude of the electron mobilities achievable by selective doping is shown[18] in Fig. 9. In this case, the mobility increases from 8000 cm^2/V-s at room temperature to 200,000 cm^2/V-s at 77 K. At 4 K the mobility reached a value of 1.5×10^6 cm^2/V-s.

Traditionally molecular beam epitaxy (MBE) has been used to grow the layer sequences for heterostructure devices. In this case the undoped GaAs is p-type with a carrier concentration of $\sim 10^{14}$ cm^{-3}. This minimizes Coulombic scattering of electrons in the 2DEG; and additionally, electrostatic screening by the high electron density further reduces ionized impurity scattering. The 2DEG is confined to within ~ 100 Å of the interface, with a sheet carrier density of around 10^{12} cm^{-2}. This corresponds to a peak electron concentration of $\sim 10^{18}$ cm^{-3}. The electron mobility can be further increased by inserting an undoped AlGaAs spacer layer between the doped AlGaAs layer and the undoped GaAs in which the 2DEG resides. This leads to a reduction in the electron density with a trade-off between the lower 2DEG density and increased electron mobility. The relationship between the spacer layer thickness, AlGaAs doping density and 2DEG electron density is shown[19] in Fig. 10, and the corresponding band diagram for this structure is shown in Fig. 11. In the HFET structure, a large spacer layer results in greater mobility but a lower sheet charge in the channel. Decreased gate to 2DEG spacing results in a higher device transconductance. Typically, device structures utilize spacer-layer thicknesses in the range of 20 to 70 Å. The electron mobility and density are also functions of the conduction-band discontinuity between the AlGaAs and the GaAs. This increases with increasing Al mole fraction in the AlGaAs. The trade-off in this

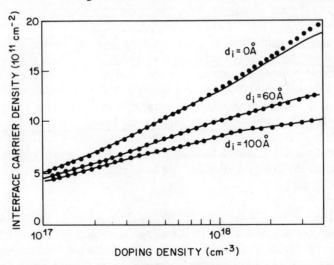

Fig. 10 2DEG carrier density as a function of doping density in the AlGaAs, for various spacer layer thicknesses d_i. (After Ref. 19) The solid and dotted lines correspond to analytical or exact numerical solutions to change density equations.

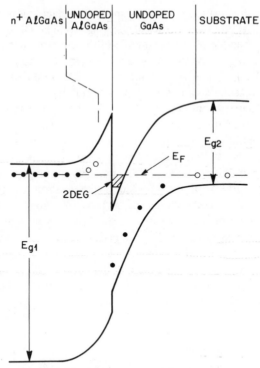

Fig. 11 Band diagram for a HFET structure.

case is that, for high values of Al composition, the material quality of the AlGaAs degrades, trapping by DX centers increases, and the contact resistance to such a wide-bandgap semiconductor is rather high for device applications. For these reasons the Al mole fraction is normally in the range of 0.25 to 0.33.

The trade-off between the transconductance and the low-field mobility illustrates that, although the high low-field mobility is an important characteristic of the HFET, it is not the most important parameter in determining the performance of the HFET. Even though electrons in the lowest subband in the 2DEG have a velocity of 2×10^7 cm/s, the average velocity of all electrons is 1.7×10^7 cm/s at 77 K, with a field of 500 V/cm.[20] At higher fields, the electron mobility of the 2DEG is reduced, due to scattering of electrons into lower-mobility subbands.

Scattering Mechanisms in the HFET.   When electrons in an HFET are accelerated in the presence of a field, they undergo scattering events, which determine their mean free path and their velocity. The scattering mechanisms affecting the 2DEG in HFETs are Coulomb scattering due to ionized donors and deep levels primarily in the doped AlGaAs region, electron-phonon scattering due to the polar optical phonons in the channel, scattering at heterointerfaces, intervalley scattering, impact ionization, and real-space transfer. The scattering of electrons not only limits the low-field mobility but also causes the removal of carriers from the lowest subband of the 2DEG. Real-space transfer of electrons from the 2DEG into the AlGaAs layer occurs by thermionic emission or thermionic field emission. The electrons may be trapped in immobile states on the interface or in the AlGaAs near the interface. They may also be scattered into the GaAs buffer layer or the first excited subband.

Carrier Velocity. Figure 12 is the plot of electron velocity in GaAs and various compositions of $In_yGa_{1-y}As$. The typical peak velocity is approximately 2×10^7 cm/s in GaAs. These velocities are the velocities that can be measured in a constant electric field. It should be noted that two other velocities are often quoted in the literature. One is the "effective" carrier velocity. This velocity is the averaged velocity of carriers in an FET structure. Since the field in the channel is not constant, this value is less than the "peak" velocity. The peak velocity has been calculated using a Monte Carlo method.[21] A high peak velocity ($>2 \times 10^7$ cm/s), well above the values in the curves, can be achieved. At

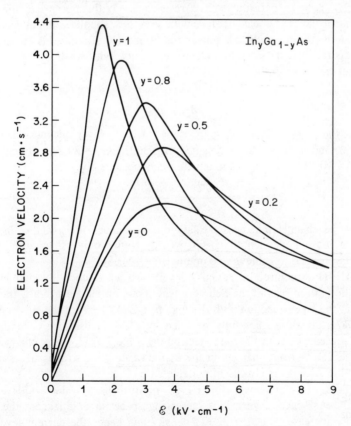

Fig. 12 Velocity-field characteristics for $In_yGa_{1-y}As$.

high fields where non-equilibrium transport is important, carrier velocities higher than the values shown on the curves can be reached.

The electron transit time τ under the gate is given by

$$\tau = \frac{C_g}{g_m} = \frac{L_g}{v}, \tag{24}$$

where C_g is the gate capacitance, g_m is the intrinsic transconductance, L_g is the gate length, and v is the average electron velocity under the gate. The electron velocity appears to be similar to that in undoped GaAs ($\sim 2 \times 10^7$ cm^2/s at 300 K). Using this relation,

$$f_T = \frac{g_m}{2\pi C_g} \propto \frac{v}{L_g} \tag{25}$$

for an HFET. This is the same equation for a conventional GaAs MESFET with velocity saturation in the channel region. The drain voltage at which velocity saturation and high transconductance are observed in the HFET is lower than in the MESFET since the higher carrier density and mobility in the 2DEG of the HFET reduces parasitic source resistance. With appropriate threshold voltage control the HFET can operate with very low supply voltages and logic swings, and, therefore, with low power requirements. Taking this further, for an HFET operating at its saturation electron velocity, with small gate length, the transconductance is given by

$$g_m = \frac{\epsilon v W}{d_d + \Delta d}, \tag{26}$$

where W is the channel width, d_d is the doped AlGaAs layer thickness (and ϵ its permittivity), and Δd is a scaling distance for the 2DEG (~ 80 Å). By comparison, a GaAs MESFET has a maximum transconductance given by

$$g_m = \frac{\epsilon v W}{a} = \frac{\epsilon \mu W V_D}{a L_g}, \tag{27}$$

where a is the channel thickness. Using reliable values for μ and v in these equations, a 0.5-μm gate-length HFET would require only 10 to 20% of the voltage swing needed to achieve maximum f_T in a MESFET.

In the HFET structure the AlGaAs layers are acted upon by two built-in voltages—the built-in bias of the Schottky gate metal on the surface and the voltage offset at the heterojunction with the undoped GaAs beneath the spacer layer. If the AlGaAs layers are not fully depleted by these fields, then a conducting channel is present between the gate electrode and the 2DEG. This channel acts to shield the 2DEG from changes in the applied gate bias, so it is necessary to ensure that the AlGaAs is always fully depleted.

One of the key parameters for HFET performance is the width of the undoped AlGaAs spacer layer. For increasing spacer-layer thickness the electron mobility increases, but the 2DEG density and hence the current level and transconductance are decreased. Figure 13 shows the maximum transconduct-

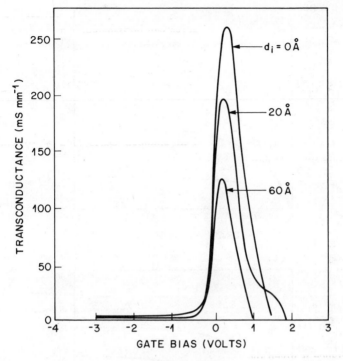

Fig. 13   Transconductance of HFET with different spacer thicknesses (d_i).

ance of an HFET as a function of applied gate bias for different spacer-layer thicknesses. The g_m falls off rapidly under forward biases where charge conduction moves from the 2DEG into the AlGaAs donor layer.[22] A spacer-layer thickness of 20 Å is widely used as a compromise thickness to balance the requirements of high mobility and high current driving capability. It is possible to achieve good HFET performance even with no spacer layer, and Fig. 14 shows drain I-V characteristics at both 300 and 77 K for a 1-μm gate-length device.[23] The maximum transconductances at these temperatures are 250 and 400 mS/mm, respectively.

HFETs offer performance advantages over more conventional MESFETs even at room temperature, but they would be expected to display even better characteristics at cryogenic temperatures. The 2DEG mobility rises rapidly from $\sim$ 8000 cm^2/V-s at 300 K to more than $>$ 100,000 cm^2/V-s at 77 K, and higher transconductance results. For a standard 1-μm gate-length HFET structure this increase in g_m was from 195 mS/mm at 300 K to 305 mS/mm at 77 K.[24] An improvement in operating speed for medium-scale integrated circuits has also been reported upon cooling to 77 K.[25] There are difficulties, of course, with operation of devices at low temperatures because of the change of threshold voltage, and cryogenic wafer probing will be required to optimize performance.

The use of an AlAs-n^+GaAs short-period superlattice to replace the AlGaAs donor layer is necessary for low-temperature HFET operation to circumvent the

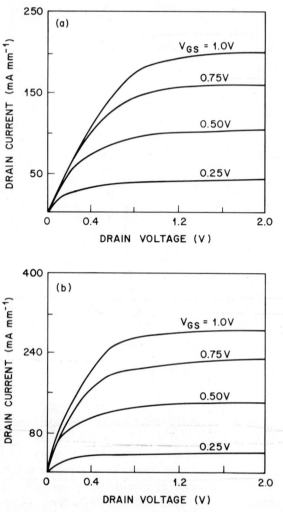

Fig. 14 Drain *I-V* characteristics of HFET with 0 Å spacer at (a) 300 K and (b) 77 K. (After Ref. 23)

collapse of the drain *I-V* characteristics discussed earlier. A further alternative is to replace Si as the dopant in the AlGaAs with Se or possibly S. The DX centers associated with these elements have slightly different energy levels than do those of the Si related centers, and, hence, the trapping and emission time constants are also different. Finally, one can prevent DX center formation entirely by lowering the AlAs mole fraction in the AlGaAs to less than 0.22. The penalty for this is a lower conductance-band offset in the heterostructure, with consequently poorer charge localization in the 2DEG.

Backgating or crosstalk between devices is often a problem in MBE growth HFET structures, and is usually related to the presence of a conducting layer at the substrate-epi interface. This can be circumvented by in situ thermal etching of the substrate prior to growth to ensure removal of interfacial contaminants.

HFETs are particularly suited to low-noise applications. Experimentally, HFET devices show lower noise figures at microwave frequencies than GaAs MESFETs. This lower noise ascribed to the high electron mobility in the HFET, which enhances the carrier velocity overshoot at these frequencies. This leads to higher f_T values from

$$f_T \propto \frac{v}{L_g}. \tag{28}$$

The quantity $1/2\pi f_T$ is the transit time for an FET. Figure 15 is a comparison of GaAs MESFET and AlGaAs/GaAs HFETs for transit time as a function of gate length, and shows higher effective velocity in the HFET. The associated noise figure (NF) is given by [26]

$$NF = 1 + K\frac{f}{f_T} \sqrt{g_m} \ (R_s + R_g) \ , \tag{29}$$

where K is a fitting parameter from the Fukui analysis, f is the frequency of oscillation, g_m is the intrinsic transconductance, R_s is the source resistance, and R_g is the gate resistance. Experimentally it has been found that $K \sim 2.5$. The noise figure is smaller for high values of f_T and therefore lower for HFETs than for MESFETs. Figure 16 shows a comparison of the minimum noise figure for both types of devices over the range 8 to 60 GHz, for room-temperature operation.[27] The gate lengths in both cases were 0.25 μm. By cooling to cryogenic temperature (12.5 K), an associated noise temperature of 5.3 K at 8.5 HGz was obtained.

Gate Barrier on HFETs. The forward *I-V* characteristics of the gate contact on the HFET determine the operating characteristics of the enhancement mode HFET. The fabrication of enhancement (E) mode HFETs requires that a large

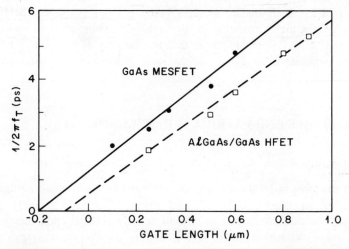

Fig. 15 Transit time for electrons versus gate length for GaAs MESFETs and GaAs-AlGaAs HFETs.

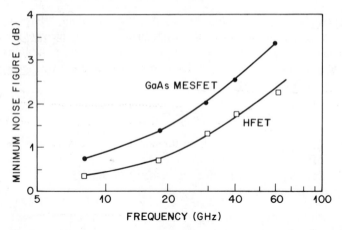

Fig. 16 Minimum noise figure versus frequency for 0.25-μm gate-length GaAs MESFETs and GaAs-AlGaAs HFETs. (After Ref. 27)

enough barrier height is maintained at the Schottky metal-to-semiconductor interface such that the gate does not conduct at bias voltages used in circuit operation. The study of the gate current behavior across the Schottky barrier in HFETs has indicated methods to improve the gate turn-on of the device.[28] The mechanism of gate current was identified as being controlled by two barriers: One was the metal-semiconductor junction and the other the AlGaAs-GaAs junction. The former is the controlling mechanism at low gate bias but, at high forward bias, the latter determined the gate current. Therefore, incorporation of an AlAs spacer layer at the heterojunction between the n^+-AlGaAs layer and the GaAs improves the gate characteristics at forward bias. The typical gate turn-on voltage for HFETs is measured to be ~0.1 V higher than that for GaAs MESFETs, that is, 0.8 V. The enhancement of gate barrier height has been studied for HFETs by growing p^+ layers between the n-AlGaAs and the metal, resulting in barrier-height enhancement of up to 0.8 V. The importance of this technique is that the turn-on of the gate for E-mode devices is suppressed, so that a larger logic swing can be sustained in the devices.

5.4 MATERIALS AND FABRICATION TECHNOLOGY

5.4.1 Materials

The rapid advances in HFET devices and circuits have been based on the exceptional control of doping densities, and layer thicknesses afforded by two epitaxial-growth methods, molecular beam epitaxy (MBE)[29] and metal-organic chemical vapor deposition (MOCVD).[30] In this section we will review these growth technologies, as well as the processing steps required to fabricate typical

HFET structures. A third technique, metal-organic molecular beam epitaxy (MOMBE),[31] a hybrid of the other two methods, has also demonstrated the capability of growing high-quality heterostructures. We will also review the characteristics of this growth technique, as it promises to play an important role in providing a high-throughput capability for HFET structures. For completeness it is worth mentioning that the other common methods for achieving doped layers of GaAs, namely liquid-phase or vapor-phase epitaxy, or ion implantation, are not capable of the type of control needed to produce a typical HFET structure. Liquid-phase epitaxy (LPE) suffers from relatively poor thickness control, non-abrupt interfaces, rough surface morphology, and inadequate layer uniformity. Vapor-phase epitaxy (VPE) is capable of growing very high purity GaAs, but its growth rates are too high for the thicknesses used in HFET structure, and the usual chloride-VPE method cannot produce high-quality AlGaAs. Ion implantation also suffers from non-abrupt doping profiles, and it does not have the ability to produce multi-layered heterostructures consisting of two or more materials.

Molecular Beam Epitaxy. The concept behind MBE is quite simple: atomic or molecular beams of the lattice constituents (Ga, As, Al) produced by heating high-purity, solid sources of these elements in a high vacuum ($\sim 10^{-10}$ torr) chamber, are directed onto a heated substrate. The group III atoms have a high sticking coefficient on the substrate surface at the growth temperature (typically $\sim 550\,^{\circ}C$), and, together with the As atoms, they migrate short distances on the surface, producing layer-by-layer stoichiometric epitaxial growth. The rate of this growth is determined by the arrival rate of the Ga (and Al for AlGaAs) atoms on the substrate surface, and for HFET structures is on the order of 1 μm per hour. A typical MBE system is shown[32] in Fig. 17. Shutters, in front of

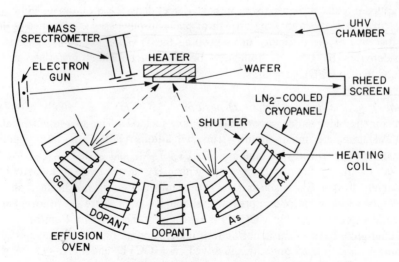

Fig. 17 Typical MBE growth chamber (courtesy M. B. Panish).

the small ovens that heat the source chemicals, allow switching in of Al beams for the growth of AlGaAs, or dopant beams; typically, Si for n-type doping and Be for p-type doping. The growth chamber also contains several in situ diagnostic techniques. Residual gas analyzers are used to check for the presence of impurities prior to growth. Reflection high-energy electron diffraction (RHEED) patterns provide an extremely accurate measure of the growth rate and of the quality of the surface reconstruction during in situ cleaning steps and at the nucleation stage. Finally, ion gauges are used to measure the impinging beam fluxes. Doping calibrations are usually performed on separate, thick-layered structures.

Typical free-electron concentrations in highly doped GaAs grown by MBE may exceed 5×10^{18} cm^{-3}, while in n^+-AlGaAs, the electron density is generally limited to 2×10^{18} cm^{-3}. The run-to-run reproducibility of this doping is $\pm 5\%$, while the run-to-run variations in thickness can be as little as $\pm 1\%$ with the use of RHEED oscillation control. Since the elemental As, Ga, and Al used as sources can be refined to very high purities (seven nines or better), and the growth chamber is itself an inherently clean environment, undoped MBE material has very low carrier concentrations. Undoped GaAs is typically p-type with hole densities in the range 10^{13} to 10^{15} cm^{-3}, due to residual C contamination. At the present time there are three basic limitations when the MBE process and materials are used for HFETs. The first is that the throughput of wafers is only one per growth run. Multiwafer MBE machines will be introduced shortly to overcome this problem. The second limitation is the rather high surface-defect density caused by Ga spitting during growth and by particulates on the wafer itself. This surface-defect density is usually between 200 and 2000 cm^{-2}, although numbers as low as 100 cm^{-2} have been achieved. It is generally accepted that, when the gate of an HFET is located in the vicinity of these surface defects, the device will be inoperative. Therefore, to fabricate high-density integrated circuits, it is necessary to have defect densities ≤ 50 cm^{-2}. The third problem with MBE is that the growth of phosphorous based compounds is difficult because of the difficulty in controlling the P vapor pressure. This is not considered to be a severe limitation because P containing materials are not widely used in HFET structures.

Metal-Organic Chemical Vapor Deposition.   MOCVD is essentially a replacement for the older halide vapor-phase epitaxy method. The source chemicals in MOCVD are the metal-organic trimethylgallium (TMG) or triethylgallium (TEG) for Ga, trimethylaluminum (TEAL) for Al, and arsine (AsH$_3$) for As. Growth occurs by flowing these gases with a H$_2$ carrier over an inductivity or radiatively heated GaAs substrate. The chamber itself is not heated, and the growth occurs at atmospheric pressure, although low pressure (~ 76 torr) can be used. The growth rate for MOCVD is typically of the order of 4 μm/min.

The growth mechanism is not as well understood as for MBE, predominantly because of the higher pressures involved in MOCVD and the more complex source chemicals. At the gas-substrate interface a significant amount of gas-phase pyrolysis of the source materials occurs, and one can consider that these species diffuse across a stagnant boundary layer to the substrate. A combination of gas-phase pyrolysis and surface pyrolysis produces a population of Ga and As (and Al if AlGaAs is being grown) that incorporate into the growing layer. As in MBE, excess As evaporates as As_2 dimers, and stoichiometric GaAs is produced by altering the TMG-to-AsH_3 ratio to give an excess of As at the substrate surface at a given growth temperature. The typical chemical reaction producing GaAs from TMGa and AsH_3 can be written as

$$(CH_3)_3Ga + AsH_3 \rightarrow GaAs + 3CH_4 \tag{30}$$

while for AlGaAs, it is

$$x(CH_3)_3Al + (1-x)(CH_3)_3Ga + AsH_3 \rightarrow Al_xGaAs_{1-x} + 3CH_4 \tag{31}$$

In HFET structures, the GaAs is usually grown at a temperature of around 650°C, and AlGaAs at 675 to 700°C. The quality of AlGaAs grown by MOCVD is extremely high, and typically has much better photoluminescent intensity compared to MBE material. A typical layout for an MOCVD reactor is shown in Fig. 18. A major part of such a system is the gas-handling manifold, which allows control of switching the source gases. The substrate is usually loaded onto a graphite susceptor that is rotated to improve the growth uniformity. For growth rate and composition control, MOCVD makes use of high-precision electronic mass-flow controllers and fast switching valves, which can provide rapid switching of the growth reagents to either the reactor or vent.

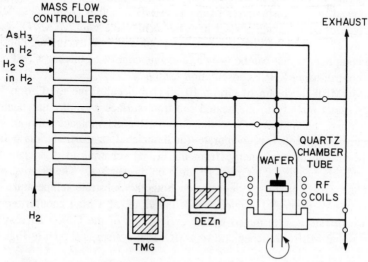

Fig. 18 Schematic of MOCVD reactor. (After Ref. 30) The trimethylgallium and diethylzinc sources are labeled TMG and DEZn, respectively, while the gaseous source chemicals enter through mass flow controllers in the gas-handling manifold at left. The wafer or substrate to be grown upon sits on a heated holder within the chamber tube.

In its ultimate limit, alternate layers of Ga or As can be grown by alternately switching the gases in and out of the system. This variation of MOCVD is known as atomic layer epitaxy (ALE). Since there is no solid Ga source in MOCVD, the problem of surface defects is limited to the more easily controlled particulates.

Typically, three to five wafers are grown in each MOCVD run, giving it a much higher throughput than MBE. Basically this occurs because it is easier to achieve a uniform reactant flux with gaseous sources than with solid sources. The run-to-run and across-wafer doping and thickness uniformities are generally about $\pm 7\%$ for MOCVD grown layers. The purity of undoped GaAs grown by MOCVD is also somewhat worse at this point compared with MBE material and is typically in the range 10^{14} to 10^{15} cm^{-3} n-type because of residual Si and Ge impurities in the source chemicals. There have, however, been demonstrations of the capability of MOCVD to produce high-quality HFET structures with acceptable thickness and doping uniformities. Safety issues are more of a concern with MOCVD than with MBE because of the large volume of toxic gases used in the former. The typical dopants in MOCVD are Si (derived from disilane) for n-type materials, and Zn (derived from trimethylzinc) for p-type materials.

Metal-Organic Molecular Beam Epitaxy.   MOMBE is a hybrid technique combining the gaseous sources of MOCVD with the high-vacuum environment of MBE. In principle, this method is scalable to multiple, large-diameter (3 in.) wafer growth, that have highly abrupt interfaces. Since the source chemicals are external to the growth chamber, operation of those systems is not interrupted by the need to replenish source ovens as in MBE. Other advantages include the same type of low surface- defect densities achievable with MOCVD, versatility in terms of the number of different types of material (including phosphorous based compounds) that can be grown, and fine control of thickness and compositional control using RHEED oscillation feedback.

In MOMBE the beam of Group III alkyl molecules and the Group V dimer species travel directly to the heated substrate surface as molecular beams, with no gas-phase interaction. Once the Group III alkyl molecule impinges on the substrate it can either re-evaporate (undissociated or partially dissociated) or acquire enough thermal energy from the heated substrate to dissociate its three alkyl radicals, leaving the Group III atom on the surface. The relative probability of these two processes depends on both the substrate temperature and the arrival rate of the metal-organics, and, therefore, at a high enough temperature the growth rate is determined by the arrival rate of the Group III alkyls, as in conventional MBE. A schematic of a MOMBE system is shown in Fig. 19.

5.4.2   Fabrication of HFET Devices and Circuits

The uniformity of the thickness and the doping of the AlGaAs layers in the HFET structure and the uniformity of the subsequent selective etches for defin-

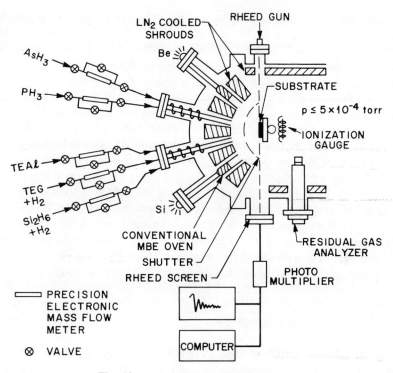

Fig. 19 Schematic of MOMBE system.

ing the driver transistor and load devices are two of the critical features in fabricating HFET circuits.

Ohmic Contacts to HFET Structures. Ohmic contact to the 2DEG is achieved by depositing AuGeNi eutectic, which is subsequently patterned by lift off. The contacts are then alloyed at a temperature around 420 °C for 20 to 30 s. The current understanding of the role of the Au is that it selectively gathers Ga from the underlying semiconductor material, allowing its replacement by Ge to produce a degenerately doped n^+ layer. This allows ohmic flow of current through the doped region. The Ni is used to prevent "balling-up" of the contact upon alloying. Typical contact resistances to the HFET are in the range 0.15 to 0.35 Ω-mm, as measured by the transmission line method. There are many limitations, in terms of fabrication yield and reliability associated with alloyed ohmic contacts to GaAs and AlGaAs. There is considerable interest in fabricating non-alloyed ohmic contacts to these materials, based on epitaxial growth of low-bandgap material such as InGaAs, for the contact layer.[33]

Gate Technology. To achieve a high-yield, high-performance IC technology based on HFETs it is necessary to have uniform and reproducible threshold-voltage control, and to reduce the source resistance of the transistor. Wet-chemical etching is adequate only for discrete-device fabrication so, as the drive to develop HFET circuits continues, much more emphasis has been placed on controlling the threshold voltage by using selective dry etching of GaAs over AlGaAs. To utilize this method to its fullest advantage, it is necessary to incorporate a thin $Al_xGa_{1-x}As$ etch-stop layer in the HFET structure, to allow fabrication of enhancement (E) and depletion mode (D) devices on the same wafer. Such a structure is shown[32] schematically in Fig. 20. Using the same resist mask, selective dry etching is used to remove the top GaAs layer for D-HFETs and also to remove the GaAs layer under the thin AlGaAs etch stop-layer for E-HFETs. As mentioned before, the selective dry etching CCl_2F_2-based chemistry has a very high selectivity for GaAs over AlGaAs. A typical dry etch rate for GaAs is ~5000 Å /min, but upon etching through to the AlGaAs, a thin, involatile film of AlF is formed on the surface, and the AlGaAs etch rate is only ~20 Å/min. With this method the threshold voltages are largely determined by the thickness and doping of the as-grown layers rather than by the processing as in the wet-chemical approach. The control of uniformity and yield afforded by the dry-etch process has allowed fabrication of medium-to-large scale HFET

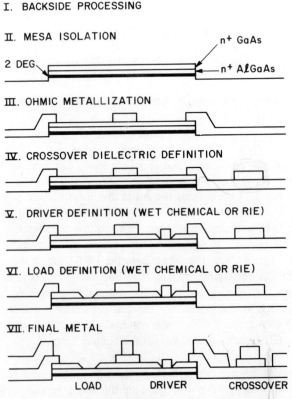

Fig. 20   Schematic of the processing sequence for load and driver HFET devices.

integrated circuits. The most complex circuit reported to date is a 4K SRAM using a 0.5-μm gate length.[33] The anistropy of dry etching makes it readily applicable to submicron features for fabricating complex circuits. The only concern with the use of dry etching is the introduction of damage into the near-surface region by ion bombardment during the etch. The minimization of this damage usually requires self-biases of ≤ 50 V on the cathode of the etch system.

The performance of HFETs can be improved by reducing the source resistance of the transistors, so there are advantages in having a fully self-aligned-gate process. Ion implantation into the source and drain regions reduces both the contact resistance of the ohmic metallization and the resistance between the source and gate and the drain and gate. Both of these effects reduce the knee voltage of drain I-V curves, and increase the extrinsic transconductance of the device. A schematic of a self-aligned-gate process for HFETs is shown in Fig. 21. Once again the mesa isolation step is done by ion milling, although ion implant isolation is also used, followed by deposition of the Schottky metallization, and patterning of this metal by dry etching. The most commonly used metallization is sputtered WSi, which has stable Schottky characteristics at the annealing temperature needed to activate Si ions implanted into the GaAs ($\sim 900\,°C$). This self-aligned implantation step requires that the Si ions be acti-

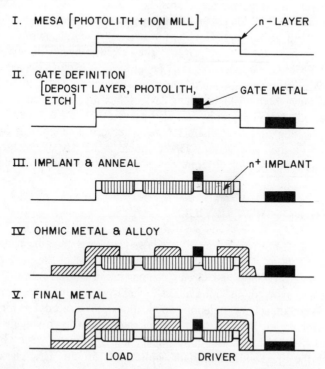

Fig. 21 Schematic of the processing sequence for a self-aligned refractory gate ion-implanted load and driver HFET device.

vated while at the same time both the surface- and heterostructure-interface quality must be preserved.

This technology is planar, allowing a high integration density, and it eliminates the threshold voltage non-uniformity that may be caused by variations in the gate-to-channel spacing as a result of the gate-etching process in the mesa technique. Therefore, the device threshold voltage is determined largely by the uniformity, thickness, and doping of the epitaxial layers, rather than by the processing steps. A limitation of the self-aligned technology is that it allows only one threshold voltage for the HFETs. Therefore, for circuits that need load devices, selective area, ion-implanted saturated resistors are used. For short gate lengths, lateral straggling of the implanted Si ions under the gate metal can affect the threshold voltage of the HFET. These effects can be reduced by spacing the implant away from the edge of the gate by depositing a dielectric sidewall on the gate metallization.

Submicron Lithography for Gates. Much of the lithography for the fabrication of HFET devices and circuits has been performed by optical techniques. However, the application of advanced electron-beam techniques to GaAs devices has a long history based on the advantages in performance to be gained from submicron geometries. The principle application has been for microwave devices, with the development of multilevel resists.[34] These resists are used to make short gates with a "T" shape to minimize gate resistance and yet maintain a small gate footprint to optimize performance. A large number of HFET devices[35] have been made using direct-write, electron-beam lithography with a single-layer PMMA. This technique is being used to explore device and small-scale circuit performance down to short gate lengths. The use of multilevel resists to achieve high yield on such devices has also emerged as a common tool.[36] Electron-beam lithography has also proven to be an indispensable tool to explore the scaling of device characteristics, because a range of gate sizes can be fabricated on the same wafer. One example is a study of voltage gain as a function of gate length[36] for different device structures as shown in Fig. 22.

5.5 FAMILY OF HFET DEVICES

The simple single-heterojunction n^+-AlGaAs/GaAs FET has been described in the previous sections. This device is the basis of most of the pioneering work on device characterization and circuit demonstrations. However, as the HFET technology has evolved, the motivation to improve on this structure has come from the requirements for potential applications. The need for high current-drive FET for digital ICs,[37] a high current, high breakdown-voltage device for power FETs, and low-noise microwave and millimeter-wave FETs have emerged as the three most important applications for this new class of FETs. For digital circuits and for the power FETs, a high extrinsic transconductance

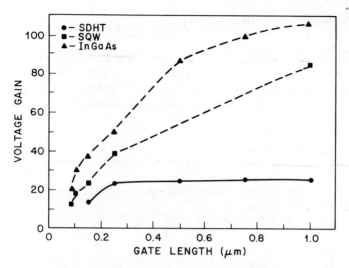

Fig. 22 Voltage gain as a function of gate length for HFETs with different types of channels: Single quantum well (SQW), conventional selectively doped heterostructure transistor (SDHT) or InGaAs channel.

over a large range of gate bias is desirable. In the case of microwave devices, the peak transconductance is important, and in all cases, a minimum gate-channel capacitance is desired for effective modulation of the charge in the channel. The confinement of carriers to a well-defined channel eliminates two unwanted effects. One is the real-space transfer of carriers from the channel to the higher bandgap confining layers, which leads to the formation of a parallel MESFET in the AlGaAs. The second is the space-charge injection of carriers into the substrate, which leads to an increase in the drain output conductance, especially in short channel devices. Finally, the need to have good device behavior at short channel lengths, where the greatest speed is required, is critical for a good HFET technology.

With these simple guidelines, there are two basic variants of the HFET. One is a device whose structure is optimized to achieve the best channel confinement. This type of HFET requires an adjustment of the layer thicknesses and compositions, that is, engineering the band structure of the device. The second variant is an HFET in which doping is distributed to allow the charge to populate the device. Doping may be done from a bulk-doped layer, a δ-doped layer, a superlattice, by doping the conducting channel itself, or by the creation of an inversion channel, with no doping. The family tree of HFETs is shown in Table 2, and conveniently categorizes the large variety of HFET structures that have been developed. Particular examples of each of these devices are now reviewed. Figure 23 is a schematic of the band diagrams of the range of HFETs discussed in this section.

TABLE 2 Family Tree of HFET Devices

Donor Layer	Selective doping	Bulk doped donor layer (SDHT, MODFET, TEGFET, HEMT, etc.) δ-doped donor layer Superlattice donor layer
	Insulated gate	MISFET SISFET
Channel confinements	Quantum-well channel (SQW, etc.) Inverted structure (I-HEMT, I^2-HEMT, etc.)	
Channel doping	Undoped Doped (DMT, etc.)	

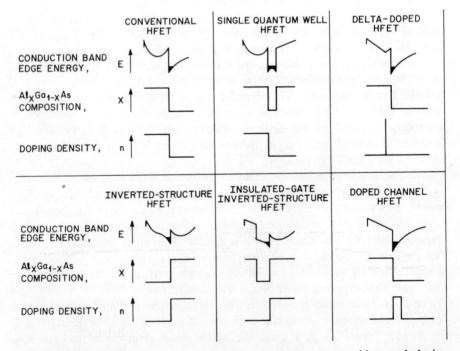

Fig. 23 Schematic diagrams of the conduction-band edge, composition, and doping structures of different HFETs.

5.5.1 HFETs with Improved Channel Confinement

The single quantum well HFET is a structure where the GaAs conducting channel is bounded on each side by AlGaAs confinement layers. This is also referred to as the double-heterostructure (DH) FET. The effect of the confinement is to increase the effective carrier velocity in the channel, and to improve the drain conductance and short-channel properties of the FET, by improving the charge confinement in the channel. In most cases, both of the confining layers are doped to maximize the charge in the intervening channel layer. In addition, the sheet-electron concentration is twice that of the conventional single-heterojunction device,[38] which made this an attractive alternative to the conventional HFET. Since a good-quality GaAs layer must be grown on AlGaAs, a suitable heterointerface for device fabrication[39] is achieved by controlling the growth rate and temperature. The diffusion of the impurities from the lower AlGaAs layer into the channel also demands control of the substrate temperature during growth.

There has been an attempt to make "inverted structure" HFETs (I-HEMT), where the doping lies not between the channel and the gate but below the channel. Much work has also been done on the I^2-HEMT where the gate is placed on an undoped AlGaAs layer, under which is the GaAs channel and then the n^+-AlGaAs donor layer. The significant feature of this device is that the gate leakage is small under forward bias up to 1.4 V, compared to the Schottky turn-on voltage of 0.8 V for the conventional HFET. Comparisons between the I-HEMT and I^2-HEMT have been reported.[40] The I-HEMT has no doping between the gate and the channel, so the threshold voltage sensitivity to the recess depth in the I-HEMT is less than that in the conventional HFET. Also, it does not have the problems of the parallel MESFET due to the population of the n^+-AlGaAs donor layer in the conventional structure. Whereas the gate of the I-HEMT is placed on undoped GaAs, the I^2-HEMT has undoped AlGaAs as the barrier under the gate. The improved barrier height at the metal-semiconductor interface and the better confinement of carriers are seen as advantages of this device over the I-HEMT. The I^2-HEMT has remarkable short-channel behavior. An I^2-HEMT with 0.8-μm gates has a drain conductance of only 2 mS/mm at 300 K, and a peak transconductance of 230 mS/mm.

5.5.2 HFETs with Different Donor Layer

The charge control in a superlattice donor structure has been discussed by a number of authors.[41,42] In the short-period superlattice, there is no parallel conduction in the AlGaAs layers in the superlattices, since they are undoped and have a high bandgap. The compositions reported are in the range of $Al_{0.6}Ga_{0.4}As$ to AlAs. The narrow doped GaAs layers also have high eigenenergy states, so they do not contribute to parasitic parallel conduction under forward bias, which is important for the E-mode devices used in digital circuits. The superlattice donor has been used extensively for the fabrication of devices and circuits by the self-aligned ion implantation process.[43]

Delta-doping, also referred to as pulse or atomic layer or spike doping consists of growth interruption,[44,45] and the incorporation of a doping layer of density $> 1 \times 10^{13}$ cm^{-2}. Low-noise microwave GaAs MESFETs with an AlGaAs buffer layer and a δ-doped channel were fabricated with a gate length of 0.1 μm.[46] The device had a minimum noise figure of 0.8 dB at 18 GHz, and reduced short-channel effects compared to conventional MESFETs. δ-doping has been adopted for a number of devices as a desirable alternative to bulk doped donor layers.[45]

5.5.3 Insulated Gate HFETs

The heterostructure insulated-gate FET (HIGFET), also known as the heterostructure MISFET, is similar to the conventional HFET except that the epitaxial layer structure is undoped. The attraction of this device is that there is no doped AlGaAs layer that would have trapping behavior, and that the threshold voltage of the device is governed by the material and its composition, not its thickness and doping, as is the case for the n^+-AlGaAs/GaAs HFET. Because the layers are undoped, however, there is a need for a self-aligned process rather than the simpler recess-gate technology to reduce parasitic source resistance. These devices have evolved in two forms, one where the gate consists of a metal (HIGFET) and one where the gate is a semiconductor (SISFET).

The GaAs-gate FET is analogous to the polysilicon gate FET in Si MOS devices.[47,48] The threshold voltage of the GaAs-gate HFET is determined by the work function difference between the gate material and the channel, which in this case is close to zero. This threshold voltage is also almost independent of the AlGaAs layer thickness, in contrast to the n^+-AlGaAs/GaAs HFET, where there is a strong sensitivity to doping and thickness. In addition, the Ge can be deposited in situ in the MBE system, and then selectively etched to form the gate pattern. This is a SISFET, employing Ge rather than GaAs as the semiconductor gate, and has been implemented into static frequency dividers operating at 11.3 GHz.[49]

The use of an undoped epitaxial layer for an n-channel device has also been attempted.[50] The uniformity of these MIS-like HFETs was good for a threshold voltage of 0.25 V, the standard deviation was 11 mV for a quadrant of a 2-in. wafer. The control was attributed to the lack of dependence of doping and thickness on the threshold voltage of this device.

In contrast to the SISFETs, the W-Si refractory metal gate HIGFET has a threshold voltage of +0.9 V, and the metallization is performed after the epitaxial growth rather than in situ. A number of studies of the HIGFET device operation have been made.[51]

A number of authors have investigated AlGaAs/GaAs heterostructure FETs that contain a two-dimensional hole gas (2DHG), using p-type doping of the AlGaAs acceptor layer. The motivation was not to implement p-type logic only, but to determine the suitability of the 2DHG device as the p-channel device in complementary HFET technology. The room temperature transconductance of these devices was poor, but due to the enhancement of hole mobility at 77 K, a peak transconductance of 28 mS/mm was measured for a 2-μm FET.[52] It is interesting to note that the p-channel device does not exhibit "collapse" of the drain I-V characteristics, in contrast to the n-channel device.[53] The absence of doping in the active layers of the HIGFET and SISFET devices allows the creation of n-channel and p-channel devices by the choice of implantation species for the self-aligned implant. The complementary HFET technology has speed comparable to the n-channel HFET, at one-tenth of the power.[54]

5.5.4 Doped-Channel HFET

The doped-channel HFET is a significant departure from the conventional HFET in which the GaAs channel is not doped to minimize impurity scattering and, therefore, to maximize the mobility. The excellent performance of the doped-channel device confirmed that the confinement of carriers at a heterointerface in a large concentration is more important than their mobility, at least for transistor action.[55] In the case of the doped-channel device, the gate is deposited on undoped AlGaAs that acts as an insulator and as a confinement barrier for the underlying layer of n^+-GaAs, which is the channel. A self-aligned ion-implanted HFET with W-Al as the refractory metal was made with both a conventional and doped-channel structure. The doping was 2×10^{18} cm^{-3} in the 60-Å-thick GaAs channel.[56] This device is commonly called the doped-channel MIS-like FET, or DMT. The gate-drain breakdown voltage was superior to the MESFET or conventional HFET, because the gate contact is on undoped AlGaAs. A high transconductance was sustained over a gate voltage range of 2.5 V with a peak value of 310 mS/mm and saturation current of 650 mA/mm at 300 K for a 0.5-μm gate, which is four times that of the conventional HFET. The absence of doping in the AlGaAs resulted in the removal of the usual persistent photoconductivity effects. The room temperature electron saturation velocity was calculated as 1.5×10^7 cm/s, which rose to 2×10^7 cm/s at 77 K.

5.5.5 Summary of Heterostructure FET Structures

The simple single-heterojunction n^+-AlGaAs/GaAs HFET spurred many fundamental studies and applications of the high-speed capabilities of the device. A number of significant advances have been made to improve on these structures. These advances are summarized here. The introduction of the quantum well for carrier confinement and the insulated gate for the reduction of gate current have made significant improvements in the device behavior. The use of δ doping, the doped channel, and the superlattice donor have all proven to be better than n^+-AlGaAs for the tailoring of the band structure and material properties to improve charge control in the channel. The innovations in the SISFET and HIGFET, and their applications to n-channel and complementary devices have opened a new regime for HFET device operation.

The single-quantum-well (inverted and insulated-gate inverted) HFET, the heterojunction insulated-gate FET, and the doped-channel devices, each with their own characteristic advantages, are merging into a new class of HFETs with low drain conductance, high gate-drain breakdown voltage, low gate conduction under forward bias, and a higher channel sheet-charge density than the conventional n^+-AlGaAs/GaAs HFET.

5.5.6 HFETs in Different Materials Systems

The small effective electron mass (m^*), high low-field mobility (μ_0), high electron saturation velocity (v_{sat}), and large energy gap between the Γ- and L-valleys make InP and In$_{1-x}$Ga$_x$As promising materials for high-speed electronic applications. A comparison of principal physical parameters of GaAs, InP, InAs, and In$_{0.53}$Ga$_{0.47}$As was given earlier.[57] In particular, In$_{0.53}$Ga$_{0.47}$As, which is lattice matched to InP, is a semiconductor of great interest for electronic as well as optoelectronic applications. Experimentally,[58] mobilities as high as 13,800 cm^2/V-s an effective saturation velocity up to 2.95×10^7 cm/s,[59] and an energy separation between the Γ- and L-valley as large as 0.55 eV have been reported[60] for In$_{0.53}$Ga$_{0.47}$As at room temperature. InAlAs/InGaAs HFETs, which take advantage of the high electron mobility of the 2DEG in InGaAs, have been demonstrated to have superior device characteristics compared with conventional GaAs/AlGaAs devices. One example is the achievement of an f_T of over 200 GHz with 0.1-μm gate length HFETs.[1]

5.5.7 Lattice-Matched HFETs

Due to the intrinsically higher electron mobility, velocity, and larger discontinuity of the conduction-band edge, the InGaAs HFET potentially offers performance improvements over those of AlGaAs/GaAs HFETs or other InGaAs FETs. At room temperature, mobility of $\sim 10^4$ cm^2/V-s has been demonstrated in a selectively doped n-InAlAs/InGaAs structure.[61]

5.5.8 Strained-Layer n-HFETs

Early predictions[62] and later experimental confirmation[63] showed that the lattice mismatch in layers thinner than the critical thickness can be totally accommodated by coherent layer strain.[64] More recent work has resulted in improvements in the understanding of the stability of strained-layer structures.[65] These studies were motivated by the need to reduce or eliminate some of the problems associated with deep-level traps. The conduction-band edge drops below the DX centers if the aluminum mole fraction is less than 0.22, reducing the deep-level density. Unfortunately, the use of low AlAs mole fractions also degrades the performance of the FETs by reducing the conduction-band-edge discontinuity, which results in degradation in the carrier confinement. The strained-layer AlGaAs/InGaAs structure maintains sufficient conduction-bandgap edge discontinuity for 2DEG confinement by using a lower bandgap material in the channel. A mobility of 4210 cm^2/V-s with a sheet-charge density of 4.5×10^{12} cm^{-2} has been demonstrated at 300 K and 18,640 cm^2/V-s and 2.4×10^{12} cm^{-2} at 77 K.[66] Figure 24 shows the conduction-band diagram of such a structure.

The indium mole fraction in this mismatched channel with a practical thickness (> 100 Å) is limited to 0.25. It is desirable to go to an even higher indium mole fraction for better performance. This is achieved with device structures grown on InP substrates. A strained-layer n-In$_{0.52-u}$Al$_{0.48+u}$As/In$_{0.53+u}$Ga$_{0.47-u}$As HFET structure on InP has been demonstrated.[67] For $u = 0.07$ both the InAlAs donor layer and the InGaAs channel have a 1% lattice mismatch, but the strains are equal and in opposite directions. The In$_{0.6}$Ga$_{0.4}$As channel offers not only higher electron velocity and mobility, but also has the largest conduction-band-edge discontinuity among all HFET structures reported in the literature. The higher aluminum mole fraction in the InAlAs layer also leads to larger bandgap and higher Schottky barrier height than those in the n-InAlAs/InGaAs HFETS.[68] The measured unity-current-gain frequencies of various GaAs- and InP-based HFETs are compared in Fig. 25. Figure 26 is a comparison of the transit times of such devices as a function of

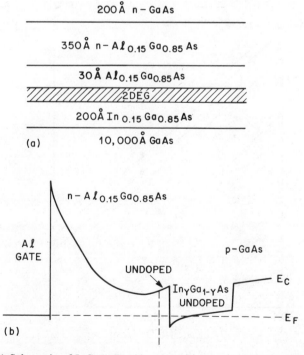

Fig. 24 (a) Schematic of InGaAs/InAlAs pseudomorphic HFET (b) Conduction-band diagram for this structure. (After Ref. 73)

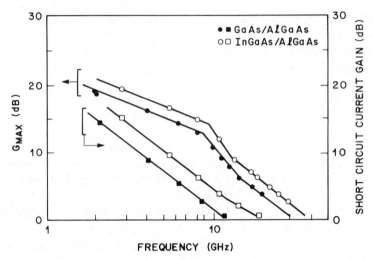

Fig. 25 Maximum available gain (G_{max}) shown as circles and short circuit gain shown as squares of conventional GaAs/AlGaAs HFETs and $In_{0.15}Ga_{0.85}As/Al_{0.15}Ga_{0.85}As$ HFETs. (After Ref. 73)

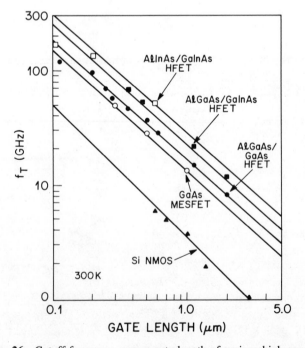

Fig. 26 Cutoff frequency versus gate length of various high-speed devices.

gate length. As expected, the data clearly demonstrates that the speed of the InP-based HFETs are comparable to or better than those of the best AlGaAs/GaAs HFETs.

5.6   DEVICE AND CIRCUIT PERFORMANCE

There are several possibilities for the logic family used for the implementation of HFET circuits. These alternatives determine the type of devices to be incorporated in the circuit. The most widely used logic family is direct-coupled FET logic (DCFL), which incorporates series pairs of depletion mode devices and enhancement mode loads. DCFL has the advantage of high-speed operation, low power requirements, and simpler and more compact designs compared to alternatives such as Schottky-diode FET logic (SDFL) or buffered FET logic.[17,69,70] Both of these logic families use enhancement mode devices.

One of the reasons for the high-speed capability of DCFL is the comparatively small logic swing involved. This places fairly stringent limits on the allowable noise margins in the DCFL gates, with a concurrent requirement for very good control of the uniformity of device threshold voltages. Typical noise margins of HFET gates using DCFL are on the order of 200 mV. Fortunately with the control offered by MBE growth and dry-etching techniques, this is readily achievable, allowing fabrication of circuits up to 16K gates.

5.6.1   Device Options

In the DCFL design, one of the critical choices is that of the width of the driver gates. Increased widths give rise to higher transconductance and current driving capability, while at the same time increasing the power requirements in the circuit.[19,70] Typical gate widths are in the range of 10 to 50 μm, which appears to be an optimum compromise. For the highest speed operation it is also necessary to operate the load devices, either ungated HFETs or saturated resistors, at approximately half the current of the driver devices. The minimum supply voltage for this type of operation is given by the sum of the knee voltages of the driver and load device. The maximum supply voltage is limited both by the gate turn-on voltage (~ 1.5 V) and the sheet carrier density in the 2DEG. The logic swing for both MESFET and HFET DCFL circuits is limited, in general, to less than one volt.

The potential applications for HFETs have motivated the development of a number of different device structures. There are high-current devices for digital ICs and power FETs, and low-noise microwave and millimeter-wave FETs for communication applications.[19] The high-current devices require high transconductance over a wide range of applied gate bias, whereas the microwave FETs require the highest peak transconductance. These different applications have led to the development of two basic HFET structures. The first strives for the best

2DEG confinement by adjustment of the layer thicknesses and AlGaAs composition. The second type of structure varies the mode by which doping is introduced into the device. The conventional structure uses a uniformly doped AlGaAs layer, although, as we have seen earlier, an AlAs-GaAs superlattice may be used in its place. It is also possible to planar-dope the GaAs channel to achieve a 2DEG. In this method Si doping is limited to basically one monolayer during the growth of the structure. This layer is grown by actually stopping the growth of the GaAs (or AlGaAs), allowing one monolayer (~ 2.6 Å) of Si atoms to settle on the surface and then resuming the growth of undoped GaAs.[71] The final variants of the doping method for HFETs include doping the channel layer itself and also creating an inversion layer, with no doping. Each of these particular structures within the two basic types of HFETs has been reviewed previously.[72] The formation of two-dimensional hole gases (2DHG) using p-type doping of the AlGaAs allows consideration of complementary logic. The valence-band offset for GaAs-AlGaAs is only about 20% that of the conduction band, so that, in general the Al fraction in the AlGaAs is increased to improve confinement of the 2DHG. Hole mobilities above $250,00$ cm^2/V-s at 2 K have been reported, and p-channel devices do not suffer from the drain I-V collapse problem because holes are not trapped effectively by DX centers. Complementary logic is also possible in insulated-gate HFET structures using selective, self-aligned implantation of both n- and p-type dopants.

5.6.2   Circuit Operation

In terms of logic circuits, most work has been carried out with ring oscillators. These simple circuits have provided ample proof of the high-speed, lower power capabilities of HFET devices.[73] Ring oscillator propagation delays below 5 ps with sub-micron gate length HFETs have been demonstrated. A comparison of the scaling properties of HFETs and MESFETs showed reduced short-channel effects in the former because of better confinement of the 2DEG. The speed of HFETs improves with smaller gate length, making it suitable for scaling to sub-micron feature sizes. Ring oscillators are often used as demonstration circuits to check material and processing quality, but they are not regarded as entirely practical because loading effects are not considered. Figure 27 is a comparison of published ring oscillator data for HFETs and MESFETs.

To evaluate HFETs in more complex logic circuits, frequency dividers and multipliers are the most common test vehicles. The frequency dividers have been of various types—single clocked divide-by-two circuits consisting of eight DCFL NOR gates, an inverter, and four output buffers, which operated up to 8.9 GHz at 77 K and up to 5.5 GHz at 300 K, and circuits based on four 0.7 μm DCFL AND-NOR gates, which operated up to 13 GHz at 77 K.[74] The 77 K results are the fastest, low-power frequency dividers in any technology. Silicon bipolar circuits have approached the operating speed of HFETs, but at much higher power levels (approximately an order of magnitude higher). GaAs MESFETs also have higher power dissipation than HFETs. Multipliers of type 4×4, 5×5, and 8×8 have also been demonstrated with HFETs.

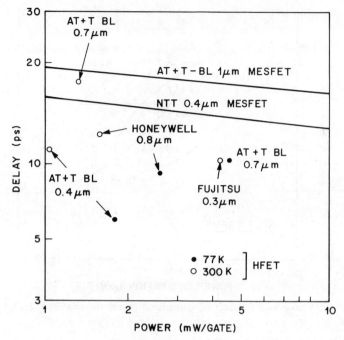

Fig. 27 Comparison of ring oscillator delays of HFETs and MESFETs. (After Ref. 72)

Memory circuits are another potential application for HFET technology. The most complex design to date is a 16 × 16K static random-access memory (SRAM), with an access time of 3.4 ns at 77 K. The power dissipation was 1.34 W at this temperature. In general, for SRAMs the access times are less than a nanosecond, which is faster than comparable memories using MESFETs. Figure 28 is a comparison of HFET SRAM speeds. These memory circuits are the most complicated HFET designs employing depletion mode devices for loads, and enhancement-depletion DCFL circuits. Dry processing is a critical part of the fabrication of these large-scale circuits, particularly with DCFL designs. In addition, memories may be fabricated with complementary logic using the insulated-gate approach, which is more tolerant to threshold voltage variations than DCFL. This is inherently a low-power approach. Remember, that low operating voltages are necessary for low-power operation, and, therefore, very fine control over the turn-on characteristics is required. Reduced gate lengths are also important for improving circuit performance. As the gate lengths are reduced below one micron, the gate capacitance is also reduced and the device transconductance improves, leading to better current-gain cutoff frequencies and faster switching times. To keep power consumption low, it is usually necessary to use enhancement mode devices, as much as possible, in the circuits. Enhancement mode devices are used as the switches in the circuits, whereas the loads are depletion mode HFETs with the gate shorted to the source. In the enhancement mode switches, a positive gate voltage is applied to

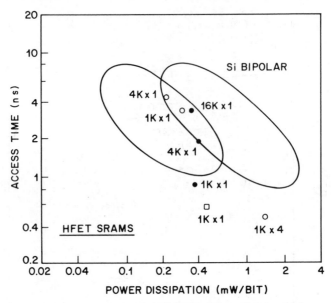

Fig. 28 Access time as a function of normalized power dissipation of HFET SRAMS. (After Ref. 72)

turn the device on. The ultimate speed of such a switch is determined by its transconductance divided by the associated capacitance, comprising the gate capacitance and interconnect capacitances.

For VLSI complexity, it appears necessary to achieve switching delays below 10^{-11} s, with power dissipation of $\leq 10^{-4}$ W/stage using 1-μm design rules. There do not appear to be any fundamental restrictions on achieving this type of performance.

5.6.3   Practical Applications for Heterojunction FETs

HFET devices and circuits are best suited to high-speed, low-noise, and low-power applications, even though they may, obviously, be used in any situation where conventional GaAs MESFETs are now used.[70,73] Some of these areas include communications, such as radar links, direct broadcast satellite television, cellular telephones, cable television converters, and data processing applications such as supercomputers, where high-speed and low-power consumption are critical. There are also a large number of military applications, particularly phased-array antenna radar. This basically consists of a large number of individual transistors connected in an IC that operates at microwave frequencies. The overriding key to success of antenna radar is minaturization, and this requires compromises in the adaption of the transistors with the other elements in the microwave circuit. The transistors usually cannot be operated at their optimum conditions, and, therefore, to achieve acceptable performance, very high quality devices are necessary so that, even with degraded performance, their performance is still adequate.

A key future application is also likely to be in high-definition television, where the increased complexity of the electronics requires minimum levels of power dissipation from individual transistors.

5.6.4 Future Directions

Although the GaAs-AlGaAs system has been the most widely studied and is a mature technology, there are advantages from the use of InGaAs as the channel layer. InGaAs has a smaller bandgap than GaAs, which leads to a higher electron mobility and a concomitantly higher saturated-electron velocity. This, in turn, leads to a higher transconductance and a faster device than conventional GaAs-AlGaAs. Similar performance improvements are achieved with InAlAs-InGaAs, in part because of the improved Schottky barrier height of InAlAs (~ 0.6 eV) relative to InGaAs (~ 0.2 eV) and InP ($0.4 - 0.5$ eV).

5.7 SUMMARY AND FUTURE TRENDS

The heterostructure is the basic building block for high-speed FETs. It combines stringent requirements on the control of layer thicknesses, doping, and composition with exacting processing steps. The result is a truly novel charge-transfer device in which the motion of charge carriers is confined to two dimensions by the quantum well formed at a GaAs-AlGaAs heterostructure. HFET devices and circuits already have practical applications, and with the move to more "Si-like" processing it is likely that such circuits will be applied aggressively wherever low-noise and high-speed performance are critical.

HFETs have inherent performance advantages over more conventional GaAs MESFETs, predominantly because of the mobility enhancement in selectively doped heterostructures. Further advantages accrue with the use of other materials systems such as InGaAs and InAlAs. These materials are also interesting because of their compatibility with InP-based lightwave devices such as lasers and photodetectors. The use of HFETs in monolithic microwave integrated circuits appears to be an attractive combination for use in microwave communication systems. The large enhancement in electron mobility in HFETs at cryogenic temperatures, provides another relatively unexplored area for the exploitation of their speed. One obvious application is in the next generation of supercomputers.

PROBLEMS

1. The structures of the conventional and delta-doped HFETs are shown in Figure 23. Evaluate the sensitivity of the threshold voltage of the two

devices as a function of the AlGaAs layer thickness. Comment on the implications of using these two types of structures for HFET circuits. The threshold voltage is estimated as

$$\Phi_m - \frac{\Delta E_c}{q} - \frac{q}{\epsilon} \int N_D(z) z \, dz \; ,$$

where Φ_m is the Schottky barrier height, ΔE_c is the band edge discontinuity, and $N_D(z)$ is the net doping density profile.

2. When the HFET structures experience a high temperature, the Si dopants can diffuse. The usual design for a conventional HFET requires an undoped AlGaAs spacer layer of 25 Å to be maintained between the n^+-AlGaAs and the undoped GaAs. Calculate the diffusion of Si impurities during a typical growth run, which should be allowed for in the device design. (The typical growth rate is 1 μm per hour, and the thickness to be grown is 500 Å.) What are the implications if this device has to be ion-implanted and annealed at 900°C for 10 s? The diffusion length (assuming a Gaussian distribution is estimated using the diffusion length formula, $L_D = \sqrt{(Dt)}$. The value of D is given by $D = D_0 \exp(-E_a/k_B T)$, where $D_0 = 4 \times 10^{-8} \text{cm}^2/\text{s}$ and $E_a = 1.3$ eV.

3. Carefully study Figures 20 and 21, the fabrication sequences for the non-self-aligned (recess-gate) and self-aligned ion-implanted HFET processes. Describe all the associated process steps, and identify the critical steps. Also, comment on the advantages and limitations of each of the processes for integrated circuit fabrication.

4. Plot the transit time as a function of gate length for HFETs with a saturation velocity of 1.6×10^7 cm/s. What is the cutoff frequency at 0.5 and 0.1 μm gate length? Given that the typical g_m of an HFET is 300 mS/mm, what is the estimated C_g for a 0.5-μm gate?

5. The typical growth rate during MBE is 1 μm/hour. Calculate the flux rate of atoms arriving at the wafer surface. Compare this to the flux rate of gaseous atoms arriving at the surface given the background pressure in the typical MBE system is better than 10^{-7} Pa.

6. The oval defect density of an MBE growth run can be as low as 100 defects/cm. What chip area can be fabricated such that the defect density is below one defect per chip? Discuss the implications for VLSI electronics based on molecular beam epitaxially grown wafers.

7. Compare the characteristics of a GaAs MESFET, the conventional HFET, and the strained-layer (InGaAs channel) HFET. Concentrate your comments on figures of merit such as g_m and f_T and the impact of the ΔE_C of the heterostructures created in these devices.

REFERENCES

1. U. K. Mishra, A. S. Brown, S. E. Rosenbaum, C. E. Hooper, M. W. Pierce, M. J. Delaney, S. Vaughn, and K. White, "Microwave Performance of AllnAs-GaInAs HEMT's with 0.2 and 0.1 μm Gate Length," *IEEE Electron Dev. Lett.*, **EDL-12**, 647 (1988).

2. N. J. Shah, S. S. Pei, C. W. Tu, and R. C. Tiberio, "Gate Length Dependence of the Speed of SSI Circuits Using Submicrometer Selectively Doped Heterostructure Transistor technology," *IEEE Trans. Electron Dev.*, **ED-33**, 543 (1986).

3. S. Notomi, Y. Awano, M. Kosugi, T. Nagata, K. Kosammura, M. Ono, N. Kobyashi, H. Ishiwara, K. Odani, T. Mimura, and M. Abe, "A High Speed 1 K × 4-bit Static RAM Using 0.5 Micron Gate HEMT," IEEE GaAs IC Symposium, p. 177 (1987).

4. L. Esaki and R. Tsu, "Superlattice and Negative Conductivity in Semiconductors," IBM Research, Internal Report RC2418, March 26, 1969.

5. R. Dingle, H. L. Stormer, A. C. Gossard, and W. Wiegmann, "Electron Mobilities in Modulation-doped GaAs-AlGaAs Heterojunction Superlattices," *Appl. Phys. Lett.*, **33**, 665 (1978).

6. F. Capasso, S. Sen, F. Beltram, L. Lunardi, A. S. Vengurlekar, P. R. Smith, N. J. Shah, R. J. Malik, and A. Y. Cho, "Quantum Functional Devices: Resonant Tunnelling Transistors, Memory Devices, Multiple Valued Logic and Circuits with Reduced Complexity," *IEEE Trans. Electron Dev.*, **ED-36**, 2065 (1989).

7. H. Kroemer, "Critique of Two Recent Theories of Heterojunction Lineups," *IEEE Electron Dev. Lett.*, **EDL-4**, 25 (1983).

8. R. I. Anderson, "Experiments on Ge-GaAs Heterojunctions," *Solid State Electron.*, **5**, 341 (1962).

9. T. Baba, T. Mizutani, and M. Ogawa, "Elimination of Persistent Photoconductivity by Al Spatial Separation from Ga and Si—A Novel, Short Period AlAs-n GaAs Superlattice," *Jap. J. Appl. Phys.*, **22**, L627 (1983).

10. H. C. Casey and M. B. Panish, *Heterostructure Lasers, Part A: Fundamental Principles*, Academic, New York, 1978.

11. See, for example, W. I. Wang and F. Stern, "Valence Band Offset in AlAs/GaAs Heterojunctions and the Empirical Relation for Band Alignment," *J. Vac. Sci. Technol. B*, **3**, 1280 (1985) and the references therein.

12. A. J. Hill and P. H. Ladbrooke, "Dependence of Conduction Band Discontinuity on Al Mole Fraction in GaAs/AlGaAs Heterojunctions," *Electronics Lett.*, **22**, 218 (9186).

13. H. Stormer, R. Dingle, A. C. Gossard, and W. Weigmann, "Two Dimensional Electron Gas at a Semiconductor-Semiconductor Interface," *Solid State Commun.*, **29**, 705 (1979).

14. S. Tiwari, "Threshold and Sheet Concentration Sensitivity of HEMTs," *IEEE Trans. Electron. Dev.* **ED-31**, 879 (1984).

15. J. V. Dilorenzo, R. Dingle, M. D. Feuer, A. C. Gossard, R. H. Hendel, J. C. M. Hwang, A. Kastalsky, R. A. Kiehl, and P. O'Connor, "Selectively Doped Heterostructure FETs," *IEDM Tech. Dig.*, p. 578 (1982).

16. M. C. Foisy, P. J. Tasker, B. Hughes, and L. F. Eastman, "The Role of Inefficient Charge Modulation in Limiting the Current Gain Cutoff Frequency of the MODFET," *IEEE Trans. Electron Dev.*, **ED-35**, 871 (1988).

17. R. Dingle, M. D. Feuer, and C. W. Tu, in *The Selectively Doped Heterostructure Transistor, Materials, Devices and Circuits VLSI Electronics Microstructure Science*, Vol. 11, N. G. Einspruch, Ed., Academic, New York, 1985, Chap. 6.

18. M. Abe, T. Mimura, K. Nishiuchi, A. Shibatomi, M. Kobyashi, and T. Misugi, in *Ultra High Speed HEMT Integrated Circuits, Semiconductors and Semimetals*, Vol. 24, R. Dingle, Ed., Academic, New York, 1988, Chap. 4.

19. N. Y. Linh, in *Two Dimensional Electron GaAs FETs: Microwave Applications Semiconductors and Semimetals*, Vol. 24, R. Dingle, Ed., Academic, New York, 1988, pp. 203–247.

20. E. F. Schubert and K. Ploog, "Selectively Doped *n*-AlGaAs/GaAs Heterostructures with High Mobility Two-Dimensional Electron Gas for Field Effect Transistors," *Appl. Phys.*, **A33**, 193, (1984).

21. I. C. Kizilyalli, K. Hess, J. L. Larson, and D. J. Wildiger, "Scaling Properties of High Electron Mobility Transistors," *IEEE Trans. Electron Dev.*, **ED-33**, 1427 (1986).

22. T. J.Drummond, R. Fischer, S. L. Su, W. G. Lyons, H. Morkoc, K. Lee, and M. S. Shur, "Characteristics of Modulation-doped AlGaAs-GaAs FETs-effect of Donor-Electron Separation" *Appl. Phys. Lett.*, **62**, 262, 1983.

23. L. H. Kamnitz, P. J. Tasker, H. Lee, D. van der Merwe, and L. F. Eastman, "High Performance Modulation-doped FETs," *Tech. Dig. Int. Electron. Dev. Meeting*, p. 360, 1984.

24. S. S. Pei, N. J. Shah, R. H. Hendel, C. W. Tu, and R. Dingle, "Logic Circuits with SDHTs," *Tech. Dig. of 1984 IEEE GaAs IC Symp.*, p. 129 (1984).

25. D. K. Arch, B. K. Bebz, P. J. Vold, J. K. Abrokwah, and N. C. Cirillo, "A Self-Aligned Gate Superlattice AlGaAs-*n*$^+$ GaAs MODFET 5 × 5 Bit Parallel Multiplier," *IEEE Electron Dev. Lett.*, **EDL-7**, 700 (1986).

26. H. Fukui, "Optimized Noise Figure of Microwave GaAs MESFETs," *IEEE Trans. Electron Dev.*, **ED-26**, 1032 (1979).

27. K. H. G. Duh, M. W. Pospeiszalski, W. F. Kopp, P. Ho, A. A. Jabra, P. C.Chao, P. M. Smith, L. F. Eastman, J. M. Ballingall, and S. Weinreb,

"Ultra Low Noise Cryogenic HEMTs," *IEEE Trans. Electron Dev.*, **ED-35**, 249 (1988).

28. P. P. Ruden, C. J. Han, C. H. Chen, S. Baier, and D. K. Arch, "Improved MODFET Performance through Gate Current Control," 45th Annual Dev. Res. Conf., 1987.

29. For example, see A. Y. Cho and J. R. Arthur, "Molecular Beam Epitaxy," *Prog. Solid State Chem.*, **10**, 157 (1975); A. C. Gossard, "Molecular Beam Epitaxy of III–V Semiconductors," *Treat. Mater. Sci. Technol.*, **24**, 13 (1982); K. Ploog, in *MBE of III–V Materials Crystals; Growth, Properties and Applications*, H. C. Freyhardt, Ed., Springer-Verlag, Berlin, 1980, Chap. 3, p. 75.

30. P. D. Dapkus, "Metalorganic Chemical Vapor Deposition," *Ann. Rev. Mater. Sci.*, **12**, 243 (1983).

31. M. B. Panish, Gas Source MBE of GaInAsP:GaS Sources, "Single Quantum Wells, Superlattice *p-i-n*'s and Bipolar Transistors," *J. Cryst. Growth*, **81**, 249 (1987).

32. C. W. Tu, R. Dingle, and R. H. Hendel, in *MBE and the Technology of SDHTs, Gallium Arsenide Technology*, D. K. Ferry, Ed., SAMS, Indianapolis, IN, 1985, Chap. 4.

33. For example, S. Kuroda, N. Harada, T. Katakami, T. Mimura, and M. Abe, "HEMT VLSI Technology using Non-Alloyed Ohmic Contacts," *Tech. Dig. IEDM*, p. 680 (1988).

34. M. Hatzakis, "PMMA Copolymers as High Sensitivity Electron Resists," *J. Vac. Sci. Technol.*, **16**, 1984 (1979).

35. See, for example, N. J. Shah, S. S. Pei, C. W. Tu, and R. C. Tiberio, "Gate Length Dependence of the Speed of SSI Circuits Using Submicrometer Selectively Doped Heterostructure Transistor Technology," *IEEE Trans. Electron. Dev.*, **ED-33**, 543 (1986).

36. D. J. Resnick, D. K. Atwood, T. Y. Juo, N. J. Shah, and F. Ren, "0.50 μm Direct Write Gate Lithography for Selectively Doped Heterostructure Transistor Devices," Proceedings of the SPIE Meeting, March 1988.

37. P. M. Solomon and H. Morkoc, "Modulation Doped GaAs/AlGaAs Heterostructure FETs (MODFETs)-Ultra High Speed Devices for Supercomputers," *IEEE Trans. Electron. Dev.*, **ED-31**, 1015 (1984).

38. K. Inoue, H. Sakaki, and J. Yoshino, "MBE Growth and Properties of AlGaAs/GaAs/AlGaAs Selectively Doped Double Heterostructures with Very High Conductivity," *Jpn. J. Appl. Phys.*, **23**, L767 (1984).

39. F. Alexandre, L. Goldstein, G. Levoux, M. C. Jancour, H. Thibierge, and E. V. K. Rao, "Investigation of Surface Roughness of MBE Grown GaAlAs Layers and Its Consequences on GaAs/AlGaAs Heterostructures," *J. Vac. Sci. Tech.* **B3**, **950** (1985).

40. H. Kinoshita, T. Ishida, H. Inomata, M. Akiyama, and K. Kaminishi, "Submicrometer Insulated Gate Inverted Structure HEMT for High Speed

Large Logic Swing DCFL Gate," *IEEE Trans. Electron. Dev.*, **ED-33**, 608 (1986).

41. T. Baba, T. Mizutani, M. Ogawa, and K. Ohata, "High Performance AlAs/*n*-GaAs Superlattice GaAs 2DEG FETs with Stabilized Threshold Voltage," *Jpn. J. Appl. Phys.*, **22**, L654 (1984).

42. C. W. Tu, W. L. Jones, R. F. Kopf, L. D. Urbanek, and S. S. Pei, "Properties of Selectively Doped Heterostructure Transistors Incorporating a Superlattice Donor Layer," *IEEE Electron. Dev. Lett.*, **EDL-7**, 552 (1986).

43. J. K. Abrokwah, N. C. Cirillo, D. Arch, R. R. Daniels, M. Hibbs-Brenner, A. Fraasch, P. Vold, and P. Joslyn, "Superlattice Donor Modulation Doped FETs Using Self-Aligned Gate Technology," *J. Vac. Sci. Tech.*, **B4**, 615 (1986).

44. E. F. Schubert, A. Fischer, and K. Ploog, "The Delta-doped Field Effect Transistor," *IEEE Trans. Electron. Dev.*, **ED-33**, 625 (1986).

45. M. Hueshen, N. Moll, E. Gowen, and J. Miller, "Pulse Doped Modulation Doped FETs," *IEDM Tech. Dig.*, p. 348 (1984).

46. U. Mishra, R. Beaubien, M. Delaney, A. Brown, and L. Hackett, "MBE Grown GaAs MESFETs with Ultra High g_m and f_T," *IEDM Tech. Dig.*, p. 829 (1986).

47. P. M. Solomon, C. M. Knoedler, and S. L. Wright, "A GaAs Gate Heterojunction FET," *IEEE Electron Dev. Lett.*, **EDL-5**, 379 (1984).

48. K. Matsumoto, M. Ogara, T. Wada, N. Hashizume, T. Yao, and Y. Hayashi, "n^+ GaAs/Undoped GaAs/AlGaAs FET," *Electron. Lett.*, **20**, 462 (1984).

49. S. Fujita, M. Hirano, K. Maezawa, and T. Mizutani, "A High Speed Frequency Divider Using n^+ Ge-gate AlGaAs/GaAs MISFETs," *IEEE Electron. Dev. Lett.*, **EDL-8**, 226 (1987).

50. Y. Katayama, M. Morioka, Y. Sawada, K. Ueyanagi, T. Mishima, Y. Ono, T. Usagawa, and Y. Shiroki, "A New 2DEG FET Fabricated on Undoped AlGaAs-GaAs Heterojunction," *Jpn. J. Appl. Phys.*, **23**, L150 (1984).

51. M. S. Shur, D. K. Arch, R. R. Daniels, and J. K. Abrokwah, "New Negative Resistance Regime of Heterostructure Insulated Gate Transistor (HIGFET) Operation," *IEEE Electron Dev. Lett.*, **EDL-7**, 78 (1986).

52. H. L. Stormer, K. Baldwin, A. C. Gossard, and W. Weigmann, "Modulation Doped Field Effect Transistor Based on a 2-Dimensional Hole Gas," *Appl. Phys. Lett.*, **44(11)**, 1061 (1984).

53. S. Tiwari and W. I. Wang, "*p*-Channel MODFET Using a GaAlAs/GaAs Two-Dimensional Hole Gate," *IEEE Electron Dev. Lett.*, **EDL-5**, 333 (1984).

54. R. A. Kiehl, M. A. Scontras, D. J. Widiger, and W. M. Kwapien, "Poten-

tial of Complementary Heterostructure FET ICs," *IEEE Trans. Electron. Dev.*, **ED-34**, 2412 (1987).

55. F. Hasegawa, "Role of Carrier Confinement in MODFET Performance," Proc. 43rd Annual Device Research Conference, IIA-1, 1985.

56. H. Inomata, S. Nishi, S. Takahashi, and K. Kaminishi, "Improved Transconductance of AlGaAs/GaAs Heterostructure FET with Si-doped Channel," *Jpn. J. Appl. Phys.*, **25**, L731 (1986).

57. A. Cappy. B. Carnez, R. Fauquembergues, G. Salmer, and E. Constant, "Comparative Potential Performance of Si, GaAs, InGaAs, InAs Sub-Micron FETs," *IEEE Trans. Electron Dev.*, **ED-27**, 2158 (1980).

58. J. D. Oliver, Jr. and L. F. Eastman, "LPE Growth and Characterization of High Purity Lattice Matched InGaAs on InP," *J. Electron. Mat.*, **9**, 693 (1980).

59. S. Bandy, C. Nishimoto, S. Hyder, and C. Hooper, "Saturation Velocity Determination for InGaAs FETs," *Appl. Phys. Lett.*, **38**, 817 (1981).

60. K. Y. Chen, A. Y. Cho, S. B. Christman, T. P. Pearsall, and J. E. Rowe, "Measurement of the Γ-L Separation in InGaAs by UV Photoemission," *Appl. Phys. Lett.*, **40**, 423 (1982).

61. K. Y. Chang, A. Y. Cho, T. J. Drummond, and H. Morkoc, "Electron Mobilities in Modulation Doped InGaAs/AlInAs Heterojunctions Grown by MBE," *Appl. Phys. Lett.*, **40**, 147 (1982); U. K. Mishra, A. S. Brown, L. M. Jelloian, L. H. Hackett, and M. J. Delaney, reported in 45th Annual Device Research Conference, IIA-6, Santa Barbara, CA, June 22–24, 1987.

62. J. H. van der Merwe, "Crystal Interfaces Part II: Finite Overgrowths," *J. Appl. Phys.*, **34**, 123 (1963).

63. Y. Sugita and M. Tamura, "Misfit Dislocations in Bicrystals of Epitaxially Grown Si on Boron-doped Si Substrates," *J. Appl. Phys.*, **40**, 3089 (1969).

64. J. W. Matthews and A. E. Blakeslee, "Coherent Strain in Epitaxially Grown Films," *J. Cryst. Growth*, **27**, 118 (1974).

65. T. Zipperian, E. Jones, B. Dodson, J. Klein, and P. Gourley "Stable and Metastable AlGaAs/InGaAs n-channel strained QW FETs," GaAs IC Symposium, *Tech. Dig.*, **10**, 251 (1988).

66. T. Henderson, M. I. Aksun, C. K. Peng, H. Morkoc, P. C. Chao, P. M. Smith, K. H. G. Duh, and L. F. Lester, "Power and Noise Performance of the Pseudomorphic MODFET at 60 GHz," *IEDM Tech. Dig.* (1986).

67. J. M. Kuo, B. Lalevic, and T. Y. Change, "New Pseudomorphic MODFETs Utilizing InGaAs/AlInAs Heterostructures," *IEDM Tech. Dig.*, (1986); J. M. Kuo, T. Y. Chang, and B. Lalevic, "GaInAs/AlInAs pseudomorphic modulation-doped FETs," *IEEE Electron Dev. Lett.*, **EDL-8**, 380 (1987).

68. P. Chu, C. L. Lin, and H. H. Wieder, "Schottky Barrier Height of InAlAs," *Electron. Lett.*, **22**, 890 (1986); C. L. Lin, P. Chu, A. L.

Kellner, and Wieder, "Composition Dependence of Au/InAlAs Schottky Barrier Heights," *Appl. Phys. Lett.*, **49**, 1593 (1986).

69. R. J. Van Tuyl, C. A. Liechti, R. E. Lee, and E. McGowan, "HEMT Circuits," *IEEE J. Solid State Circuits*, **SC-12**, 485 (1977).

70. M. Abe, T. Mimura, K. Nishiuchi, A. Shibatomi, M. Kobyashi, and T. Misugi, in *Ultra High Speed HEMT Integrated Circuits, Semiconductors and Semimetals*, Vol. 24, R. Dingle, Ed., Academic, New York, 1988, Chap. 4.

71. K. Ploog, in *MBE of III–V Materials Crystals; Growth, Properties and Applications*, H. C. Freyhardt, Ed., Springer-Verlag, Berlin, 1980, Chap. 3, p. 75.

72. S. S. Pei, N. J. Shah, and S. J. Pearton, in *Gallium Arsenide Technology* Plenum, New York, in press.

73. H. Morkoc and H. Unlu, in *Factors Affecting the Performance of AlGaAs/GaAs and AlGaAs/InGaAs MODFETs: Microwave and Digital Applications, Semiconductors and Semimetals*, Vol. 24, R. Dingle, Ed., Academic, New York, 1987, Chap. 2.

74. R. H. Hendel, S. S. Pei, C. W. Tu, B. J. Roman, and N. J. Shah, "Realization of Sub-10 Picosecond Switching Times in SDHTs," *IEDM Tech. Dig.*, p. 867 (1984).

6 Bipolar Transistors

P. M. Asbeck
Rockwell International Science Center
Thousand Oaks, California

6.1 INTRODUCTION

Bipolar transistors were the first three-terminal semiconductor devices with useful amplifying properties, and traditionally have been pre-eminent in high-speed performance. In recent years, this pre-eminence has been severely challenged by metal-oxide-semiconductor field-effect transistor (MOSFET) technology. The area of application of MOSFETs has grown enormously, but bipolar transistors still provide higher speed in most circuit applications. Further, bipolar technology is presently undergoing rapid transformations that promise to maintain or extend its leadership in circuit operating speed.

Figure 1 illustrates the schematic band diagram in the direction of electron travel for a representative bipolar transistor. An *n-p-n* device is pictured, and will continue to be used as an example throughout the chapter (although most of the results apply equally to *p-n-p* transistors). The bipolar structure provides a number of natural advantages:

1. Electrons travel from the emitter to the collector, perpendicular to the wafer surface. The dimensions that control transit time are established by processes with excellent dimensional control, such as diffusion, ion implantation, and epitaxy. It is, therefore, easy to produce structures in which the electron transit time is short (corresponding to high cutoff frequency, f_T). Moreover, this can be accomplished without particularly stringent demands on lithography (as is typically required with FETs).

2. The entire emitter area conducts current (unlike the case of FETs, in which only a thin surface channel conducts current). It is, therefore, possible to provide large amounts of output current per unit chip area, and maintain high circuit density.

High-Speed Semiconductor Devices, Edited by S.M. Sze. ISBN 0-471-62307-5
© 1990 John Wiley & Sons, Inc.

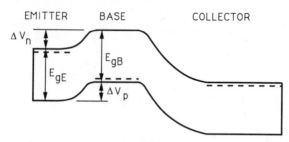

Fig. 1 Energy-band diagram of an *n-p-n* bipolar transistor along the direction of electron travel.

3. The input voltage directly controls the density of the carriers that provide the output current, I_C, leading to output current variation of the form $\exp(qV/nKT)$ with n close to or equal to 1. This leads, in turn, to a transconductance, g_m, of the form $g_m = qI_C/kT$. This transconductance is the highest obtainable in any three-terminal device in which current is controlled through voltage modulation of carrier density (all devices used today). The high transconductance of bipolar transistors allows circuit operation with small input voltage swings, which is central to low power-delay products in logic applications.

4. The turn-on voltage of bipolar transistors, V_{BE}, is relatively independent of device size and process variations, since it corresponds to the built-in potential of a *p-n* junction. This minimizes variations in the turn-on voltage across a wafer, as required in logic circuits with small voltage swings.

5. The input capacitance of bipolar transistors tends to scale with the operating current to the extent that it is dominated by the diffusion capacitance. As a result, bipolar devices can be operated at high or low current, and the input capacitance will adjust accordingly, so that fanout delays do not vary inordinately. Typically, in bipolar circuitry it is not necessary to tailor device sizes to correspond to the load being driven.

Bipolar transistors also suffer from several disadvantages in comparison with FETs. Included among these are:

1. Finite input current, even at dc.

2. When operated in the saturation regime (forward-biased base-collector junction), base current increases and excess charge is stored.

3. Turn-on voltage cannot be used as a design variable within a circuit. Only enhancement mode transistors are producible at present.

4. Fabrication processes are more complex and the resultant circuit areas tend to be larger than with FETs.

Continuous progress has been made in strengthening the advantages of bipolar transistors and minimizing their shortcomings. The use of heterojunctions, for example, adds a new degree of freedom in transistor design, and enables higher speeds. Where ultra-high input impedance is required and/or where transistors must be used as transfer gates, bipolars have been combined with FETs (e.g., in the BiCMOS technology).

This chapter discusses the underlying principles of recent advances in bipolar transistor design and performance. Principles of bipolar transistor operation are first discussed, including primary limits on performance. The approaches taken in advanced Si homojunction transistor fabrication and in heterojunction bipolar transistor technology are then described. Results for heterojunction transistors in various material systems, including Si, GaAlAs/GaAs, and InGaAs/InAlAs/InP, are then presented.

6.2 PRINCIPLES OF BIPOLAR TRANSISTOR OPERATION

6.2.1 Transistor Gain Mechanism

The application of a forward-bias input voltage to the base-emitter junction of a bipolar transistor (as represented in Fig. 1) reduces the conduction-band energy barrier for electron flow from emitter to base. Electrons that surmount this barrier travel across the base (by drift and diffusion), and when they reach the collector-base junction, are swept into the collector by the high fields within the junction region. At the same time, the input voltage leads to the flow of holes into the base. The hole current has a number of different components, corresponding to recombination of holes with electrons in different regions of the device, as pictured in Fig. 2. There is unavoidable (but generally small) excess recombination in the quasineutral base, J_{b1}. Significant numbers of holes typically recombine at deep levels in the emitter-base space-charge region, J_{b2}. Frequently there exists recombination at the emitter periphery (a component J_{b3} not shown in the one-dimensional plot of Fig. 2). Finally, holes from the base surmount the energy barrier in the valence band at the emitter-base junction and flow into the emitter. These holes recombine with electrons in the emitter body (J_{b4}) or at the emitter surface (J_{b5}).

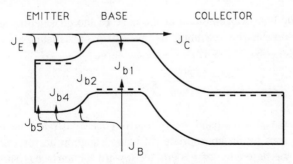

Fig. 2 Components of base current in *n-p-n* transistors.

For most applications, dc current gain on the order of 10 to 100 is required. A major driver in transistor design is the need to minimize the base current to maintain this gain. The key determinants of current gain are discussed in the following simplified analysis.

The collector current density J_C (for the homojunction transistor case) typically is limited by the rate at which electrons travel across the base (rather than by the supply of electrons from the emitter). Then

$$J_C = q\, n_{B0} v_e \,, \tag{1}$$

where n_{B0} is the density of electrons injected at the emitter edge of the base, and v_e is their effective net velocity toward the collector at the base edge. From Boltzmann statistics,

$$n_{B0} = \frac{n_E N_{CB}}{N_{CE}} \exp\left(-\frac{\Delta V_n}{kT}\right) \,, \tag{2}$$

where n_E is the emitter free-electron concentration, N_{CB} and N_{CE} are densities of states in the conduction band for base and emitter (which in general may be slightly different), and ΔV_n is the conduction-band energy difference between quasineutral emitter and base regions as shown in Fig. 1. For the case of simple diffusion, v_e is approximately given by[1]

$$v_e = \frac{D_n}{L_B \sinh(W_B/L_B)} \simeq \frac{D_n}{W_B} \,, \tag{3}$$

where D_n is the electron diffusivity in the base region, $L_B = (D_n \tau_B)^{1/2}$ is their diffusion length, τ_B is their recombination lifetime, and W_B is the thickness of the quasineutral base. For more realistic cases, in which an effective electric field is established in the base due to gradients in acceptor doping, an appropriate transport equation with drift and diffusion terms governs electron flow.

Similarly, the hole current density flowing from base to emitter is expressible as

$$J_B = q p_{E0} v_h \,, \tag{4}$$

where p_{E0} is the hole concentration at the edge of the quasineutral emitter, with v_h their effective velocity. Assuming again Boltzmann statistics,

$$p_{E0} = \frac{p_B N_{VE}}{N_{VB}} \exp\left(-\frac{\Delta V_p}{kT}\right) \,, \tag{5}$$

where ΔV_p is the valence-band energy difference illustrated in Fig. 1. The velocity v_h is given by an appropriate diffusion equation corresponding to hole transport from the base edge of the emitter toward the surface. The recombination velocity, s_0, at the emitter surface is a key boundary condition. The effective hole velocity for uniform emitter doping is given by

$$v_h = v_0 \frac{1 - \dfrac{v_0 - s_o}{v_0 + s_0} \exp(2W_E/L_E)}{1 + \dfrac{v_0 - s_0}{v_0 + s_0} \exp(2W_E/L_E)}. \tag{6}$$

Here, W_E is the emitter thickness, D_h is the hole diffusivity, L_E the hole diffusion length and $v_0 = D_h/L_E$ is the velocity for an infinitely thick emitter. From Eqs. 1 and 3, the current gain as limited by emitter injection efficiency is given by

$$h_{fe} = \frac{n_E v_E N_{CB} N_{VB}}{p_B v_h N_{CE} N_{VE}} \exp\frac{\Delta E_g}{kT}. \tag{7}$$

In Eq. 7, $\Delta E_g = E_{ge} - E_{gb} = \Delta V_p - \Delta V_n$ has been used. This relation is evident in Fig. 1 (where it may be seen $E_{ge} + \Delta V_n = E_{gb} + \Delta V_p$). More exact expressions for current may be obtained by using Fermi statistics (rather than the Boltzmann approximation), considering detailed doping profiles within emitter and base to calculate electric fields that modify carrier transport, using realistic boundary conditions at the collector edge of the base (corresponding to limiting velocity of electrons), and so on. The more exact calculations are conveniently done numerically, and 1-, 2- and 3-dimensional codes have been developed. Important factors for transistor design are nonetheless evident in Eq. 7:

1. Current gain is directly proportional to the ratio n_E/p_B of emitter to base doping concentration. In order to maintain suitably high values of h_{fe}, this ratio typically is chosen to be on the order of 100 to 1000. Figure 3 illus-

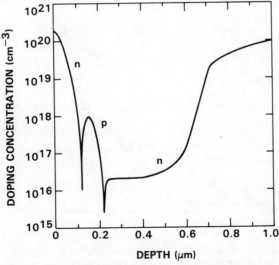

Fig. 3 Representative doping profile for a Si homojunction bipolar transistor.

trates a representative doping profile for a Si homojunction transistor. Emitter doping levels of 10^{20}cm^{-3} and base doping levels on the order of 10^{17} to 10^{18}cm^{-3} are typically used.

2. Emitter injection efficiency is inversely related to base thickness, W_B, from Eq. 7 (W_B also affects current gain through an additional slight dependence associated with base transport factor—the fraction of electrons injected into the base that reaches the collector without recombining). The product $G_B = p_B w_B$ is denoted the base Gummel number; h_{fe} varies inversely with G_B.

3. The emitter surface-recombination velocity is a critical parameter, particularly for thin emitter regions. This velocity is highly technology-dependent. For conventional approaches with emitter-contact metallization deposited directly on Si surfaces, the velocity tends to be very high. For surfaces covered with thin oxide, the velocity is low.

4. The difference in bandgap energy between emitter and base has a major effect. In homojunction transistors, the principal bandgap difference stems from the bandgap shrinkage from heavy doping on the emitter side of the device. For large doping levels, the effective bandgap energy is reduced because of the effects of the potentials of the donor or acceptor atoms and their associated fluctuations (bandtailing effect) and because of the binding energy of the associated electron and hole gases. The extent of bandgap shrinkage in Si has been extensively studied theoretically and experimentally. Figure 4 shows the bandgap shrinkage as a function of doping concentration.[2] As a result of the reduction in emitter bandgap, the current gain is reduced according to the factor $\exp(-\Delta E_{gj}/kT)$, where ΔE_{gj} is the emitter bandgap narrowing at the emitter-base junction edge. A further reduction in gain can occur through the v_h factor in Eq. 7, since the variation of

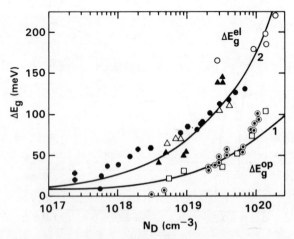

Fig. 4 Theoretical and experimental results for bandgap narrowing in n-type silicon as a function of doping concentration. (After Ref. 2) Variations in bandgap determined in electrical measurements and in optical absorption measurements are contrasted.

carrier concentration and associated bandgap narrowing within the emitter often provide quasielectric fields that drive holes toward the emitter contact and increase the hole current J_{b5}.

6.2.2 Figures of Merit for Transistor Performance

Current-Gain Cutoff Frequency. The frequency at which the transistor incremental current gain drops to unity is called the current-gain cutoff frequency f_T. It can be estimated through the relation[1]

$$\frac{1}{2\pi f_T} = \tau_{EC} = \tau_E + \tau_B + \tau_{CSCL} + \tau_C , \tag{8}$$

where τ_{EC} is the emitter-to-collector transit time, made up of delay components corresponding to different regions of the device. These delay components are discussed in the following paragraphs.

Within the charge control framework for estimating the transistor dynamic performance, the incremental current gain h_{fe} at high frequency is given by

$$h_{fe} = \left| \frac{dI_C}{dI_B} \right| = \left| \frac{dI_C}{dQ_B} \frac{1}{j\omega} \right| = \frac{f_T}{f} . \tag{9}$$

Then

$$\tau_{EC} = \frac{dQ_B}{dI_C} , \tag{10}$$

where dQ_B is the base charge associated with an increment input voltage and dI_C is the corresponding increment in output current. It may be seen that the delay components in Eq. 8 correspond directly to charges stored in different regions of the transistor.

τ_E is the emitter delay. It contains a component associated with charges Q_{Ej} stored at the emitter-base depletion region edges, needed to establish the electrostatic fields at the junction. This part is given by $dQ_{Ej}/dI_C = C_{BEj} \, dV_{BE}/dI_C = C_{BE}/g_{mo}$ where C_{BE} is the base-emitter junction capacitance, and g_{mo} is the intrinsic transconductance of the device, typically qI_C/kT. An additional part of τ_E is associated with holes that are stored in the quasineutral emitter, $Q_{En} = \int p(x) \, dx$. The component Q_{En} increases for decreasing emitter efficiency, decreasing emitter surface-recombination velocity, increasing emitter width, and increasing emitter bandgap narrowing. In homojunction transistors it accounts for a major fraction (10 to 30%) of the total delay τ_{EC}.

τ_B is the base delay due to the hole charge stored in the quasineutral base to neutralize the charge associated with the electrons traversing this region. In the limit of electron flow by diffusion only (for devices with negligible doping gradients or energy gap variations in the base), τ_B is given by

$$\tau_B = \frac{W_B^2}{2D_n} + \frac{W_B}{v_m} , \tag{11}$$

where W_B is the base thickness, D_n the electron diffusivity, and v_m is the velocity at which the electrons exit the base at the collector edge. v_m is typically given by the effusion or thermionic emission velocity of electrons, $v_m = (kT/2\pi m^*)^{1/2}$.

In situations where significant electrostatic fields are present in the base, electron flow proceeds by drift as well as by diffusion, and τ_B is customarily represented as

$$\tau_B = \frac{W_B^2}{\gamma D_n} ,$$ (12)

where γ is an adjustment factor that depends on the magnitude of the electric field present. For a constant field $\mathscr{E}_0$, $\gamma \approx 2[1 + (q\mathscr{E}_0 W_B/2kT)^{3/2}]$.

τ_{CSCL} is the delay due to the hole charge stored in the base that is required to neutralize the charge of electrons traversing the base-collector depletion region, given by

$$\tau_{CSCL} = \frac{1}{2} \int \frac{dx}{v(x)}$$ (13)

(where the integration is carried out over the base-collector depletion region, of width W_C). For electron travel at saturated velocity, $\tau_{CSCL} = W_C/(2v_{sat})$. The factor of two in this expression arises because only half the charge needed to neutralize the traveling carriers corresponds to hole charge in the base; the remainder is neutralized in the collector.

τ_C is the RC time constant for charging the base-collector capacitance. For an incremental input voltage dV_{BE}, the base-collector junction voltage change (for the condition where the output terminals are short-circuited) is given by $dI_C(R_E + R_C + kT/qI_C)$ where the resistances correspond to emitter, collector, and dynamic junction resistances. From this, τ_C is found to be

$$\tau_c = C_{BC}\left[R_E + R_C + \frac{kT}{qI_C}\right].$$ (14)

Maximum Frequency of Oscillation. The frequency at which the maximum available power gain of the transistor drops to unity is called the maximum frequency of oscillation. f_{max} is widely used to estimate power gain since, over a wide range of frequencies, maximum available power gain follows the relation

$$G_p = \left[\frac{f_{max}}{f}\right]^2.$$ (15)

For bipolar transistors a convenient approximate expression for f_{max} is[1]

$$f_{max} = \left[\frac{f_T}{8\pi R_B C_{BC}}\right]^{1/2},$$ (16)

where R_B is the parasitic base resistance and C_{BC} the base collector capacitance.

More exact expressions have been given, where the components of R_B and C_{BC} are properly weighted and phase variations are taken into account.

In order to realize the power gain inferred from Eq. 15, it is necessary to match input and output impedances of the transistor.

Logic Switching Delay.   The figures of merit f_T and f_{max} do not correspond closely to the maximum possible frequency of operation of logic circuits, and are not necessarily the best guides to transistor design for digital circuits. Attempts have been made to determine simple expressions to quantify the behavior of transistors in logic circuits, although the variety of possible circuits and the intrinsic nonlinearity of the problem make this difficult.[3–6]

The power-transfer cutoff frequency f_{pT} is a figure of merit closely tied to digital structures.[4] It is the frequency at which power gain drops to unity in a chain of single-stage identical amplifiers constructed with only resistors and capacitors. This construction does not allow, in general, impedance matching to realize the maximum available power gain. Simple analytical expressions for f_{pT} are not available. In the limit of vanishing base resistance, f_{pT} coincides with f_T. For non-zero base resistance, $f_{pT} < f_T$.

In a somewhat less general fashion, expressions can be developed to represent the propagation delay per gate for specific logic families. Among the most used circuit approaches for high-speed operation are emitter-coupled logic (ECL) and current-mode logic (CML). Figure 5 represents the appropriate circuit schematic for both structures (in CML, the emitter-followers at the outputs

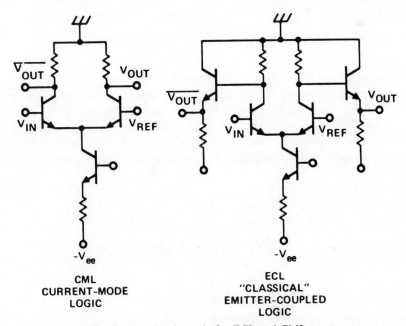

Fig. 5 Circuit schematic for ECL and CML gates.

are omitted). To determine the delay per stage, a particularly simple approximation is to consider a small-signal representation of the circuit (using, e.g., the classical hybrid-π transistor model shown in Fig. 6). After determining the appropriate voltage transfer $F(\omega)$ between output and input in the frequency domain, the associated delay t_d in the time domain is estimated by the relation[6]

$$t_d = \frac{\int_0^\infty t f(t)\, dt}{\int_0^\infty f(t)\, dt} = \frac{i\,[dF(\omega)/d\omega]}{F(\omega)}\bigg|_{\omega = 0}. \tag{17}$$

The application of this technique to CML gates leads to the relation

$$t_d = (1 + g_m R_L) R_B C_{BC} + R_B(C_{BE} + C_d) + \frac{2 C_{BC} + \tfrac{1}{2} C_{BE} + \tfrac{1}{2} C_d}{g_m} \tag{18}$$

for a simplified case with no fanout loading.[6] Here, $g_m = q I_C / kT$ is the transistor transconductance, R_L is the load resistance, C_{BE} is the base-emitter space-charge layer capacitance, and C_d is the diffusion capacitance, with $C_d = g_m(\tau_B + \tau_{CSCL})$. This technique is limited because of its reliance on a linearized model and associated small-signal parameters, while in switching circuits the appropriate element values vary over wide ranges (e.g., g_m varies by orders of magnitude during switching transients). More detailed descriptions of CML and ECL switching have been given, including the recent analysis by Stork.[5] Over a limited range of parameters, the delay of CML can be represented by

$$t_d = [k\tau_{EC}(R_L + 2R_B)(2C_{BC} + C_L)]^{1/2}, \tag{19}$$

where R_L is the load resistance used, C_L is parasitic capacitance associated with the output node, and k is a constant of order unity.

A widespread technique for present device design is to use numerical techniques to arrive at accurate estimates of delay, as a function of device parameters. A useful approach to obtain insight into underlying physical mechanisms is to represent the numerically calculated delays as linear combinations of elementary time constants, with coefficients that are valid only over a range of device

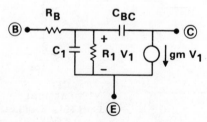

Fig. 6 Basic hybrid π model of bipolar transistor. Here, $C_1 = C_{BE} + C_d$, as defined in the text, and $R_1 = h_{fe}/g_m$.

and circuit parameters. A representative expression developed by Tang and Solomon for the delay of ECL gates is[3]

$$t_d = k_1 R_L C_{BC} + k_2 R_L C_{BE} + k_3 R_L C_S + k_4 R_L C_L + \frac{k_5 C_d}{g_m}$$

$$+ k_6 R_{Bi} C_{BC} + k_7 R_{Bi} C_d + k_8 R_{Bx} C_d , \tag{20}$$

where k_i are coefficients of order unity. In this expression, C_S is the collector-substrate capacitance (ignored in Eqs. 18 and 19), and the base resistance R_B is separated into intrinsic and extrinsic parts, R_{Bi} and R_{Bx}, for greater accuracy. The different elementary time constants vary in general differently with current per gate. Figure 7 illustrates, on a log-log scale, the current dependence of the components. The resistance R_L varies with current because it is necessary to maintain a fixed value of the logic swing, $V_L = I_C R_L$ (of the order of 0.25 to 0.6 V depending upon the noise margin that can be tolerated). It is evident from these curves that the effect of improvement in different transistor parameters is strongly dependent on the desired current range of operation. For a given device, it is also clear that there is a current for which delay is minimized.

Logic Power-Delay Product.   The propagation delay per gate varies inversely with the power per gate over a substantial range of operating current choices (constant power-delay product). For VLSI circuits, the maximum power that can be dissipated in the chip package or that can be accommodated in the system power budget frequently limits the power per gate, and therefore the delay per gate. Consideration of the gate power-delay product is therefore of fundamental importance. In ECL the power dissipated is virtually independent of the rate of switching of the gate. In other technologies [notably complementary MOS (CMOS)] there is a clear difference between static power consumption (in the

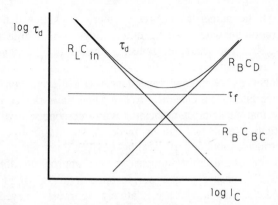

Fig. 7 Schematic dependence of various components of ECL gate propagation delay on transistor current.

absence of switching) and dynamic power consumption, which is given by the switching rate times an associated energy per cycle. Insight into the dynamic switching energy can be obtained by simple consideration of charge flow. During one switching cycle, a charge Q greater than or equal to $C_0 V_L$ (where C_0 is the total capacitance loading the output and V_L is the logic swing) is transported from the positive power supply to the load, and subsequently transported to the negative supply. A lower bound for the energy dissipated per cycle is therefore

$$E_d = C_0 \dot{V}_L V_S ,   \tag{21}$$

where V_S is the voltage difference between positive and negative supplies. To minimize this quantity, it is apparent that low supply voltages, low logic swings, and low capacitance should be pursued. The high transconductance and turn-on voltage uniformity of bipolar transistors make it particularly easy to reduce V_L down to values of 5 to $10kT$, which is probably a practical limit. The value of C_0 includes the stray capacitance associated with the output node, the input capacitance of following devices, and the wiring capacitance associated with interconnects. For bipolar transistors, the capacitances are strong functions of terminal voltages and currents. Equation 21 can be used, however, by considering large-signal average capacitances, given by the total difference in the charge stored between logic high and logic low, divided by the voltage swing. In this sense, the capacitance associated with a CML gate with unity fan-out is

$$C = \frac{C_{BE}}{2} + 2C_{BC} + C_S + \frac{\tau_B + \tau_{CSCL}}{R_L} ,   \tag{22}$$

where C_S is the collector-substrate capacitance. The factor of 2 weighting C_{BC} is due to the Miller effect, and the factor of ½ for C_{BE} is due to the division of the input swing between the two transistors of the differential pair.

6.2.3 Scaling Relations

Consideration of the numerous figures of merit illustrates that the path to higher-speed operation includes, among other things, minimizing the different components of τ_{ec} by, for example, reducing the vertical dimensions of the device, reducing base resistance, and, for highly integrated logic circuits, reducing junction capacitances. Reducing capacitances is achieved, conveniently, by reductions in device lateral dimensions. At the same time, device packing density is reduced, so that capacitance associated with interconnects will be reduced also. Semi-quantitative estimates can be made of the improvements in device and circuit performance obtained by scaling device dimensions. Coordinated scaling of vertical and lateral device dimensions, and appropriate doping changes are typically needed to ensure proper device operation. In MOSFET technology, the associated "scaling laws" are relatively rigorously obeyed in technology development. With bipolar transistors, the scaling laws have been less

TABLE 1 Scaling Relations for Bipolar Transistor (Constant Voltage and Current Case)

Parameter	Scaling Rule
Horizontal Dimension	k
Base Width, W_B	$k^{0.5}D_n^{0.5}$
Base Doping, p_B	W_B^{-2}
Collector Current density, J_C	k^{-2}
Collector Doping Level, N_C	J_C
Supply Voltage	Constant
Depletion Capacitance	k
Depletion Time Constant	k
Diffusion Time Constant	k
$R_B C_{BC}$	k^2/μ_p
Circuit Delay	k

meaningful, and successive generations of transistors have had significant changes in structure that violate simple scaling. One set of proposed scaling relations[7] is shown in Table 1. According to this scaling scenario, the propagation delay decreases linearly with the scaling factor for device size, while the current and voltage remain essentially constant. Collector current density rises rapidly with device scaling. Alternative scaling possibilities can be implemented to avoid power dissipation constraints on circuit density.

6.2.4 Limiting Factors in Transistor Performance

Base Pushout. An important limit on collector current density for bipolar transistors is base pushout or the Kirk effect. When the electron density injected into the base-collector depletion region becomes comparable to the background donor density, the fields in the base-collector depletion region become distorted. When the electron density ($n = J_C/qv_s$, where v_s is the saturation velocity) attains a critical value of

$$n_C = N_D + \frac{2\epsilon(V_{CB} + V_{Bi})}{qW_C^2} \tag{23}$$

(where W_C is the n-collector layer width, V_{CB} the base-collector bias, and V_{Bi} the base-collector built-in potential), then the electrostatic field at the edge of the base vanishes. Above this current density, holes penetrate into the collector, leading to increased base transit time and reduced current gain. As a result of this effect, increasing values of collector doping are required as devices are scaled at constant current density.

Base Punchthrough. As base regions are made thinner in an attempt to shorten transit time, a limit is established by the fact that at high collector-base biases the collector depletion region may extend all the way across the base and reach the emitter depletion region. This "punchthrough" leads to a reduction in the barrier for current flow from the emitter, and uncontrolled current injection. Figure 8 illustrates the limits on base doping and width to avoid punchthrough at a representative voltage of $V_{CE} = 3$ V. In small base regions, it is important to take into account the statistical nature of the dopant atom distribution. If the number of acceptors within the base in any square of dimension W_C^2 is smaller than that needed to terminate the collector field, there will be localized punchthrough that will cause circuit failure. The number of impurities in such a small square may be small enough that its standard deviation is significant compared to its mean, and there can be an appreciable probability of punchthrough even if the mean base doping is acceptably high.

Current Crowding. The ac and dc currents flowing through the base layer produce *I-R* drops that tend to reduce the forward-bias V_{BE} of the base-emitter junction. As a result, the center of emitter regions ceases to conduct current during transistor operation at high current levels and high frequencies (or at short times after the application of an input pulse), reducing the device transconductance. This complicates the theoretical picture of transistor performance at high frequencies (as well as leading to the interesting result that the base resistance decreases at high frequencies from its intrinsic value to its extrinsic value).

Parasitic Elements. To this point the discussion has focused primarily on the intrinsic device, viewed as a one-dimensional structure. However, bipolar transistor performance is affected significantly by parasitic elements surrounding the device. To make contact to the base layer, extrinsic base regions are formed on

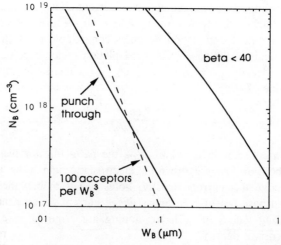

Fig. 8 Limits on base doping established by base punchthrough.

one or both sides of the emitter. The resistance associated with the base contacts and with the regions between the base contacts and the intrinsic device can limit device performance significantly. More important, the capacitances between these extrinsic base regions and the collector is often the dominant part of C_{BC}. The intrinsic base-collector junction area is typically only $1/3$ to $1/6$ of the overall junction area. It must be noted, however, that in some circuit contexts, the extrinsic part of C_{BC} is not as detrimental as the intrinsic part, because the charging base resistance associated with it is lower.

The extrinsic base also borders the emitter and forms sidewalls whose specific capacitance is high because of the high doping levels on both sides of the junction. These sidewalls are also susceptible to leakage currents and premature breakdown. Capacitance between collector and substrate is another parasitic that can significantly load the circuit output. It is somewhat less problematic than the feedback capacitance C_{BC}, since it is not affected by Miller multiplication.

Another parasitic element of significance is emitter resistance, R_E. The area of the emitter contact is equal to the area of the intrinsic device. Values of intrinsic g_m achievable with bipolar transistors are very high (up to 40 mS/μm^2 of emitter), but the extrinsic g_m may be limited if R_E is not maintained below 10 to 50 Ω-μm^2.

6.3   ADVANCED Si BIPOLAR TRANSISTORS

In recent years, dramatic advances have been made in the fabrication technology of Si bipolar transistors, which have led to considerable reductions in vertical and lateral device dimensions, and corresponding increases in performance. There have been improvements in the technique of forming the emitter and extrinsic base junctions, in isolation methods, and in the degree of self-alignment between the different device regions. This section discusses several of the common themes of present technology.[7-12]

6.3.1   Self-Aligned Emitter and Base Contacts

"Conventional" bipolar transistor fabrication procedures during the 1960s and 1970s employed individual diffusion or implant steps to produce emitters and extrinsic bases, which were subsequently contacted by metal. A representative "conventional" device cross section is shown in Fig. 9. Significant improvements were shown to be possible by employing emitters and extrinsic bases diffused from polysilicon layers, which remain on the wafer forming the respective contacts. The polysilicon layers withstand high process temperatures and can be oxidized, so they lend themselves to a variety of sophisticated processing techniques. In particular, the separation between the emitter contact and the base contact can be self-aligned, using a dielectric sidewall spacer, whose width can be accurately controlled over the range of several thousand angstroms. The resultant double-polysilicon, self-aligned device is pictured in Fig. 10. Repre-

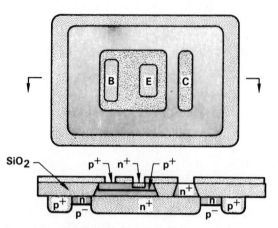

Fig. 9 Cross section of representative "conventional" Si bipolar transistor and device layout. (After Ref. 8)

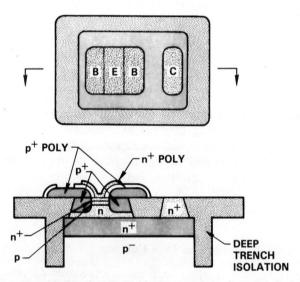

Fig. 10 Cross section of "advanced" Si bipolar transistor featuring double polysilicon self-aligned process and trench isolation and the associated device layout. (After Ref. 8)

sentative process steps to produce the structure are shown in Fig. 11. The width of the emitter is equal to the width of the opening in the base polysilicon (defined photolithographically) less the sidewall dimensions. By repeated application of sidewall spacers, emitter widths as small as 0.35 μm have been made, beginning with 1-μm lithographic patterns. The emitter polysilicon can extend over a considerably wider region than the area of contact, reducing its series resistance. The base-collector junction area is kept at a minimum (typically 3

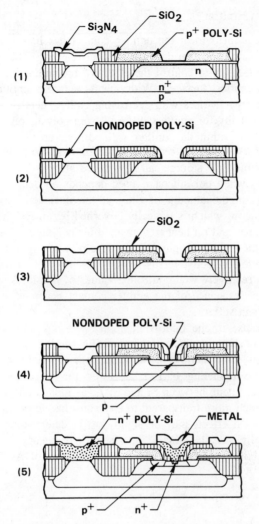

Fig. 11 Outline of double polysilicon self-aligned process steps. (After Ref. 11)

times the emitter area, even for the very narrow emitters). Dramatic savings in device area are made over non-self-aligned devices, which require wide separations between base and emitter to allow for photolithographic alignment tolerances.

The principal limits associated with this technique are related to the region of overlap between the n^+ and p^+ doping distributions produced in the single-crystal silicon by diffusion from the closely spaced polysilicon areas. If the heavily doped regions overlap too much, then tunneling can occur. Hot-electron production can also occur, due to enhancement of the electric field, which leads to surface leakage.

6.3.2   Polysilicon Emitter

The use of polysilicon to form the emitter contact has had a major impact on transistor current gain, and therefore on vertical scaling. The principal role of the polysilicon process is to control the effective recombination velocity, s_0, at the single-crystal-emitter surface. With metal-contacted emitters, s_0 is very high, so that base current rises with decreasing emitter depth. Lower s_0 values are found experimentally for emitters diffused from polysilicon. As a result, the distribution of holes within the emitter is modified, as illustrated in Fig. 12. The associated base current is reduced, and for a sufficiently narrow single-crystal-emitter region, hole storage can be reduced also.

The value of s_0 when polysilicon is used depends on detailed process conditions, particularly on the amount of SiO_2 at the interface between the single crystal and polysilicon (which is typically governed by an HF dip prior to deposition of polysilicon), and on heat treatments after polysilicon deposition (which tend to break up the oxide). Figure 13 shows a plot of experimentally obtained surface-recombination velocity versus oxide thickness. It is necessary to produce devices with relatively well controlled values of this thickness; if the oxide is too thick, series resistance of the emitter contact becomes excessive, if it is too thin, current gain suffers.

The physical cause of the improvement in s_0 is not well understood. For structures with a continuous SiO_2 film separating single-crystal Si and polysilicon emitter regions, current flow of both electrons and holes occurs by tunneling. The effective tunneling barrier is lower for electrons than for holes because of the band lineup between Si and SiO_2, as illustrated in Fig. 14. Thus, the SiO_2 film is much more transparent to electrons than to holes.

Additional physical effects may be significant in determining current flow at the polysilicon–single-crystal-silicon interface, particularly in devices with incomplete oxide coverage of the interface. Segregation of As impurities at the

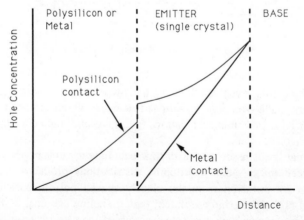

Fig. 12 Distribution of holes within the emitter of *n-p-n* transistors fabricated with (a) metal-contacted single-crystal Si emitter; and (b) polysilicon-contacted Si emitter.

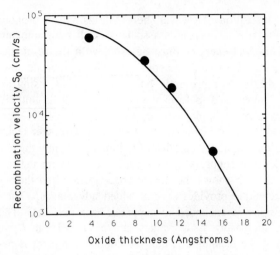

Fig. 13 Emitter surface recombination velocity as a function of SiO_2 thickness separating single crystalline emitter and polysilicon contact. (After Ref. 10)

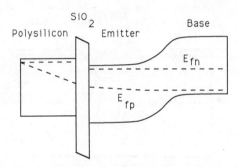

Fig. 14 Band diagram of polysilicon-contacted emitter in the presence of an SiO_2 interfacial layer.

interface has been observed. This impurity pileup could lead to band-bending at the interface, which provides a barrier for hole transport out of the single-crystal region, leading to improved current gain.

With the use of polysilicon, thin emitters can be produced with acceptable current gain. Thin emitters limit hole storage and allow thin base regions to be formed with adequate control and with higher base doping. The use of polysilicon also provides benefits from the standpoint of processing technology. Doping is accomplished by As implantation of the polysilicon, followed by a drive-in anneal. Implant damage is kept away from the single-crystal material, and by appropriate control over the anneal, extremely shallow emitter depths can be reproduced. The formation of the base regions is also a critical fabrication step. The conventional method of forming the base layer is by boron implantation prior to emitter deposition. However, boron atoms have a pronounced tendency to channel, giving rise to deep tails associated with their implant distributions. Thus, very low implant energies must be used, with the result that base charge becomes very sensitive to emitter depth. To obtain reproducible shallow base profiles, boron diffusion is presently being investigated, using as a source the same polysilicon layer used to form the emitter (which is initially made p-type and subsequently arsenic implanted), or by using a layer of borosilicate glass deposited prior to the polysilicon and then stripped. Another technique is to use conventional boron implants, and later to implant P into the collector region underneath the base to suppress the doping associated with the tail of the boron distribution. Figure 15 shows the device cross section and the resultant doping profile. This technique provides several added advantages. The doping in the collector drift region is enhanced to suppress base pushout. This allows high current density, increasing f_T to values above 25 GHz at $J_C = 10^5$ A/cm^2. Another feature of the technique is that the collector implant is patterned by the emitter opening, so that it is formed only in the intrinsic device region. Collector doping in the extrinsic areas remains low to minimize base-collector capacitance.

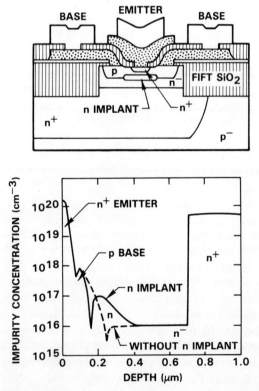

Fig. 15 Cross section of transistor with self-aligned collector implant, and resulting doping profile. (After Ref. 11)

Although polysilicon emitters have allowed shallower emitters and thinner base layers to be produced, there are limits to the vertical scaling that can be obtained. (1) As the base is thinned, base resistance is increased unless doping is also increased. This limit can be counteracted to a considerable extent by making narrow emitters, so that the distance between base contact and center of the emitter stripe is kept to a minimum. Moreover, for logic applications there is a tendency to reduce overall transistor current to control power dissipation. Therefore, load resistances are increased and the tolerable values of base resistance increase correspondingly. (2) It is necessary to avoid base punchthrough at reasonable operating voltages. This requires increasing values of base doping, p_B, as the structure is scaled. The requirement roughly corresponds to base pinch resistances of the order of 30 kΩ or lower to prevent punchthrough at $V_{CB} = 3$ V (for base widths in the vicinity of 500 Å). (3) For high p_B values, charge storage in the emitter is increased, decreasing f_T. Current gain drops due to both increasing hole injection into the emitter, and to the appearance of tunneling current at the base-emitter junction (particularly at low bias voltages). Consideration of the trade-offs required for p_B has let to the estimate that base thicknesses of 500 Å should be achievable, with doping levels of mid 10^{18} cm^{-3}. The maximum f_T, under these conditions, is on the order of 40 to 50 GHz, limited to a considerable extent by hole storage in the emitter.

6.3.3 Trench Isolation

For complex integrated circuits, dense packing of transistors is a key concern, because it minimizes the distances needed to interconnect devices, and hence reduces the wiring capacitance loading the logic gates. A major advance in packing density can be implemented with trench isolation. The advanced transistor structure shown in Fig. 10 incorporates trench isolating; the conventional device, shown for comparison in Fig. 9, employs junction isolation. Trench isolation is based on narrow grooves etched into the silicon from the surface down to the p-substrate. The grooves are lined with SiO$_2$, filled with polysilicon, and planarized. The trench width is of the order of lithographic linewidths (1 to 2 μm) allowing closely spaced devices. The capacitance associated with the isolation regions is greatly reduced because the high specific capacitance associated with the junction isolation is eliminated. Furthermore, the technique can be readily combined with localized-oxidation technology (LOCOS), with which the entire wafer surface outside of active device areas can be coated with oxide. This tends to decrease the capacitance per unit length of interconnect metal.

Trench formation is based on the use of anisotropic plasma etching, typically reactive-ion etching (RIE), which permits formation of high aspect ratio grooves. P implants typically are done at the bottom of the trenches to provide a channel stop. Sidewalls are oxidized to provide isolation, and the grooves are subsequently conformally filled with undoped polysilicon. The surface polysilicon is then etched using anisotropic (vertical) etching, with well-controlled etch-stop based on a suitable endpoint detection system. This leaves a planar surface that can be oxidized in the same step as the formation of the field oxide (at the same time filling any voids that might be present because of the volume expansion during oxidation).

In the design of a deep-trench process, it is desirable to increase the SiO_2 thickness as far as possible to decrease the sidewall capacitance. It is also necessary to avoid formation of high-stress areas, which can lead to the production of dislocations. Sharp corners both at the bottom and top of the trench should be avoided. Secondary concerns are related to the controllability of the etch profile and damage associated with the plasma etching.

6.3.4   Sidewall Contact Process

The sidewall base contact structure (SICOS) transistor[12] makes innovative use of polysilicon contacts to achieve considerable improvement in device performance and to increase the possible circuit uses of the device. A SICOS transistor is shown in Fig. 16. Thick oxide layers separate the base polysilicon from the collector, so that extrinsic base-collector capacitance is dramatically reduced. The device structure becomes a nearly perfect embodiment of the one-dimensional model. With appropriate doping profiles, a nearly identical geometry can be obtained for current flow downward from emitter to collector and upward from collector to emitter. Figure 17 shows experimental current gain measurements in upward and downward directions. The availability of high-quality upward transistors can be used in circuit applications to simplify layout. It is particularly important in the implementation of integrated injection logic (I^2L) circuits in which compact collectors to increase packing density are desirable. With the SICOS process, monolithic integration of ECL and I^2L circuits has been demonstrated, in order to combine high-density and low-power circuits with ultra-high-speed circuits.

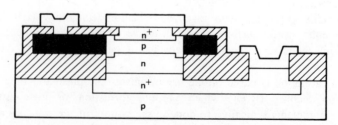

Fig. 16   Cross section of sidewall contacted structure (SICOS) transistor. (After Ref. 12)

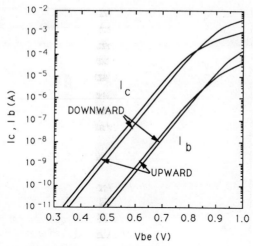

Fig. 17 Collector and base current versus V_{BE} for SICOS transistor operated in upward and downward modes. (After Ref. 12)

6.3.5   Advanced Si Bipolar Circuit Results

The reductions in device size with associated minimization of parasitic capacitances, the increases in f_T from vertical scaling, and the improved isolation have led to greater switching speed in digital circuits. Key device characteristics can be conveniently summarized in a table of parameter values such as those used in the SPICE circuit simulation program. Table 2 shows representative results for minimum-geometry switching transistors. The corresponding delay time measured in ring oscillators has shown steady improvement. Figure 18 illustrates the measured delay time as a function of emitter width, showing nearly linear progression as the devices are scaled.

**TABLE 2   Representative Device Characteristics
(SPICE Parameters) for Double Polysilicon
Self-Aligned Bipolar Transistors**

Emitter Area	$1 \times 2\,\mu m^2$
Base-Collector Area	$3 \times 2\,\mu m^2$
Collector-Substrate Area	$3 \times 7\,\mu m^2$
R_E	$40\,\Omega$
R_B	$80\,\Omega$
C_{BE}	10 fF
C_{BC}	3 fF
C_{CS}	5 fF
f_T	18 GHz

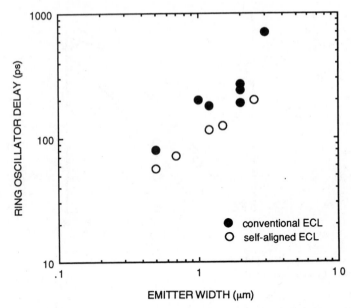

Fig. 18   Delay time of Si bipolar ring oscillators as a function of emitter width. (After Ref. 8)

6.4   HETEROJUNCTION BIPOLAR TRANSISTORS

6.4.1   Bandgap Engineering in Bipolar Transistors

The constraints on vertical scaling of bipolar transistors discussed in Sections 6.2 and 6.3.2 may be overcome if semiconductor bandgaps can be changed appropriately. The effect on current gain of bandgap differences between emitter and base was examined in Section 6.2.1, where the bandgap shrinkage arising from heavy emitter doping was seen to decrease current gain. Shockley and Kroemer as long ago as the late 1950s envisioned that considerable improvement in transistor performance could be obtained by intentionally changing the semiconductor material composition within the device to produce a bandgap that is wider in the emitter than in the base.[13] More generally, the importance of bandgap variation stems from the ability to engineer forces on electrons and holes separately. In semiconductors of uniform composition, electric fields produce forces on electrons and holes that act in opposite directions, as illustrated in the band diagram of Fig. 19. By providing a progressive change in the semiconductor bandgap, forces are established that drive electrons and holes in the same direction as shown in Fig. 19b. With appropriate combinations of electric fields and bandgap variations, the forces on the individual carriers may be established separately. The ability to tailor these forces provides a powerful new degree of freedom in the design of bipolar devices.

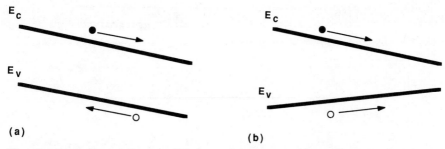

Fig. 19   Conduction- and valence-band energies versus position with (a) varying electrostatic potential, and (b) varying bandgap. (After Ref. 14)

The implementation of heterojunction approaches in bipolar devices was delayed by decades from the time of the initial concepts. The delay was caused by the technological problem of providing interfaces between dissimilar materials that were free of imperfections, either impurities or structural defects as a result of mismatch in lattice constant. Even now, high-performance heterojunction bipolar transistors (HBTs) are limited to relatively few materials systems. The systems explored furthest are those involving semiconductors that have identical lattice constants such as in the GaAlAs/GaAs and InGaAs/InAlAs/InP systems.

6.4.2   Bandgap Variations in HBTs

Abrupt and Graded Junctions.   At abrupt junctions between different semiconductors, in general, there are discontinuities in the energy of the conduction-band minima and valence-band maxima. As detailed in Chapter 1, the change in energy gap ΔE_g between the materials is made up of a conduction-band energy step, ΔE_c, and a valence-band energy step, ΔE_v (with $\Delta E_g = \Delta E_c + \Delta E_v$). A representative band diagram for an *n-p-n* HBT with a wide-bandgap emitter and an abrupt emitter-base junction is shown in Fig. 20. As is clear from the figure, an energy barrier appears in the conduction band at the emitter-base junction, which tends to retard the flow of electrons from emitter to base. This decreases the emitter injection efficiency, but it provides benefits for high-speed operation because the electrons that surmount the energy barrier are injected into the base with high forward velocities, reducing base transit time. To improve the current gain of the HBTs, the composition of the materials may be graded over distances of several hundred angstroms. The result of grading is shown in the band diagram of Fig. 21. Detailed behavior of the band edges depend on the doping profile and the grading profile. As illustrated in Fig. 22, the conduction-band energy $E_C(x)$ can be readily determined from its components, the electrostatic potential variation $\phi(x)$ and the variation of the electron affinity of the material, $\chi(x)$. To avoid the formation of any maxima in conduction-band energy, grading of alloy composition [hence $\chi(x)$], that is quad-

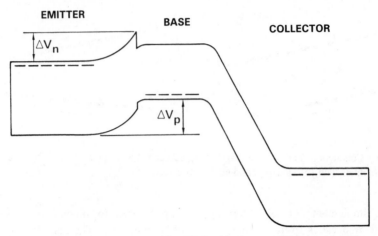

Fig. 20   Energy-band diagram for HBT with abrupt emitter-base junction.

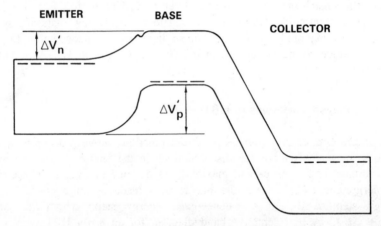

Fig. 21   Energy-band diagram with graded alloy composition at emitter-base junction.

ratic in distance from the junction may be used, for the case where doping is uniform (and thus electrostatic potential also varies quadratically with distance).

Wide-Bandgap Emitter.   With the band diagram of Fig. 21, following the analysis of Eqs. 1 to 7, the current gain, as limited by emitter injection efficiency is

$$h_{\mathrm{fe}} = \frac{n_E}{p_B} \frac{v_E}{v_h} \frac{N_{\mathrm{CB}} N_{\mathrm{VB}}}{N_{\mathrm{CE}} N_{\mathrm{VE}}} \exp \frac{\Delta E_g}{kT} . \tag{24}$$

The difference in bandgap between emitter and base, ΔE_g, is typically chosen to be greater than 250 meV ($10kT$), providing a factor of 10^4 improvement over the homojunction case. It is, therefore, possible to ensure adequate injection efficiency with very high levels of base doping, and very low levels of emitter

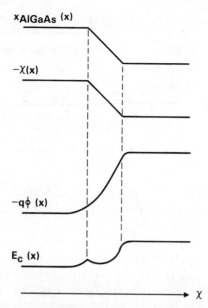

Fig. 22   Relationship between conduction-band energy versus position, and the variation of composition, electrostatic-potential, and electron affinity at a heterojunction.

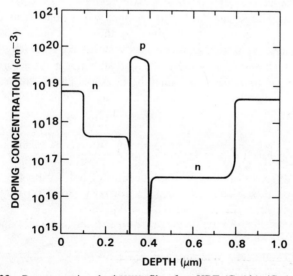

Fig. 23   Representative doping profile of an HBT (GaAlAs/GaAs case).

doping. Representative doping concentrations in current HBTs fabricated in the GaAlAs/GaAs system are shown in Fig. 23, which should be contrasted with those of Fig. 3 for the homojunction case. In HBTs, doping levels of 10^{20} cm^{-3} have been used in the base. As a result, base sheet resistance can be greatly decreased, even with ultranarrow base regions. Transistor f_{max} is thereby greatly increased. The limitations to device scaling due to base punch-through disappear as a result of the higher doping. Current-gain reduction from high injection effects at high current density are similarly avoided. The Early voltage of the device increases considerably, because the base charge is insensitive to output voltage V_{CB}. At the same time, emitter doping levels can be reduced dramatically. This allows the emitter-base space-charge region to broaden considerably on the emitter side of the junction, and leads to a reduction in emitter junction capacitance. The storage of holes in the emitter region vanishes for abrupt heterojunctions, and is greatly minimized for graded heterojunctions, increasing f_T.

Graded Base. With the ability to control semiconductor bandgap, it is worthwhile also to establish a gradual change in bandgap across the base layer from E_{g0} near the emitter to $E_{g0} - \delta E_g$ near the collector, as pictured in Fig. 24. In such a situation, the high hole conductivity ensures that the valence band is effectively flat, and the bandgap shifts establish a conduction-band energy-gradient equal to $\delta E_g / W_B$ (where W_B is the base thickness). This energy gradient constitutes a quasi-electric field that drives electrons across the base by drift as well as diffusion. In homojunction devices, drift fields are of the order of $kT/qW_B\ln(N_{Ao}/N_{Af})$, where N_{Ao} and N_{Af} are the accepter concentrations at the beginning and the end of the quasineutral base. This field is typically of the order of $2\,kT/q$ across 500 to 1000-Å base regions (2 kV/cm). With HBTs, fields 2 to 5 times larger are easily implemented. It is noteworthy that in group III-V compounds, electrons can be driven by these high quasi-electric fields to velocities in excess of the value predicted by steady-state velocity-field curves. These high velocities occur because the overall voltage drop across the base generally is limited to values lower than the threshold energy for scattering from the gamma conduction band valley to satellite valleys. The result is a considerable improvement in f_T

Wide-Bandgap Collector. Another possibility offered by bandgap engineering is to increase the bandgap of the collector, as illustrated in Fig. 25. This has the beneficial effect of eliminating the injection of holes from the base into the collector when the base-collector becomes forward biased (in a manner analo-

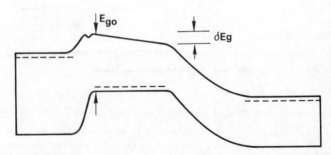

Fig. 24   Band diagram of HBT with graded composition base region.

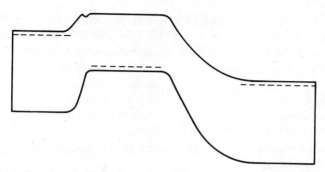

Fig. 25 Band diagram of double heterojunction HBT (with wide-bandgap emitter and collector).

gous to the wide-gap emitter effect). Increasing the bandgap of the collector greatly diminishes the saturation stored-charge density and speeds up device turn-off after the device is biased in the saturation regime. The saturation stored charge, it must be noted, is not fully eliminated, inasmuch as there remains a component of electrons injected into the base from the collector. With double-heterojunction transistors that have wide-gap emitters and collectors, grading of the material composition at the base-collector junction is critical, inasmuch as the appearance of a potential barrier can suppress collector current. Even with grading, it has been shown that an effect analogous to base pushout can occur at lower collector-current densities than expected from Eq. 23. With double-heterojunction devices, symmetrical operation in upward and downward direction can be established, which leads to circuit flexibility.

Additional advantages of wide-bandgap collector devices are the increase in breakdown voltage and the reduction in leakage current. It is possible to employ narrow-bandgap base materials, yet not suffer from the problems of leakage and low breakdown voltage that occur in homojunction devices with similarly low bandgaps (such as Ge transistors).

6.4.3 Modeling of HBTs

For structures such as HBTs with variable semiconductor composition, the classical semiconductor device analysis equations must be supplemented with new terms to account for the variations in conduction- and valence-band characteristics. The modified transport equations may be obtained by considering that, for small departures from equilibrium, carrier flows are linearly dependent on the corresponding gradients in the electrochemical potential, $\psi(x)$, so that, for example, for electrons

$$J_n = \frac{n\,\mu_n\,d\psi(x)}{dx},$$
(25)

where μ_n is the electron mobility. $\psi(x)$ is made up of terms corresponding to (a) electrostatic potential, $\phi(x)$; (b) the energy of the conduction-band minimum measured with respect to a convenient reference, such as the energy of an electron in vacuum outside the semiconductor (electron affinity); and (c) chemical potential, given for the case of Boltzmann statistics by $kT \ln(n/N_c)$ where n is the local electron density and N_c is the effective density of states in the conduction band. With this starting point, the semiconductor device equations of Table 3 may be derived.[15] The carrier-transport equations that describe the electron and hole current densities in Table 3 involve two new terms: a drift-like term that includes the gradient in conduction- or valence-band energy and a diffusion-like term that includes the gradient in effective density of conduction- or valence-band states.

Solution of the equations of Table 3 for geometries representative of HBTs can be easily carried out with numerical techniques. Codes that predict HBT performance in one- and two-dimensional approximations have been written, and are used extensively for device design. The accuracy of these codes is typically limited by the above assumption of small departures from equilibrium. For example, implicit in the simulations is the assumption of infinite effusion (or thermionic emission) velocity, and of steady-state velocity-field characteristics (that ignore velocity overshoot effects). Analytical results for HBT characteristics may be obtained in various situations.[16,17] In general in HBTs, collector current may be limited by (a) transport of electrons across the base (as it is in homojunction transistors); (b) transport of carriers across the conduction-band barrier (as it is for abrupt junctions); or (c) supply of electrons from the emitter (if the emitter is very lightly doped). For case (a), it has been shown that collector current, J_C, is given by a simple generalization of the corresponding homojunction equation[17]:

$$J_C = \frac{q \exp(qV_{BE}/kT)}{\int [p \, dx / D_n n_i^2(x)]} .$$

(26)

TABLE 3 Equations for Semiconductor Device Analysis for the Case of Spatially Varying Composition

$$\frac{d^2V}{dx^2} = \frac{q}{\epsilon_S}(n - p + N_A^- - N_D^+)$$

$$\frac{\partial n}{\partial t} = G_n - U_n + \frac{1}{q}\frac{\partial J_n}{\partial x}$$

$$\frac{\partial p}{\partial t} = G_p - U_p - \frac{1}{q}\frac{\partial J_p}{\partial x}$$

$$J_n = q\mu_n n\left[-\frac{dV}{dx} + \frac{1}{q}\frac{d\Delta E_C}{dx}\right] + qD_n\left[\frac{dn}{dx} - \frac{n}{N_C}\frac{dN_C}{dx}\right]$$

$$J_p = q\mu_p p\left[\frac{dV}{dx} + \frac{1}{q}\frac{d\Delta E_V}{dx}\right] + qD_p\left[\frac{dp}{dx} - \frac{p}{N_V}\frac{dN_V}{dx}\right]$$

This equation, the Moll-Ross-Kroemer relation, illustrates that J_C depends solely on the characteristics of the base region, and that the effect of variable base bandgap is taken into account simply by the variation of intrinsic carrier concentration, n_i. The ideality factor associated with J_C is unity.

The more general case where current flow is limited by mechanisms (b) and (c) as well as (a) has been considered.[16] J_C can be expressed in the form

$$J_C = \frac{1}{1/J_1 + 1/J_2 + 1/J_3} \exp\frac{qV_{BE}}{kT} , \qquad (27)$$

where J_1, J_2, and J_3 are, in general, voltage-dependent current densities governed by the different mechanisms. The turn-on voltage is increased from the value that applies in the base transport limited case. The ideality factor for J_C is somewhat greater than unity because the height of the conduction-band barrier increases somewhat with increasing V_{BE}. For single heterojunction transistors it is possible to determine experimentally if a given HBT is significantly limited by factors other than base transport by comparing forward and reverse characteristics. Equation 26 predicts that collector current as a function of V_{BE} should be the same as emitter current as a function of V_{BC} (although the corresponding values of base current will differ by orders of magnitude). If there is a significant barrier in the conduction band at the base-emitter junction, then the reverse current at a given voltage will be higher because the conduction-band barrier will be lower under zero-bias conditions than under base-emitter forward bias. Experimental results for the two cases are shown in Fig. 26. While at high cur-

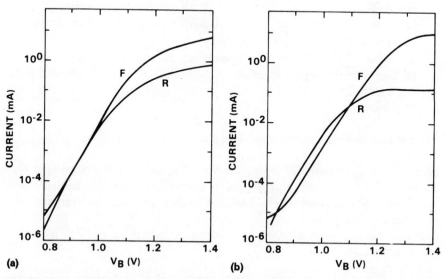

Fig. 26 Experimental measurements of I_C as a function of V_{BE}, and of I_E as a function of V_{BC}, for two devices: (a) device for which base transport limits electron flow; (b) device for which potential barrier at emitter-base junction limits current flow.

rent levels IR drops dominate the measured characteristics, the predicted result is observed over several orders of magnitude of current.

Equation 24 can be used to calculate current gain as limited by emitter-injection effects for the case of base-emitter junctions with no potential barrier [case (a)]. A noteworthy feature of this expression is that current gain decreases with increasing temperature. This is a natural consequence of the fact that the energy barrier for hole flow is greater than that for electron flow, and hence its thermal activation energy is greater.

6.5 SILICON-BASED HBTS

To obtain the benefits of heterojunctions combined with standard Si bipolar technology, considerable efforts have been made to identify a suitable wide-bandgap semiconductor to be used as the emitter, or, more recently, a suitable narrow-bandgap semiconductor to be used as the base within Si devices.

6.5.1 Wide-Gap Emitter Approaches

A variety of candidate materials have been tried as emitters in Si-based bipolar transistors. To date, no single material system has emerged as leading contender for future development. Several generic problems affect progress:

1. Wide-gap emitter materials frequently lead to high emitter resistance, associated with either the material itself, with the emitter/Si interface, or with contacts to the wide-gap material. If the resulting total emitter specific resistance is greater than 20 to 50 Ω-μm^2, however, the characteristics of the transistors are significantly degraded.
2. Diffusion of donors into the single-crystal silicon must be suppressed. If the single-crystal region becomes n-type, it will become the effective emitter, and the same stored-hole charge of the polysilicon emitter structure will be present.

Candidate materials include the following:

SIPOS (Semi-Insulating Polycrystalline Si). The bandgap of this material is of the order of 1.5 eV. High current-gain transistors have been obtained even with high base Gummel numbers, as indicated in Fig. 27. Emitter resistance has been high. The improvement in transistor characteristics may, in fact, be related to the formation of oxide layers at the surface of the Si, with an attendant series resistance.

Amorphous Silicon. The bandgap (mobility edge) of amorphous Si is on the order of 1.6 eV. High current-gain devices have been produced, but again, these have had large series resistance that is believed to be associated with the contacts to the amorphous Si rather than with interfacial oxides.

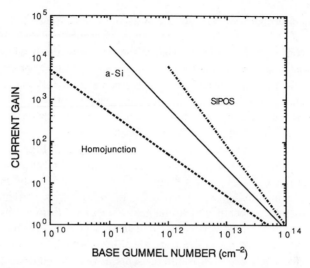

Fig. 27 Measured current gain as a function of base Gummel number for representative Si bipolar transistors with wide-bandgap emitters.

Microcrystalline Silicon. Phosphorus-doped hydrogenated microcrystalline Si with grain size on the order of 50 Å can be made to have low resistivity (0.01 to 0.2 Ω-cm) and a wide optical gap (1.5 to 1.9 eV). The material is deposited at low temperature ($<450\,°$C) by plasma chemical vapor deposition (CVD) techniques. HBTs have been produced with adequate current gain with base sheet resistances down to 2 kΩ/square. At present, the stability of the hydrogen incorporated in the structure is a problem. Contact resistance is also a problem, with specific resistance typically near 1000 Ω-μm^2; this can be tolerated only with a strategy of making the emitter contact significantly larger than the actual emitter.

β-SiC. Epitaxial growth of this wide gap (2.2 eV) semiconductor on Si has been reported. Despite the considerable lattice mismatch with the Si substrate (20%), current gains up to 800 have been obtained.

GaP. Although lattice match of this widegap (2.26 eV) semiconductor is good with respect to Si (0.4% mismatch), difficulties arise from cross-doping, and from the generation of antiphase domains during GaP growth.

6.5.2 $Si_{1-x}Ge_x$ HBTs

Band Structure. Addition of Ge reduces the bandgap of Si, leading to an alloy that can be used to form the base region of HBTs. The lattice constant of the alloy differs considerably from that of Si, but as shown in Chapter 1, if the thickness of epitaxial layers is kept below a critical thickness, then the mismatch between the alloy and the Si substrate is accommodated elastically and no misfit dislocations form (pseudomorphic growth). The resulting SiGe pseudomorphic layers are considerably strained. Due to the details of the band structure of Si, this strain provides major benefits for transistor operation. The conduction-band minima in unstrained Si and SiGe correspond to 6-fold degenerate valleys located along {100} directions in k-space (delta points). In the presence of strain, the degeneracy of the valleys is lifted. For growth on (100) substrates, the 4 valleys oriented in the plane of the heterojunction are lowered in energy, while the remaining two valleys are raised in energy (see Fig. 28). As a result, the overall bandgap shrinkage is increased over that of unstrained material with the same Ge content. At the same time, electrons preferentially populate valleys for which the effective mass in the direction of travel is particularly low. In fact, for transport perpendicular to the heterojunction, electron mass corresponds approximately to the transverse mass $(0.19m_0)$, which represents a reduction of the order of 60% from the unstrained case. Similarly, the degeneracy of the top of the valence band of SiGe is lifted. This also contributes to increasing the shift in bandgap energy for a given Ge mole fraction. The energetically favored valence band corresponds to light hole mass for transport in the plane of the heterojunction, as desired to obtain low base resistance. The effective mass changes from the strain anisotropy more than compensate for the decrease in mobility expected from alloy scattering.[18]

The variation of bandgap of $Si_{1-x}Ge_x$ pseudomorphically grown on Si is shown in Fig. 29. The bandgap difference between SiGe and Si appears principally as a valence-band step, a situation of considerable benefit for the formation of n-p-n HBTs.

SiGe HBTs have achieved average Ge mole fraction of about 20%, corresponding to a bandgap difference on the order of 200 meV ($8kT$). This allows an appreciable increase in injection efficiency, from Eq. 24.

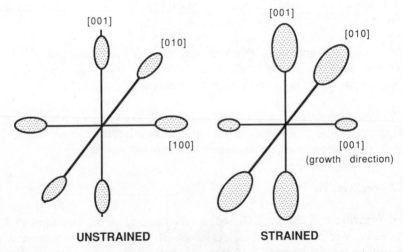

Fig. 28 Constant energy ellipsoids for the different conduction-band valleys of SiGe alloys, illustrating the splitting of the degeneracy due to strain.

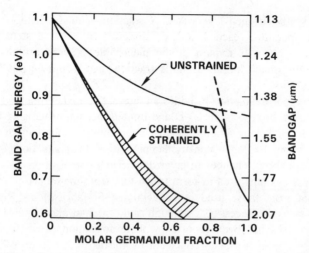

Fig. 29 Variation of energy gap for pseudomorphic $Si_{1-x}Ge_x$ on Si, and unstrained alloy, as a function of alloy composition.

HBT Fabrication. Several experimental demonstrations of HBTs employing SiGe bases have been reported.[19-21] A representative layer structure and device cross section are shown in Fig. 30. The epitaxial growth technique used to form the alloy has been molecular beam epitaxy (MBE) in most work, which allows growth at relatively low temperatures (usually in the range 550 to 800°C). As described in Chapter 1, the low temperature process allows pseudomorphic growth of materials that under equilibrium conditions would tend to

Layer	Composition	Doping (cm^{-3})	Thickness (μm)
Contact	Si	n=1x10^{20}	0.1
Emitter	Si	n=5x10^{17}	0.2
Base	Si$_{0.8}$Ge$_{0.2}$	p=2x10^{18}	0.08
Collector	Si	n=3x10^{16}	0.3
Substrate	Si	n=2x10^{19}	

```
                        Emitter
          Base        ▓▓▓▓▓▓▓▓▓        Base
                      n+   Si
        ▓▓▓▓▓▓▓       n-   Si        ▓▓▓▓▓▓▓
                      p+  SiGe

                      n-   Si

                   n+ Si substrate
        ▓▓▓▓▓▓▓▓▓▓▓▓▓▓▓▓▓▓▓▓▓▓▓▓▓▓▓▓▓▓▓
                      Collector
```

Fig. 30 Layer structure of Si/SiGe HBT and corresponding device cross section.

form incoherently strained layers (with formation of misfit dislocations). If the maximum temperature reached during subsequent processing steps is kept low enough, the layers will remain in the metastable pseudomorphic state. This allows achieving much larger Ge mole fraction than would be possible in equilibrium.

Chemical vapor deposition provides an alternative growth method for Si/SiGe structures. Ultra-high-vacuum CVD and limited thermal processing CVD[21] have been used for HBT fabrication. In this last technique, rapid thermal switching (based on heating by high-intensity lamps) rather than gas flow switching, is used to achieve abrupt changes in layer doping and composition.

The processing steps used to form transistors from the epitaxial structures differ somewhat from those used in conventional Si fabrication. Relatively low temperature steps are necessary, although rapid thermal annealing of implants up to 850 °C for 10 s has been reported without formation of misfit dislocations. The thin base region has been contacted by means of low-energy Ga implants, or by means of selective etching. Wet etches have been developed that are strain sensitive, and will stop when they reach the SiGe base. Low-temperature dielectrics have typically been deposited for passivation. Isolation has been accomplished with mesa etching in work to date, although considerable effort is being devoted to develop structures closer to present Si homojunction devices.

Device Characteristics. The decrease in energy gap of the base leads to changes in the $I_C - V_{BE}$ relation toward lower turn-on voltage as expected from Eq. 26. Figure 31 shows experimental results for a series of devices with constant base Gummel number ($2 \times 10^{13} \mathrm{cm}^{-2}$) and varying allow content in the base. The base current is significantly decreased from the homojunction case at a given base doping. Figure 32 shows the variation of base current with base

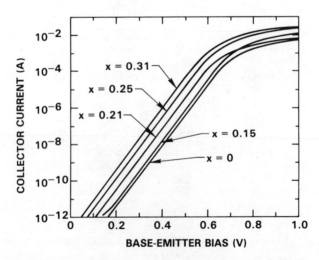

Fig. 31 I_C versus V_{BE} for SiGe HBTs with varying Ge mole fraction x. The HBTs have approximately constant base Gummel number $2 \times 10^{13} \mathrm{cm}^{-2}$, while the homojunction device (with $x = 0$) has Gummel number $10^{12} \mathrm{cm}^{-2}$. (After Ref. 21)

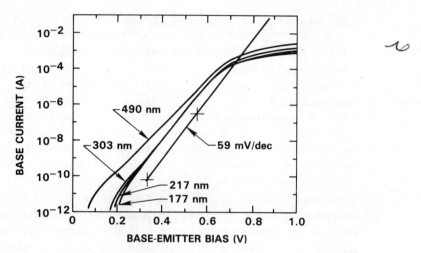

Fig. 32 I_B versus V_{BE} for SiGe HBTs with fixed Ge mole fraction $x = 0.215$, and varying thickness W_B. (After Ref. 21)

width, W_B, at a constant base alloy composition ($x = 0.215$). For small W_B, ideality factor is near unity, and base current is independent of W_B (as expected for emitter injection efficiency limited current gain). For W_B greater than 2000 Å, the ideality factor and magnitude of base current increase, as expected from the formation of misfit dislocations. From the dc results, inferences about the electron diffusivity in the SiGe base layers indicate that it is not as high as in single-crystal Si. As a result of the excellent prospects of SiGe HBTs for ultra-high-speed operation, the structures are currently the topic of very active research. In recent work, f_t up to 75 GHz has been demonstrated with a Si/SiGe HBT.[41] The Ge concentration in the base was graded up to a peak concentration of 7%, providing a built-in drift field for electrons. The intrinsic base sheet resistance was 17 KΩ/square, achieved with a nominal base thickness of 45 nm.

6.6 GaAlAs/GaAs HBTs

6.6.1 Advantages of GaAlAs/GaAs

1. The GaAlAs/GaAs material system has an excellent lattice match. The difference in lattice constant between AlAs and GaAs is on the order of 0.14% at room temperature, and due to slight differences in thermal expansion coefficient, is even less at representative crystal growth temperatures. Matching may therefore be obtained for any choice of Al mole fraction. This simplifies demands on composition control during epitaxial growth.

2. Advanced material-growth techniques are available, such as molecular beam epitaxy (MBE) and metal-organic chemical vapor deposition (MOCVD), as discussed in Chapter 1. These techniques permit growth of ultra-thin device layers with excellent control.

3. Significant bandgap differences may be obtained. The bandgap of the alloy system as a function of AlAs mole fraction x, is given by $E_g = 1.424 + 1.247x$ for $x < 0.45$, and $E_g = 1.424 + 1.247x + 1.147(x - 0.45)^2$ for higher x values. Below the critical composition $x = 0.45$, the bandgap is direct. Above this composition, the conduction-band minima are at X points of the Brillouin zone, while the valence-band maximum remains at gamma. The majority of the bandgap difference (approximately 62%) corresponds to a difference in conduction-band energy, with the remainder (38%) corresponding to the valence-band energy difference. As a result of this, abrupt Np heterojunctions have significant conduction-band energy barriers.

4. The electron mobility is high in $Al_xGa_{1-x}As$ for $x < 0.45$ (the direct-bandgap composition range) as a result of the small effective mass associated with the gamma minimum ($m^* = 0.065$). In pure material, electron mobility reaches 8000 cm^2/V-s. The high mobility tends to decrease base-transit time, decrease charge storage at the base-emitter junction, and increase conductivity of the undepleted n-collector. Also of importance in bipolar transistors is the hole mobility since f_{max} varies approximately as $(\mu_n\mu_p)^{1/2}$. Hole mobility is not particularly high, due to the presence of light holes and heavy holes, with respective effective masses $m_{lh}^* = 0.075$ and $m_{hh}^* = 0.50$. The effective hole mobility in pure GaAs (that is free of impurity scattering) is 380 cm^2/V-s.

5. At high electric fields (> 3 kV/cm), electrons in GaAs display negative differential mobility under steady-state conditions, and their velocity decreases with increasing electric field till it saturates at a value of 0.6 to 0.8×10^7 cm/s. When electrons enter high field regions their velocity considerably exceeds this limit for short times. This phenomenon, known as velocity overshoot, has been presented in previous chapters. Velocity overshoot is of considerable importance in decreasing the transit time of carriers across the base-collector depletion region.[22,23]

6. As a result of the wide bandgap of GaAs, the intrinsic carrier concentration is low, $n_i = 2.3 \times 10^6$ cm^{-3} at room temperature. Intrinsic material has a resistivity on the order of 5×10^8 Ω-cm (frequently described as "semi-insulating"). Substrates can be manufactured conveniently with resistivities approaching this limit because of easily produced deep levels that can pin the Fermi level near mid gap. The semi-insulating nature of the substrate simplifies the task of isolating devices and interconnects. The capacitance between devices or interconnects and the substrate, a major factor in silicon-based designs, is reduced to negligible values. The virtual elimination of collector-substrate capacitance is a major benefit. In the case of digital circuits, the decreased cross-capacitance between interconnects achievable with GaAs tends to be less significant, because interconnect lines in most alternative technologies are narrow (1 to 2 μm), and can be placed on dielectrics (SiO_2 or polyimide) of thickness nearly comparable to the inter-

connect separation. As a result, the influence of a conducting substrate is slight. In microwave applications, interconnects tend to be much wider in order to limit attenuation and to control impedance, and the effect of the substrate is of major significance. Monolithic microwave-integrated circuits are well developed in GaAs, and have not proven successful with Si yet.

7. GaAlAs/GaAs have been used to fabricate a variety of optoelectronic devices including light-emitting diodes (LED)s, lasers, modulators, and detectors of various types. These devices can be monolithically integrated with circuits based on GaAlAs/GaAs HBTs.

6.6.2 Device Structure

Figure 33 shows a representative device cross section and associated layer structure. Emitter layers consist of $Ga_{1-x}Al_xAs$ with AlAs mole fraction x chosen to be of the order of 0.25. With greater x values, deep donors, known as DX cen-

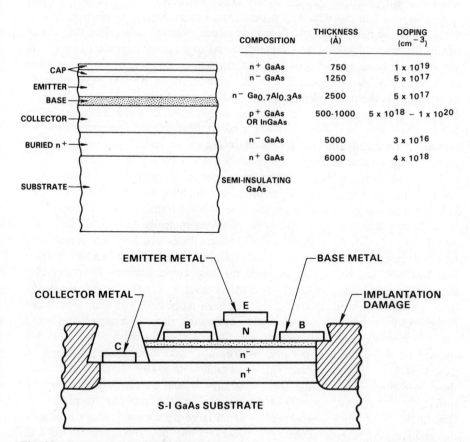

Fig. 33 Representative layer structure and device cross section for GaAlAs/GaAs HBTs.

ters, begin to appear in n-type GaAlAs, which increase the capacitance of the emitter depletion region, and may contribute to trapping effects. With $x = 0.25$, the bandgap of the emitter is 0.30 eV wider than that of the base, allowing huge increases in injection efficiency. The majority of the energy gap difference corresponds to a conduction-band step (0.2 eV). Accordingly, in much of the work the alloy composition is graded at the base-emitter junction. Base layers are typically made with thicknesses of 0.05 to 0.1 μm, with values of doping N_A from 5×10^{18} to 10^{20}cm^{-3}. Extremely high values of base doping may be used without incurring hole injection into the emitter (although other sources of base current are present, and these also tend to increase with N_A). Resulting values of base sheet resistance are in the range 100 to 600 Ω/square. The practical limit to N_A is established by metallurgical considerations. The diffusivity of the most frequently used acceptors (Be in MBE and Zn in MOCVD) increases with concentration. Unless care is exercised during crystal growth, and high-temperature processing is eliminated subsequent to growth, the acceptors will penetrate into the lightly doped GaAlAs. When the p-n junction is thus moved into the wide-gap material, an energy barrier to electron flow is formed and the barrier to hole flow is reduced, reducing gain. For the case of drastic acceptor diffusion, the structure reverts to a homojunction transistor with variable-base bandgap and poor injection efficiency. Grading of the composition of the base has been demonstrated, with variation of the Al content by as much as 10% across the base. With excessive Al content at the top of the base, however, diffusion of acceptors is enhanced, etches used in subsequent fabrication are less selective, and base contacts tend to have higher specific resistance. Structures frequently include layers of InGaAs (not lattice matched to GaAs) grown on the emitter surface to decrease the emitter contact resistance.

Many variations in HBT structure have been demonstrated. (1) Additions of InAs can be made to the base layer producing the alloy GaInAs, whose bandgap is lower than that of GaAs. As in the case of SiGe transistors, if the base layer is sufficiently thin, the layer will grow pseudomorphically (coherently strained). Advantages of GaInAs include wider-bandgap difference between emitter and base, increased electron mobility, strain splitting of the valence band to increase hole mobility, and the ability to grade the base composition without incurring the above-mentioned problems of GaAlAs bases. (2) Double-heterojunction devices with wide-bandgap collectors have been fabricated, and with appropriate processing have been shown to lead to symmetric device characteristics when operated in upwards and downwards directions. (3) While most of the work has focused on n-p-n devices, p-n-p transistors also have been made. Theoretical expectations are that the f_T and f_{max} achievable with p-n-ps should nearly equal those of n-p-ns, although the layer structures should be configured somewhat differently.[34,35] To maximize f_T in a p-n-p device, it is important to reduce the base thickness as much as possible, to avoid large base transit times (in light of the relatively poor hole mobility). However, the base resistance penalty in using ultra-thin bases should not be very great, because of the very high electron mobility in GaAs. (4) In work oriented toward I^{2}L circuitry, transistors optimized for operation in the upward direction have been made, with a wide-bandgap emitter layer underneath the base and collector. In this approach,[31] the base doping is typically produced by ion implantation rather than epitaxy.

6.6.3 Fabrication Technology

Epitaxial layer structures have been typically produced with MBE or MOCVD techniques, which have been extensively developed for this material system. Devices are fabricated using a mesa geometry for isolation, or by using ion implantation damage to make the epitaxial regions outside the device semi-insulating. In order to contact the base and collector, vias are etched to the appropriate layers as pictured in Fig. 33, or the material above the layer to be contacted is type-converted by implantation or diffusion (for instance, Zn diffusion or Be implantation has been used to contact the base). The process of etching down from the surface to the thin base layer is facilitated by the availability of composition-selective etches, whose etch rate can be made to slow considerably when they reach the desired layer. As a result of the mesa structure and the deep vias to the buried collector, there can be concerns about the step coverage of metallizations. To alleviate the problem, planarization of the wafer surface can be done, or orientation-selective etching can be used so that etched sidewalls with a specific crystallographic orientation become sloped at tolerable angles (55° from the surface). A major objective of process development is to produce a self-aligned structure in which the edge of the emitter, emitter contact, and base contact are produced with the same photolithographic pattern. Various self-alignment techniques have been reported for GaAlAs/GaAs HBTs.[24,25,27] Figure 34 illustrates a number of the device cross sections. Central themes of the fabrication approaches are the use of sidewall spacers at the edges of the emitter contact (as in structures a and b in Fig. 34), and the use of refractory emitter contacts based on InAs cap layers (employed in structures a, b and d in Fig. 34). In the dual-liftoff approach (structure c), a dielectric layer is deposited on top of the base contact metal, and both are patterned simultaneously by liftoff.[24] A fabrication technique to minimize extrinsic base-collector capacitance has been developed, based on the use of implants of oxygen or protons into the regions of the n-collector layer underneath the base contacts, as pictured in Fig. 35. The implants render the material semi-insulating, and decrease capacitance per unit area by widening the effective separation between base and collector quasineutral regions. The implants can be carried out through the base without significantly altering its conductivity, because of the high base doping.

6.6.4 Device Characteristics

Figure 36 shows I-V characteristics of a representative single heterojunction GaAlAs/GaAs HBT. Distinctive features of the dc characteristics include the output conductance, the offset voltage, and the current gain variation.

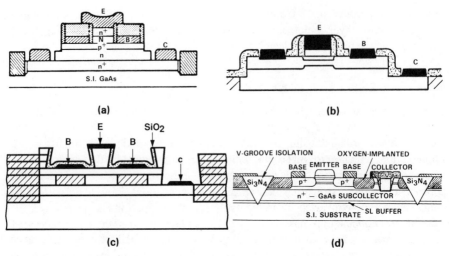

Fig. 34 Various approaches to the fabrication of fully self-aligned GaAlAs/GaAs HBTs.

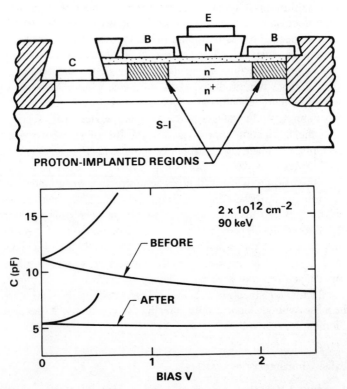

Fig. 35 Proton implantation technique to reduce the extrinsic base-collector capacitance in GaAlAs/GaAs HBTs. Capacitance-voltage data for test diodes with and without the implants are shown.

Output Conductance.   The curves display high positive output resistance, or even differential negative output resistance. High-output resistance (high Early voltage, V_A) stems from the high doping of the base region, so that the base Gummel number, $G_B = p_B W_B$, is not significantly changed with the application of a base-collector bias. The induced change in G_B is $dG_B = -C_{BC} dV_{BC}$, so that from Eq. 26,

$$\frac{dI_C}{I_C} = \frac{-dG_B}{G_B} = \frac{C_{BC} dV_{BC}}{p_B W_B} .$$

(28)

Corresponding values of $V_A = I_C/(dI_C/dV_{BC})$ are in the range of 100 to 200 V. The negative resistance is obtained only on a slow time scale, and is due to the decreasing current gain with increasing temperature, as described in Section 6.4.2.

Offset Voltage.   Figure 36 shows a significant offset voltage, V_{CEsat}, that must be applied between collector and emitter before positive I_c will flow. The offset voltage appears because the turn-on voltage for current flow for the base-emitter junction is greater than that for the base-collector junction. For single hetero-junction abrupt HBTs, a portion of the difference in turn-on voltage is attributa-ble to the conduction band barrier at the EB junction. For HBTs with graded EB junctions, this contribution is absent and electron flow for both EB and CB junctions is equal at equal forward bias voltages (as described in Section 6.4.3). Even for these transistors, however, there is a difference in turn-on voltage between EB and CB junctions associated with the difference in hole current at a given bias. Hole flow is typically negligible for the EB junction, but it typically

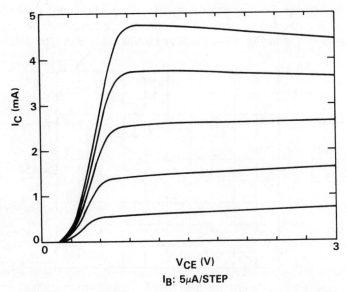

Fig. 36 I_C versus V_{CE} for a representative GaAlAs/GaAs HBT.

is orders of magnitude greater than electron flow in forward biased CB homojunctions as a result of the doping ratio of the collector and base. The off-set voltage can be decreased substantially by the use of a wide bandgap collector.

Current Gain.   The *I-V* curves of Fig. 36 and the representative Gummel plot of Fig. 37 illustrate that current gain is a function of current, and decreases to less than unity for very low I_c values. In most GaAlAs/GaAs HBTs, hole recombination in the quasineutral emitter is effectively suppressed by the emitter bandgap step. The base current arises, in part, from recombination at the BE space charge region due to deep levels in the GaAlAs and GaAs, whose concentration is dependent on the growth technique. Deep-level recombination can be suppressed to some extent by the technique of moving the depletion region into the wide bandgap GaAlAs region, because the magnitude of the Shockley–Read–Hall recombination depends directly on the intrinsic carrier concentration, n_i. Another important contribution to base current, dominant in small devices, is recombination associated with the emitter periphery. The surface recombination velocity of GaAs is very high (of the order of 10^6 cm/s). As a result, electrons injected into the base that diffuse to the exposed surface areas of the base near the emitter edges have a high probability of recombining at the base surface. Current gain in such circumstances decreases with the emitter periphery/area ratio, and can be a critical limitation for emitters of the order of 1 μm wide. The effect may be suppressed (a) by using thin base regions with built-in fields from composition grading, which reduces the number of injected electrons that diffuse to the surface;[27] (b) by using a device structure which has the surface of the base covered with GaAlAs, so that there is a potential barrier for electrons between the base and the surface. This can be accomplished if the emitter layer is converted to *p*-type by diffusion or implantation in the regions of the base contact[25] (as in Fig. 34) or by using a thin layer of *n*-type AlGaAs cov-

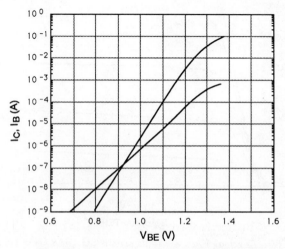

Fig. 37   Gummel plot for GaAlAs/GaAs HBT.

ering the emitter edges, which is pinched off from surface depletion so that it is an effective-insulating passivation layer;[26] or (c) by decreasing the surface recombination velocity of GaAs by chemical treatments such as the deposition of $Na_2S_{0.9}H_2O$. The physical mechanism for the change is still under study. Unfortunately, the effect disappears after a period of the order of hours-to-days.

RF Characteristics.   Figure 38 shows various measures of the gain of a representative microwave HBT as a function of frequency. The gain values are calculated from S-parameters measured with microwave network analyzers. The small-signal current-gain, h_{21}, at low frequency has the value of dc current-gain h_{210}. Above frequency of the order of f_T/h_{210}, the current-gain begins to roll-off with a limiting roll-off rate of 6 dB/octave ($1/f$ behavior). f_T is estimated for the device of Fig. 38 to be 50 GHz by extrapolating the data assuming the 6 dB/octave dependence. Other measures of gain correspond to maximum available gain, MAG (the maximum power gain obtainable by conjugate matching of input and output ports); and unilateral or Mason gain, U (the maximum gain obtainable by conjugate matching of inputs and outputs, together with the use of a lossless feedback network to tune-out any internal device feedback). MAG and U both reach unity at the same frequency, f_{max}, which can be extrapolated conveniently for most devices from the behavior of U (which more closely corresponds to 6 dB/octave roll-off). For some combination of input and output match, device are potentially unstable below a frequency f_1, while above f_1 the devices tend to be unconditionally stable. In the regime of potential instability, MAG is not defined, and maximum stable gain, MSG, is used as a measure of gain.

From measured f_T values, the emitter-to-collector transit time τ_{EC} has values down to 1 to 2 ps. The contributions associated with τ_E and τ_C (emitter and col-

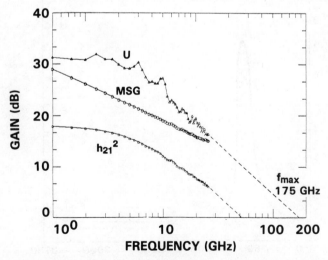

Fig. 38   Gain variation with frequency measured for a GaAlAs/GaAs HBT. Here h_{21} is the current gain, MAG the maximum available power gain, MSG the maximum stable power gain, and U is the unilatral gain or Mason gain invariant.

lector delays, as defined in Section 6.2.2) decrease with current density J_C, and reach values of the order of 0.4 ps for J_C of 5×10^4 A/cm² or above (J_C is limited typically by base pushout to $0.5 - 2 \times 10^5$ A/cm²). The contribution to τ_E from hole storage in the emitter is entirely absent in abrupt base-emitter heterojunction devices, and is small in graded heterojunction structures. The contribution for base transit time, τ_B, for thin bases is influenced by the thermionic-emission velocity or effusion velocity of electrons (2×10^7 cm/s in GaAs) as well as the diffusion time, and for 600 Å heavily doped bases has a value on the order of 0.6ps. The collector-space-charge layer transit time, τ_{CSCL}, is dependent on collector-base voltage, V_{CB}. As in Si devices there is an effect of widening of the space-charge layer with increasing V_{CB}, which increases τ_{CSCL}. In addition, there are strong effects on electron velocity in GaAs. The saturation velocity, v_{sat} [at high electric fields (> 10 kV/cm)] is of the order of 10^7 cm/s. However, when electrons enter regions of high field from low-field regions, they are accelerated to velocities much higher the v_{sat} before eventually slowing down due to scattering to the satellite valleys of the conduction band. Figure 39 shows calculated results for the velocity of electrons as a function of position near the base-collector junction of a GaAlAs/GaAs HBT, obtained by Monte Carlo techniques. Velocities of more than 7×10^7 cm/s are reached, although the high velocities only persist over distances of the order of 500 Å. The detailed behavior is dependent on the applied bias V_{CB}. To obtain an estimate of the distance of significant velocity overshoot, one can calculate the distance required for electrons to pick up an energy of 0.36 eV from the

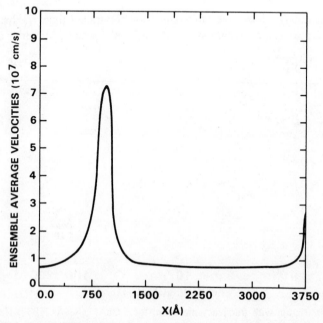

Fig. 39 Calculated electron velocity as a function of position near the base-collector junction of a GaAlAs/GaAs HBT (using the Monte Carlo technique). (After Ref. 22)

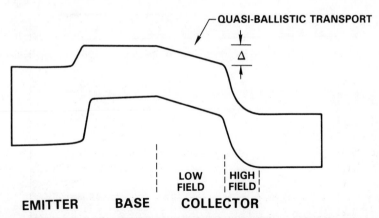

Fig. 40 Band diagram for the ballistic collection transistor, in which the distance of significant velocity overshoot is increased. (After Ref. 28)

electrostatic potential. With this energy they can scatter to the satellite L valleys with the emission of an optical phonon. In specially designed structures, the overshoot can be increased considerably over the conventional case by decreasing the field at the base-collector depletion edge with an appropriate acceptor doping of the collector-depletion region.[22,28] Figure 40 shows the band diagram corresponding to a representative structure of this type, termed the ballistic collection transistor. With the BCT, an f_T value of 105 GHz has been obtained.[28]

6.6.5 Applications of GaAlAs/GaAs HBTs

Microwave and mm-Wave Circuits. The high values of power gain obtainable with HBTs make them usable for amplification over a frequency range that extends well into the mm-wave regime. The f_{max} values are considerably higher than obtained with Si bipolar transistors, because of the much lower base resistivity. An added benefit is that the current crowding effect is significantly reduced. Emitter current uniformity may be described quantitatively through the emitter utilization factor $\gamma = \int j(x)dx/J_p W_E$. Figure 41 shows calculations of γ as a function of emitter width W_E, for various frequencies of operation, using representative HBT parameters. Up to 94 GHz, 1 μm emitters are adequate. This allows a considerable simplification in processing compared to competing mm-wave transistors (for example, heterojunction FETs and permeable base transistors).

Another advantage of HBTs in microwave power applications is the high degree of control over breakdown voltage.[29,30] Breakdown characteristics are determined by the doping and thickness of the n-collector region. For high-power applications, efficiency is increased with higher voltage operation. This can be done with HBTs with thick collector-depletion regions, although there is an associated falloff of f_T due to increasing τ_{CSCL}.

Digital Circuits. With low values of R_B and τ_{EC} (equivalent to low diffusion capacitance C_d), the logic propagation-delay time can be very short. Logic cir-

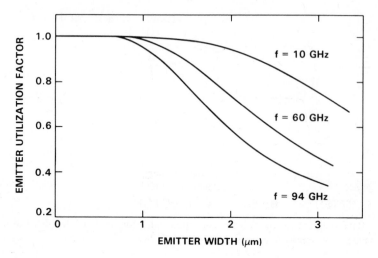

Fig. 41 Calculated emitter utilization factor as a function of emitter width for representative microwave HBTs ($f_T = 70$ GHz and base sheet resistance of 200 Ω were assumed).

cuits based on ECL and CML have been demonstrated with ultrafast operation. Gate delays measured in ring oscillators are below 10 ps. Frequency dividers, configured from master-slave flip-flops, are used to measure the maximum toggle rate of flip-flops. The maximum reported speed of such frequency dividers has reached 30 GHz. Many of the highest speed circuits attained to-date have been based GaAlAs/GaAs HBTs.

Large logic circuits of various types have been demonstrated. An ECL gate array containing about 1200 equivalent NOR gates has been reported, with flip-flop toggle rates up to 7.7 GHz. Substantially larger gate arrays and logic circuits (including a 32-bit microprocessor) have been reported using the heterojunction integrated injection logic (HI^{2}L) approach. The circuit diagram corresponding to this approach is shown in Fig. 42. Switching is accomplished with Schottky diodes (which can be very fast in GaAs, with characteristic frequencies given by the product of on-resistance and off-capacitance up to 1 THz). Transistors operate in the upward mode with the collector on the wafer surface and emitter buried below. The transistors saturate during the gate transitions, but the saturation stored charge is eliminated rapidly because of recombination at the GaAs surface.

The power-delay product attained in GaAlAs/GaAs HBTs has not been significantly lower than for other digital technologies to date. In terms of the factors in dynamic switching energy described in Eq. 21, the HBTs benefit from significant reductions in device-diffusion capacitance. Other contributors to device-input capacitance (C_{BE}, C_{BC}) have not been particularly low because device sizes used to-date have not been aggressively scaled as they have been for advanced Si-bipolar transistors. This will occur as the technology develops, although fabrication techniques must deal with the emitter-size effect on current gain.

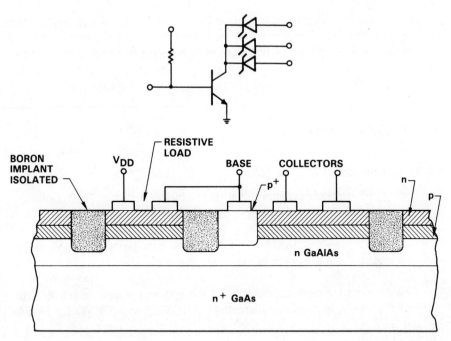

Fig. 42   (a) Circuit diagram and (b) device cross section for heterojunction integrated injection logic circuits. (After Ref. 31)

Another factor affecting power-delay product is the somewhat larger power-supply voltages needed with GaAlAs /GaAs HBTs because the turn-on value of V_{BE} is higher than in Si (V_{BE} = 1.4 V for current densities above 10^4 A/cm^2, determined by the high bandgap of GaAs together with the high base Gummel numbers used).

Analog Circuits.   GaAlAs/GaAs HBTs have a variety of advantages for use in broadband dc-coupled analog circuits, including dc-coupled linear amplifiers, variable-gain amplifiers, logarithmic amplifiers, analog-to-digital converters, sample-and-hold circuits and digital-to-analog converters. Their advantages compared with FETs stem from the reproducible turn-on values of V_{BE} (which reduces offset errors), the absence of trapping effects and hysteresis often found in GaAs FETs, low $1/f$ noise, and high transconductance. With respect to Si bipolars, GaAlAs/GaAs HBTs benefit from lower base resistance achievable with relatively large, easily controllable dimensions, from higher output resistance (high Early voltage), from higher f_T and f_{max} values, from the absence of current crowding, and from the availability of semi-insulating substrates to form low-capacitance interconnects. The importance of low-base resistivity is greater in analog circuits, which are required to conduct significant amounts of current, than in digital circuits, which are evolving towards very low currents per device. The significance of achieving high f_T and low R_B with large dimensions is more evident in analog circuits, since matching of devices is significantly better with larger devices. As a result of these advantages, a variety of analog circuits based on GaAlAs/GaAs HBTs have been reported,[32,33] including 4- and 6-bit flash analog-to-digital converters, 8-bit digital-to-analog converters, 2-Giga sample-per-second sample-and-hold circuits, logarithmic amplifiers, etc.

6.7 THE InGaAs HBT

6.7.1 Advantages of InGaAs HBTs

The set of III-V semiconductors that are lattice-matched to InP substrates includes $In_{0.53}Ga_{0.47}As$ (with energy gap 0.75 eV) and $In_{0.52}Al_{0.48}As$ (with energy gap 1.5 eV). HBTs configured using $In_{0.53}Ga_{0.47}As$ (abbreviated InGaAs) as the base and InAlAs or InP (whose bandgap is 1.34 eV) as the wide-gap emitter have a number of attractive features:

1. Well-developed growth technology for the epitaxial layers, based on MBE for InGaAs and In AlAs, or alternatively, MOMBE or MOCVD for InGaAs and InP.

2. High electron mobility in InGaAs, 1.6 times higher than GaAs and 9 times higher than Si in the case of pure material. The extent of transient electron velocity overshoot is also greater in InGaAs, InP, and InAlAs than in GaAs, inasmuch as the separation between the gamma conduction, band minimum, and the satellite valleys is considerably higher than in the case of GaAs. Table 4 summarizes values for band structure and electron and holes transport parameters. As a result, higher f_T values can be obtained with InGaAs HBTs than with GaAs devices.

TABLE 4 Band Structure and Carrier Transport Parameters of GaAs and $In_{0.53}Ga_{0.47}As$

	GaAs	$In_{0.53}Ga_{0.47}As$
$E_g(eV)$	1.42	0.75
$\Delta E(\Gamma - L)(eV)$	0.3	0.55
$\Delta E(\Gamma - X)(eV)$	0.48	1.0
$m_e^*(m_0)$	0.067	0.045
$\mu_e(cm^2/V\text{-}s)$	8,000	13,000
$v_{peak}(cm/s)$	2×10^7	2.7×10^7
$m_{lh}^*(m_0)$	0.07	0.05
$m_{hh}^*(m_0)$	0.62	0.65

3. The bandgap of InGaAs is smaller than that of GaAs or Si. The resulting turn-on V_{BE} of HBTs is correspondingly smaller (with graded heterojunction HBTs), and as a result, the power-supply voltage and power dissipation can be lower. This improves the power-delay product in logic circuits (as detailed in Section 6.2).

4. The recombination velocity at surfaces of InGaAs is much smaller than that of GaAs surfaces (10^3 cm/s versus 10^6 cm/s). As a result, there is less base current caused by emitter periphery recombination, and scaling to small device dimensions is easier.

5. Semi-insulating substrates are available, made of InP doped with Fe (which produces a deep acceptor near midgap).

6. Higher substrate thermal conductivity than in the case of GaAs (0.7 versus 0.46 W-cm/K).

7. Direct compatibility with sources (lasers and light-emitting diodes) and detectors (pin diodes) for 1.3-μm radiation, which corresponds to the wavelength of lowest frequency dispersion in silica-based fibers, and the wavelength of semiconductor lasers with the fastest modulation capabilities.

6.7.2 Device Structure and Fabrication

Figure 43 illustrates the representative epitaxial layer structure and cross section of an InGaAs HBT. As noted above, a choice between InAlAs and InP must be made for the semiconductor to be used in emitter (and collector regions). The

Layer	Composition	Doping (cm-3)	Thickness (μ m)
Contact	GaInAs	n=1x10^{19}	0.15
Emitter	AlInAs	n=1x10^{19}	0.1
Emitter	AlInAs	n=5x10^{17}	0.15
Spacer	GaInAs	n=5x10^{17}	0.02
Base	GaInAs	p=5x10^{18}	0.15
Collector	GaInAs	n=1x10^{16}	0.6
Subcollector	GaInAs	n=1x10^{19}	0.7
Substrate	InP	semi-insulating	

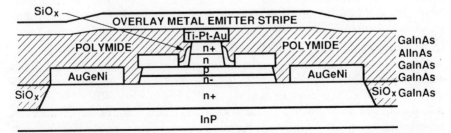

Fig. 43 Epitaxial layer structure and device cross section for representative InGaAs HBT. (After Ref. 40)

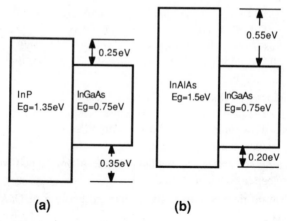

(a)   **(b)**

Fig. 44   (a) Band lineups between InP and InGaAs and (b) between InAlAs and InGaAs.

band lineups of these materials with respect to InGaAs are shown in Fig. 44. Abrupt InP junctions exhibit greater valence-band steps than the corresponding InAlAs abrupt junctions. For both systems it is necessary to grade the BE junction to obtain the benefit of low turn-on voltage. But, in both cases, if all or part of the conduction-band step is left, then "ballistic launching ramps" that inject electrons into the base with high forward momentum can be formed. This will reduce base transit time considerably, if bases are thin enough. The momentum achievable is much higher in InGaAs than in GaAs because of the low electron mass, and the fact that the upper limit on energy (energy separation between gamma and satellite valleys) is greater.

The low value of bandgap in InGaAs can lead to low values of base-collector breakdown voltage, and to relatively high leakage currents at the base-collector junction. To reduce these problems, it is possible to use wide-bandgap collector regions. As discussed in Section 6.4.2, it is crucial to grade the base-collector junction of such structures to avoid the formation of a conduction-band barrier, which can diminish the collection efficiency for electrons.

High-performance devices have been made with both InP and InAlAs wide-bandgap materials.[36-40] InP is difficult to grow with conventional MBE, because of the difficulty of handling solid phosphorus and its high vapor pressure. The technique of MOMBE conveniently solves this problem by providing a gas source (PH$_3$), as does MOCVD. For graded composition structures, quaternary alloys can be used, which combine In, Ga, Al, and As or In, Ga, As, and P. It is necessary to maintain reasonable lattice-match to the substrate, which restricts the composition range that can be used to a one-parameter family. The required degree of composition control is relatively difficult to achieve. It is possible as an alternative to use "quasi-quaternary" alloys, superlattices of short period (10 to 20 Å) in which the composition is changed abruptly between the endpoints, with a duty cycle that is progressively changed.

   Fabrication approaches for InGaAs HBTs are similar to those used for GaAs
HBTs. A difference exists in regard to isolation techniques: it is not as straightforward to use implant damage to convert epitaxial layers to semi-insulating
material. Therefore, deep mesa or trench isolation is preferred. Self-aligned
fabrication techniques have been developed, using sidewall dielectric spacers to
define separations between base and emitter.

6.7.3 Device Characteristics

The *I-V* characteristics of an InGaAs HBT with an abrupt, single-heterojunction
structure are shown in Fig. 45. The high offset voltage is the result of the significant difference in turn-on voltage between BE and BC junctions. Breakdown
is relatively low, on account of the low InGaAs bandgap. Both features can be

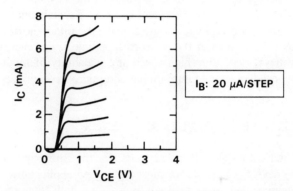

Fig. 45 *I-V* characteristics of representative InAlAs/InGaAs HBT with abrupt BE junction, and InGaAs collector region.

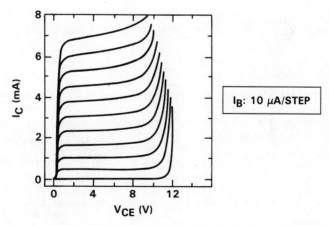

Fig. 46 *I-V* characteristics of InAlAs/InGaAs HBT with graded BE junction, and
InAlAs collector region.

remedied with the use of wide-bandgap collectors and graded junctions, as shown in Fig. 46. Current gain in the InGaAs HBTs can be very high, even in structures with narrow emitters, because of the low surface-recombination velocity. Devices with emitters as narrow as 0.3 μm have been reported, with gains as high as 115.

The high-frequency performance of InGaAs HBTs is very impressive. Figure 47 shows the incremental current gain versus frequency for an InP/InGaAs/InGaAs HBT, demonstrating f_T of 165 GHz, among the highest of any semiconductor device.[39] The associated value of τ_{EC} is 1 ps. Each of the components of τ_{EC} is minimized in the device. Emitter charging time is reduced because there is no hole storage at the heterojunction, and by operation at high current density (2×10^5 A/cm^2). Base transit time is reduced by injection of electrons with high forward momentum as a result of an abrupt BE heterojunction. Collector space-charge layer transit time is reduced by electron travel at very high velocity, in the overshoot regime. To maintain this condition, narrow collector depletion regions are used, and the voltage is restricted to a range of V_{CE} bias of the order of 0.6 V, corresponding to the energy separating gamma and L valleys. The ultra-high speed and scalability of these devices make them attractive candidates for future digital and microwave circuits.

6.8 SUMMARY AND FUTURE TRENDS

Bipolar transistors have long been and, for the foreseeable future, will continue to be the devices of choice for most high-speed applications. Key characteristics in maintaining their advantage are high f_T and high transconductance. Both features of bipolar transistors can be achieved with relatively large lateral dimensions, so lithography requirements are not severe.

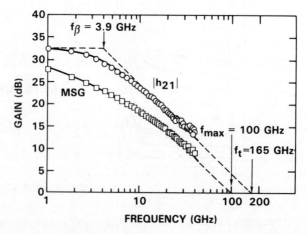

Fig. 47 Incremental current gain and maximum stable power gain versus frequency measured for an InP/InGaAs HBT. (After Ref. 39)

For digital applications, a major technology objective over recent years has been to increase circuit density. Lateral scaling has been carried out aggressively, with corresponding decreases in device parasitics, in interconnect capacitance (from higher packing density), and in the power dissipation per device. Already emitter widths of 0.35 μm are in use in experimental integrated circuits. This trend will continue as lithography techniques improve. Corresponding scaling in the vertical dimension is limited in homojunction devices. Present circuit approaches (and the constant V_{BE} of transistors as devices are scaled) dictate that power-supply voltages should remain constant. From this, lower limits on base doping are determined to prevent base punchthrough. Emitter doping must increase to maintain current gain, and the emitter stored charge rises, limiting f_T. Predictions indicate that f_T of the order of 40 to 50 GHz is the maximum achievable with homojunction devices. HBTs provide a way for further improvement of operating speed, by minimizing the emitter stored charge while avoiding base punchthrough. Already f_T above 100 GHz has been achieved in HBTs.

It is interesting to note that in very large, dense digital circuits the importance of lowering base sheet resistance is not great. As emitter width W_E is decreased, typically collector currents are reduced with logic swings kept constant. Circuit load resistances increase as W_E decreases, and thus the tolerable values of R_B also increase. At the same time, the base resistance for a given base sheet resistance decreases as W_E decreases. The permissible base sheet resistances increase as W_E^2.

Additional problems for bipolar transistors that will arise from continued dimensional scaling are associated with breakdown, tunnel currents in the base-emitter junctions, and hot-electron effects due to increasing electric fields. Many of these problems could be minimized if the operating voltages could be scaled down with transistor size. Additional opportunities exist from the physics of small dimensions, notably the velocity overshoot effects discussed below. Quantum size effects are still fairly remote at dimensions foreseeable in the near future.

Inherent problems of the present bipolar logic approaches remain even with scaling. These include considerable static power dissipation (along with the dynamic power) and saturation charge storage.

For analog and microwave applications, the trend to smaller devices is not as clear-cut. Power levels used in circuits are dictated by the impedance of free space (or of conveniently fabricated transmission lines), together with noise considerations. Therefore currents used in transistors cannot be reduced. With impedance-matching circuits, there is continued gain improvement and noise reduction from decreasing R_B to very low values. Current crowding is a central concern. For analog circuits, matching of devices is frequently of key importance. This is more readily achieved with large lateral device dimensions. Therefore, the advantages of HBTs (with low base sheet resistances and high gain achievable with large emitter dimensions) over homojunction transistors are even more evident.

These trends are based on direct extrapolation of present technology. In addition, a variety of new themes will be explored in the future with bipolar transistors. These new themes will include increased use of velocity overshoot, new materials, and integration of bipolars with different devices.

Velocity Overshoot Effects. As discussed in Sections 6.6.3 and 6.7.3, the velocity of electrons in semiconductors, particularly in III-V compounds, may attain high values (well above 10^7 cm/s) for brief periods after the electrons enter regions with high electric fields. This effect is particularly important in HBTs for electrons traversing the base-collector depletion region. This phenomenon can be exploited to increase the f_T of the device, by reducing the transit time of electrons across the base-collector depletion region. It also has a significant role in increasing the maximum current density, J_{max}, that can be carried by an HBT without incurring base pushout. Velocity overshoot also contributes to reduced base-collector capacitance at a fixed value of f_T (because it increases the distance traveled by electrons in a given amount of time). HBTs can take advantage of the velocity overshoot phenomenon because the dimensions involved can be made small and are reproducible. A major drawback is that the voltage range over which velocity overshoot persists is not very great. To take best advantage of it, logic circuits will have to be configured to keep voltage swings small, and associated values of V_{CE} relatively fixed. For microwave power applications, it will be necessary to seek ways to achieve efficient power combining from low-voltage, high-current sources.

New Materials. Exploration of new materials systems is an important avenue to improve devices. Material systems will be tailored to specific applications. For microwave power generation and amplification, high output power density is desirable, as well as operation at convenient (high) impedance levels. A useful figure of merit for this application is $v_{sat}\mathcal{E}_b^2$, where $\mathcal{E}_b$ is the breakdown electric field. The semiconductors used most extensively for bipolar transistors at present do not differ significantly in this figure of merit. Different semiconductor materials, including diamond, offer the possibility of dramatic increases in this value.

For use in dense logic circuits, choice of new materials will be dictated by entirely different considerations. It has been argued in Section 6.2.4 that the power-delay product that limits very large circuits ("wire-limited chips") is remarkably insensitive to transistor parameters. Key parameters are layout size, which can determine interconnect lengths, and threshold voltage V_{th} for current flow. One theme of future device evolution will be to minimize V_{th}, which for bipolar transistors corresponds to the turn-on V_{BE} value. For FET technologies, the threshold voltage for switching is relatively easily varied by changing channel doping and thickness. FETs, however, tend to have non-uniform V_{th} and low transconductance, and as a result are difficult to operate with low voltage swings. With bipolar technologies, V_{th} is fixed by the material system. For this reason, it is relatively uniform across wafers, but it cannot be readily changed to minimize power dissipation. From Eq. 26, V_{th} in HBTs with graded junctions is dependent on the bandgap of the material used in the base, E_{gB}, approximately according to $V_{th} = E_{gB}/q - 10kT/q$. To minimize dynamic power-delay product, it is important to choose materials providing a low value of E_{gB}. Figure 48 illustrates the measured dependence of collector current on V_{BE} for Si bipolar transistors, GaAs graded junction HBTs, and InAlAs/InGaAs HBTs (lattice matched to InP substrates) with graded and abrupt emitter-base junctions. The considerable shift in V_{th} between the materials is clear from the figure. The value of supply voltage depends on the detailed circuit approach (ECL, I²L, NTL (nonthreshold logic), etc.) but scales roughly as $V_{th} + V_L$ (where V_L is the logic swing voltage). For AlGaAs/GaAs, Si, InGaAs/InP, the corresponding quantities are approximately in the ratio 1 : 0.8 : 0.5. A further decrease to 0.3 can be expected by using InAs as the base material. Corresponding reductions in dynamic power-delay product may be expected with the narrower bandgap materials. This trend can be expected to continue until the leakage currents associated with thermal generation of carriers in the small-bandgap materials become excessive. The critical leakage of the base-collector depletion region can be minimized by the use of a wide-bandgap material in this regions (double-heterojunction device).

Integration of Bipolar Transistors with Different Devices.   While bipolar transistors do not have all the characteristics needed for optimal circuits, they can be combined monolithically with other devices that supplement their strengths. One well-known example of this is BiCMOS technology, in which *n-p-n* bipolar

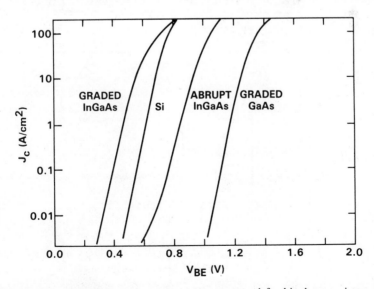

Fig. 48   Collector current density versus V_{BE} measured for bipolar transistors in a number of material systems: Si transistors; GaAlAs/GaAs HBTs with graded BE junction; and InAlAs/InGaAs HBTs with graded and abrupt BE junctions.

transistors are combined with *n*-channel and *p*-channel MOSFETs. Figure 49 shows the cross section of a representative BiCMOS structure, and the circuit diagram of a BiCMOS gate. The logic tends to combine the best features of CMOS and bipolar technologies, low static-power dissipation and high-drive capability (from high transconductance output transistors). Although the process is relatively complex, the use of BiCMOS technology is increasing.

Related combinations of devices may be expected in III-V materials. Combinations of HBTs and heterostructure FETs will be developed, with the objective of combining high-speed and low-power benefits for logic circuits, to achieve combinations of high-power and low-noise characteristics in the microwave domain, and to add devices with high dc input impedance to bipolar circuits in analog applications. A major theme within bipolar technology itself is the combination of *n-p-n* and *p-n-p* devices. This chapter has not given extensive coverage to *p-n-p* transistor development, since *p-n-p* performance has traditionally lagged that of the *n-p-n*. Particular synergy occurs, however, with combinations of *n-p-n* and *p-n-p* transistors on the same chip. This combination has long been practiced in analog circuits, where *p-n-p* devices form active loads (current mirrors) in high-gain stages, and are used in output drivers (complementary push-

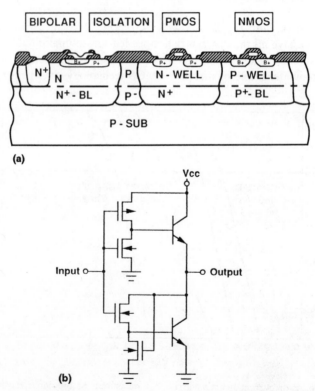

Fig. 49 (a) Schematic cross section of representative BiCMOS structure. (b) Circuit diagram of representative BiCMOS logic gate.

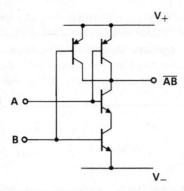

Fig. 50   Circuit diagram for complementary bipolar logic.

pull amplifiers). The I^2L approach is also based on merged n-p-n/p-n-p transistor combinations. The p-n-p most often implemented has been a lateral structure with limited f_T. More recently, a high f_T p-n-p has been developed and incorporated on the same chip with the fast n-p-n. The resulting structures can provide significant performance improvements in circuits. A long-range objective is to develop a logic approach that minimizes static power dissipation, the bipolar analog of CMOS. The most straightforward implementation of such circuits, illustrated in Fig. 50, requires transistors that have adequate current gain when operated in saturation and do not store excess charge under this condition. Such devices have not yet been developed. Alternative approaches have been suggested, in which logic is carried out by diode switches or by n-p-n devices only and the output drivers are implemented with complementary circuits.

On a long-term basis, it can be envisioned that logic approaches will be developed based on complementary HBTs that will virtually eliminate static power dissipation and that will operate with the low logic swings characteristic of bipolar circuits. These circuits will minimize power supply voltage by appropriate choice of materials to achieve low V_{BE}. By combining such structures with an advanced interconnect technology to ensure low wiring capacitance, an "optimal" logic approach will have been established.

PROBLEMS

1. According to a simple model of the single-crystal Si/polysilicon boundary, the hole current J^+ traversing the boundary in the direction of the contact is $J^+ = qp_s v_s$, where p_s is the hole concentration in the single-crystal silicon at the boundary, and v_s is an effective velocity. A corresponding current $J^- = qp_p v_s$ flows from the polysilicon to the single-crystal material, where p_p is the hole concentration in the poly material at the boundary. What is the effective surface recombination velocity s_0 for the single-crystal emitter? Discuss how s_0 varies with polysilicon thickness.

2. If a transistor has two base contact stripes, one on either side of an emitter stripe, find the component of base resistance associated with the flow of current underneath the emitter. Assume the transistor length is L, the emitter stripe width is W, and the base sheet resistance is R_s. Consider that R_s is low enough that the emitter current is effectively uniform.

3. Determine the value of collector current density such that $\tau_E + \tau_C$ contribute less than half of τ_{EC} in a heterojunction transistor. Assume $C_{BE} = 5 \text{ } fF/\mu m^2$, $C_{BC} = 0.2 \text{ } fF/\mu m^2$, $\tau_B = 2 \text{ ps}$, $\tau_{cscl} = 3 \text{ ps}$, $R_E = 10^{-6} \text{ } \Omega\text{-cm}^2$, $R_C = 8 \times 10^{-6} \text{ } \Omega\text{-cm}^2$. What limit is set on N_C to avoid base pushout at this current density?

4. Contrast the electric field built into the base region due to doping gradients for the case where the base majority carriers are non-degenerate and for the case where they are degenerate.

5. Determine the value of base thickness that maximizes f_{max} of a bipolar transistor. Assume that minority carrier transport across the base is by diffusion only.

6. Consider a SiGe/Si HBT with $x_{Ge} = 10\%$ in the base region (and 0% in emitter and collector regions). If the base current is due to emitter injection efficiency only, what is the expected change in current gain between 0 and 100°C?

7. Estimate the intrinsic offset voltage V_{CEsat} for a single heterojunction AlGaAs/GaAs HBT with graded BE junction, and $p_B = 10^{19} \text{ cm}^{-3}$, $n_E = 5 \times 10^{17} \text{ cm}^{-3}$, and $n_C = 5 \times 10^{16} \text{ cm}^{-3}$. Assume the recombination lifetime for electrons and holes is 1 ns.

8. Discuss how f_T varies as V_{CE} is changed (around a nominal value of 2 V) in a homojunction transistor ($p_B = 5 \times 10^{17} \text{ cm}^{-3}$, $W_B = 800 \text{ Å}$, $N_C = 5 \times 10^{16} \text{ cm}^{-3}$, $W_C = 1.0 \text{ } \mu m$), and in a heterojunction transistor (which differs by having $p_B = 5 \times 10^{19} \text{ cm}^{-3}$)? Assume $J_C = 10^4 \text{ A/cm}^2$.

9. Determine the velocity of electrons injected into the base region from abrupt emitter-base heterojunctions of InP/InGaAs and InAlAs/InGaAs. Assume parabolic bands for the InGaAs (although in reality non-parabolicity affects the answer significantly). Determine the angular distribution of electron velocity within the base near the emitter.

10. The circuit model for a bipolar transistor typically takes the form of a T-connected circuit, with a current source $\alpha_0 e^{-j\omega\tau} i_e$ between base and collector, controlled by the emitter current i_e; or a π-connected circuit, as shown in Fig. 6, with a current source $g_{m0} e^{-j\omega\tau} v_1$ between collector and emitter, controlled by the voltage v_1 between base and emitter. Find the relationship between the elements of the two models.

REFERENCES

1. S. M. Sze, *Physics of Semiconductor Devices,* 2nd ed., Wiley, New York, 1981.

2. D. S. Lee and J. G. Fossum, "Energy Band Distortion in Highly Doped Silicon," *IEEE Trans. Electron Dev.,* **ED-30**, 626 (1983).

3. D. D. Tang and P. M. Solomon, "Bipolar Transistor Design for Optimized Power-Delay Logic Circuits," *IEEE J. Solid-State Circ.,* **SC-14,** 679 (1979).

4. G. W. Taylor and J. G. Simmons, "Figure of Merit for Integrated Bipolar Transistors," *Solid-State Electron.,* **29,** 941 (1986).

5. J. M. C. Stork, "Bipolar Transistor Scaling for Minimum Switching Delay and Energy Dissipation," *Tech. Dig. IEDM,* p. 550 (1988).

6. K. U. Ashar, "The Method of Estimating Delay in Switching Circuits and the Figure of Merit of a Switching Transistor,"*IEEE Trans. Electron Dev.,* **ED-13,** 497 (1964).

7. P. M. Solomon and D. D. Tang, "Bipolar Circuit Scaling," *ISSCC Dig. Tech. Pap.,* p. 86 (1979).

8. T. H. Ning and D. D. Tang, "Bipolar Trends," *Proc. IEEE,* **74,** 1669 (1986).

9. D. D. Tang, P. M. Solomon, T. H. Ning, R. D. Isaac, and R. E. Burger, "1.25 μm Deep-Grove-isolated Self-aligned Bipolar Circuits," *IEEE J. Solid State Circ.,* **SC-17,** 925 (1982).

10. H. Schaber, J. Bieger, B. Benna, and T. Meister, "Vertical Scaling Considerations for Polysilicon-Emitter Bipolar Transistors," in *Solid State Devices,* G. Soncini and P. U. Calzolari, Eds., Elsevier Science, North-Holland, 1988, p. 685.

11. S. Konaka, Y. Amemiya, K. Sakuma, and T. Sakai, "A 20 pS/G Si Bipolar IC Using Advanced SST with Collector Ion Implantation," *Ext. Abstr. 19th Conf. Solid-State Dev. Matls.,* Tokyo, 1987, p. 331.

12. T. Nakamura, T. Miyazaki, S. Takahashi, T. Kure, T. Okabe, and M. Nagata, "Self-Aligned Transistor with Sidewall Base Electrodes," *IEEE Trans. Electron Dev.,* **ED-29,** 596 (1982).

13. H. Kroemer, "Theory of Wide-Gap Emitter for Transistors," *Proc. IRE,* **45,** 1535 (1957).

14. H. Kroemer, "Heterostructure Bipolar Transistors and Integrated Circuits," *Proc. IEEE,* **70,** 13 (1982).

15. J. E. Sutherland and J. R. Hauser, "A Computer Analysis of Heterojunction and Graded Composition Solar Cells," *IEEE Trans. Electron Dev.,* **ED-24,** 363 (1977).

16. A. Marty, G. E. Rey, and J. P. Bailbe, "Electrical Behavior of an NPN GaAlAs/GaAs Heterojunction Transistor," *Solid-State Electron.*, **22**, 549 (1979).

17. H. Kroemer, "Two Integral Relations Pertaining to the Electron Transport through a Bipolar Transistor with a Nonuniform Energy Gap in the Base Region," *Solid-State Electron.*, **28**, 1101 (1985).

18. C. Smith and A. D. Welbourn, "Prospects for a Heterostructure Bipolar Transistor Using a Silicon-Germanium Alloy," *Tech. Dig. IEEE Bip. Circ. Tech. Mtg.*, p. 57 (1987).

19. G. L. Patton, S. S. Iyer, S. L. Delage, S. Tiwari, and J. M. C. Stork, "Silicon-Germanium Base Heterojunction Bipolar Transistors by Molecular Beam Epitaxy," *IEEE Electron Dev. Lett.*, **9**, 165 (1988).

20. H. Temkin, J. C. Bean, A. Antreasyan, and R. Leibenguth, "Ge_xSi_{1-x} Strained-Layer Heterostructure Bipolar Transistors," *Appl. Phys. Lett.*, **52**, 1089 (1988).

21. J. F. Gibbons, C. A. King, J. L. Hoyt, D. B. Noble, C. M. Gronet, M. P. Scott, S. J. Rosner, G. Reid, S. Laderman, K. Nauka, J. Turner, and T. I. Kamins, "Si/SiGe Heterojunction Bipolar Transistors Fabricated by Limited Reaction Processing," *Tech. Dig. IEDM*, p. 566 (1988).

22. C. M. Maziar, M. E. Klausmeier-Brown, and M. Lundstrom, "A Proposed Structure for Collector Transit-Time Reduction in AlGaAs/GaAs Bipolar Transistors," *IEEE Electron Dev. Lett.*, **EDL-7**, 483 (1986).

23. D. Ankri and L. F. Eastman, "GaAlAs/GaAs Ballistic Heterojunction Bipolar Transistor," *Electron. Lett.*, **18**, 750 (1982).

24. M. F. Chang, P. M. Asbeck, K. C. Wang, G. J. Sullivan, N. H. Sheng, J. A. Higgins, and D. L. Miller, "AlGaAs/GaAs Heterojunction Bipolar Transistors Fabricated Using a Self-Aligned Dual-Liftoff Process," *IEEE Electron Dev. Lett.*, **EDL-8**, 7 (1987).

25. S. Tiwari, "GaAlAs/GaAs Heterostructure Bipolar Transistors: Experiment and Theory," *Tech. Dig. IEDM*, p. 262 (1986).

26. H. H. Lin and S. C. Lee, "Super-Gain AlGaAs/GaAs Heterojunction Bipolar Transistors Using an Emitter Edge-thinning Design," *Appl. Phys. Lett.*, **47**, 839 (1985).

27. T. Isawa, T. Ishibashi, and T. Sugeta, "AlGaAs/GaAs Heterojunction Bipolar Transistors," *Tech. Dig. IEDM*, p. 328 (1985).

28. T. Ishibashi and Y. Yamauchi, "A Possible Near-Ballistic Collection in an AlGaAs/GaAs HBT with a Modified Collector Structure," *IEEE Trans. Electron Dev.*, **ED-35**, 401 (1988).

29. B. Bayraktaroglu, N. Camilieri, H. D. Shih, and H. Q. Tserng, "AlGaAs/GaAs Heterojunction Bipolar Transistors with 4 W/mm Power Density at X-Band," *Tech. Dig. MTT-S*, p. 969 (1987).

30. N. H. Sheng, M. F. Chang, P. M. Asbeck, K. C. Wang, G. J. Sullivan, D. L. Miller, J. A. Higgins, E. Sovero, and H. Basit, "High Power

GaAlAs/GaAs HBTs for Microwave Applications," *Tech. Dig. IEDM* (1987).

31. H. T. Yuan, W. V. McLevige, H. D. Shih, and A. S. Hearn, "GaAs Heterojunction Bipolar 1K Gate Array," *ISSCC Tech. Dig.*, p. 42 (1984).

32. K. C. Wang, P. M. Asbeck, M. F. Chang, G. J. Sullivan, and D. L. Miller, "A 4-bit Quantizer Implemented with Heterojunction Bipolar Transistors," *Tech. Dig. GaAs IC Symp.*, p. 83 (1987).

33. A. K. Oki, M. E. Kim, J. B. Camou, C. L. Robertson, G. M. Gorman, K. B. Weber, L. M. Holbrock, S. W. Southwell, and B. Y. Oyama, "High Performance GaAs/AlGaAs Heterojunction Bipolar Transistor 4-Bit and 2-Bit A/D Converters and 8-Bit D/A Converter," *Tech. Dig. GaAs IC Symp.*, p. 137 (1987).

34. N. Chand, T. Henderson, R. Fischer, W. Kopp, H. Morkoc, and L. J. Giacoletto, "A *pnp* AlGaAs/GaAs Heterojunction Bipolar Transistor," *Appl. Phys. Lett.*, **46,** 302 (1985).

35. D. A. Sunderland and P. D. Dapkus, "The Performance Potential of *p-n-p* Heterojunction Bipolar Transistors," *IEEE Electron Dev. Lett.*, **EDL-6,** 648 (1985); see also J. Hutchby, "High Performance Pnp AlGaAs/GaAs Heterojunction Bipolar Transistors: A Theoretical Analysis," *IEEE Electron Dev. Lett.*, **EDL-7,** 108 (1986).

36. R. J. Malik, J. R. Hayes, F. Capasso, K. Alavi, and A. Y. Cho, "High-Gain $Al_{0.48}In_{0.52}As/Ga_{0.47}In_{0.53}As$ Vertical *npn* Heterojunction Bipolar Transistors Grown by Molecular Beam Epitaxy," *IEEE Electron Dev. Lett.*, **EDL-4,** 383 (1983).

37. H. Kanbe, J. C. Vlcek, and C. G. Fonstad, "(In, Ga)As/InP *npn* Heterojunction Bipolar Transistors Grown by Liquid Phase Epitaxy with High Current Gain," *IEEE Electron Dev. Lett.*, **EDL-5,** 172 (1984).

38. R. N. Nottenburg, H. Temkin, M. B. Panish, R. Bhat, and J. C. Bischoff, "InGaAs/InP Double-Heterostructure Bipolar Transistors with Near-Ideal Beta versus I_C Characteristics," *IEEE Electron Dev. Lett.*, **EDL-7,** 643 (1986).

39. Y. K. Chen, R. N. Nottenburg, M. B. Panish, R. A. Hamm, and D. A. Humphrey, "Subpicosecond InP/InGaAs Heterostructure Bipolar Transistors," *IEEE Electron Dev. Lett.*, **EDL-10,** 267 (1989).

40. U. K. Mishra, J. F. Jensen, D. B. Rensch, A. S. Brown, M. W. Pierce, L. G. McGray, T. V. Kargodorian, W. S. Hoefer, and R. E. Kastris, "48 GHz AlInAs/GaInAs Heterojunction Bipolar Transistors," *Tech. Dig. IEDM,* p. 873 (1988).

41. G. L. Patton, J. H. Comfort, B. S. Meyerson, D. F. Crabbe, G. J. Scilla, E. DeFresart, J. M. C. Stork, J. Y.-C. Sun, D. L. Harame, and J. N. burghartz, "75-GhZ f_T SiGe-Base Heterojunction Bipolar Transistors," *IEEE Electron Dev. Lett.*, **EDL-11**, 171 (1990).

7 Hot-Electron Transistors

S. Luryi

AT&T Bell Laboratories
Murray Hill, New Jersey

7.1 INTRODUCTION

As the dimensions of semiconductor devices shrink and the internal fields rise, a large fraction of carriers in the active regions of the device during its operation are in states of high kinetic energy.[1] At a given point in space and time the velocity distribution of carriers may be narrowly peaked, in which case one speaks about "ballistic" electron packets. At other times and locations, the non-equilibrium electron ensemble can have a broad velocity distribution—usually taken to be Maxwellian and parameterized by an effective electron temperature $T_e > T$, where T is the lattice temperature. Hot-electron phenomena have become important for the understanding of all modern semiconductor devices.[2] Moreover, a number of devices have been proposed whose principle is based on such effects. This group of devices will be considered here. We shall be concerned only with three-terminal hot-electron *injection* devices, that is, transistors, based on hot carriers that are physically transferred between adjacent semiconductor layers.

Although the first hot-electron injection devices were proposed a long time ago, their full potential has become realizable only with the advent of such epitaxial techniques as the molecular beam epitaxy (MBE) and metal-organic chemical vapor deposition (MOCVD). These techniques can provide abrupt heterojunction interfaces and abrupt changes in the doping profile—on the scale of a nanometer—that are essential for the implementation of hot-electron devices.

7.1.1 Field-Effect and Potential-Effect Transistors

By the physical principle involved, most of the three-terminal semiconductor devices can be classified into either of the two groups: field-effect transistors

High-Speed Semiconductor Devices, Edited by S.M. Sze. ISBN 0-471-62307-5

(FET) and potential-effect transistors (PET). In the first group, containing a great variety of FETs, the transistor action results from screening of an input electric field by the modulation of charge in a conducting channel.† In the second group, which includes the bipolar transistor, the so-called analog transistors,[3-6] and a number of ballistic-injection, hot-electron transistors discussed in this chapter, the transistor action results from *modulating by an input electrode the height of a potential barrier for carrier injection.* Of course, both the field-screening and the potential-modulation effects are at work in every transistor. Nevertheless, this classification based on the physical principle of the transistor control action, is instructive because it allows us to abstract certain generalizations about each group of devices.[7]

Consider a field-effect transistor, Fig. 1a. In the "on" state the channel contains a charge Q moving toward the drain with the speed v. This means that the channel current equals

$$I = \frac{Q\,v}{L}\,,\tag{1}$$

where L is the source-to-drain separation. Because the channel charge has been induced by the field effect, the input circuit must have placed on the gate a charge of opposite polarity and at least of the same magnitude Q. How long would it take for the output current to charge the gate of the next-stage identical transistor to the same level? The answer is clear:

$$\tau = \frac{Q}{I} = \frac{L}{v}\,.\tag{2}$$

This "time-of-flight" delay represents the characteristic speed limitation of an FET. It is an idealized limit (cf. Problem 1). In reality, besides the useful channel charge Q, the output current must supply an additional charge Q_p equal

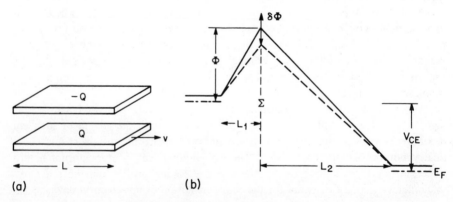

(a)   (b)

Fig. 1   Schematic illustration of (a) field-effect transistor and (b) potential-effect transistor.

---

†This is the precise meaning of the widely used though rarely defined term "field effect".

in magnitude to the combined parasitic charges that the gate induces outside the channel. Another factor that further degrades the transistor speed compared to the idealization, Eq. (2), is the non-vanishing source resistance, that is, the series resistance between the modulated end of the channel and the true source ground electrode. To enhance the performance of field-effect transistors one seeks to shrink the channel length L and, at the same time, maintain or improve the useful fraction of the modulated charge and minimize the source resistance.

At first sight, factors limiting the speed of potential-effect transistors are quite different. Consider the operation of an idealized PET, Fig. 1b. It represents a triangular-barrier diode with some (unspecified) means to modulate the "planar-doped" charge Σ, which forms the barrier. For a fixed bias between the emitter and the collector, the height of the barrier will be an approximately linear function of Σ, while the current will be an exponential function of the barrier height. This implies that the speed of a potential-effect transistor increases with its output current. Such an enhancement, however, cannot be continued indefinitely, because the exponential regime will be limited. In a triangular-barrier diode the exponential I-V characteristics saturate at sufficiently high current densities and are eventually replaced by a linear dependence (Section 2.4.2). From the point of view of enhancing the speed, it makes no sense to further increase the current, because the required input charge (as well as the charge stored in all parasitic capacitances) will rise proportionately.†

Saturation of the exponential characteristics occurs when the mobile charge associated with the increasing current becomes sufficiently large to screen internal fields in the device. The fact that this is the physical origin of the limitation of speed in potential-effect transistors, can be seen clearly from a small-signal analysis (cf. Problem 2). One obtains the characteristic speed limitations of a potential-effect transistor:

$$\tau_E = \frac{L_1}{v_T}, \qquad \tau_C = \frac{L_2}{v_S} \tag{3}$$

called, respectively, the emitter and the collector charging times. Here v_S is the saturated drift velocity of carriers on the downhill slope of the barrier and v_T their effective diffusion velocity on the uphill slope (Section 2.4.2). Another important limitation is associated with the finite time it takes to modulate the controlling charge Σ. This has to be done by feeding an input current that charges or discharges the base-emitter and base-collector capacitances through a

---

†Most FETs also possess a range of biases where the output current is an exponential function of the input voltage. The name notwithstanding, in these regimes the device control is due to the potential—not field—effect. For example, a metal-oxide-semiconductor field-effect transistor (MOSFET) below threshold could be classified as a PET. In this regime, the device speed rises with the current, much like in a bipolar transistor. Above threshold, the output current becomes a linear function of the gate bias V_G (for sufficiently short-channel devices) and further increase of V_G does not enhance the speed. It should be noted that MOSFETs are typically operated with the "on" state chosen quite far from the threshold—for reasons associated with the circuit stability, threshold variation margins, etc.

finite lateral base resistance. The resultant RC delay,

$$\tau_B = R_B (C_{\text{BE}} + C_{\text{BC}}) , \tag{4}$$

is illustrated in Problem 2c for the model PET of Fig. 1b.

Different types of potential-effect transistors employ different ways of supplying charge to the base. Of course, the most important property of the base is that it must be "transparent" for carriers injected from the emitter. It is a notable feature of all PETs that carriers participating in the output current are distinct from those that control the barrier height. Any mixing between these groups of carriers leads to a degradation of the transistor performance. The classic example of a PET is the bipolar transistor where the identity of the two groups is determined by the band in which the carriers move.† In a vacuum triode, which is the prototype of all potential-effect transistors, the two groups of carriers move in different regions of the "real space"; similar situation occurs in most analog transistors.

In ballistic hot-electron transistors, each group of carriers is maintained in a distinct narrow energy range within the same band. One group of carriers is injected from the emitter at high energy and travels across the base ballistically; the other group of carriers are "native" to the base and are distributed there in an equilibrium fashion. As in a bipolar transistor, one can say that the controlling and the injected carriers move in different portions of the momentum space. Base transparency for ballistic carriers relies on the short time required to cross the narrow base compared to the energy relaxation time, which is the time required for the two groups of carriers in the base to lose their identity. Ballistic hot-electron transistors will be discussed in Section 7.3.

Another group of hot-electron devices are the so-called real-space transfer transistors, which we shall discuss in Section 7.4. Interestingly, these devices are based *neither* on the potential-effect *nor* on the field-effect principle. One can illustrate their operational principle by an analogy with the glow cathode of a vacuum diode. The anode current of a vacuum diode saturates as function of the applied anode voltage when the latter is high enough to sweep all the electrons that are thermally emitted by the heated cathode. The higher the cathode temperature the higher is the level of current saturation. One can imagine an amplifying device in which the input circuit would be the cathode heater, Fig. 2. In this hypothetical device the cathode provides carriers by thermionic emission over the barrier of fixed height (the cathode's work function) with the temperature of the cathode modulated in a controlled fashion. Needless to say this would be a slow amplifier, as the material temperature cannot be modulated rapidly.‡ In the real-space-transfer transistors, one modulates the effective hot-

---

†Transparency of the base for the minority carriers relies on the fact that their lifetime is substantially longer than the time they need to travel across the base and form the output current.

‡The characteristic times of such processes are of the form RC, where C is the heat capacity of a material and R the thermal resistance of the link to a heat bath. The reader is invited to make estimates.

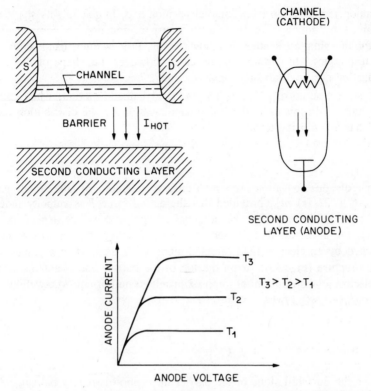

Fig. 2 Illustration of the principle of real-space transfer transistors. The channel serves as a cathode whose effective electron temperature T_e is controlled by the source-to-drain field. The second conducting layer, separated by a potential barrier, serves as an anode and is biased positively. To the extent that the barrier lowering by the anode field can be neglected, the anode current as a function of the anode voltage exhibits quasi-saturation at a value determined by T_e.

electron temperature, and this can be done *very* rapidly with only picosecond delays associated with the electron energy-relaxation time.

7.1.2 Hot Electrons

The term "hot electrons" purports a non-equilibrium ensemble of high-energy carriers. It is often possible to pump external energy (e.g., by shining light or applying an electric field) directly into the system of carriers. If the power input of the electronic system exceeds the rate of energy loss by that system to the lattice, then the carriers "heat up," that is, their velocity distribution deviates significantly from the equilibrium Maxwellian form.† In a steady state, this

---

†Note that one can also *cool* the carrier system by making it do work against an external field at a fast rate compared to power replenished by the lattice.

distribution is a function of the spatial position and, in general, its shape can be determined by solving the Boltzmann transport equation, a complicated task even for the simplest scattering models. This task is often handled by Monte Carlo techniques.[8] In certain limiting cases, however, the shape of the distribution function can be established approximately.

One such case corresponds to the so-called ballistic motion, when collisions are negligible. In this case, the Boltzmann equation for the distribution function $f(\mathbf{r}, \mathbf{v})$ has no collision term:

$$\mathbf{v} \cdot \frac{\partial f}{\partial \mathbf{r}} + \frac{\mathbf{F}}{m} \cdot \frac{\partial f}{\partial \mathbf{v}} = 0 .$$

Solution of this equation is an arbitrary function of the total energy, $f = f[\mathbf{v}^2 + 2U(\mathbf{r})/m]$, provided that the acting force F is velocity independent, $F \equiv -\partial U/\partial \mathbf{r}$, and that the potential energy U does not depend on time. Suppose, for example, that carriers are injected into a region where they are accelerated by an electric field directed along the z axis, as in a planar-doped-barrier structure (Fig. 3a). If on the top of the barrier one maintains a steady-state electronic distribution of (approximately) equilibrium Maxwellian form, $f_0(\mathbf{v}) = f_0(v_x) f_0(v_y) f_0(v_z)$, where

$$f_0(v_j) = \left[\frac{m}{2\pi kT} \right]^{1/2} e^{-mv_j^2/2kT} , \quad j = x, y, z \tag{5}$$

and if on the downhill slope of the barrier the motion is purely ballistic, at least down to a point z where the potential is $V(z)$, then at that point the distribution function is given by

$$f(\mathbf{r}, \mathbf{v}) = f_0([v_z^2 - 2qV(z)/m]^{1/2}) f_0(v_x) f_0(v_y) , \quad (v_z > 0) . \tag{6}$$

For $qV(z) >> kT$ the energy distribution function dN/dE at the point z is sharply peaked. Ballistic motion can be launched without an accelerating electric field: electrons injected over a "cliff" in a heterojunction (Fig. 3b) begin their travel with a high kinetic energy equal to the conduction-band discontinuity. Of course, these narrow high-energy electrons can have distributions other than Eq. 6. For example, an electron distribution injected by tunneling from a Fermi-degenerate emitter, Fig. 3c, would have a shape determined by the energy dependence of the barrier transmission coefficient. Such "filtered" distributions can be transported ballistically across a semiconductor layer.[9]

Another simple limit is realized when the carrier density is high (of order $10^{18}\,\text{cm}^{-3}$ or higher). Due to the carrier-carrier interaction, the extra kinetic energy is redistributed among carriers much faster than it is dissipated to the crystal lattice. Typically, multiple collisions result in a Gaussian distribution of carrier velocities, which can be interpreted by analogy with the Maxwellian ensemble in Eq. 5 as a quasi-equilibrium distribution—but with an effective electron temperature T_e. If such an ensemble is established in the vicinity of a potential barrier of height Φ, then the carrier flux over this barrier is propor-

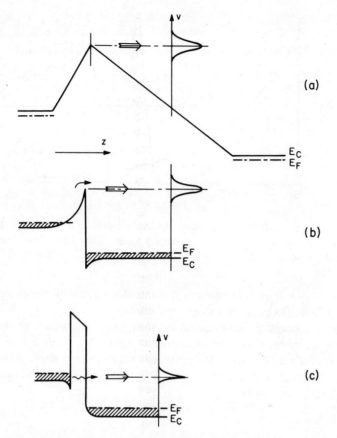

Fig. 3 Ballistic motion of electrons (a) accelerated by an electric field; (b) thermionically injected from a wider-gap material; (c) injected by tunneling into a state of high kinetic energy.

tional to $\exp(-\Phi/kT_e)$. This process is quite similar to the usual thermionic emission. It is usually referred to as the real-space transfer.

Energy distributions, corresponding to these simplest hot-electron regimes (the ballistic and the T_e limits) are illustrated schematically in Fig. 4.

Two distinct classes of hot-electron-injection devices can be identified depending on which of the two hot-electron regimes is employed. In the *ballistic* devices, electrons are injected into a narrow base layer at a high initial energy in the direction normal to the plane of the layer. Performance of these devices is limited by various energy-loss mechanisms in the base and by the finite probability of a reflection at the base-collector barrier. In the *real-space transfer* (RST) devices the heating electric field is applied parallel to the semiconductor layers with hot electrons then spilling over to the adjacent layers over an energy barrier. Even though only a small fraction of electrons, those in the high-energy tail of the hot-carrier distribution function, can participate in this

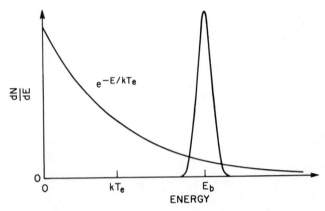

Fig. 4 Hot-electron distribution functions. In ballistic devices dN/dE is sharply peaked at $E_b \approx mv_b^2/2$, with v_b mainly directed perpendicular to the base layer. In the real-space transfer devices, the distribution is, approximately, $dN/dE \propto \exp(-E/kT_e)$.

flux, their number is replenished at a fast rate determined by the energy relaxation time, so that the injection can be very efficient.

Our main conceptual classification is, thus, made according to the type of hot-electron ensemble employed in the device operation. Although all injection devices involve a real-space transfer of hot electrons, we shall, following the established terminology, reserve this term for devices operating in the electron-temperature regime.

7.2   ELECTRON TRANSMISSION ACROSS AN INTERFACE OF DIFFERENT CRYSTALS

Hot-electron transistors are usually implemented in a heterojunction system. What happens when an electron crosses a heterointerface? One would like to consider this problem as if electrons were free in both materials (albeit with an effective, rather than the free electron mass), and treat it as a barrier transmission problem. The key advantage in using this approach, called the envelope function approximation, is that the estimates of the barrier transmission coefficient are in terms of such parameters of the structure as the effective electron masses in both constituent crystals and the conduction-band discontinuity at the interface. Without extensive numerical band structure calculations, these are the only parameters that are known reasonably well.

The simplest envelope function description of the transmission problem consists in assuming the wave function Φ in the form of a combination of plane-wave solutions to the effective-mass Hamiltonian,

$$H^{(i)} = E_C^{(i)} - \frac{\hbar^2 \nabla^2}{2m_i} \ , \qquad i = 1, 2 \tag{7}$$

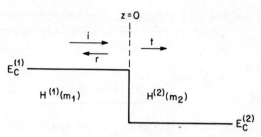

Fig. 5 Envelope function description of a heterjunction. The interface separates regions where the electronic motion is governed by the effective Hamiltonians $H^{(1)}$ and $H^{(2)}$. The potential step at the junction equals $E_C^{(1)} - E_C^{(2)}$. With the eigenfuction of the form (Eq. 8) the electron group velocities in each material ($i = 1, 2$) are $v_i = \hbar k_i / m_i$.

describing electronic motion near the band edges $E_C^{(i)}$ in each crystal, Fig. 5. Using the reflection (r) and the transmission (t) amplitudes, one can write Φ as follows:

$$\Phi(\mathbf{r}) = \begin{cases} e^{j\mathbf{k}_1 \cdot \mathbf{r}} + re^{-j\mathbf{k}_1 \cdot \mathbf{r}}, & z \leq 0, \\[2mm] te^{j\mathbf{k}_2 \cdot \mathbf{r}}, & z \geq 0. \end{cases} \tag{8}$$

However, the envelope function Φ and its normal derivatives need not be continuous across the interface, and the appropriate matching conditions are not known. Without the matching conditions at the interface, the amplitudes r and t are indeterminate and the wave function of Eq. 8 is pretty much useless.

Of course, the exact wave function Ψ and its derivative must be continuous. This continuity is intimately related to the existence of the quantum-mechanical current-density operator,[10]

$$J_z = -\frac{\hbar}{2jm_0} \left[\Psi^* \frac{\partial(\Psi)}{\partial z} - \Psi \frac{\partial(\Psi^*)}{\partial z} \right], \tag{9}$$

and it automatically guarantees conservation of the particle flux across the boundary (m_0 is the free-electron mass). In principle, matching conditions for envelope functions would follow from an expression for the current operator in terms of Φ and Φ', which would be valid in regions where the band structure is rapidly varying (or, in the limit of interest to us, changing abruptly across an interface).†

---

†Such an expression, if it exists, depends on the proper form of the effective kinetic-energy operator acting on the envelope functions—which would generalize Eq. 7 to the regions of variable band structure. Although a number of self-consistent choices for this operator have been discussed in the literature, it has never been satisfactorily derived from the first principles.

Exercise 1: The simplest such expression, parameterized solely in terms of the effective mass $m(z)$, corresponds to a kinetic energy operator of the form

$$\hat{K}\Phi = \frac{\hbar^2}{2}\frac{\partial}{\partial z}\left[\frac{1}{m}\frac{\partial\Phi}{\partial z}\right]. \tag{10}$$

Write down the corresponding expression for the current density. Show that it leads to the following matching condition:

$$\Phi(0^-) = \Phi(0^+); \tag{11}$$

$$\frac{1}{m_1}\frac{\partial\Phi(0^-)}{\partial z} = \frac{1}{m_2}\frac{\partial\Phi(0^+)}{\partial z},$$

which ensures the conservation of flux since the group velocities of electrons in the effective-mass approximation are given by $\hbar k_i/m_i$.

This is a widely employed matching condition; its chief merit lies in the fact that it is practical to use in the absence of relevant information specifying the discontinuity in Φ, which would characterize a given crystal interface.

Exercise 2: Show that the matching conditions as in Eq. 11 lead to a reflection amplitude of the form

$$r = \frac{(k_1/m_1) - (k_2/m_2)}{(k_1/m_1) + (k_2/m_2)}. \tag{12}$$

In particular, if the edges of the two crystal bands coincide ($V = 0$, cf. Fig. 5), so that $k_1^2/m_1 = k_2^2/m_2$, then the reflection amplitude,

$$r = -\frac{\sqrt{m_1} - \sqrt{m_2}}{\sqrt{m_1} + \sqrt{m_2}} = -\frac{k_1 - k_2}{k_1 + k_2}, \tag{13}$$

is different by the sign only from (and the reflection coefficient is identical to) the well-known expression for a free-electron incident on a step potential barrier.

A more consistent approach[11] is to use two parameters, α and β, and let *both* Φ and Φ' be discontinuous at the interface:

$$\beta_1\Phi(0^-) = \beta_2\Phi(0^+); \tag{14}$$

$$\alpha_1\frac{\partial\Phi(0^-)}{\partial z} = \alpha_2\frac{\partial\Phi(0^+)}{\partial z}.$$

Exercise 3: Show that these matching conditions are consistent with the following expression for the current density in a region of continuously varying band structure:

$$J_z = - \frac{\hbar}{2jm_0} \left[\Phi^* \frac{\alpha\, \partial(\beta\Phi)}{\partial z} - \Phi \frac{\alpha\, \partial(\beta\Phi^*)}{\partial z} \right] , \tag{15}$$

and write down a related expression for the kinetic-energy operator acting on the envelope functions.

If one assumes the current density in the form of Eq. 15 and uses the effective-mass approximation for the electron energy spectrum, then the particle conservation requires that $\alpha\beta m$ be continuous across the interface. Matching conditions of Eq. 14 can therefore be expressed as the continuity at the interface of

$$\zeta\, \Phi \quad - \quad \text{continuous} , \tag{16}$$

$$\frac{1}{m\zeta} \frac{\partial\Phi}{\partial z} \quad - \quad \text{continuous} ,$$

introducing, besides the ratio m_1/m_2, another (usually unknown) parameter ζ_1/ζ_2, characterizing the interface. These matching conditions result in the following expression for the reflection amplitude:

$$r = \frac{(k_1/\zeta_1 m_1) - (k_2/\zeta_2 m_2)}{(k_1/\zeta_1 m_1) + (k_2/\zeta_2 m_2)} . \tag{17}$$

If $\zeta_1/\zeta_2 = 1$, then Eqs. 17 and 12 coincide. Since the group velocities of incident and reflected electrons are equal, the intensity reflection coefficient is simply given by

$$R = |r|^2 , \tag{18}$$

and for the transmission coefficient $T = 1 - R$ one obtains the following formula:

$$T = \frac{4(k_1/\zeta_1 m_1)(k_2/\zeta_2 m_2)}{[(k_1/\zeta_1 m_1) + (k_2/\zeta_2 m_2)]^2} . \tag{19}$$

7.2.1   An Exactly Soluble One-Dimensional Model

It may appear that in order to obtain a rigorous solution of the heterojunction transmission problem, it is sufficient to replace the plane waves $e^{j\mathbf{k}_i \cdot \mathbf{r}}$ in Eq. 8 by exact Bloch waves $\Psi_{\mathbf{k}_i}^{(i)}$ in each material ($i = 1, 2$) and use the standard quantum mechanical requirement of continuity for the wave function and its derivative as the matching condition at the interface. Unfortunately, this program fails not only because it is extremely difficult to compute the Bloch functions but because even the exact knowledge of these functions in constituent bulk crystals is insufficient. The problem is that for a given energy in the allowed band,

Bloch solutions do not form a complete set in the double crystal. A rigorous solution must also include electronic states localized in the vicinity of the interface.

These states, often called the evanescent states, owe their existence to the violation of the crystal periodicity by the interface. The presence of a boundary allows the existence of states with a complex Bloch wave vector. The imaginary part of this vector describes an exponential decay of the interface-state wave function into the bulk of the constituent crystals. In semiconductors, these states can have energies in the forbidden gaps, in which case they do not affect the transmission of the bulk states incident on the interface (unless inelastic processes are also considered). On the other hand, those evanescent states whose energies lie within the allowed bands of the constituent crystals, are indispensable for correctly matching the wave functions at the interface. Indeed, in three dimensions it is impossible to match the Bloch solutions of two different crystals across a continuous boundary surface.

A similar situation occurs if one considers a two-dimensional (2D) double-crystal model with the transmission across a linear boundary: Bloch solutions of each individual crystal cannot be matched along the entire boundary. However, in one dimension (1D) this becomes possible because the boundary is no longer a laterally extended object. In the 1D case, the evanescent states do not mix with the Bloch states (because their energies fall strictly within the forbidden gaps) and do not come into play in the transmission problem for band electrons. As a step toward an understanding and possible justification of the envelope function approximation, it is therefore instructive to consider the heterointerface transmission problem in a 1D model. It turns out that this model admits of an exact analytic solution.[12]

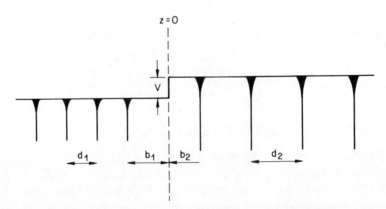

Fig. 6 Illustration of the 1D model of a heterojunction. Two crystals, adjacent at $z = 0$, are formed by periodic arrays of δ-function potentials of the form, $-P_i \delta [z \pm (b_i + nd_i)]$, $n = 0, 1, 2, \cdots$, where the $+$ sign refers to the left crystal $(i = 1)$ and the $-$ sign to the right crystal $(i = 2)$. A potential step, V, at the interface can be thought of as produced by a dipole layer of infinitesimal extent.

Consider two contiguous Kronig-Penney crystals, as illustrated in Fig. 6. The left "crystal" is labeled 1, the right 2. The crystals represent arrays of δ-functions of strength P_i and periodicity d_i, $i = 1, 2$. An important characteristic of the boundary is the size $b \equiv b_1 + b_2$ of the cell containing "atoms" of both crystals; for simplicity of presentation we shall assume that $b < d_1 + d_2$. Then we can choose the unit cells in both crystals so that they abut at the interface without a gap or overlap. The Schrödinger equation in these abutting cells is of the form

$$\frac{d^2 \Psi_{k_i}^{(i)}}{dz^2} + \left[q_i^2 + \frac{2m_0 P_i}{\hbar^2} \delta(z \pm b_i) \right] \Psi_{k_i}^{(i)} = 0 , \qquad i = 1, 2 \qquad (20)$$

where

$$\hbar^2 q_1^2 = 2m_0 E_{k_1} \quad \text{and} \quad \hbar^2 q_2^2 = 2m_0(E_{k_2} - V) , \qquad (21)$$

m_0 is the free-electron mass, E_{k_i} the electron energy, and V a potential barrier at the interface. The + sign in the δ-function argument is used when $i = 1$ and − when $i = 2$.

The Bloch solutions of Eq. 20 are of the form

$$\Psi_{k_i}^{(i)}(z) = \frac{1 - e^{-j(q_i + k_i)d_i}}{1 - e^{-j(q_i - k_i)d_i}} e^{j q_i (z \pm b_i)}$$

$$+ e^{-j q_i (z + b_i)} - [1 - e^{-j(q_i + k_i)d_i}] e^{j q_i |z \pm b_i|} , \qquad (22)$$

where the + sign corresponds to $i = 1$ and − to $i = 2$.

Exercise 4: Show that the eigenfunctions $\Psi_{k_1}^{(1)}$ and $\Psi_{k_2}^{(2)}$ satisfy the Bloch relations

$$\Psi_{k_1}^{(1)}(0) = e^{jk_1 d_1} \Psi_{k_1}^{(1)}(-d_1) , \qquad \Psi_{k_1}^{(1)'}(0) = e^{jk_1 d_1} \Psi_{k_1}^{(1)'}(-d_1) ; \qquad (23a)$$

$$\Psi_{k_2}^{(2)}(d_2) = e^{jk_2 d_2} \Psi_{k_2}^{(2)}(0) , \qquad \Psi_{k_2}^{(2)'}(d_2) = e^{jk_2 d_2} \Psi_{k_2}^{(2)'}(0) . \qquad (23b)$$

Show that the corresponding energy eigenvalues $E(k_i)$ satisfy the Kronig-Penney dispersion equations:

$$\cos(k_i d_i) = \cos(q_i d_i) - U_i \frac{\sin(q_i d_i)}{q_i d_i} , \qquad i = 1, 2 \qquad (24)$$

$$U_i \equiv \frac{m_0 P_i d_i}{\hbar^2} . \qquad (25)$$

The eigenfunctions $\Psi_{k_1}^{(1)}$ and $\Psi_{k_2}^{(2)}$ correspond to the band-electron motion with group velocities v_{k_1} and v_{k_2} in the left and the right crystals, respectively,[†] where $v_{k_i} = (1/\hbar) \partial E_{k_i}/\partial k_i$. Considering the reflection of a band-electron incident on the interface from the left, the total wave function can be written in the form

$$
\Psi(z) = \begin{cases} \Psi_{k_1}^{(1)} + r \Psi_{-k_1}^{(1)}, & z \leq 0 \\[2mm] t \Psi_{k_2}^{(2)}, & z \geq 0 \end{cases}
\tag{26}
$$

where r and t are the reflection and the transmission amplitudes, respectively; these can be determined by matching the wave functions and their derivatives at the interface:

$$
\Psi_{k_1}^{(1)}(0) + r \Psi_{-k_1}^{(1)}(0) = t \Psi_{k_2}^{(2)}(0) ,
\tag{27}
$$

$$
\Psi_{k_1}^{(1)\,'}(0) + r \Psi_{-k_1}^{(1)\,'}(0) = t \Psi_{k_2}^{(2)\,'}(0) .
$$

The resultant analytic expression for t and r in the general case is rather cumbersome and will not be presented here (see Ref. 12).

Exercise 5: Show that for zero potential barrier at the interface ($V = 0$ and hence $q_1 = q_2 \equiv q$), in the limit of small $k_1 d_1$ and $k_2 d_2$ the transmission coefficient $T \equiv 1 - R$ is given by

$$
T = \frac{4(k_1 d_1)(k_2 d_2)}{D}
\tag{28a}
$$

with

$$
D = \tan\left(\frac{qd_1}{2}\right) \tan\left(\frac{qd_2}{2}\right) \left[\frac{k_1 d_1}{\tan(qd_1/2)} + \frac{k_2 d_2}{\tan(qd_2/2)} \right]^2
$$
$$
+ U_1 U_2 \frac{\sin(qd_1)\sin(qd_2)}{(qd_1)(qd_2)} \sin^2 \gamma ,
\tag{28b}
$$

where

$$
\gamma = q\left(\frac{d_1 + d_2}{2} - b\right) .
\tag{29}
$$

[†]Evanescent-wave solutions of Eq. 20, exponentially localized near the interface, have been considered in Ref. 13. These states correspond to electron energies in the bands forbidden by Eq. 24 and do not affect the electron motion in the allowed bands. We stress that this is a special feature of the one-dimensional case.

The case when $V = 0$ occurs rarely in heterostructure transmission problems, but it will allow us to discuss the questions of principle regarding the effective-mass and the envelope function approximations—without introducing additional complications, peripheral to these questions.

7.2.2 The Effective-Mass Approximation

The effective-mass description of electronic motion in each of the constituent crystals is applicable near the band edges. Consider those edges that correspond in the reduced-zone picture to $k \rightarrow 0$. For $k = 0$ the dispersion equation 24 has two types of solution, defining the band edges $E_n \equiv \hbar^2 q_n^2 / 2 m_0$ (here we shall suppress the indices $i = 1, 2$ of the crystals):

$$\tan \frac{q_n d}{2} = - \frac{U}{q_n d} , \qquad n = 1, 3, \cdots ; \qquad (30a)$$

$$\sin \frac{q_n d}{2} = 0 , \qquad n = 2, 4, \cdots . \qquad (30b)$$

Near the $k = 0$ band edges the electron energy is given by $E_n + \hbar^2 k^2 / 2 m_n$. The odd bands, whose extrema are described by Eq. 30a, possess a positive effective mass m_n for a repulsive δ-potential ($U < 0$), and a negative m_n for $U > 0$. The opposite is true for the even bands described by Eq. 30b.

Exercise 6: Using the dispersion equation 24, show that the effective masses at $k = 0$ band edges are

$$\frac{m_n}{m_0} = - \frac{U}{(q_n d)^2} \left[1 - \frac{2U}{(q_n d)^2 + U^2} \right] \qquad \text{for } n = 1, 3, \cdots \qquad (31a)$$

$$\frac{m_n}{m_0} = \frac{U}{(\pi n)^2} \qquad \text{for } n = 2, 4, \cdots \qquad (31b)$$

Derive analogous results for the band edges at $k \rightarrow \pi / d$.

Considering electron transmission at $V = 0$ and small k_1 and k_2, we can express the exact result for T in terms of the effective masses m_1 and m_2. [Here and below we shall suppress the band index n in the effective mass $m_i \equiv (m_n)_i$, $i = 1, 2$.] To do this we replace the energy parameter q by q_n in Eq. 28 and use the dispersion relation. For the case when the bands of both crystals are of the type described by Eq. 30a, we find

$$T = \frac{4 (k_1 / \zeta_1 m_1)(k_2 / \zeta_2 m_2)}{[(k_1 / \zeta_1 m_1) + (k_2 / \zeta_2 m_2)]^2 + (q_n^2 / m_0^2) \sin^2 \gamma} , \qquad (32)$$

where

$$\varsigma_{1,2} \equiv \left[1 - \frac{2U_{1,2}}{U_{1,2}^2 + (q_n d_{1,2})^2} \right]^{-1} \qquad (33)$$

Comparing Eqs. 32 and 19 we see that these equations coincide at $\gamma = 0$. Thus, for this special type of boundary, the phenomenological coefficients $\varsigma_{1,2}$ introduced within the envelope function matching procedure, acquire a rigorous meaning and a concrete form, given by Eq. 33. For higher-lying bands, where $q_n d_{1,2} >> |U_{1,2}|$, these parameters tend to unity and even the simplest envelope function approximation, Eq. 12, becomes applicable.

However, there is nothing special about the $\gamma = 0$ boundary from the crystallographic point of view. If the size b of the interfacial cell deviates from the arithmetic mean of the lattice constants d_1 and d_2, then Eq. 32 is not modeled by Eq. 19. This shows that the interface of two one-dimensional crystals can be such that the matching conditions of Eq. 14 are inadequate. It turns out that the exact result near the band edges can be reproduced in an extended phenomenological model.

Exercise 7: Consider the transmission of a particle from a medium governed by the Hamiltonian $H^{(1)}$ to that governed by $H^{(2)}$ (Eq. 7) *through a "vacuum" gap* of thickness $\tilde{b}$, where the particle is assumed to have the free-electron mass m_0, see Fig. 7. Assume that the "band edges" not only coincide but also have a vanishing affinity, $E_n^{(1)} = E_n^{(2)} = 0$, so that there are no potential steps at the vacuum interfaces. Apply Eq. 16 at each interface (taking—without a loss of generality— $\varsigma_0 = 1$ for the "vacuum") and show that the resultant transmission coefficient is of the form (32), with $\gamma \equiv q\tilde{b}$ and $q \equiv (2m_0 E/\hbar)^2$.

From this exercise we conclude that in the effective-mass approximation near a band edge, envelope function matching in the general one-dimensional case can be accomplished with *three* parameters: m, ς, and γ.

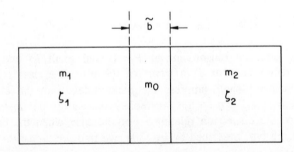

Fig. 7 Illustration of an envelope function model that results in the same transmission coefficient as that given by the rigorous Eq. 32 if the matching conditions (Eq. 16) are used at each interface.

7.2.3 Numerical Examples; Impedance Matching

The one-dimensional model described above, admits of an analytic solution in the entire range of energies (not only near the band edges) and also in the presence of a conduction-band discontinuity at the interface. A misalignment of band edges can be produced either by varying the crystal parameters (U and/or d) or by introducing a potential step V at the $z = 0$ plane defining the interface, as shown in Fig. 6. Examples below (Figs. 8 and 9) correspond to $V = 0$.

Figures 8 illustrate the situation when the "conduction" band edges of two crystals are arranged to coincide by adjusting the periods and the strength of the δ-potentials. The Kronig-Penney dispersion relations are plotted side by side for both crystals along with the exactly calculated reflection coefficient for arbitrary incident Bloch wave vectors. Transmission near the band edges is faithfully reproduced by the analytic formulas 32 and 33. A more general situation, with a conduction-band discontinuity at the interface, is illustrated in Fig. 9—where the structure parameters are chosen similar to those in Fig. 8, respectively, except for the strength of the δ-potential—which is adjusted so as to produce an upward step in the band edges.

An interesting feature seen in Fig. 9a is the perfect transmission ($R = 0$) at a certain "resonant" energy for an "ideal" interface, corresponding to $\gamma = 0$. The possibility that quantum reflections at an abrupt heterojunction interface may vanish for some energy has been discussed[14] in connection with the fundamental limits of hot-electron transistors; this situation is analogous to impedance matching in microwave transmission lines. In the simplest plane-wave effective-mass approximation the impedance-matched condition arises when the electron velocities k_i/m_i are the same on either side of the junction (cf. Eq. 12). In the example of Fig. 9b, when the transmission occurs from a higher-lying band of crystal 1 to the ground band of crystal 2, there is no perfect impedance match for any value of γ. That this was going to be the case, could be surmised from the band structure diagram: at all coincident energy levels in the allowed bands of the two crystals, the electron group velocity is manifestly higher in the left crystal.

7.2.4 Word of Caution

Using the simplest soluble Kronig-Penney model of two one-dimensional crystals, we were able to compare and assess the widely used envelope function (plane-wave) approximations of different degrees of refinement. It turns out that neither the simplest such approximation, corresponding to a continuous envelope function Φ and a discontinuous gradient Φ' (with a single-parameter matching of the electron flux $m^{-1}\Phi'$ across the interface) nor the refined two-parameter matching procedure, in which *both* Φ and Φ' are discontinuous, produce, in general, an adequate description. Even near the band edges, where the effective-mass approximation in the constituent crystals is excellent, a rigorous plane-wave approximation requires *three* parameters: coefficients α and β of Eq. 14

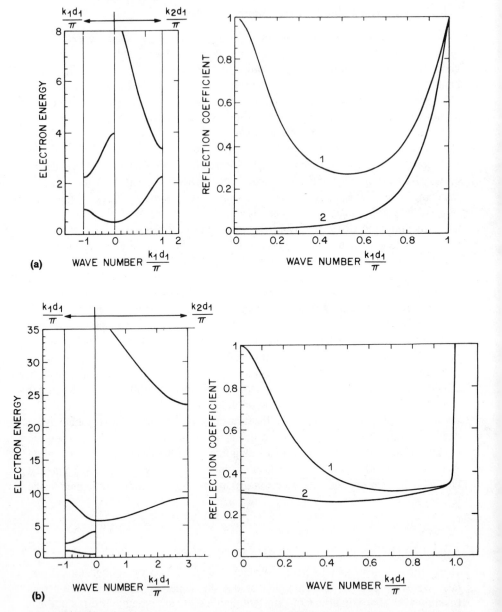

Fig. 8 Electron transmission across the interface of two Kronig-Penney crystals with coincident conduction-band edges. Left side of the figures shows the dispersion relations in the two crystals with the energy plotted in units of $\pi^2\hbar^2/2m_0 d_1^2$. On the right side the reflection coefficient is plotted for $b = d_1$ (curve 1) and $b = (d_1 + d_2)/2$ (curve 2, corresponds to $\gamma = 0$). (a) Case when the coincident band edges are lowest in both crystals. Structure parameters: $d_2/d_1 = 1.5$, $U_1 = -3\pi/2$ and $U_2 \approx 0.3\,U_1$. (b) Case when the bottom of the third band in the left crystal coincides with that of the first band in the right crystal. Structure parameters: $d_1/d_2 = 3$ and $U_1 = -3\pi/2$, $U_2 \approx 3U_1/2$.

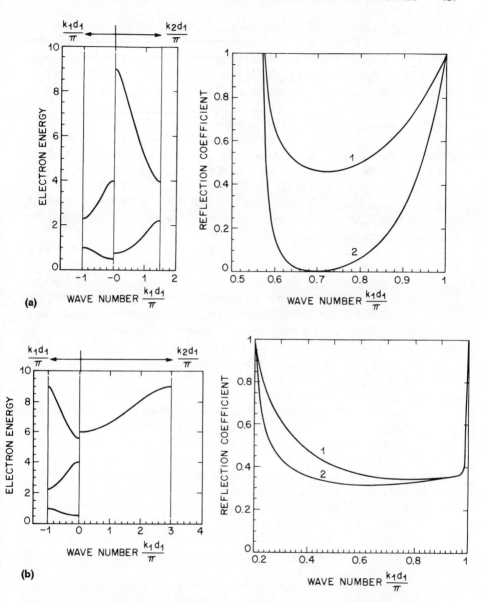

Fig. 9 Electron transmission across a heterointerface with an upward step in the conduction band. Arrangement of the figure and axes labeling is similar to those in Fig. 8. (a) Case when incident and transmitted electrons are moving in the ground bands of respective crystals. The structure parameters: as in Fig. 8 except for $U_2 = 0.5 U_1$. (b) Case when incident electrons are moving in a higher-lying band and transmitted electrons in the ground band. Structure parameters: as in Fig. 8 except for $U_2 = 1.8 U_1$.

plus another coefficient γ describing the size of the crystal cell at the interface. Only for a special type of boundary, corresponding to $\gamma = 0$, does the envelope function description provide a reasonable approximation. Moreover, an exact expression derived for α and β shows that in this case even the simpler one-parameter effective-mass matching of Φ' becomes adequate for higher-lying bands.

To the extent that an essentially one-dimensional analysis can shed light on these problems, we can expect that lattice-matched heterostructures do correspond to a special type of the boundary with $\gamma = 0$ and consequently the envelope function approach is applicable. This approximation works reasonably well (perhaps, even surprisingly well) for GaAs/Al$_x$Ga$_{1-x}$As heterojunctions, with $x \lesssim 0.4$, that is, when electrons in both materials move in the same Γ valley of the conduction band. For a general heterojunction transmission problem, however, the envelope function approach is hopelessly inadequate.

In order to appreciate the difficulty, consider a coherent transmission of electrons between indirect semiconductors, whose conduction-band minima $\mathbf{k}_C$ are located in different parts of the Brillouin zone. The problem is that the total crystal momentum $\hbar\mathbf{k}$ of a conduction-band electron ($\mathbf{k} \equiv \mathbf{k}_C + \mathbf{q}$ being its Bloch wave vector) is very different from its quasiparticle momentum $\hbar\mathbf{q}$, which enters the envelope function description. Imagine an ideal interface between, say, Si and Ge, assuming for simplicity that the interface imparts no parallel momentum to an electron passing across.† In this case, conservation of the crystal momentum $\mathbf{k}_\| \equiv (k_x, k_y)$ parallel to the interface requires that both the initial and final states of the electronic motion lie in the Brillouin zones of both materials on the same perpendicular to the interface plane, (k_x, k_y) (cf. Fig. 10). If the initial electron state is near the bottom of the conduction band, it will not have an allowed counterpart in the other material. Except for a few special orientations of the interface, the ellipsoids of Si and Ge do not project onto one another. Consequently, the transmission probability will vanish, a result hardly contained in any envelope function description.

One runs into a similar difficulty trying to describe, within an envelope function approximation, the process of electronic transmission across a polycrystalline grain boundary. A simple (inadequate, but perhaps illustrative) model of such a boundary is a mathematical plane separating crystal regions rotated with respect to one another. For an indirect semiconductor, like Si or Ge, the coherent electronic transmission through such a boundary will vanish, because in gen-

---

†This would happen if the boundary is along a crystallographic plane containing two primitive vectors of the bulk lattice. For a general boundary, there is an additional complication analogous to diffraction in optics. Regarding the interface as a 2D crystal, it is clear that its reciprocal lattice is spanned by the primitive vectors that can be shorter than those that span the reciprocal lattice of the bulk crystal. (This complication does not arise when two non-parallel vectors of the latter happen to lie in the plane of the interface.) Consequently, the parallel crystal momentum of a coherently transmitted electron can be changed by a non-trivial amount equal to a reciprocal lattice vector of the 2D crystal at the interface.

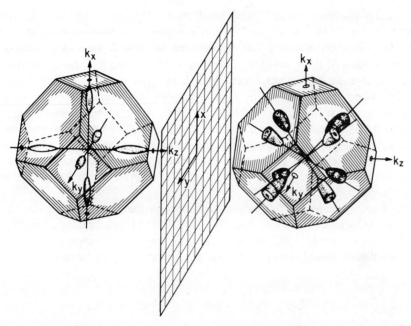

Fig. 10 Illustration of the absence of coherent transmission of conduction-band electrons across an ideal {100} interface between Si and Ge. Electronic states corresponding to a given in-plane crystal momentum must lie on the same perpendicular to the interface (*xy*) plane. However, all allowed low-energy conduction-band states in each semiconductor are in small ellipsoidal pockets, often referred to as the valleys. In the example shown, it is evident that no two valleys lie on the same perpendicular to the interface plane.

eral the band-edge ellipsoids will correspond to different values of $\mathbf{k}_\parallel$ in the rotated crystal.†

Thus, the envelope-function description does not appear promising for modeling electronic transmission across boundaries between indirect semiconductors.

7.3 BALLISTIC-INJECTION DEVICES

Ballistic-injection transistors differ by the materials employed and by the physical mechanism of hot-electron injection into the base. The first proposal of a

---

†One could hope that for special boundary orientations when the initial and the rotated ellipsoids do project onto one another, the use of the envelope functions and the effective-mass approximation may be justified. This, however, does not seem to be the case even for the simplest case of a Σ_3 twin boundary, which is the lowest energy and therefore the most commonly occurring grain boundary in polycrystalline Si. In a Σ_3 twin the two crystals are mirror images of one another in a {111} crystallographic plane. A naive model based on the effective-mass approximation would predict no reflection in this case, whereas a sophisticated numerical calculation[15] indicates that at low energies near the conduction band edge it is the transmission coefficient that vanishes.

hot-electron injection device[16] was based on electron tunneling from a metal emitter through a thin oxide barrier into a high-energy state in a metal base, Fig. 11a. Another insulating barrier separated the base from a metal collector electrode and the whole structure was called the MOMOM (metal-oxide-metal-oxide-metal) transistor. Subsequent versions of this device[17] had the second MOM replaced by a metal-semiconductor junction, resulting in a transistor structure called the MOMS (Fig. 11b). Attempts have also been made to employ a vacuum collector barrier (MOMVM). Metal-base transistors (MBTs), which employ thermionic rather than tunneling injection of hot carriers into the base, were first proposed in the form of a semiconductor-metal-semi-conductor (SMS) structure.[18, 19] Schematic band diagram of an SMS transistor is illustrated in Fig. 11c.

The basic principles of all ballistic hot-electron transistors can be discussed using the SMS as an example. The SMS is a direct unipolar analog of the bipolar junction transistor. The emitter-base and the base-collector junctions are Schottky diodes biased, respectively, in the forward and the reverse direction.

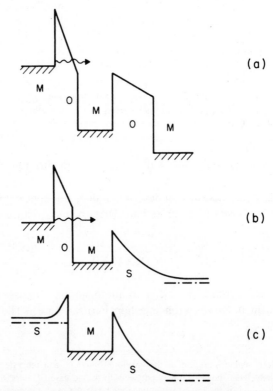

Fig. 11 Metal-base transistors: (a) MOMOM; (b) MOMS; (c) SMS. Each of the M (metal) or S (doped-semiconductor) electrodes is contacted independently and can be biased with respect to the other electrodes. The figure shows the energy-band diagrams of metal-base transistors under operating bias conditions.

As V_{BE} increases, the emitter current rises exponentially. The hope is that electrons injected into the metal base traverse it without losing too much energy, so that at the plane of the base-collector junction the injected electron energy is sufficiently high to clear the metal-semiconductor barrier. Following the bipolar terminology, the fraction of injected electrons that make it to the collector (the transfer ratio) is referred to as the common-base current gain α. If α is close to unity, then the transistor power gain is approximately equal to the ratio of the large output resistance to the small input resistance, which becomes progressively smaller as the emitter-base Schottky diode is forward biased.

The tunnel-emitter (MOMOM, MOMS) structures operate somewhat differently. The emitter current is not and does not have to be a strong function of the input bias. What is being controlled is the electron injection energy relative to the top of the collector barrier. Depending on that energy, the injected electrons end up mostly in the base or in the collector. In principle, these transistors are devices with a controlled transfer ratio α—switching a relatively constant emitter current between the base and the collector circuits. There has been little development of tunnel-emitter metal-base transistor concepts in recent years. However, their all-semiconductor analogs,[26] implemented with the help of various heterostructures, are currently under intense investigation (Section 7.3.2).

7.3.1 Metal-Base Transistors

Experimental studies of the SMS device are being actively pursued. Recent advances have been associated with the development of epitaxial techniques for the growth of single-crystal silicon-silicide-silicon structures, using metallic silicides, such as $CoSi_2$ and $NiSi_2$. This continued interest is explained not only by the scientific usefulness of the SMS structure (it is an excellent tool for studying fundamental properties of hot-electron transport through thin films), but also by lingering hopes of producing a transistor that is faster than bipolar or FET devices.

The problem that has plagued the SMS (and all other metal-base) transistors is their poor transfer ratio α. Even assuming an ideal monocrystalline SMS structure and extrapolating the base thickness to zero, the typical calculated values of α are unacceptably low—mainly due to the quantum-mechanical (QM) reflection of electrons at the base-collector interface.[20] Exact calculations carried out recently[21] for the case of electron transmission through $NiSi_2$-Si interfaces, indicate that over 50% of electrons are reflected even in the most ideal situation. Nevertheless, there have been experimental reports[22, 23] of a transistor action in monocrystalline $Si/CoSi_2/Si$ structures with α as high as 0.6. One cannot rule out some "accidental" resonance that aids the QM transmission of hot electrons in these devices. Such an interpretation, however, is not very likely. A more probable explanation is related to the existence of pinholes in the base metal film, that is, continuous silicon "pipes" between the emitter and the collector. A careful analysis[24] of the correlation between the pinhole sizes and the device characteristics revealed no evidence for a hot-electron component of the current through the base. On the other hand, the pinhole conduction in some cases gives an α as high as 0.95. The device, therefore, works like a permeable base transistor (PBT),[5] which in our classification is not a hot-electron device. This "natural" version of the permeable base transistor had, in fact, been proposed, and experimentally studied[25] prior to the invention of the PBT.

If the area of each pinhole is small, it is often difficult to tell whether the PBT or the MBT mechanism has contributed to the observed *I-V* characteristics. Thermionic emission through a permeable base has, in our opinion, a greater potential for use in transistors than the hot-electron transport through a metal base. Thin silicide films may offer an attractive way of fabricating the PBT—if one learns how to control the distribution of pinhole sizes, making it sharply peaked at a desired area scale.

7.3.2   Doped-Base Transistors

The interest in ballistic hot-electron transistors was revived[26] by the tremendous progress in the epitaxial growth of semiconductor heterojunctions. The problem of QM reflections can be largely avoided in such structures—provided the carrier transport occurs in similar Brillouin-zone points on both sides of the interface. Even when the base is degenerately doped, the Fermi energy is typically less than 0.1 eV. It is possible, therefore, to arrange an injection energy so that, say, $\Phi/E \leq 1/2$ and $R \leq 0.03$ (cf. Problem 3). A number of such devices have been manufactured recently, using ion-implanted "camel" barriers,[27] MBE grown planar-doped barriers,[28, 29] GaAs/AlGaAs heterostructure barriers grown by MBE[30–32] or MOCVD,[33] MBE-grown InGaAs/InAlAs barriers,[34] and pseudomorphic MBE grown structures with an InGaAs base on GaAs substrates.[35, 36] Both tunnel-emitter and thermionic-emitter versions of the ballistic hot-electron transistor have been implemented. Figure 12 shows the schematic energy-band diagrams of these devices.

It is important to understand the trade-off involved in the design of all hot-electron transistors with a doped base: cooling of hot-electrons by phonon emission and other inelastic processes (minimized by thin base layers) against the increasing base resistance for thinner layers. It is easy to estimate the *RC* delay associated with charging the working base-emitter capacitance and the parasitic base-collector capacitance through the lateral base resistance:

$$RC = \tau_B = \frac{\epsilon_s L^2}{L_i \mu \sigma}, \tag{34}$$

where L_i is the thickness of the emitter or the collector barriers, $L_i \sim 10^{-5}$ cm, L the characteristic lateral base dimension, $L \sim 10^{-4}$ cm, μ the electron mobility in the base, σ the mobile charge density per unit base area, and ϵ_s the dielectric permittivity. For a hot-electron transistor to be competitive, it must have $\tau_B \approx 1$ ps, which means that the sheet resistance in the base must be $(\mu \sigma)^{-1} \leq 1$ kΩ/$\square$. On the other hand, making the base too thick, say thicker

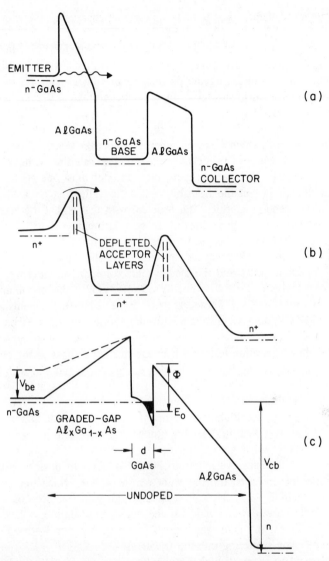

Fig. 12 Unipolar ballistic transistors with a monolithic all-semiconductor structure. (a) tunneling hot-electron transfer amplifier (THETA), a tunnel-emitter transistor, (b) planar-doped barrier (PDB) transistor, (c) induced-base (IBT) transistor. (After Refs. 26, 28, 45)

than 1000 Å, would lead to a degradation in α due to various energy-loss mechanisms. For example, hot electrons in GaAs lose energy at the rate of about 0.16 eV/ps due to the emission of optical phonons.[37] The limitation expressed by Eq. 34 is rather severe. The minimum value of L is governed by the lithographic resolution. One cannot really make the barrier thicknesses L_i much larger than 1000 Å, since this would introduce the emitter and the collector delays of more than 1 ps. It may appear that increasing the base doping can resolve all the difficulties; however, if it is increased much beyond $10^{18}\,\mathrm{cm}^{-3}$, then we can expect a strong degradation in the transfer ratio due to plasmon scattering.[38] Monte Carlo simulations of a GaAs/AlGaAs ballistic hot-electron transistor at 77 K, including the plasmon scattering in addition to conventional scattering mechanisms, showed[39] an α no better than 0.9 even with a 100-Å-thick heavily doped base.

An important limitation in these estimates originates from the intervalley scattering mechanism, which turns on when the electron kinetic energy exceeds the satellite valley energy (in GaAs the lowest satellite valleys are L and $\Delta E_{\Gamma L} \approx 0.3\,\mathrm{eV}$). For the best gain, the transistor must operate at low injection energies, that is, when the transfer to satellite valleys is reduced. This limits the possible injection energy. The problem with that is that the elastic Coulomb scattering from ionized impurities in the base is minimized with increased injection energy. The mean free path of ballistic electrons in a doped GaAs base—limited at high doping levels by the inelastic interaction with coupled plasmon/phonon modes and the elastic ionized-impurity scattering—was considered theoretically in Ref. 40. For $N_D \geq 10^{18}\,\mathrm{cm}^{-3}$ the maximum mean free path is of order 300 Å. These estimates have led to the conclusion[40] that the GaAs/AlGaAs system is unsuitable for the fabrication of a viable doped-base ballistic transistor. This type of device requires either a material with a higher satellite-valley separation to take advantage of the decrease in the elastic scattering rate with increasing injected-electron energy, or a material with lower effective mass, which corresponds to a lower density of states and therefore also a lower scattering rate.

Experimentally, the highest gain in a ballistic GaAs/AlGaAs heterostructure transistor was reported in THETA (tunneling hot-electron transfer amplifier) devices.[41] These MBE grown devices, had relatively low-doped n-type GaAs base layers of thickness varied from 300 to 800 Å. In devices with narrowest base widths the observed transfer ratio was $\alpha = 0.9$, which implies a differential current gain $\beta \approx 9$. However, such a high gain was seen only at injection energies just below the threshold for the transfer to satellite valleys. At higher injection energies, the gain was reduced from 9 to about 3—due to the energy losses in electron scattering into the L valleys.†

It is natural to expect a higher transport efficiency in ballistic devices implemented using heterostructures with larger separation to satellite valleys, such as InGaAs. This conclusion is supported by Monte Carlo simulations[44] of hot-electron transport through the base of an InGaAs/InP heterojunction ballistic transistor. Experimentally, the highest gain for a hot-electron transistor with a doped base was demonstrated[36] in a pseudomorphic THETA device containing a

---

†The fact that intervalley transfer caused the gain degradation was elegantly confirmed by experiments[42] in which the variation of α was measured at different hydrostatic pressures. It is known[43] that the Γ-L separation decreases with pressure at the rate $\sim 6.3\,\mathrm{meV/kbar}$. Application of a hydrostatic pressure in the kilobar range can, therefore, significantly lower the maximum α.

GaAs emitter, an AlGaAs tunneling emitter barrier, and a strained 200-Å-thick $In_{0.12}Ga_{0.88}As$ base. The high current gain achieved ($\beta \approx 41$ at 4.2 K and 27 at 77 K) was attributed mainly to the larger Γ-L energy separation ($\approx 380\,mV$) in the base. As a function of the injection energy (controlled by the base-emitter voltage), the gain increases, reaches its maximum, and then drops sharply at the onset of the transfer to L-valleys.

7.3.3 Induced-Base Transistor (IBT)

In this device, illustrated in Fig. 12c, the base conductivity is provided by a 2-dimensional (2D) electron gas induced by the collector field at an undoped heterointerface. The density of the induced charge is limited by a dielectric breakdown in the collector barrier. For a GaAs/AlGaAs system this means $\sigma/q \lesssim 2 \times 10^{12}\ cm^{-2}$. In the IBT operation the lateral electric field in the base is low and hence the device can take a direct advantage of the high electron mobility in a 2D electron gas at an undoped heterojunction interface.† At room temperature μ is limited by phonon scattering, $\mu \lesssim 8000\ cm^2/V$-s, giving $(\mu\sigma)^{-1} \approx 400\ \Omega/\square$ at the highest sheet concentrations in the base. The base sheet resistance is much lower at 77 K.

The IBT represents an attempt to circumvent the limitation of Eq. 35. The induced-base conductivity is virtually independent of its thickness down to $d \lesssim 100\,Å$. At such short distances the loss of hot electrons due to scattering is small. Injected hot electrons, traveling across the base with a ballistic velocity of order $10^8\ cm/s$, lose their energy mainly through the emission of polar optic phonons. For $d = 100\ Å$ the attendant decrease in α is estimated to be about 1%. Energy losses to the collective and single-electron excitations of the two-dimensional electron gas are negligible. The IBT can be regarded as a metal-base transistor—with the notable difference that the base "metal" is two-dimensional. This permits a significant improvement in the transfer ratio α—due, mainly, to a lower quantum-mechanical reflection coefficient R at the collector barrier interface.

Let us mention some alternative possibilities for an implementation of the IBT. An injecting emitter barrier of triangular shape can be organized using either the graded-gap technique, as illustrated in Fig. 12c, or using planar-doped barriers. The undoped, graded-gap version has the advantage in that it avoids the adverse effects of doping fluctuations both on the barrier injection and the base conductivity. The latter is particularly important for shrinking the *lateral* device dimensions.[46] The second alternative, illustrated in Fig. 13, is attractive when one wishes to employ heterostructure materials in which lattice-matched

---

†In such electronic systems, the low-field electron mobility parallel to the layers is greatly enhanced (especially at lower temperatures) because of the suppressed Coulomb scattering of electrons by ionized impurities—due to (i) spatial separation from the scatterers and (ii) higher than thermal electron Fermi velocity in a degenerate 2D electron gas. The latter reduces the scattering cross section in accordance with the Rutherford formula.

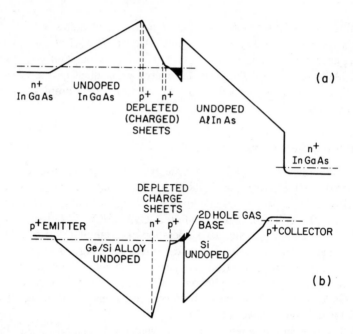

Fig. 13 Possible induced-base transistor designs with a planar-doped emitter barrier in (a) $Al_{0.48}In_{0.52}As/In_{0.53}Ga_{0.47}As$ heterostructure and (b) Ge/Si heterostructure. The structures must contain two built-in charge sheets—acceptors and donors—designed to have both sheets depleted of mobile carriers to provide a desired barrier height. The 2-dimensional carrier gas induced at the heterointerface by the collector field should be separated from the nearest planar-doped sheet by an undoped setback layer of ~ 50 Å to maintain the benefit of enhanced mobility in the base.

grading of the gap is difficult to achieve. If implementation of IBT in a Ge/Si hetero-system is desired (provided, of course, one can achieve a high-quality interface of these lattice-mismatched materials), it makes sense to use the injection of hot holes because, unlike the conduction-band minima, the valence-band maxima are located at the same $k = 0$ point in both semiconductors (cf. Section 7.2.4). For a discussion of other exotic possibilities, related to the IBT concept, see Ref. 47.

The first experimental implementation of the IBT was demonstrated[48] in an AlGaAs/GaAs graded-gap heterostructure grown by MBE. The device had a maximum sheet carrier concentration $n_s \approx 3 \times 10^{12} \, cm^{-2}$ induced in a 100-Å quantum-well base layer, with a minimum base resistivity $\rho_\square \approx 310 \, \Omega$. In a relatively narrow range of applied biases the IBT showed a common-base current gain of $\alpha \approx 0.96$ at room temperature. This value of gain approaches the estimated limit[45] for a 100-Å-thick GaAs base.

7.3.4   Speed of Ballistic Transistors

Like the bipolar, the FET, and most other transistors, hot-electron transistors have a regime in which their output current I rises exponentially with the input (base-emitter) voltage. In this regime, the maximum speed of operation is proportional to I. Eventually, the exponential dependence saturates and goes over into a linear law. One gains no further advantage in speed by increasing I, because the charge stored in all the input capacitances will rise proportionally with the current. Ultimately, the speed of any transistor is determined by the current level at which one has a crossover between the exponential and the linear regimes.[7] The crossover generally occurs when the amount of injected charge associated with the output current $I = I_{EC}$ exceeds the amount of charge required to control the current.

Let us trace how this crossover limits the transistor speed. Consider a small-signal switching of a transistor in which the output current has been changed from I to $I + \delta I$. The new state differs by the amount of charge Q_α stored in various regions of the device (α labels different contributions to the overall modulated charge $\delta Q \equiv \Sigma \, \delta Q_\alpha$). One can choose different ways of separating the charge into components Q_α; let us write schematically $Q = Q_{ctl} + Q_i$, where Q_{ctl} denotes the "controlling" charge (whose modulation δQ_{ctl} has caused the variation δI in the first place; this charge is stored in the input capacitance), and Q_i represents the modulated mobile charge,† whose variation is proportional to δI.

One way of describing the transistor speed is to calculate the delay τ necessary for the variation in the output current δI to deliver the total amount of modulated charge. This is equivalent to calculating the cutoff frequency $f_T = 1/2\pi\tau$ for short-circuit unity current gain.[2]

At low current levels the output current I depends exponentially on the control charge Q_{ctl} because the latter is linearly related to the barrier height for thermionic emission.‡ In this regime the delay τ_{ctl} is proportional to $1/I$. The other delay, associated with $Q_i = \int I \, dx / v(x)$, is approximately independent of the current level and is equal to the total time of flight of injected carriers across the structure:

$$\tau = \int_E^C \frac{dx}{v(x)} \, . \tag{35}$$

---

†Mobile charge is proportional to the current in those parts of the device where the carrier velocity is independent of the small current variations. This is the case, at least approximately, for the injected carrier transport on the slopes of the potential barrier, and across the base of the transistor. In heavily doped contact layers, the mobile charge concentration is not modulated and current variations are accommodated by small changes in the carrier drift velocity.

‡This situation is common to any potential effect transistor, based on thermionic emission of carriers. For example, a MOSFET below threshold is such a transistor. In the subthreshold regime, both the channel charge and the output current are exponential functions of the controlling charge on the gate.

As the output current increases, the delay τ_{ctl}, which dominates at lower currents, becomes progressively less important. Strictly speaking, one can enter the regime with delay dominated by Eq. 35, while the dependence of I on the control charge (and the input voltage) is still exponential. Such a situation often occurs in bipolar transistors, but rarely in unipolar ballistic transistors. The difference is rooted in the physics of the injection process. In a forward-biased p-n junction, the depletion region shrinks with increasing current—introducing no unwelcome space-charge effects, except at the highest current densities. In contrast, the current in a unipolar triangular-barrier diode suffers from a space-charge limitation on the uphill slope (cf. Chapter 2). As a result, the exponential injection characteristics in unipolar transistors may saturate at lower currents than in bipolar devices where the exponential regime can persist to very high current densities. Conceptually, the distinction is not very deep: it relies on the rather arbitrary assignment of the charge due to carriers on the slope of the emitter-base barrier to the control category. One can lump it, instead, with Q_i by introducing an effective diffusion velocity $(1/n)D\,\nabla n$ on the uphill slope.† This allows one to write the total intrinsic delay in the form of Eq. 35. The length of the uphill slope contributes a non-negligible delay. It should be emphasized that the disadvantage of lower crossover currents of unipolar transistors does not reduce to this additional time-of-flight delay. An often important part of the "control" charge resides in parasitic capacitances C_p parallel to the input. As long as the output current rises exponentially, the delay associated with charging the parasitic capacitances decreases proportionally to $1/I$, stabilizing at a value $C_p/g_m \propto I_{max}^{-1}$ when the transistor input characteristics become linear. The higher the crossover current I_{max}, the lower the dynamic emitter resistance (g_m^{-1}) one can achieve in a given transistor. Therefore, by enhancing I_{max}, one gets closer to the fundamental limit, Eq. 35.

It may appear that all one needs to do to increase the speed of a potential-effect transistor is to shorten the time of flight. That, however, cannot be done indefinitely because of the complementary limitation (Eq. 34) associated with charging the input capacitances through the finite base spreading resistance. Indeed, if the time of flight is reduced by narrowing the physical dimensions (the barrier slopes L_i), the improvement in τ comes with increasing the base-emitter and the base-collector capacitances and hence is at the expense of a degradation in τ_B. The best way of improving the speed performance of a transistor is to reduce the time-of-flight delay *through increasing the carrier speed*. Here is where ballistic devices have an edge over conventional bipolar transistors. This advantage has been demonstrated most convincingly by the recent progress of heterojunction bipolar transistors with a ballistic injector, Section 7.3.5.

Speed limitations of tunnel-injection ballistic transistors may be somewhat different from those of thermionic-injection devices. The difference, again rooted

---

†Magnitude of this velocity cannot exceed v_T (cf. Chapter 2, Section 2.4.2).

in the physics of the injection process, manifests itself in the fact that the emitter current in THETA devices is only a weak function of the base-emitter bias; it varies only to the extent of the variation in the average tunneling-barrier phase area. Operation of these transistors consists in *switching* of the emitter current between the base and the collector circuits. Consequently, the maximum transconductance achievable in THETA depends on the degree of collimation of the injected electrons in a narrow energy range, the preservation of this narrow range in transit through the base, and on the selectivity of the collector-barrier analyzer. On the other hand, the fact that a tunneling barrier must be thin ($\leq 100 \, \text{Å}$) puts a more stringent requirement on the base resistance, since THETA transistors are still subject to the limitation of Eq. 34. To operate such transistors at picosecond speeds one must achieve a base sheet resistance $(\mu \sigma)^{-1} \leq 100 \, \Omega / \square$.

7.3.5 Hot-Electron Bipolar Transistors

Speed limitations of thermionic-injection ballistic transistors are rather similar to those of another important heterojunction device, the heterojunction bipolar transistor or HBT. The idea of a HBT with a wide-gap emitter was first proposed by Shockley[49] and developed theoretically by a number of workers, most notably Kroemer[50]—at the time when no practical technology existed for its implementation. With the advent of modern crystal-growth techniques, a number of HBT designs have been demonstrated experimentally (see Chapter 6). The advantage of the wide-gap emitter concept is that the minority-carrier injection into the emitter can be practically suppressed. This allows the use of heavily doped base layers without degrading the gain. At the same time, the transport of injected carriers across the base can be improved through a creative use of what Kroemer had called the *central design principle* for heterostructure devices: separate and independent control of the forces acting on electrons and holes.[51] For example, by incorporating a quasi-electric field in the base (by grading its bandgap), one can replace a minority-carrier diffusion transport across the base by a relatively faster drift transport.

Recently, great progress was achieved in the implementation and perfection of the hot-electron HBT—a device that employs "non-equilibrium" ballistic transport in the base.† In these devices, carriers are launched at high energies into a small angular cone perpendicular to the base-emitter junction. The fact that the velocity distribution is sharply peaked in the direction perpendicular to the base layer is beneficial for downscaling the device area. In bipolar transis-

---

†The word "non-equilibrium" is in quotes to stress the fact that any minority-carrier transport is non-equilibrium. Minority carriers are always "hot" with respect to the possibility of recombination. However, because the recombination times are typically much longer than the kinetic energy relaxation time, one can talk about a quasi-equilibrium energy distribution of minority carriers—and deviations from that equilibrium of minority-carrier packets injected at high energies and traveling ballistically across the base.

tors the lateral scaling is limited mostly by recombination in the extrinsic base region (both in the base bulk and on the surface adjacent to the emitter stripe). The high injection velocity in HBT—compared with a characteristic lateral diffusion velocity D/L, where L is the lateral dimension of the intrinsic base, and D the diffusion coefficient—results in a better spatial confinement of injected carriers to the intrinsic base area.

For the successful implementation of hot-electron HBTs, some of the material requirements are similar to those for unipolar hot-electron devices. For example, the base layer should be beneficially made of a semiconductor with high-lying satellite valleys. At the same time, for reduced scattering of minority electrons, one should maximize the heavy-hole mass in the base.[52] From these considerations, InAs and InSb were proposed as the best candidates for the base-layer material. For example, in p-type InAs doped to $2 \times 10^{18} \, cm^{-3}$, the total mean free path of a hot electron 400 meV above the conduction band edge was calculated[52] to be around $1\,000 \, \text{Å}$.

Experimentally, the most impressive hot-electron HBTs were obtained in heterostructures, lattice matched to InP—the base layer being $In_{0.53}Ga_{0.47}As$ and the emitter-launcher either InP itself[53, 54] or $Al_{0.48}In_{0.52}As$. The latter structure, illustrated in Fig. 14, provides a higher injection energy ($\approx 0.5 \, eV$) and leads to a near-ideal lateral scaling to submicron emitter-stripe dimensions.[55] Hot-electron HBTs show both high-gain and exceptionally high-speed operation. Because of the low base resistivity and the improved scalability of these devices the base delay, expressed by Eq. 34, can be minimized. The dominant delay in hot-electron HBTs is the time of transit of hot minority carriers across the base and the base-collector depletion layer. Recently, an InGaAs/InP transistor was demonstrated[53] with a record f_T of 165 GHz at room temperature. The current progress is so rapid that this record will not last very long. Hot-electron HBT appears to be the first transistor to reach subpicosecond speeds.

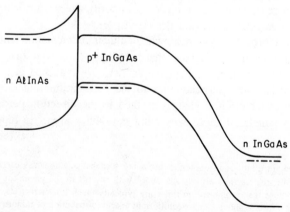

Fig. 14 Energy-band diagram for a heterojunction bipolar transistor in an $In_{0.53}Ga_{0.47}As/Al_{0.48}In_{0.52}As$ material system lattice matched to InP.

Finally, let us mention the idea of the so-called Auger transistor—a bipolar transistor operating in the extreme case of high-energy electron injection.[51, 56] If the base is made of a narrow-gap material—narrower than the conduction-band discontinuity at the base-emitter junction—then the injection energy may exceed the threshold for the Auger impact ionization process. The Auger process creates an electron-hole pair by the excitation of an additional electron from the valence band into the conduction band and results in the multiplication of electrons in the base. Both the primary and the Auger electrons can be swept by the collector-junction field. Consequently the Auger transistor can exhibit a common-base current gain α exceeding unity. Whether or not this will be a useful property is unclear.

7.3.6 Hot-Electron Spectrometers

Ballistic transistor structures allow fundamental studies of the dynamics of non-equilibrium carriers in semiconductors. The idea of hot-electron spectroscopy[57] consists in the following: one measures the dependence of the collector current I_C on the collector-base bias V_{CB} at a fixed emitter-base bias V_{BE} and plots $(\partial I_C/\partial V_{CB})_{V_{BE}}$ versus V_{CB}. If certain important conditions are met, the resultant curve is proportional to the number of carriers arriving at the collector barrier with a normal component of the kinetic energy (i.e., portion of the energy corresponding to the motion normal to the barrier) equal to the barrier height Φ_C. For this to be true, one must ensure that the collector barrier height linearly depends on the bias, $\delta\Phi_C \propto \delta V_{CB}$, and that the collector bias does not affect the hot-electron energy distribution in the base, the emitter injection efficiency, and the above-barrier QM reflection.

Exercise 8: Write the collector current I_C over a barrier of height Φ as an integral over the distribution function $f(v_z)$. Neglect both the above-barrier reflection and the under-barrier transmission. Calculate the derivative $\partial I_C/\partial V_{CB}$, assuming an approximately linear $\Phi_C(V_{CB})$ dependence characteristic of a planar-doped-barrier diode (Section 2.1.3).

Figure 15 illustrates a spectrometer based on the planar-doped-barrier transistor structure. With this device the authors of Ref. 58 were able to observe ballistic transport of hot electrons through the GaAs base and obtain information about the dynamics of the electron energy loss further interpreted in terms of various electron-scattering mechanisms.[59, 60] The planar-doped barrier spectrometer has the drawback that the width of the electron distribution injected into the base and the breadth of the analyzer filter function are both adversely affected by fluctuations in the barrier doping. These effects have been clearly shown by a Monte Carlo analysis[61] of the planar-doped spectrometer structure. In a related ballistic spectrometer, the planar-doped emitter was replaced by a graded-gap triangular-barrier structure.[62] This structure was used to study the effects of the width and the doping level in the base.

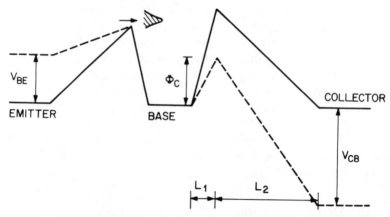

Fig. 15 Schematic diagram of a hot-electron spectrometer based on a planar-doped barrier structure. To a reasonable approximation, $\delta\Phi_C \approx e\, \delta V_{CB}\, L_1/(L_1 + L_2)$. (After Ref. 58)

Another type of a hot-electron spectrometer is based on the THETA device.[63, 9] As illustrated in Fig. 16, these devices have a tunnel emitter and an abrupt heterojunction collector barrier. The required linear $\Phi_C(V_{CB})$ dependence occurs at *negative* collector biases—with the analyzer plane located at the collector edge of the base-collector heterojunction barrier. The heterojunction spectrometer[9] enabled an unambiguous observation of the ballistic electron transport, which is particularly remarkable because electrons had to traverse (without scattering) not only the GaAs base but also the entire length of the AlGaAs collector barrier (~ 1000 Å). Examples of the measured hot-electron energy distribution are shown in Fig. 16b. The main peak increases with the collector current and occurs at the values of V_{CB} that are consistent with the theoretically expected values, assuming the barrier height $\Phi_C(V_{CB})$, calculated from the structure parameters. The majority of electrons contributing to the peak arrive at the analyzer without a single scattering event—otherwise, the peak would be displaced by at least 36 mV (the optical-phonon energy) to lower voltages, which could be accounted for, with the data of Fig. 16b, only by postulating an unrealistically low barrier height. The sharper distributions obtained in heterojunction spectrometers are, at least in part, due to the absence of the dopant-fluctuation broadening mentioned above in connection with the planar-doped-barrier devices.

Figure 17 shows a *lateral* hot-electron device,[64, 65] fabricated in the plane of a two-dimensional electron gas. Two potential barriers, illustrated in Fig. 17c, were induced by 500-Å-wide metal gates deposited on a GaAs/AlGaAs selectively doped heterostructure. Hot electrons with narrow energy distributions (~ 5 meV wide at $T = 4.2$ K) were observed to traverse the 2D electron gas base (region between the gates, $d \approx 1700$ Å wide) without scattering. From the measured ballistic fractions α the authors[64] deduced a mean free path $\lambda \approx 4800$ Å (defined by $\alpha = e^{-d/\lambda}$), which is considerably longer than that

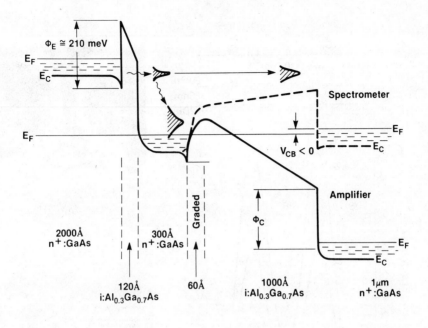

(a)

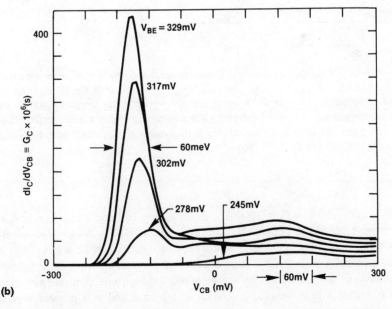

(b)

Fig. 16 Ballistic electron spectroscopy in a THETA device. (a) Band diagram of the structure. Spectrometer configuration corresponds to $V_{CB} < 0$ when $\delta\Phi_C \approx e\,\delta V_{CB}$. (b) Example of a measured energy spectrum. The main peak is due to electrons arriving at the analyzer plane without a single scattering event. (After Ref. 9)

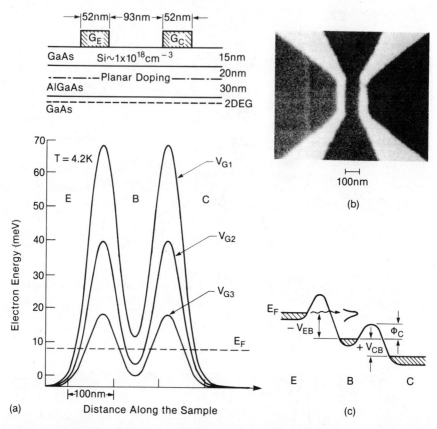

Fig. 17 Lateral hot-electron device. (a) Cross-sectional diagram showing the selectively doped structure and the gates on top. Potential shape in the plane of the 2D electron gas is plotted underneath, showing the injector and the spectrometer barriers. As V_G becomes more negative, the barriers increase. (b) A scanning electron micrograph (SEM) showing the gates configuration. (c) Schematic illustration of the lateral potential distribution under bias. (After Refs. 64, 65)

found in vertical devices ($\lambda \approx 1000$ Å). Such a long ballistic mean free path of 2D electrons enabled the implementation of a lateral THETA device with exceptionally high gain.[65] The maximum β observed in a lateral device was as high as 105, corresponding to $\alpha > 0.99$ (this does not contradict the previously quoted value of λ since, as pointed out by the authors,[65] at high injection energies electrons may suffer a few inelastic collisions in the base and still possess enough energy to pass over the collector barrier). This is the highest current gain reported for a hot-electron device. Because it is possible to tune the emitter and collector barrier heights, the lateral THETA device is extremely powerful for studying ballistic transport in small systems.

7.4   REAL-SPACE TRANSFER DEVICES

The term "real-space transfer" (RST) was coined[66] to describe a new mechanism for negative differential resistance (NDR) in layered heterostructures.[66, 67] The original RST structure is illustrated in Fig. 18. In equilibrium the mobile electrons, which determine the conductivity of the system in an external electric field parallel to the layers, reside in undoped GaAs quantum wells and are spatially separated from their parent donors in AlGaAs layers. If the power input into the electronic system exceeds the rate of energy loss by that system to the lattice, then the carriers "heat up" and undergo partial transfer into the wide-gap layers where they may have a different mobility. If the mobility in layers 2 is much lower, an NDR will occur in the two-terminal circuit. There is a strong analogy to the Gunn effect, based on the momentum-space intervalley transfer.[66−68] In fact, in a simple model, which neglects polarization of the structure arising from electron redistribution between the layers in RST, the problem is mathematically equivalent to that of a two-valley model of the Gunn effect (see Problem 5).

Experimentally, the RST effect was discovered[69] in a multilayer modulation-doped $GaAs/Al_xGa_{1-x}As$ heterostructure; microwave generation in an RST diode was subsequently demonstrated.[70] If the device is used as an oscillator, electrons must cycle back and forth between the high and low mobility layers. The maximum oscillation frequency is limited by the delay due to "cold" elec-

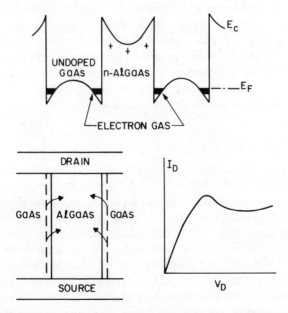

Fig. 18 The real-space transfer diode. Electrons, heated by an applied electric field, transfer into the wide-gap layers, where their mobility is substantially lower, giving rise to a negative differential conductivity. (After Ref. 66)

trons returning from the potential "pockets" in layer 2. This process occurs mainly by thermionic emission over the potential barrier created by the space-charge of ionized donors. For a modulation-doped AlGaAs/GaAs hetero-structure at room temperature the return time can be estimated to be at least 10^{-11} s and still longer at lower temperatures. On the other hand, the time constants involved in the initial transfer of hot electrons are considerably shorter.[71]

Exercise 9: Make simple estimates. Assume that the potential "pockets" in layer 2 are due to uniformly distributed positively charged donors of surface density $10^{12}\,cm^{-2}$ and that $2 \times 10^{11}\,cm^{-2}$ transferred electrons are in these pockets. Calculate the return flux by a thermionic formula and estimate the time required for the escape of half of the "trapped" carriers. Calculate the rate of another restoring proces—electrons in the wide-gap layers drifting to the drain contact while the high-mobility layers are being refilled from the source contact—by estimating the effective capacitance and resistance involved in such a recharging process. Which of these processes becomes more efficient as the temperature is lowered?

Transistor applications of real-space transfer began with the proposal[72] of a three-terminal hot-electron device structure, in which the RST effect gives rise to charge injection between two conducting layers isolated by a potential barrier and contacted separately. Transistor action results from the control of the electron temperature T_e in one of the conducting layers resulting in a modulation of the current into the other layer. Based on this principle, several new device concepts were proposed and experimentally demonstrated: the charge injection transistor or CHINT,[73] the negative resistance field-effect transistor or NERFET,[74] and the hot-electron memory element.[75] In the next sections we shall review these devices and some of their possible circuit applications, emphasizing the physics of the device operation.

7.4.1   Device Structures

The basic structures used for three-terminal RST devices are illustrated in Fig. 19. In the original structure[73-75] the second conducting layer was implemented as a conducting GaAs substrate separated by a graded-gap AlGaAs barrier from the channel of a modulation-doped FET with source (S) and drain (D) contacts, Fig. 19a. This device had an auxiliary fourth electrode (gate) that concentrated the lateral electric field under a 1-μm-wide notch. In the more recent work[76] both the gate electrode and the modulation-doping were eliminated, Fig. 19b, and the channel was induced at the undoped heterointerface by a back-gate action of the second conducting layer. Also in the new structure the rectangular potential barrier provides better insulation between the two conducting layers. Still more recently, a substantial progress was achieved[77,78] using a structure similar to that in Fig. 19b but grown by MOCVD instead of MBE.†

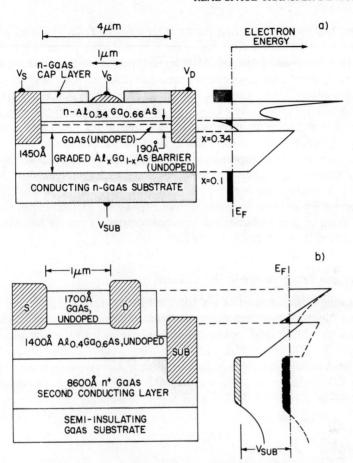

Fig. 19 Cross section and the energy band diagram of three-terminal RST devices. (a) Type 1 MBE-grown structure had a MODFET-like channel separated from the second conducting layer (the SUB electrode) by a graded $Al_xGa_{1-x}As$ barrier. Two-dimensional electron gas is present in the channel even at $V_{SUB} = 0$ as well as with a floating SUB. Furthermore, the quasi-electric field in the barrier aids the drift of injected electrons toward the second conducting layer. However, the barrier breakdown occurs in this structure at relatively low values of V_G, which limits the operating range. (b) Type 2 structure has no electrons in the "channel"—until it is induced by a positive $V_{SUB} > V_T$. The barrier is ungraded, which helps provide a better insulation of the second conducting layer from the channel. This structure was implemented by MBE and MOCVD. (After Refs. 73, 74, 76, 77, 78)

---

†Even though in these structures the second conducting layer is implemented as a heavily doped *n*-GaAs layer on a semi-insulating substrate, we shall keep the designation SUB for this electrode. In the literature, it is sometimes referred to as the collector or the anode.

A critical step in manufacturing the three-terminal RST devices is to provide ohmic contacts to the 2D electron gas in the channel, while preserving the insulation between the channel and the SUB layer. This is no easy task when the two conducting layers are separated by less than 1500 Å. One way of implementing the contacts[73] was to use the Au/Ge-Ag alloy which has an abrupt edge of penetration into $Al_xGa_{1-x}As$. Figure 20 shows the characteristics of a diode formed between the collector electrode and the source and drain terminals tied together and grounded. It is evident that devices with a rectangular barrier have better insulation at 300K, especially in MOCVD grown structures. For the graded-gap structure the diode characteristics were strongly asymmetric[79] with the forward direction corresponding to $V_{SUB} < 0$ (not shown in Fig. 20), as can be expected for thermionic emission over a triangular barrier. The observed current at $V_{SUB} > 0$ is probably due to a combination of barrier lowering and tunneling, especially at lower temperatures.

7.4.2   Electron Temperature in the Channel

Electron heating in the channel of a CHINT/NERFET device involves an interplay between the real-space and momentum-space transfer effects, formation of domains of high electric field, redistribution of electrons both laterally and vertically within the channel and the effect of such redistribution on the self-consistent electric field, transient ballistic transport, and various quantum effects due to the 2D confinement. A comprehensive theory that can encompass all these effects—in a realistic device geometry and including an accurate band

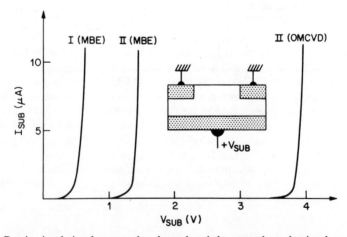

Fig. 20   Barrier insulation between the channel and the second conducting layer at room temperature. Characteristics shown correspond to the positive voltage applied to the SUB electrode, which is the situation in the operating regime of both CHINT and NERFET. For the opposite bias polarity, the characteristics were approximately symmetric in the case of a rectangular-barrier type 2 structure and strongly asymmetric for the triangular-barrier type 1 structures, as can be expected from the band diagrams in Fig. 19.

structure of Γ, X and L valleys—is available only using large computational resources and Monte Carlo techniques, self-consistently coupled with Poisson's equation. Such a large-scale numerical analysis of the CHINT/NERFET structure, including both steady-state and transient effects, has been reported recently.[80, 81]

Our treatment of most of the "complications" will not go beyond naming them. A qualitative picture of hot-electron injection can be based on the assumption that the relevant non-equilibrium properties of the ensemble of channel electrons can be described by postulating a local electron temperature $T_e(x)$ that is a function of the channel position between the surface electrodes S and D. The surface density $J(x)$ of the hot-electron current is then given by a thermionic formula

$$ J = \frac{\sigma v}{\Delta} e^{-\Phi/kT_e}, \tag{36} $$

where $\sigma(x) \equiv qn_s(x)$ is the channel charge density per unit area, Δ is the channel thickness, and $v = v(T_e) = (kT_e/2\pi m)^{1/2}$.

Typical hot-electron injection characteristics, taken at 77 K, are shown in Fig. 21. One of the surface electrodes (D) was grounded and the heating voltage V_{SD} of both polarities was applied to the other electrode. The substrate was kept at a fixed positive voltage V_{SUB}. We see that the substrate current I_{SUB} exhibits a sharp minimum when $V_{SD} \rightarrow 0$. As the heating voltage increases *in either polarity*, the current rises by many orders of magnitude—its polarity corresponding to electrons going into the substrate. This is direct evidence of the hot-electron nature of I_{SUB}. Analysis of the $V_{SD} < 0$ branch of these curves allows an approximate determination of the electron temperature in the channel.[73]

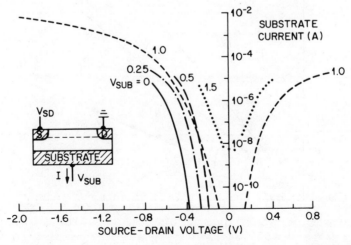

Fig. 21 Hot-electron injection curves for different biases V_{SUB}. (After Ref. 73)

Exercise 10: Assume that the barrier height $\Phi(V_{SUB})$ is independent of V_{SD}. Since most of the charge injection occurs near the drain, this may be approximately true in the $V_{SD} < 0$ branch. Assume further that T_e has a power law dependence on V_{SD}: $(T_e - T)/T \propto (V_{SD})^m$. Under these assumptions and neglecting variations of the pre-exponential factor in Eq. 36 show that

$$f \equiv \left[V_{SD} \frac{d \ln I_{SUB}}{dV_{SD}} \right]^{-1} = \frac{kT_e}{m\Phi} \frac{T_e}{T_e - T} . \tag{37}$$

Note that a similar analysis cannot be applied to the $V_{SD} > 0$ branch, which corresponds to the grounded electrode D being the source. If most of the injection occurs near the drain, then the relevant barrier height (which depends on the SUB to channel voltage) for that branch is affected by V_{SD}, for example, $\Phi \approx \Phi(V_{SUB} - V_{SD})$

By plotting f versus $(V_{SD})^m$ for different m in the limit $T_e \gg T$, one can determine both the power law m (if the dependence turns out to be linear for some m) and the electron temperature in units of $m\Phi$, as in Eq. 37. Analyzing the data of Fig. 21 in this way it was found that $m \approx 2$ for low values of V_{SUB}; this allows a plot of the approximate dependence of T_e on the heating voltage, as shown in Fig. 22. It is clear that charge-injection curves contain valuable information about the electron heating in a transistor channel. Examples of such information (magnetic-field dependence, domain formation, etc.) have been reported in Ref. 79.

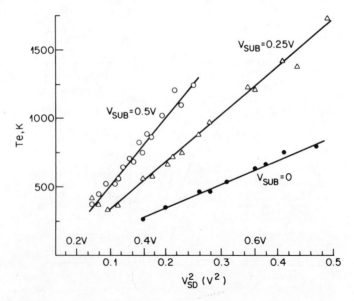

Fig. 22 Electron temperature T_e as a function of the heating voltage V_{SD} for several low values of V_{SUB}. For higher V_{SUB} this simple dependence no longer holds.

7.4.3   Various Bias Configurations and I-V Characteristics

The charge injection transistor (CHINT) is a solid-state analog of the hypothetical vacuum diode with a controlled cathode temperature, that was discussed in connection with Fig. 2. Application of a voltage V_{SD} produces a lateral electric field that heats the channel electrons and leads to an exponential enhancement of charge injection over the barrier. Figure 23 displays the collector characteristics of CHINT as a function of the heating voltage V_{SD} with the collector voltage V_{SUB} as a parameter.[78] In the operating regime, this device had a transconductance,

$$g_m \equiv \left[\frac{\partial I_{SUB}}{\partial V_{SD}} \right]_{V_{SUB}}, \qquad (38)$$

of more than 1000 mS/mm at room temperature.

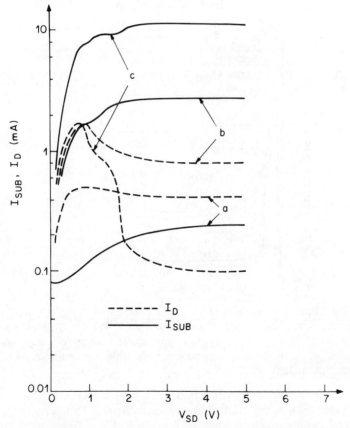

Fig. 23   Typical experimental current-voltage characteristics, I_{SUB}-V_{SD} and I_D-V_{SD} at room temperature and different collector voltages $V_{SUB} - V_T$ in MOCVD-grown CHINT/NERFET devices. (a) $V_{SUB} - V_T = 1\,V$; (b) $V_{SUB} - V_T = 2\,V$; (c) $V_{SUB} - V_T = 2.7\,V$. The threshold voltage $V_T \approx 5$ V. With increasing heating voltage V_{SD}, one clearly sees a rapid rise and subsequent saturation of the collector current I_{SUB}, accompanied by an increase and subsequent sharp drop in the drain current I_D. (After Ref. 78)

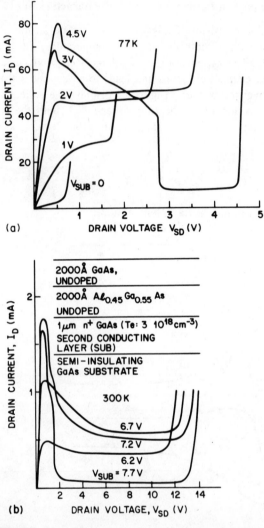

Fig. 24 The NERFET characteristics. (a) Type 1 structure (cf. Fig. 19) at 77 K. Gate dimensions: $1 \, \mu m \times 250 \, \mu m$. (b) Type 2 structure at 300 K. Channel dimensions: $2 \, \mu m \times 100 \, \mu m$. Note the strongly pronounced *negative transconductance* $g_m \equiv (\partial I_D / \partial V_{SD})|_{V_{SUB}}$ in the quasi-saturation regions of the characteristics for both types of structures. (After Refs. 73, 77)

As evident from Fig. 23 (dashed lines), the hot-electron injection in CHINT is accompanied by a strong NDR in the channel circuit. This permits the implementation of the negative-resistance FET or NERFET.[74] The typical NERFET characteristics are shown in Fig. 24. Note that the NDR is strongly affected by V_{SUB}. It is clear that higher V_{SUB} enhances the electron concentration in the channel (acting as a back gate), but it also affects the magnitude of the hot-electron flux corresponding to a given T_e (by lowering the collector barrier).

The highest peak-to-valley current ratio reported in these devices was 160 at room temperature, observed in MOCVD devices.[77] The NDR device can work as an efficient generator and amplifier of electromagnetic oscillations. The main advantage of NERFET over other microwave generators lies in the possibility of controlling the oscillations by a third electrode.

This advantage can be used in logic applications. When any two negative-resistance devices (e.g., Esaki or Gunn diodes) are connected in series and the total applied voltage V_{DD} exceeds roughly twice the critical voltage for the onset of NDR in the single device, then an instability occurs in which one of the devices takes most of the applied voltage, while the other is in the low-field mode. This is illustrated by the usual load-line graphical construct, Fig. 25. The operating points A and C are stable, while B is unstable. Which of the two devices is in the high-field regime is determined by a fluctuation or the history of preparation. Because it has a third electrode, NERFET offers new possibilities for controlling this bistability. A simple circuit[76] that does this is illustrated in Fig. 26. Two NERFETs with nearly identical characteristics are connected in series. One of the controlling voltages is fixed, $V_{SUB2} = 2.5$ V, and the output voltage V_{OUT} is measured as a function of V_{SUB1}. As the controlling voltage V_{SUB1} is varied, the system smoothly approaches the switch points (sharply defined and repetitive within 1 mV at room temperature), at which V_{OUT} jumps between the low and the high values. Such a behavior is reminiscent of a phase transition.

Two types of logic operation can be thought of in this configuration. First, the input voltage can be set to a dc value in the middle of the hysteretic loop, say $V_{SUB1} = 2.5$ V. When a control signal $\Delta V_{SUB1}(t)$ in the form of a short low-amplitude ($|\Delta V| \geq 0.15$ V) pulse of alternate polarity is applied, the system will switch. The result is a *bistable element*: the system will "remember" the sign of the last pulse, namely V_{OUT} = high for $\Delta V < 0$ and V_{OUT} = low for $\Delta V > 0$. Second, an *amplifying inverter* can be implemented by setting V_{SUB1} to a dc value high enough ($V_{SUB} \geq 2.7$ V) to ensure a stable low state.

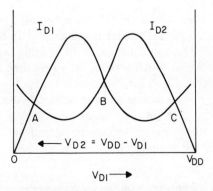

Fig. 25 Graphical construct for determining the operating points of a circuit formed by two identical NDR elements in series. Points A and C are stable, point B unstable.

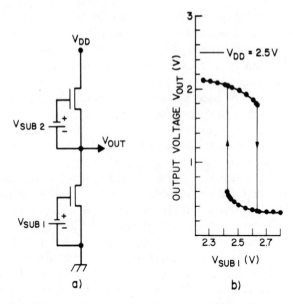

Fig. 26 Simplest NERFET logic circuit. (a) Schematic diagram. Conventional FET circuit symbols are used with the understanding that the SUB electrode plays the role of a gate. (b) Logic transitions at room temperature. The output voltage was measured at fixed $V_{SUB2} = 2.5\,V$ as a function of V_{SUB1} slowly varied in the direction shown by the arrows. An interesting observation was that the switch points were sharply defined and repetitive to within less than 1 mV (i.e., to within $<< kT/q$) in a given device pair. For the displayed example of a fixed $V_{SUB2} = 2.5\,V$, the transition from the low to the high state always occurred at $V_{SUB1} = 2.424\,V$ independent of the value of V_{DD}, provided that the latter is $\leq 3.5\,V$, at which value the new feature of a third stable state appeared, characterized by V_{OUT} being in a range near $V_{DD}/2$. (After Ref. 76)

The system will then switch to its high state only during a pulse of negative polarity $|\Delta V_{SUB1}| \geq 0.3$ V. Both operations have been demonstrated using pulse-mode experiments.

Another type of a logic device (Fig. 27) can be based on the memory effect, which is observed when the SUB layer is unbiased.[75] In this case, hot-electron injection leads to a charge accumulation in the floating layer and a drop in its electrostatic potential Ψ_{SUB}, which persists for a long time after the heating voltage V_{SD} is set to zero, Fig. 27a. The negative Ψ_{SUB} depletes the channel. Transferred electrons remain mobile and can be rapidly discharged by grounding the second conducting layer (in contrast to the situation with the non-volatile memory devices). As the heating voltage is ramped up for the first time one observes an NDR in the channel circuit.† A memory device, based on this

---

†Note, however, that this is a hysteretic NDR, not capable of generating oscillations. It indicates a charge accumulation in the floating substrate—which remains charged even after the heating voltage is removed. The thermoelectric force developed between the two conducting layers[82] has a characteristic decay time determined by the ambient temperature and the barrier height.

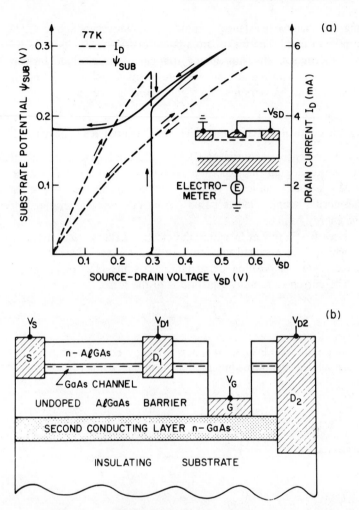

Fig. 27 Memory effect in a CHINT/NERFET structure with floating SUB electrode. (a) Substrate potential Ψ_{SUB} and the channel current I_D as functions of the heating voltage V_{SD}. Arrows indicate the direction of slow (10 mV/s) voltage ramping. (b) Schematic cross section of the proposed memory logic element. Thickness ($\sim 10^{-5}$ cm) and the doping level in the second conducting layer are chosen so that this layer can be depleted by the gate field. (After Refs. 75, 79)

effect,[79] allows a fast operation of all functions: *write, read,* and *erase.* The structure (Fig. 27b) must be grown on an insulating substrate, followed by a thin conducting GaAs layer. The key new element is the guard-gate MESFET-like structure G and the second, "deep", drain D_2, contacting both the channel and the second conducting layer. Electrically, D_2 is connected to the source S. When the guard-gate voltage is negative, the substrate conducting layer is isolated from D_2. Applying voltage to D_1 (*write*), we charge this layer by the hot-electron transfer and thus deplete the main channel. Information is *read* by probing the channel resistance. Applying a positive voltage to G, this information can be *erased*. The maximum amount of transfered charge is limited by the " cold" thermionic emission in the triangular-barrier diode forward-biased by Ψ_{SUB}. One can substantially enhance the memory effect by using thin trapezoidal or rectangular barriers instead of the triangular barrier.

7.4.4   Theoretical Description of Three-Terminal RST Devices

For a comprehensive analysis of CHINT/NERFET characteristics, that is capable of reproducing all the experimentally known features, the reader is referred to the Monte Carlo studies.[81] However, certain qualitative insight can be gained by considering simple models based on the electron temperature approximation and the energy balance.

Consider first a model in which the electric field is concentrated in a domain within which T_e is assumed uniform. In the presence of a hot-electron transfer, the current $I(x)$ is position dependent, as is the electron sheet concentration $\sigma(x)$. The current-continuity equation is of the form

$$\frac{1}{W}\frac{dI}{dx} = -J(x). \tag{39}$$

If we assume that the RST current J is given by Eq. 36 with both T_e and Φ constant, and neglect the diffusion component in the channel current,

$$\frac{I}{W} = \sigma v_S - D\frac{d\sigma}{dx} \approx \sigma v_S, \tag{40}$$

then we find that both σ and I decrease exponentially, $\sigma \propto \exp(x/\lambda)$ with the characteristic length

$$\lambda = \frac{v_S \Delta}{v(T_e)}\, e^{\Phi/kT_e}. \tag{41}$$

For $T_e \sim 1500\,\mathrm{K}$ this gives $\lambda \sim 10^{-5}\,\mathrm{cm}$. Of course, when the variation of σ is so rapid, the diffusion component is non-negligible, but this does not change the situation qualitatively.

Exercise 11: Show that inclusion of the diffusion component changes the characteristic decay length from λ to

$$\Lambda = \frac{\lambda}{2}\left[1 + \left[1 + \frac{4L_D}{\lambda}\right]^{1/2}\right],$$

where $L_D \equiv D/v_S$. Note that the limit $\lambda \ll L_D$, where this correction would be important, is hardly ever realized because at high T_e when λ is very short, the hot-electron diffusivity D drops sharply[83] and the length L_D becomes also quite short ($\leq 10^{-5}\,\mathrm{cm}$).

Estimating the electron temperature from the energy balance equation, one can derive the characteristics of CHINT/NERFET device. This has been done semi-analytically in a simplified model with a uniform electric field in the channel.[84] In the energy balance equation that determined the local T_e, it was assumed that channel electrons gain energy from the field and lose it (i) to the lattice through phonon emission and (ii) to the SUB layer through RST of hot carriers. Momentum-space transfer was neglected in the balance equation. This last assumption leads to a significant distortion of the calculated characteristics at high V_{SD} when most of the carriers in the high-field region reside in the upper valleys.† In particular, it predicts an exponential decrease of I_D without limit as V_{SD} is increased. This can already be seen in the model based on Eq. 41. If the size of the region of high T_e is much larger than λ—a situation easily realized in practice for a sufficiently high V_{SD}—then, along that region σ drops by many orders of magnitude and I_D becomes vanishingly small. However, this is not what happens experimentally. Instead, one clearly observes a saturation of I_D from below and I_{SUB} from above. Monte Carlo studies[81] indicate that this happens because of the intervening momentum-space transfer, which cuts off the heating to channel electrons and limits the I_{SUB}. At the same time, the increasing V_{SD} does not translate into a higher source current because the source is effectively decoupled from the drain by the high-field domain (the electric field near the source does not change). Since in steady state one has

$$I_S = I_D + I_{SUB} \, ,$$

the drain current must also saturate.

It is clear that the RST may lead to a much lower NERFET valley current in materials with higher satellite-valley separation. An interesting question of principle arises: what will limit the RST in the absence of momentum-space transfer? Emission of optical phonons alone can compensate the input power from the electric field only for fields $\leq 4\,kV/cm$. The extra power will have to go into the electron heating and exponential depletion of the channel by RST. The limitation will probably come from a breakdown of the electron temperature approximation.

Indeed, a T_e different from the lattice temperature T is established when the electron-electron (e-e) scattering time is substantially shorter than the relaxation time associated with the lattice. At low electron concentrations, the high-energy

---

†In the calculation of the transfer rate of hot electrons at a given T_e the neglect of satellite valleys can be justified, at least for GaAs/Al$_x$Ga$_{1-x}$As structures with $x \approx 0.4$. In this case, the barrier height Φ is approximately equal to the intervalley Γ-L energy separation. The transferred L electrons see virtually no barrier for real-space transfer and their collection efficiency is near unity. The situation with these electrons is analogous to that with minority carriers in a bipolar transistor. The energy carried away by an L electron is of the same order as that for Γ electron. Therefore, this process (transfer into a satellite valley with subsequent diffusion across the barrier) is just another real-space transfer channel, whose contribution to the overall rate of RST is not very important. What cannot be neglected, however, is the sharply lower heating efficiency of low-mobility electrons in the satellite heavy-mass valleys.

portion of the energy distribution can be strongly depressed compared to a Maxwellian curve. Such an effect has been demonstrated theoretically[85] for a two-dimensional electron gas heated by optical pumping—taking into account the e-e scattering and the optical-phonon emission. Although the total distortion of the distribution function is "integrally" weak, because it affects only the tails of the distribution (above the optical phonon threshold, $\phi \sim 0.03$ eV, where the number of electrons is small), in those very tails the distortion can be quite strong.

Similar effects can be even more important in the operation of CHINT where one is interested only in the tails of the distribution above the collector barrier height $\Phi \sim 0.3$ eV, which are constantly depleted by the injection current. Strong depletion of the channel should lead to a situation in which the hot-electron distribution function will deviate from the Maxwellian form—with the high-energy tail suppressed compared to value predicted by the electron temperature approximation. This means that one can expect a self-limitation of the RST process with the channel concentration never dropping below a critical level—determined, for a given collector barrier height, by the concentration dependence of the electron-electron interaction. This effect will also result in a saturation from below the NERFET I_D versus V_{SD} characteristic. This saturation occurs at a much lower level than the experimentally observed saturation in GaAs devices, which, as clarified by the Monte Carlo studies,[81] is controlled by momentum-space transfer.

7.4.5 Ultimate Speed of CHINT/NERFET Devices

Fundamental limitations on the intrinsic speed of three-terminal RST devices arise due to the time-of-flight delays characteristic of a space-charge-limited current and because of a finite time required for the establishment of an electron temperature.

Consider the latter limitation first. Energy relaxation of hot carriers in bulk semiconductors has been a subject of considerable number of studies.[83] The dominant mechanisms for the maxwellization of the hot-electron energy distribution function are the polar optic phonon scattering and the electron-electron interaction. The phonon mechanism is expected to be similar in the CHINT/NERFET structure compared to that in the bulk semiconductor. Monte Carlo studies indicate that the energy loss rate due to polar optic phonon emission by electrons in GaAs is nearly constant for electron energies above 0.1 eV and is of the order of 2×10^{11} eV/s. This translates into about 1 ps equilibration time for $T_e \sim 1500$ K. Monte Carlo analysis[81] of the average electron-energy transient in the CHINT channel during switching of the heating voltage V_{SD} from a low to a high value, showed that the average energy $<E>$ reaches a steady state in less than 1 ps. The influence of e-e scattering, which may be of primary importance for the establishment of the quasi-equilibrium in high-energy tails of the electron distribution, is much more difficult to take into consideration. As far as we know, there is no satisfactory treatment of this process in a two-dimensional electron gas, which would be applicable to the charge-injection problem in CHINT. It is probably safe to assume that at the device operating voltages (when the carrier concentration is high) the hot-electron ensemble equilibrates in less than 1 ps.

The second fundamental limitation of the speed arises from the space-charge capacitance associated with the mobile charge drifting in the high-field regions of the device. It reduces to the time of flight of electrons over these regions—the high-field portion of the channel and the downhill slope of the potential barrier. At high T_e both regions are of order 10^{-5} cm and the corresponding delay can be less than 1 ps.† Time-of-flight delay associated with the readjustment of mobile charge was also found to be the dominant delay in the transient Monte Carlo simulations of CHINT.[80, 81] Figure 28 illustrates the electron transfer transient obtained in a simulation where the source contact was grounded, the substrate kept at $V_{SUB} = 2.5$ V, and the drain volatge was instantaneously increased from 0.45 to 2.3 V. After about 1 ps, electrons start populating AlGaAs and a high-field region emerges near the drain—where electrons are mostly in the upper valleys. Steady state was reached in about 6 ps, which in these simulations was roughly the time of flight from the source contact to the drain contact.

Experimentally, the microwave operation of CHINT with a current gain greater than unity at room temperature and at frequencies up to 32 GHz has been demonstrated.[78] Microwave generation by NERFET in the gigahertz range had been also observed. Although in principle the CHINT and NERFET are picosecond devices, their actual speed limit, at present, arises from the *RC* delay due to large contact pads and the series channel resistance. Improvement of performance should result from minimization of the main parasitic capacitance, that between the *D* and SUB electrodes, and the channel resistance (at the peak prior to onset of the NDR), which includes the series contact resistance. Limits on the switching speed in a two-NERFET logic circuit are uncertain at present and require further study, both experimental and theoretical.

7.4.6 Concluding Remarks

We have reviewed the physical principles of three-terminal devices that employ hot-electron transfer between two conducting layers separated by a potential barrier. Their operation is based on controlling charge injection over the barrier by

---

†It should be emphasized that these time-of-flight limitations are different from the time-of-flight-under-the-gate limitation characteristic of an FET. The latter results from charging the channel by the gate field through the output resistance of a previous identical device, which necessarily gives $\tau = L/v$ with L being the gate length. In the CHINT the controlling electrode is the drain and L is the total length of the space-charge-limited current regions—which can be substantially shorter than the source-to-drain distance. Of course, the time associated with the readjustment of charge involved in the formation of the high-field domain itself cannot be shorter than the time of flight from the source contact to the drain contact, but once the domain is formed the speed with which its T_e can be modulated is not necessarily related to the transit time along the entire channel.

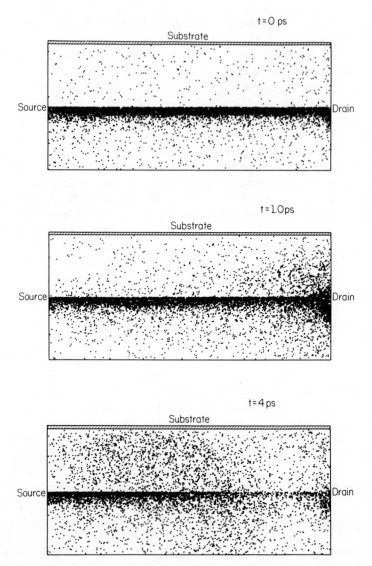

Fig. 28 Electron transfer transient obtained in Monte Carlo simulations. (After Refs. 80, 81)

modulating the electron temperature in one of the layers. The principle was illustrated by a comparison to a hypothetical vacuum diode whose cathode temperature is controlled by an input electrode.

The CHINT is a general-purpose high-speed transistor. Electrically, its operation is analogous to the bipolar transistor and the ballistic hot-electron transistors discussed in Section 7.3, if the terminals are identified as $S \equiv$ emitter, $D \equiv$ base, and SUB $\equiv$ collector. An interesting feature of CHINT is that its differential common-base current gain $\alpha \equiv (\partial I_{SUB}/\partial I_S)$ at $V_{SUB} =$ const can substantially exceed unity (due to the NDR in the S to D circuit). By the physical principle involved, the operation of CHINT is different from all previous three-terminal devices, which were based either on the potential or the field effect (Section 7.1.1). In CHINT the control of output current is accomplished by modulating the electron temperature that governs the charge injection over a barrier of fixed height. Power gain in this device had been demonstrated experimentally both in dc operation and at high frequencies. The value of the mutual conductance g_m obtained in CHINT (over 1000 mS/mm) compares favorably to the best conventional transistors.

Hot-electron injection in CHINT is accompanied by a strong negative differential resistance in the channel circuit. This gives rise to a related device, called the NERFET, which is essentially a two-terminal NDR element, controllable by the voltage on the third (SUB) electrode. The experimentally observed room-temperature NDR in NERFET had the highest peak-to-valley ratio (over 10^2) of any negative-resistance device. The NERFET can be used in several ways. First, it can be used as a controllable amplifier and generator of oscillations. Microwave generation in NERFET has been observed, with efficient dc to ac conversion and a high output power. Second, it can be applied in a variety of logic configurations, the simplest of which have been reviewed in Section 7.4.3.

All the three-terminal RST devices discussed above have been experimentally demonstrated with the help of AlGaAs/GaAs heterostructures, fabricated either by MBE or MOCVD techniques. An important feature in these structures is the interplay between the real-space and the momentum-space transfers, as clearly demonstrated by Monte Carlo simulations.[80, 81] Since the momentum-space transfer considerably slows the device operation, it had been proposed[73] to employ other heterostructure materials, such as InGaAs/InAlAs, in which the satellite valleys have a higher energy separation. A further bonus in using these materials is the lower electron effective mass, which enhances the heating effects. It may also be advantageous to use the RST transfer of holes rather than electrons.

It should be said that semiconductor heterojunctions are not the only way to implement three-terminal RST devices. An interesting possibility lies in using thin semimetal films to form a Schottky barrier with a semiconductor collector underneath. For example, Bi/Si junctions have a barrier height of 0.63 eV. Because of the reduced electron-scattering rates (due to high dielectric permittivity and low carrier concentration), bismuth is known to exhibit strong hot-electron effects. If the electric field is applied laterally to a Bi film on a silicon substrate, one can expect an efficient emission of hot electrons over the Schottky barrier. So far, this effect has not been experimentally verified.

7.5 SUMMARY AND FUTURE TRENDS

Commercial utilization of hot-electron phenomena began with the Gunn effect, based on the Hilsum-Ridley-Watkins mechanism for a negative differential resistance that results from the transfer of hot electrons from the high-mobility central valley in a direct-gap III-V compound semiconductor to its higher-lying, low-mobility satellite valleys. The Gunn diode is undoubtedly the best-known hot-electron device, for which a mature technology has developed (see Chapter 9). Another successful application of a hot-carrier effect has been made in non-volatile memory devices.[2] The floating-gate avalanche injection MOS memory device FAMOS[86] bears some conceptual similarity to the real-space-transfer memory device discussed in Section 7.4.3. The FAMOS represents a p-channel MOSFET structure with a floating gate electrode. In the process of "writing" the memory, carriers, heated by the drain field, avalanche near the drain junction with hot electrons from the avalanche plasma injected into the floating gate. As the gate is charged, its potential is lowered and the p-channel conductance increases.

In this chapter we have discussed a number of unipolar hot-electron injection transistors. These modern devices are based on the physical processes in semiconductor heterojunctions that in conventional transistors are only peripheral (or even detrimental) to their operation. The wide class of hot-electron devices were classified into two groups, the ballistic devices and the real-space-transfer devices, depending on the type of a hot-electron ensemble essentially employed in their operation. The family tree of the unipolar hot-electron injection devices is displayed in the table in Fig. 29. The family is large and its members often go under different names. In the attempt to represent only distinct ideas, we may well have overlooked some important relatives.

So far, none of these rather exotic devices have been used in electronic applications. Most of the research has concentrated on demonstrating the existence of an effect in question, proposals of new structures and effects, and studies of their potential physical limitations. Remember, however, that the main purpose

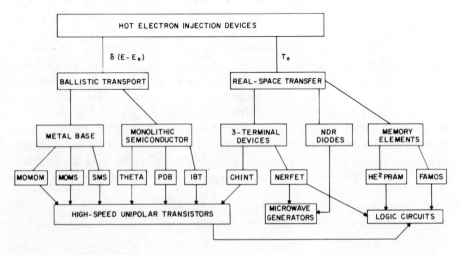

Fig. 29 The family of unipolar hot-electron injection devices and their potential applications.

of this type of work is to come up with a device that will find a practical application. Are there good reasons to hope that this will happen? We believe that the answer is in the affirmative—substantiated by the remarkable advancement in the last decade in the techniques of crystal growth (such as MBE and MOCVD) and device processing (submicron lithography, ion implantation, etc.). It is likely, in our view, that the next decade will see a commercial exploitation of hot-electron-injection devices—such as those reviewed in this chapter and those not yet invented.

PROBLEMS

1. Transconductance and small-signal delay.

    a) Write a general expression for the transconductance $g_m \equiv (\partial I_D / \partial V_G)_{V_D}$ of a field-effect transistor in terms of the gate-to-channel capacitance C_{GC}, assuming that the carrier velocity v is constant everywhere in the channel. Calculate the "small-signal" delay τ, defined by $\tau \equiv C_G / g_m$, where $C_G \equiv C_{GC} + C_p$ is the total gate capacitance and C_p is the total parasitic capacitance effectively in parallel with the "useful" capacitance C_{GC}. Show that in the limit $C_{GC} / C_G \to 1$ the small-signal delay reduces to that given by Eq. 2.

    b) (b) Show that addition of a source resistance R_s degrades the transconductance in the following manner:

    $$g_m = \frac{g_m^{(0)}}{1 + g_m^{(0)} R_s} ,$$

    where $g_m^{(0)}$ is the transconductance the transistor would have in the absence of source resistance.

2. Emitter and collector charging times. When a current is flowing, the mobile charge Q_m injected into a normally depleted region is proportional to I.

    a) Calculate the charge Q_m^{drift} accumulated on the downhill slope of length L_2 assuming drift with a saturated velocity.

    b) Estimate the charge Q_m^{diff} accumulated on the uphill slope due to the finite value of the diffusion velocity; note that the latter cannot exceed the mean thermal velocity V_T.

    c) Calculate the "drift" and "diffusion" capacitances $\partial Q_m / \partial V_{\text{in}}$, associated with the storage of the mobile charge on the downhill and the uphill barrier slopes, respectively.

d) Calculate the small-signal delay associated with charging these capacitances. Show that in the regime, when the current increases exponentially with the input voltage, they become dominant at sufficiently high currents.

3. Simplest model for the QM reflection at a metal-semiconductor barrier. Assume parabolic energy-momentum relationships in both materials. Ignore all the electronic band properties and treat the carriers as if they were free electrons impinging with the kinetic energy E on a potential step-barrier of height Φ.

a) Show that the above-barrier reflection coefficient R is given by

$$R = \left(\frac{k - q}{k + q} \right)^2 = \left(\frac{1 - \xi}{1 + \xi} \right)^2, \quad \text{where } \xi = \left[1 - \frac{\Phi}{E} \right]^{1/2},$$

$E = \hbar^2 k^2 / 2m$, and $E - \Phi = \hbar^2 q^2 / 2m$. Note that it is not the clearance $E - \Phi$ but the ratio E/Φ that enters the expression for R, and hence one must correctly choose the zero energy level.

b) In this model, the QM reflection is related to the large Fermi energy of electrons in a metal base: one typically finds that Φ/E is close to unity and the reflection is large. Estimate the probability of reflection for a ballistic electron in Al incident on the interface with GaAs at 0.4 eV above the Schottky barrier ($\Phi \approx 12 \text{ eV}$).

4. Smooth wall barrier. The abrupt barrier model assumed in Problem 3 is a somewhat pathological case: the expression for a quantum-mechanical reflection coefficient does not contain $\hbar$ (cf. Ref. 10, Section 25). Consider a barrier of height Φ, and shape $\Psi(x) = \Phi[1 + \exp(-x/a)]^{-1}$. It is an exactly soluble model of a "smooth wall" barrier graded over a distance a.

a) Show that the reflection coefficient is given by

$$R(E) = \frac{\sinh^2[\pi a (k - q)]}{\sinh^2[\pi a (k + q)]},$$

where $E = \hbar^2 k^2 / 2m$, and $E - \Phi = \hbar^2 q^2 / 2m$. Note that in this expression $\hbar$ has not dropped out in the dependence of R on energy—in contrast to the expression in Problem 3, which obtains in the limit $a \to 0$.

b) Consider the case $ka > 1$, which roughly corresponds to a being greater than the lattice constant in the metal. Show that

$$R = \left(\frac{1 - \tanh(\pi a q)}{1 + \tanh(\pi a q)} \right)^2.$$

Estimate R for $E - \Phi \geq \hbar^2/ma^2$. Note that the "smooth wall" barrier model, treated in this problem, predicts an almost complete elimination of the QM reflection at a sufficiently high electron energy. At first glance, this may suggest that the QM reflection problem is not severe, as the shape of the potential barrier (formed by the image force) for electrons leaving a metal is, typically, rather smooth on the scale of the metal lattice constant. However, estimates based on a free-electron model of reflection are certainly invalid for metals with a complicated band structure and indirect-gap semiconductors where band-structure effects further inhibit the QM transmission. For metal-semiconductor pairs studied so far,[21] the exact numerical solution gives a still lower transmission of hot electrons than that estimated from the naive expression in Problem 3.

5. Real-space transfer in a multilayer heterostructure. Consider a simple model of RST, which neglects the electric field associated with the transferred electrons and the potential "pockets" from the uncovered donor charge. Assume a periodic multilayer structure like in Fig. 18, with narrow-gap layers of thickness d_1 and wide-gap layers of thickness d_2. The effective electronic masses in the layers are m_1 and m_2. Assume a field-independent mobility μ_i in each layer ($\mu_1 > \mu_2$). Take the rate of energy loss to the lattice proportional to $(T_e - T)/\tau_E$ per electron and assume that the energy relaxation times τ_E are the same in both layers. Assume further that layers are so thin that T_e is not a local temperature but pertains to the whole electron gas and hence when layers 1 and 2 exchange electrons, the mean energy is not being transfered from one layer to another. The total density of electrons per unit area is fixed by the overall neutrality: $n_1 + n_2 = n = $ const.

a) Write down the energy balance equation.
b) Express the ratio n_1/n_2 in terms of T_e and the conduction-band discontinuity ΔE_C.
c) Derive the current-field characteristic in a parametric form

$$F = F(T_e) ; \quad J = J(T_e) ,$$

and plot $J(F)$ for several assumed values of the parameters. How are the parameters chosen to improve the peak-to-valley ratio of NDR?

REFERENCES

1. E. M. Conwell, *High Field Transport in Semiconductors*, Academic, New York, 1967.

2. S. M. Sze, *Physics of Semiconductor Devices*, 2nd ed., Wiley-Interscience, New York, 1981.

3. W. Shockley, "Transistor Electronics: Imperfections, Unipolar and Analog Transistors," *Proc. IRE* **40**, 1289 (1952).

4. J.-I. Nishizawa, T. Terasaki, and J. Shibata, "Field-Effect Transistor versus Analog Transistor (Static Induction Transistor)," *IEEE Trans. Electron Dev.* **ED-22**, 185 (1975).

5. C. O. Bozler and G. D. Alley, "The Permeable Base Transistor and Its Application to Logic Circuits," *Proc. IEEE* **70**, 46 (1982).

6. S. Luryi and R. F. Kazarinov, "On the Theory of the Thermionic Emission Transistor; TET as an Element of Logic Circuits," *Solid State Electron.* **25**, 933 (1982).

7. R. F. Kazarinov and S. Luryi, "Majority Carrier Transistor Based on Voltage-controlled Thermionic Emission," *Appl. Phys.* **A28**, 151 (1982).

8. P. J. Price, "Monte Carlo Calculation of Electron Transport in Solids," *Semiconductors and Semimetals*, Vol. 14, Academic, New York 1979, pp. 249–308.

9. M. Heiblum, M. I. Nathan, D. I. Thomas, and C. M. Knoedler, "Direct Observation of Ballistic Transport in GaAs," *Phys. Rev. Lett.* **55**, 2200 (1985).

10. L. D. Landau and E. M. Lifshitz, *Quantum Mechanics: Non-Relativistic Theory,* 3rd ed., Sect. 19, Pergamon, London, 1977.

11. W. A. Harrison, "Tunneling from an Independent Particle Point of View," *Phys. Rev.* **123**, 85 (1961).

12. A. A. Grinberg and S. Luryi, "Electron Transmission Across Interface of Different One-Dimensional Crystals," *Phys. Rev.* **B 39**, 7466 (1989).

13. M. Tomásek, "Simple Calculation of Surface States for One-Dimensional Models of Crystals," *Czech. J. Phys.* **B 12**, 159 (1962).

14. A. F. J. Levi and T. H. Chiu, "Room-Temperature Operation of Hot-Electron Transistors," *Appl. Phys. Lett.* **51**, 984 (1987).

15. M. D. Stiles and D. R. Hamann, "Ballistic Electron Transmission Through Interfaces," *Phys. Rev.* **B 38**, 2021 (1988).

16. C. A. Mead, "Tunnel-Emission Amplifiers," *Proc. IRE* **48**, 359 (1960).

17. J. P. Spratt, R. F. Schwartz, and W. M. Kane, "Hot Electrons in Metal Films: Injection and Collection," *Phys. Rev. Lett.* **6**, 341 (1961).

18. M. M. Atalla and D. Kahng, "A New Hot-Electron Triode Structure with Semiconductor-Metal Emitter," IRE-AIEE Solid State DRC, University of New Hampshire, Durham (July 1962).

19. D. V. Geppert, "A Metal-Base Transistor," *Proc. IRE* **50**, 1527 (1962).

20. C. R. Crowell and S. M. Sze, "Quantum-Mechanical Reflection of Electrons at Metal-Semiconductor Barriers: Electron Transport in Semiconductor-Metal-Semiconductor Structures," *J. Appl. Phys.* **7**, 2683 (1966).

21. M. D. Stiles and D. R. Hamann, "Electron Transmission through $NiSi_2$-Si Interfaces," *Phys. Rev.*, **B40**, 1349 (1989).

22. E. Rosencher, S. Delage, Y. Campidelli, and F. A. D'Avitaya, "Transistor Effect in Monolithic $Si/CoSi_2/Si$ Epitaxial Structures," *Electron. Lett.* **20**, 762 (1984).

23. J. C. Hensel, A. F. J. Levi, R. T. Tung, and J. M. Gibson, "Transistor Action in $Si/CoSi_2/Si$ Heterostructures," *Appl. Phys. Lett.* **47**, 151 (1985).

24. R. T. Tung, A. F. J. Levi, and J. M. Gibson, "Control of a Natural Permeable Base Transistor," *Appl. Phys. Lett.* **48**, 635 (1986).

25. J. Lindmayer, "The Metal-Gate Transistor," *Proc. IEEE* **52**, 1751 (1964).

26. M. Heiblum, "Tunneling Hot Electron Transfer Amplifiers (THETA): Amplifiers Operating up to the Infrared," *Solid-State Electron.* **24**, 343 (1981).

27. J. M. Shannon, "Hot Electron Camel Transistors," *IEE J. Solid State Electron Dev.* **3**, 142 (1979); "Hot Electron Diodes and Transistors," *Inst. Phys. Conf. Ser.* **69,** 45 (1984).

28. R. J. Malik, M. A. Hollis, L. F. Eastman, C. E. C. Wood, D. W. Woodard, and T. R. AuCoin, "GaAs Planar-Doped Barrier Transistors Grown by Molecular Beam Epitaxy," *Proc. 8th Biennial Cornell Conf. on Active Microwave Semicond. Devices and Circuits, August 1981.*

29. M. A. Hollis, S. C. Palmateer, L. F. Eastman, N. V. Dandekar, and P. M. Smith, "Importance of Electron Scattering with Coupled Plasmon-Optical Phonon Modes in GaAs Planar Doped Barrier Transistors," *IEEE Electron Dev. Lett.* **EDL-4**, 440 (1983).

30. N. Yokoyama, K. Imamura, T. Ohshima, H. Nishi, S. Muto, K. Kondo, and S. Hiyamizu, "Tunneling Hot Electron Transistor using GaAs/AlGaAs Heterojunctions," *Japan J. Appl. Phys.* **23**, L311 (1984).

31. M. Heiblum, D. C. Thomas, C. M. Knoedler, and M. I. Nathan, "Tunneling Hot-Electron Transfer Amplifier: A Hot-Electron GaAs Device with Current Gain," *Appl. Phys. Lett.* **47**, 1105 (1985).

32. S. Muto, K. Imamura, N. Yokoyama, S. Hiyamizu, and H. Nishi, "Subpicosecond Base Transit Time Observed in a Hot-Electron Transistor," *Electron. Lett.* **21**, 555 (1985).

33. I. Hase, H. Kawai, S. Imanaga, K. Kaneko, and N. Watanabe, "AlGaAs/GaAs Hot Electron Transistor Grown by MOCVD," *Inst. Phys. Conf. Ser.* **79**; 613 (1986).

34. U. K. Reddy, J. Chen, C. K. Peng, and H. Morkoç, "InGaAs/InAlAs Hot Electron Transistor," *Appl. Phys. Lett.* **48**, 1799 (1986).

35. I. Hase, K. Taira, H. Kawai, T. Watanabe, K. Kaneko, and N. Watanabe, "Strained GaInAs-Base Hot Electron Transistor," *Electron. Lett.* **24**, 279 (1988).

36. K. Seo, M. Heiblum, C. M. Knoedler, J. Oh, J. Pamulapati, and P. Bhattacharya, "High Gain Pseudomorphic InGaAs Base Ballistic Hot Electron Device," *IEEE Electron Dev. Lett.* **EDL-10**, 73 (1989).

37. T. J. Maloney, "Polar Mode Scattering in Ballistic Transport GaAs Devices," *IEEE Electron Dev. Lett.* **EDL-1**, 54 (1980).

38. P. Lugli and D. K. Ferry, "Investigation of Plasmon-induced Losses in Quasi-Ballistic Transport," *IEEE Electron Dev. Lett.* **EDL-6**, 25 (1985).

39. S. Imanaga, H. Kawai, K. Kaneko, and N. Watanabe, "Monte Carlo Simulation of AlGaAs/GaAs Hot-Electron Transistors," *J. Appl. Phys.* **59**, 3281 (1986).

40. A. F. J. Levi, J. R. Hayes, and R. Bhat, "'Ballistic' Injection in Semiconductors," *Appl. Phys. Lett.* **48**, 1609 (1986).

41. M. Heiblum, I. M. Anderson, and C. M. Knoedler, "DC Performance of Ballistic Tunneling Hot-Electron-Transfer Amplifiers," *Appl. Phys. Lett.* **49**, 207 (1986).

42. M. Heiblum, E. Calleja, I. M. Anderson, W. P. Dumke, C. M. Knoedler, and L. Osterling, "Evidence of Hot Electron Transfer into an Upper Valley in GaAs," *Phys. Rev. Lett.* **56**, 2854 (1986).

43. N. Lifshitz, A. Jayaraman, R. A. Logan, and H. C. Card, "Pressure and Compositional Dependences of the Hall Coefficient in $Al_xGa_{1-x}As$ and Their Significance," *Phys. Rev.* **B 21**, 670 (1980).

44. H. Ohnishi, N. Yokoyama, and H. Nishi, "Monte Carlo Simulation of Electron Transport Efficiency of an InGaAs/InP Hot-Electron Transistor," *IEEE Electron Dev. Lett.* **EDL-6**, 403 (1985).

45. S. Luryi, "An Induced Base Hot-Electron Transistor," *IEEE Electron Dev. Lett.* **EDL-6**, 178 (1985).

46. B. Honeisen and C. A. Mead, "Fundamental Limitations in Microelectronics—I. MOS Technology," *Solid State Electron.* **15**, 819 (1972); "II. Bipolar Technology," **15**, 981 (1972).

47. S. Luryi, "Induced Base Transistor," *Physica* **134B**, 466 (1985).

48. C. Y. Chang, W. C. Liu, M. S. Jame, Y. H. Wang, S. Luryi, and S. M. Sze, "Induced Base Transistor Fabricated by Molecular Beam Epitaxy," *IEEE Electron Dev. Lett.*, **EDL-7**, 497 (1986).

49. W. Shockley, US Patent 2,569,347 (filed 1948, issued 1951).

50. H. Kroemer, "Theory of a Wide-Gap Emitter for Transistors," *Proc. IRE* **45**, 1535 (1957).

51. H. Kroemer, "Heterostructure Bipolar Transistors and Integrated Circuits," *Proc. IEEE* **70**, 13 (1982).

52. A. F. J. Levi and Y. Yafet, "Nonequilibrium Transport in Bipolar Devices," *Appl. Phys. Lett.* **51**, 42 (1987).

53. R. N. Nottenburg, Y.-K. Chen, M. B. Panish, R. Hamm, and D. A. Humphrey, "High-Current-Gain Submicrometer InGaAs/InP Hetero-

structure Bipolar Transistor," *IEEE Electron Dev. Lett.* **EDL-9**, 524 (1988).

54. Y.-K. Chen, R. N. Nottenburg, M. B. Panish, R. Hamm, and D. A. Humphrey, "Subpicosecond InP/InGaAs Heterostructure Bipolar Transistor," *IEEE Electron Dev. Lett.* **EDL-10**, 267 (1989).

55. B. Jalali, R. N. Nottenburg, Y.-K. Chen, A. F. J. Levi, D. Sivco, A. Y. Cho, and D. A. Humphrey, "Near-Ideal Lateral Scaling in Abrupt $Al_{0.48}In_{0.52}As/In_{0.53}Ga_{0.47}As$ Heterostructure Bipolar Transistors Prepared by Molecular Beam Epitaxy," *Appl. Phys. Lett.* **54**, 2333 (1989).

56. A. G. Milnes and D. L. Feucht, *Heterojunctions and Metal Semiconductor Junctions*, Academic, New York, 1972, pp. 28–29.

57. P. Hesto, J.-F. Pone, and R. Castagne, "A Proposal and Numerical Simulation of N^+NN^+ Schottky Device for Ballistic and Quasiballistic Electron Spectroscopy," *Appl. Phys. Lett.* **40**, 405 (1982).

58. J. R. Hayes, A. F. J. Levi, and W. Wiegmann, "Hot Electron Spectroscopy," *Electron. Lett.* **20**, 851 (1984).

59. A. F. J. Levi, J. R. Hayes, P. M. Platzman, and W. Wiegmann, "Injected Hot-Electron Transport in GaAs," *Phys. Rev. Lett.* **55**, 2071 (1985).

60. J. R. Hayes and A. F. J. Levi, "Dynamics of Extreme Nonequilibrium Electron Transport in GaAs," *IEEE J. Quant. Electron.* **QE-22**, 1744 (1986).

61. T. Wang, K. Hess, and G. J. Iafrate, "Monte Carlo Simulations of Hot-Electron Spectroscopy in Planar-doped Barrier Transistors," *J. Appl. Phys.* **59**, 2125 (1986).

62. A. P. Long, P. H. Beton, and M. J. Kelly, "Hot-Electron Transport in Heavily Doped GaAs," *Semicond. Sci. Technol.* **1**, 63 (1986).

63. N. Yokoyama, K. Imamura, T. Ohshima, H. Nishi, S. Muto, K. Kondo, and S. Hiyamizu, "Characterization of Double Heterojunction GaAs/AlGaAs Hot Electron Transistors," *Int. Electron Dev. Meet. Tech. Dig.* **IEDM-84**, 532 (1984).

64. A. Palevski, M. Heiblum, C. P. Umbach, C. M. Knoedler, A. N. Broers, and R. H. Koch, "Lateral Tunneling, Ballistic Transport, and Spectroscopy in a Two-Dimensional Electron Gas," *Phys. Rev. Lett.* **62** 1776 (1989).

65. A. Palevski, C. P. Umbach, and M. Heiblum, "A High Gain Lateral Hot-Electron Device," *Appl. Phys. Lett.* **55**, 1421 (1989).

66. K. Hess, H. Morkoç, H. Shichijo, and B. G. Streetman, "Negative Differential Resistance through Real-Space Electron Transfer," *Appl. Phys. Lett.* **35**, 469 (1979).

67. Z. S. Gribnikov, "Negative Differential Conductivity in a Multilayer Heterostructure," *Fiz. Tekh. Poluprovodn.* **6**, 1380 (1972) [*Sov. Phys. Semicond.* **6**, 1204 (1973)].

68. H. Shichijo, K. Hess, and B. Streetman, "Real-Space Electron Transfer in

GaAs-Al$_x$Ga$_{1-x}$As Heterostructures: Analytical Model for Large Layer Widths," *Solid-State Electron.* **23**, 817 (1980).

69. M. Keever, H. Shichijo, K. Hess, S. Banerjee, L. Witkowski, H. Morkoç, and B. G. Streetman, "Measurements of Hot-Electron Conduction and Real-Space Transfer in GaAs/Al$_x$Ga$_{1-x}$As Heterojunction Layers," *Appl. Phys. Lett.* **38**, 36 (1981).

70. P. D. Coleman, J. Freeman, H. Morkoç, K. Hess, B. G. Streetman, and M. Keever, "Observation of a New Oscillator Based on Real-Space Transfer in Heterojunctions," *Appl. Phys. Lett.* **40**, 493 (1982).

71. K. Hess, "Principles of Hot Electron Thermionic Emission (Real Space Transfer) in Semiconductor Heterolayers and Device Applications," *Festkörperprobleme* **25**, 321 (1985).

72. A. Kastalsky and S. Luryi, "Novel Real-Space Hot-Electron Transfer Devices," *IEEE Electron Dev. Lett.* **EDL-4**, 334 (1983).

73. S. Luryi, A. Kastalsky, A. C. Gossard, and R. H. Hendel, "Charge Injection Transistor Based on Real Space Hot-Electron Transfer," *IEEE Trans. Electron Dev.* **ED-31**, 832 (1984).

74. A. Kastalsky, S. Luryi, A. C. Gossard, and R. H. Hendel, "A field-Effect Transistor with a Negative Differential Resistance," *IEEE Electron Dev. Lett.* **EDL-5**, 57 (1984).

75. S. Luryi, A. Kastalsky, A. C. Gossard, and R. H. Hendel, "Hot Electron Memory Effect in Double-Layered Heterostructures," *Appl. Phys. Lett.* **45**, 1294 (1984).

76. A. Kastalsky, S. Luryi, A. C. Gossard, and W. K. Chan "Switching in NERFET Circuits," *IEEE Electron Dev. Lett.* **EDL-6** 347 (1985).

77. A. Kastalsky, R. Bhat, W. K. Chan, and M. Koza, "Negative Resistance Field Effect Transistor Grown by Organometallic Chemical Vapor Deposition," *Solid-State Electron.* **29** 1073 (1986).

78. A. Kastalsky, J. H. Abeles, R. Bhat, W. K. Chan, and M. Koza, "High-Frequency Amplification and Generation in Charge Injection Devices," *Appl. Phys, Lett.* **48**, 71 (1986).

79. S. Luryi and A. Kastalsky, "Hot Electron Injection Devices," *Superlat. Microstruct.* **1**, 389 (1985).

80. I. C. Kizilyalli, K. Hess, T. Highman, M. Emanuel, and J. J. Coleman, "Ensemble Monte Carlo Simulation of Real Space Transfer (NERFET/CHINT) Devices," *Solid-State Electron.* **31**, 355 (1988).

81. I. C. Kizilyalli and K. Hess, "Physics of Real-Space Transfer Transistors," *J. Appl. Phys.* **65** 2005 (1989).

82. P. J. Price, "Mesostructure Electronics," *IEEE Trans. Electron Dev.* **ED-28**, 911 (1981).

83. C. Jacoboni and L. Reggiani, "Bulk Hot-Electron Properties of Cubic Semiconductors," *Adv. Phys.* **28**, 493 (1979).

84. A. Kastalsky, A. A. Grinberg, and S. Luryi, "Theory of Hot Electron Injection in CHINT/NERFET Devices," *IEEE Trans. Electron Dev.* **ED-34**, 409 (1987).

85. S. E. Esipov and I. B. Levinson, "Electron Temperature in a Two-Dimensional Gas: Energy Losses to Optical Phonons," *Zh. Eksp. Teor. Fiz.* **90**, 330 (1986) [*Sov. Phys. JETP* **63**, 191 (1986)].

86. D. Frohman-Benchkowski, "FAMOS—a New Semiconductor Charge Storage Device," *Solid-State Electron.* **17**, 517 (1974).

III QUANTUM-EFFECT, MICROWAVE, AND PHOTONIC DEVICES

8 Quantum-Effect Devices

F. Capasso, S. Sen, and F. Beltram
AT&T Bell Laboratories
Murray Hill, New Jersey

8.1 INTRODUCTION

In 1965 a remarkable and pioneering paper[1] by J. A. Morton entitled "From Physics to Function," introduced the concept of functional device. The key characteristic of such devices is that "they promise to reduce greatly the number of elements and process steps per function when their capabilities are properly matched to an old or new system function." Morton provided a few examples of functional devices, one of which was the tunnel diode.[2] The charge-coupled device (CCD),[3] invented and developed in the 1970s, is another early example of functional device, since it can perform a wide range of electronic functions including image sensing and signal processing.

Morton's vision, strongly relying on dramatic progress in growth techniques, material science, and semiconductor physics, is only now gradually becoming reality. In particular, the advent of advanced epitaxial growth techniques, such as molecular beam epitaxy (MBE) and metallorganic chemical vapor deposition (MOCVD), and of bandgap engineering[4] has made possible the development of a new class of materials and heterojunction devices with unique optical and electronic properties. The investigation of novel phenomena that arise when the layer thicknesses become comparable to the de Broglie wavelength of electrons (quantum size effect) has proceeded in parallel with the exploitation of such phenomena in novel devices such as quantum-well (QW) lasers, invented in the 1970s.

The invention of functional devices (in the sense of Morton) based on quantum confinement, however, occurred later, in the early 1980s. In the optoelectronics area an excellent example is the Self-Electrooptic Effect Device (SEED),[5] based on the quantum-confined Stark effect, which may have an important impact on photonic switching.

High-Speed Semiconductor Devices, Edited by S.M. Sze. ISBN 0-471-62307-5
© 1990 John Wiley & Sons, Inc.

Resonant-Tunneling (RT) transistors are emerging as some of the most promising electron functional devices. RT through heterojunction double barriers (DBs) was first observed in 1974.[6] However, the observed negative differential resistance (NDR) effects were too small to be useful in device applications. The impressive RT experiments at terahertz frequencies[7] in 1983 stimulated renewed interest in NDR. Oscillation frequencies in excess of 400 GHz have been demonstrated.[8] The remarkable progress in MBE during the last decade has recently made possible the observation of high peak-to-valley ratios in RT DBs.[9] The literature on RT DBs is vast. The interested reader is referred to recent reviews[10,11] covering the physics as well as the dc and high-frequency performance of RT diodes.

As early as 1963, it was suggested that the well in a unipolar RT DB could act as the control electrode of a transistor.[12,13]

In 1984 the concept of a resonant-tunneling bipolar transistor (RTBT) was proposed.[14] A similar device was discussed by independent researchers.[15] RT transistors allow the implementation of a large class of circuits (e.g., analog-to-digital converters, parity checkers, frequency multipliers, etc.) with greatly reduced complexity (i.e., fewer transistors per function compared to a circuit using conventional transistors).[14] The inherent functionality of these and other quantum electron devices has led to the projection of an intriguing scenario for the future of electronics.[16] The progress of integrated circuits has so far been marked by increased levels of miniaturization to the point that nowadays certain VLSI chips contain an average of ten million components. Owing to interconnection limitations, this scaling strategy will probably approach practical limits at a minimum lateral dimension of patterned geometries of ≈ 0.25 μm.[16] After reaching the limits of conventional scaling some time in the early twenty-first century, electronics will have to evolve along new paths in order to survive as an industry. New devices and circuit architectures will be developed. RT transistors and quantum-coupled devices may play an important role in light of their functionality and the possibility exists of direct device interconnections via tunneling.[16] It has also been pointed out that the inherent multi-state nature of an RT transistor could lead to new computer architectures that use multiple-valued logic.[14]

In 1985 the low-temperature (77 K) operation of a resonant-tunneling, hot-electron unipolar transistor (RHET) was reported.[17] Room-temperature operation of an RTBT, with a DB in the base, was demonstrated in 1986.[18] An RTBT with a single DB in the emitter was reported[19] at about the same time.

The emergence of AlInAs/GaInAs as a heterojunction ideally suited for resonant-tunneling devices, due to the light electron mass in the barrier (AlInAs) and the relatively large (direct gap) conduction-band discontinuity,[20] has further increased the pace of progress in this area, with many groups currently involved in RT transistor research. Recently, the first multiple-state RT transistor has been demonstrated, along with its circuit capabilities.[21,22]

In Section 8.2, certain RT DB structures important for their device potential or physical interest are discussed. In Section 8.3 RTBTs with a double barrier in the base region are described. Design considerations for RTBTs with ballistic injection are discussed, and the observation of minority electron ballistic RT is presented. RTBTs using thermionic injection and exhibiting high peak-to-valley ratio at room temperature in their transfer characteristics are also described. Section 8.4 deals with two- and three-terminal RT devices with multiple peaks in the current-voltage characteristics. In particular, multiple-state RTBTs with two peaks and their microwave performance are discussed; values of f_T as high as 24 GHz have been achieved. Some circuit applications of RTBTs are discussed in Section 8.5. Here we show that RTBTs allow the implementation of many analog and digital circuit functions with a greatly reduced number of transistors and show considerable promise for multiple-valued logic. Experimental results on frequency multipliers and parity-bit generators are presented. Analog-to-digital converters and memory circuits are also discussed. The structures presented in Sections 8.6 and 8.7 are of interest primarily as tools to investigate the physics of transport in lower-dimensionality (one- and two-dimensional) systems and in superlattices. RHETs, quantum-wire transistors, and gated quantum-well transistors are discussed, and the operation of two superlattice-base, negative-transconductance transistors is described. Section 8.8 presents different mechanisms of achieving negative differential resistance in superlattices and a brief discussion of quantum interference devices. For a comprehensive treaty on the physics of quantum electron devices, the interested reader is referred to the book cited in Ref. 11.

8.2 RESONANT-TUNNELING DIODES

The physics of RT has been reviewed in Chapter 2 and in a recent review article.[23] In this section we discuss specific DB structures that are of interest either as building blocks for transistors or from a physics point of view. These include AlInAs/GaInAs diodes, RT through parabolic wells, and RT spectroscopy. For circuit applications, good peak-to-valley ratios are required (typically greater than 2), together with current densities of the order of 10^4 A/cm^2 or more. The tunneling time (see later in text) should be minimized, along with the parasitic resistances and the device capacitance.

8.2.1 AlInAs/GaInAs Resonant-Tunneling Diodes

As mentioned in the introduction, this material system is particularly well suited for RT diodes because of the large ΔE_c and small electron effective mass.[24]

An RT diode in this alloy system consists of a 1-μm-thick n^+-Ga$_{0.47}$In$_{0.53}$As ($n \approx 3 \times 10^{17}$cm^{-3}) buffer layer grown on an n^+-InP substrate. On top of the buffer layer the RT DB is grown; this consists of an undoped 50-Å-wide Ga$_{0.47}$In$_{0.53}$As quantum well sandwiched between two 50-Å-wide undoped Al$_{0.48}$In$_{0.52}$As barriers. The growth ends with a 1-μm-thick Ga$_{0.47}$In$_{0.53}$As cap layer doped to $n^+ \approx 3 \times 10^{17}$cm^{-3}. The structures are etched into 50-μm-

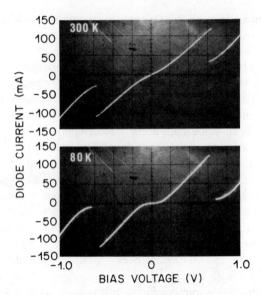

Fig. 1   Typical current-voltage characteristics of the $Al_{0.48}In_{0.52}As/Ga_{0.47}In_{0.53}As$ resonant tunneling diode at 300 K (Top) and 80 K (Bottom). Positive polarity refers to the top contact being positively biased with respect to the bottom.

diameter mesas. Figure 1 shows the current-voltage (I-V) characteristics of the diodes in both polarities measured at room temperature (top) and at 80 K (bottom). Positive polarity refers to the top contact being positively biased with respect to the bottom. The room-temperature characteristic indicates a peak-to-valley ratio of 4:1 in one polarity and 3.5:1 in the other. At low temperature (80 K), the peak-to-valley ratio increases to 15:1. It should be noted that, though the peak-to-valley ratio increases dramatically on cooling, the peak current remains the same. The peak in the I-V curve occurs at ≈ 600 mV and does not shift with temperature. An electron-tunneling transmission calculation shows that the first resonance is at $E_1 \approx 126$ meV from the bottom of the quantum well. Note that the peak in the I-V characteristic appears at a voltage greater than $2E_1/q \approx 252$ mV. This can be explained by considering the voltage circuit drop in the depletion and accumulation regions in the collector and emitter layers adjacent to the DB. Thus, a larger voltage must be applied across the entire structure to line up the first subband in the well with the bottom of the conduction band in the emitter to suppress RT. A simple calculation, taking the above effects into account, indicates that the peak should occur at ≈ 580 mV applied bias, which is in reasonable agreement with the measured value.[24]

The relatively large peak-to-valley ratio observed at room temperature makes this device suitable for many circuit applications. A circuit with a 30-Ω load resistance in series with the device and a 3.0-V supply has two stable operating points, which are measured to be 0.47 and 0.85 V, respectively, at room temperature. The corresponding load line drawn on the room-temperature I-V characteristics indicates the stable operating points at 0.46 and 0.84 V, respectively, in close agreement with the measured values. The circuit can thus be used as a static random-access memory (SRAM) cell involving only one device. Such a SRAM cell is also suitable for integration in a large memory array, as discussed in Section 8.5.3 in connection with multi-state memory.

8.2.2 Resonant Tunneling through Parabolic Quantum Wells

Parabolic QWs have interesting possibilities for device applications, because, unlike rectangular wells, the levels in such wells are equally spaced. The I-V characteristics of RT structures with parabolic wells, therefore, are expected to produce nearly equally spaced peaks in voltage. Such resonances have been observed experimentally.[25]

RT samples with parabolic quantum wells were grown by MBE on n^+ (100) GaAs substrates. Parabolically graded well compositions were produced by growth of short-period (≈ 10 Å), variable-duty-cycle, GaAs/Al_xGa_{1-x}As superlattices in which the Al content within each period of the superlattice corresponded to the Al content at the same point in a smooth parabolic well. A cross-sectional transmission electron micrograph (TEM) of one such structure is shown in Fig. 2. The structure consists of a 439-Å parabolic QW of

Fig. 2 A cross-sectional transmission electron micrograph (TEM) of a 439-Å-wide parabolic quantum well composed of Al_xGa_{1-x}As, with x varying from 0.3 at the edges to 0 at the center, sandwiched between two 34 Å AlAs barriers.

$Al_xGa_{1-x}As$, with x varying from 0.3 at the edges to 0 at the center, sandwiched between two 34-Å AlAs barriers. The parabolic part of the structure is composed of variable-gap superlattice with a period of nearly 10 Å, as discussed above. The brighter lines in the well part of the TEM picture represent $Al_{0.3}Ga_{0.7}As$ layers, while the darker lines represent GaAs layers. Notice how the relative widths of the bright and dark lines change from the edges of the well to its center. The electrons, of course, "sense" the local average composition, since their de Broglie wavelength is much greater than the superlattice period. $Al_{0.02}Ga_{0.98}As$ 1000-Å-thick layers Si-doped to $5 \times 10^{17}cm^{-3}$ (with a doping offset of 50 Å from the barriers) were used as contact regions to the RT DB. The composition of these layers was chosen in such a way that the bottom of the conduction band in the emitter is nearly lined up with (but always below) the first energy level of the well. These layers are followed by a 1000-Å region compositionally graded from $x = 0.02$ to $x = 0$ Si-doped ($n = 5 \times 10^{17}cm^{-3}$), and by 4000-Å-thick Si-doped ($n = 1 \times 10^{18}cm^{-3}$) GaAs.

The energy-band diagrams at the Γ point are shown in Fig. 3 for different bias voltages. Figure 4 shows the I-V characteristics and corresponding conductance for this sample for both bias polarities. It is interesting that the group of resonances from the fifth to the eleventh are the most pronounced ones and actually display negative differential resistance. Fourteen resonances are observed in this sample for positive polarity. In a few diodes two additional resonances were also observed. The resonances were observed up to temperatures ≈ 100 K, but were considerably less pronounced. The vertical segments near the horizontal axis indicate the calculated positions of the transmission peaks obtained by the tunneling resonance method, with the inclusion of the voltage drops across the accumulation and depletion layers adjacent to the DB. Overall good agreement with the observed minima in the conductance is found.

The overall features of the I-V characteristics can be interpreted physically by means of the band diagrams of Fig. 3 and the calculations described above. At zero bias the first six energy levels of the well are confined by a parabolic well 225 meV deep, corresponding to the grading from $x = 0$ to $x = 0.30$, and their spacing is ≈ 35 meV. When the bias is increased from 0 to 0.3 V the first four energy levels probed by RT (Fig. 3b) remain confined by the parabolic portions of the well, and their spacing is practically independent of bias. This effect is easily understood if one considers that the application of a uniform electric field to a parabolic well preserves the parabolic curvature and, therefore, the energy, level spacing. This gives rise to the calculated and observed equal spacing of the first four resonances in the I-V characteristic (Fig. 4). Consider now the higher energy levels confined by the rectangular part of the well (> 230 meV) at zero bias. When the voltage is raised above 0.3 V, these levels become increasingly confined on the emitter side by the parabolic portion of the well and on the collector side by a rectangular barrier, thus becoming progressively more separated, although retaining the nearly equal spacing (Fig. 3c). This leads to the observed gradual increase in the voltage separation of the resonances as the bias is increased from 0.3 to 1.0 V. Above 1 V the electrons

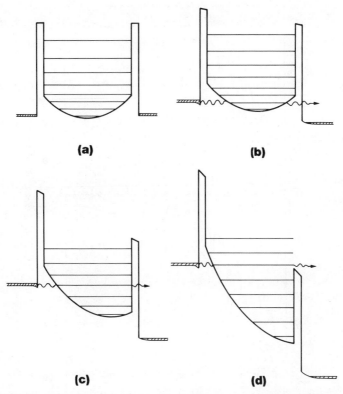

Fig. 3   Band diagrams of sample of Fig. 2 in equilibrium and under different bias conditions. The wells are drawn to scale; however, for clarity, only half the number of levels in an energy interval are shown.

injected from the emitter probe the virtual levels in the quasi-continuum above the collector barrier (Fig. 3d). These resonances result from electron interference effects associated with multiple quantum-mechanical reflections at the well-barrier interface for energies above the barrier height. It should be noted that these reflections give rise to the existence of two-dimensional (2D) quasi-eigenstates in the well region. The observed NDR is due to tunneling into these states. These interference effects produce the four resonances observed above 1 V and must be clearly distinguished from the ones occurring at lower voltages, which are due to RT through the DB.

A simple physical explanation of why the resonances above the fourth (fifth to twelfth) are the most pronounced ones, leading to negative differential resistance, can be given easily in terms of the calculated voltage dependence of the transmission. Up to 0.3 V, tunneling out of the well (Fig. 3b) occurs through the thick parabolic part of the collector barrier, and the resulting widths of the transmission resonances are very small ($< 1 \ \mu eV$). As the bias is increased above 0.3 V, not only is the barrier height further reduced, but now electrons

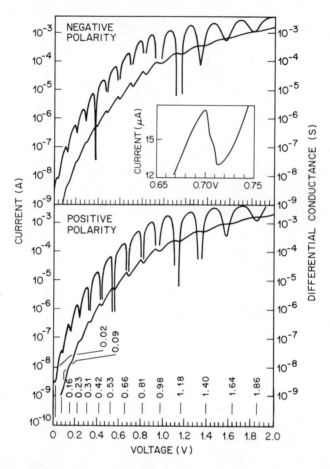

Fig. 4 Current-voltage characteristic at 7.1 K and conductance for a parabolic-well RT diode under opposite bias polarity conditions. The inset shows the eighth resonance on a linear scale. The vertical segments near the horizontal axis indicate the calculated positions of the resonances.

tunnel out of the well through the thin (20 Å) rectangular part of the barriers. This greatly enhances the barrier transmission and the resonance widths (Fig. 3c). This behavior is clearly observed in the calculation of the total transmission versus bias. For example, the calculated energy width of the level corresponding to the tenth resonance (0.9 V) is 1 meV, which is not negligible compared to the width of the incident energy distribution in the emitter. As the bias voltage is increased further, the competing effect of the decrease of the peak-to-valley ratio becomes dominant, as shown by the calculations. This explains why the highest resonances (above 1 V) become less pronounced.

8.2.3   Resonant-Tunneling Electron Spectroscopy

In this section we discuss an interesting application of RT through DBs. An RT DB can be used to analyze the energy distribution of hot electrons.[26] Compared with conventional hot-electron spectroscopy,[27] this resonant-tunneling spectroscopy technique has the advantage of not requiring derivative methods.

Figure 5 illustrates the energy-band diagram of the structures that can be used for resonant-tunneling electron spectroscopy. The first one, Fig. 5a, consists of a reverse-biased *p-i-n* heterojunction and can be used to investigate hot minority-carrier transport. Low-intensity incident light is strongly absorbed in the wide-gap p^+ layer. Photo-generated minority-carrier electrons diffuse to an adjacent low-gap layer. Upon entering this region, electrons are accelerated ballistically by the abrupt potential step and gain a kinetic energy $= \Delta E_c$ and a forward momentum $p_\perp = \sqrt{2m_e^* \Delta E_c}$. Collisions in the low-gap layer tend to randomize the injected, nearly mono-energetic distribution. Hot electrons subse-

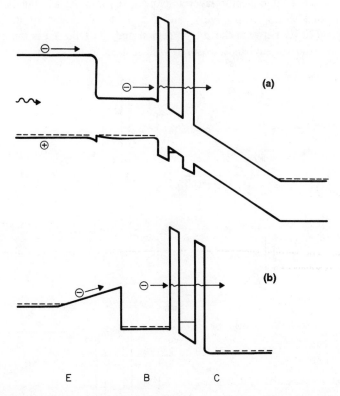

Fig. 5 (a) Band diagram of heterojunction diode used for resonant-tunneling spectroscopy of hot minority-carrier electrons. By measuring the photocurrent as a function of the reverse bias, the hot-electron energy distribution $n(E_\perp)$ can be directly probed. (b) Unipolar transistor structure for resonant tunneling spectroscopy of hot majority-carrier electrons in the n^+ base layer.

quently impinge on the DB in the collector. From simple considerations of energy and lateral-momentum $p_{||}$ conservation in the tunneling process, it can be shown that only those electrons with a perpendicular energy $E_\perp$ ($p_\perp^2/2m_e^*$ for a parabolic band) equal (within the resonance width) to the energy of the bottom of one of the subbands of the quantum well tunnel resonantly through the quantum well and give rise to a current. Thus, by varying the applied bias (i.e., changing the energy difference between the resonance of the quantum well and the bottom of the conduction band in the low-gap p^+ layer) and measuring the current, one directly probes the electron energy distribution $n(E_\perp)$ or, equivalently, the momentum distribution $n(p_\perp)$ (Fig. 6). One has, therefore,

$$E_\perp = E_n - \frac{q(V + V_{bi})(L_B + L_w/2 + L_{sp})}{L_c}, \tag{1}$$

where V is the reverse-bias voltage, V_{bi} is the built-in potential of the p-i-n diode, L_c is the total collector layer thickness, L_B and L_w are the barrier and well layer thicknesses, respectively, and L_{sp} is the thickness of the undoped spacer layer (20 Å) between the p-type region and the DB. E_n is the energy of the bottom of the nth subband measured with respect to the bottom of the center

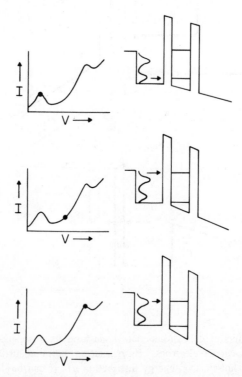

Fig. 6 Schematics of resonant-tunneling spectroscopy. By measuring the current as a function of reverse bias (left), one directly probes the energy distribution perpendicular to the layers $n(E_\perp)$ (right).

of the well. (Note that E_n is assumed to be independent of the electric field **E**, which is a good approximation as long as E_n is significantly greater than qEL_w). Identical arguments apply to the case of the unipolar transistor structure of Fig. 5b, which can be used to analyze the electron distribution in the base layer by measuring the collector current as a function of the collector-base voltage. In these arguments we have assumed that thermionic currents over the DB can be minimized. This can be done by operating the structure at sufficiently low temperature and by suitable design of the DB. To obtain the actual energy distribution $n(E_\perp)$ from the current, the latter must be properly normalized by taking into account the field dependence of the resonant-tunneling probability (integrated over the resonance width). This procedure does not alter, of course, the position of the peaks in the current-voltage characteristic, because this probability varies monotonically with the electric field, irrespective of whether electrons resonantly tunnel sequentially or coherently through the DB. Thus, the main features of the electronic transport still can be obtained directly from the current, without normalization.

The device of Fig. 5a was used to measure the hot-electron distribution of electrons launched in $p^+(\approx 3 \times 10^{18}\text{cm}^{-3})$ GaAs, of thickness ranging from 250 to 500 Å, with injection energies of ≈ 0.2 eV. It was found that the resulting energy distribution is non-Maxwellian and consists of two parts, one strongly relaxed near the bottom of the band and the other much hotter at ≈ 0.1 eV energy (Fig. 6). In this and in any other hot-electron spectroscopic technique, it is important to keep in mind that the measured energy distribution is to some extent affected by the analyzer barrier, whose height varies with applied voltage. This is particularly true for transport in thin layers (thickness $\approx$ mean free path).

8.3 RESONANT-TUNNELING BIPOLAR TRANSISTORS (RTBTS) WITH DOUBLE BARRIER IN THE BASE

The concept of an RTBT originated with the general idea of associating a logic level[14] to each state of a quantum system, for example, the energy levels of a quantum well.

This general scheme leads naturally to the idea of multiple-valued logic. Although such logic has been the subject of considerable investigations,[28] all circuits employing two-state devices are complex and cumbersome. The above correspondence (energy levels/logic states) led to the conception of a class of bipolar devices with inherent multistate operation.[14]

Figure 7 illustrates this type of device. It should be noted that, although these transistors utilize non-equilibrium injection, the underlying operating principle is at the basis of the operation of all other RTBTs with the quantum well in the base irrespective of the details of the base contact. As the base-emitter voltage is increased, RT through each subband first reaches a maximum and is then quenched as the bottom of each QW subband is lowered below the conduc-

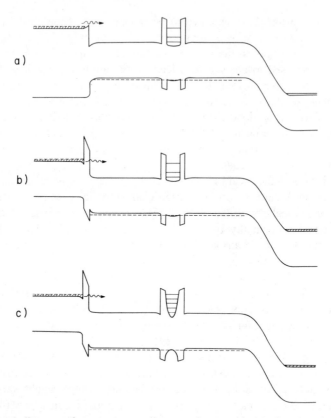

Fig. 7 Band diagram of resonant-tunneling transistors with (a) abrupt emitter and (b), (c) tunneling emitter.

tion band edge in the emitter. This produces multiple peaks in the collector current, that is, multiple negative transconductance. The tunnel emitter device with a parabolic well in the base can be used to generate equally spaced peaks. For example, using a well of width 200 Å with $Al_{0.45}Ga_{0.55}As$ barrier, one finds an energy-level separation $\approx$ 64 meV. This gives a total of five states in the well.

8.3.1 Design Considerations for RTBTs with Ballistic Injection

The design of the transistors of Fig. 7 is critical. This is due to various requirements that must be satisfied simultaneously in order to achieve acceptable current densities ($\geq 10^4 A/cm^2$), current gains ($\geq$ 10), and peak-to-valley ratios (>2:1). We assume first that the RT through the DB is coherent; this can be achieved by designing the DB so that $\hbar/\Gamma \ll \tau_\Phi$, where Γ is the resonance width at half maximum and τ_Φ is the phase relaxation time.[23] An estimate of τ_Φ can be obtained from the reciprocal of the *total* scattering rate $1/\tau_T$ (inelastic + elastic) at the energy of the incident particle.[23] As an example, in GaAs at a concentration $p = 5 \times 10^{18} cm^{-3}$, $1/\tau_T$ is in the range 2×10^{13} to

2.5×10^{13} s^{-1} for injection energies in the range $0.1 - 0.3$ eV.[29] Thus to ensure that the above condition for coherent transport is satisfied, Γ should be larger than ≈ 10 meV. This can be achieved with the ground-state resonance ($E_1 = 133$ meV) of an $Al_{0.40}Ga_{0.60}As$ (15 Å)/GaAs (30 Å) DB, for which tunneling resonance calculations show $\Gamma \approx 64$ meV. The coherence of the RT process and the lateral momentum conservation during tunneling ensure that, in symmetric DBs, incident electrons with a perpendicular energy $E_\perp$ equal to the bottom of one of the subbands of the well traverse the DB with unity transmission. However, in any experimental situation the incident perpendicular-energy distribution $n_\perp(E_\perp)$ (defined as the number of electrons per unit volume and unit energy interval) has a finite width ΔE; in order to exploit coherent RT and achieve a high base transport factor, ΔE must be smaller than Γ. If $\Delta E >> \Gamma$ (a situation commonly encountered in DB diodes), only a small fraction of the incident electrons $\sim \Gamma/\Delta E$ contributes to the RT current J_R. J_R is then approximately given by

$$J_R = qv_Rn_\perp(E_R)\Gamma T_R, \tag{2}$$

where $T_R \approx 1$, E_R is the transmission resonance energy, and v_R is the perpendicular component of the velocity corresponding to $E_\perp = E_R$. Since $\Gamma \approx E_R T_B$, where T_B is the transmission of the individual barriers (usually $<< 1$), Eq. 2 shows that, for a broad incident distribution, it is the transmission of the individual barriers and not the overall transmission of the DB that determines the current.[30] Therefore, to maximize J_R, ΔE must be smaller than Γ. To achieve this, the energy distribution in the emitter should be narrower than the resonance width and electrons should traverse the distance between the DB and the emitter quasi-ballistically. The width of the emitter energy distribution perpendicular to the barrier is approximately $k_BT + E_F$, where k_BT is the thermal energy and E_F is the quasi-Fermi energy in the emitter, which is comparable to the equilibrium Fermi level. Consider first the structure of Fig. 7a. The emitter composition should be chosen in order to have $\Delta E_c \approx E_R$, so that, under resonance conditions, the conduction band in the emitter is nearly flat to maximize the peak collector current. If ΔE_c is significantly smaller than E_R, the base-emitter junction must be biased beyond flat band to achieve resonance. In this case, the electric field in the emitter will heat the injected distribution and broaden it; an unwanted effect in light of the above discussion. On the other hand, if ΔE_c is significantly larger than E_R then at resonance the emitter current will not be large enough, because of the residual base-emitter barrier. Consider, for example, the $Al_{0.4}Ga_{0.6}As$/GaAs DB previously discussed. The emitter composition should be chosen to be approximately $Al_{0.20}Ga_{0.80}As$, which corresponds to $\Delta E_c \approx E_1 = 133$ meV. For an emitter doping density of 5×10^{17}cm^{-3} the width of the distribution ballistically launched with an energy $E_\perp = \Delta E_c$ by the abrupt emitter is ≈ 50 meV, which is close to the resonance width in the DB. One must, however, also consider the effects of scattering in the region between the DB and the emitter, which broadens the distribution. In order to achieve a high peak-to-valley ratio, scattering in this region must be minimized. Electrons are launched by the emitter with a forward velocity $(2\Delta E_c/m^*)^{1/2}$ $\approx 8 \times 10^7$ cm/s, limited by the band structure. Since the scattering rate at the injection energy is $\approx 2 \times 10^{13}s^{-1}$ (for $p = 5 \times 10^{18}$ cm^{-3} in the base), the mean free path for these electrons is $\lambda \approx 400$ Å.[29] If the distance between the DB and the emitter (L) is kept ≈ 300 Å, this implies that only half of the carriers ($\approx \exp(-L/\lambda)$) traverse this distance without collisions. Electrons that have lost a portion ($\geq$ optical phonon energy, ≈ 35 meV, for the Al$_{0.4}$Ga$_{0.6}$As (15 Å)/GaAs (30 Å) DB case under consideration) of their energy normal to the DB as a result of these collisions will see a significantly reduced transmission through the DB or "miss" the first resonance altogether if the width of the latter is too small. One way to reduce the scattering rate considerably is to dope extremely heavily ($> 10^{20}$ cm^{-3}) the region between the DB and the emitter. Recent theoretical work has shown that the inelastic scattering rate of minority carrier electrons in p-type GaAs first increases with increasing doping and rapidly decreases for doping levels well above 10^{19} cm^{-3}, due to the decreased phase space available for scattering.[29] These levels can be achieved by carbon doping,[31] which has also the advantage of a small diffusion coefficient. At the same time, elastic scattering, which increases rapidly with increasing doping, can be strongly suppressed by placing impurities in a periodic sublattice by delta doping techniques.[32] Scattering-rate calculations show that at injection energies of ≈ 0.15 eV and for $p = 2 \times 10^{20}$ cm^{-3}, mean free paths as long as 1500 Å can then be achieved in GaAs.[29] In this way one can minimize scattering in the region between the DB and the emitter. These considerations, of course, also apply to the structures of Figs. 7b and 7c.

The above discussion clearly demonstrates that the design of an RTBT with quasi-ballistic injection is an extremely difficult task. This may be somewhat easier to do in AlInAs/GaInAs and InP/GaInAs systems in light of the larger mean free path of electrons in p-type Ga$_{0.47}$In$_{0.53}$As (approximately twice that of GaAs at the same injection energy) and the smaller effective masses in the barrier and well layers. The preferred structure in this case would have a tunnel emitter of the type shown in Fig. 7b and 7c and would consist of a Ga$_{0.47}$In$_{0.53}$As layer followed by an InP or an Al$_{0.48}$In$_{0.52}$As tunnel barrier. Tunnel emitters allow one to bias the base emitter junction well beyond flat band while still maintaining a narrow incident energy distribution.

8.3.2 Observation of Quasi-Ballistic Resonant Tunneling in a Tunneling Emitter RTBT

In this section we present results on the RTBT of Fig. 7b, fabricated in the AlInAs/GaInAs system. The collector layer is 3000-Å-thick undoped Ga$_{0.47}$In$_{0.53}$As. The base layer comprises a 600-Å region on the emitter side (doped to $p = 3 \times 10^{18}$ cm^{-3}) and a 2000-Å region, doped to $p = 5 \times 10^{18}$ cm^{-3}, on the collector side, separated by an undoped Al$_{0.48}$In$_{0.52}$As (50 Å)/Ga$_{0.47}$In$_{0.53}$As (100 Å) DB. The emitter consists of an Al$_{0.48}$In$_{0.52}$As (30 Å) tunnel barrier separated from the base by a 50-Å undoped space layer and followed by an $n = 1 \times 10^{18}$cm^{-3} 3000-Å-thick Ga$_{0.47}$In$_{0.53}$As layer. The device transfer characteristics in the common-base configuration is shown in Fig. 8 at $V_{CB} = 1.0$ V, at cryogenic temperature. The collector current rises rapidly above the built-in voltage and peaks at ≈ 1.25 V. This value equals the calculated voltage required to line up the bottom of the first quantized subband in the accumulation layer on the emitter side of the tunnel barrier with the second resonance of the well ($E_2 = 193$ meV). Thus, this peak corresponds to ballistic RT of electrons injected from the emitter in the first excited state of the QW. Note that the negative transconductance region after the peak is broad (≈ 0.2 eV) and the peak-to-valley ratio is small. These features can be understood as follows. The mean free path in p^+-InGaAs for electrons with kinetic energies of the order of 100 meV can be estimated to be about 500 Å.[33] A large fraction of the electron distribution incident on the DB is therefore non-ballistic owing to scattering in the region between the emitter and the DB. The dominant scattering mechanisms for electrons in p^+-InGaAs doped to densities $> 10^{18}$ cm^{-3} are inelastic collisions with holes and coupled phonon-plasmon modes.[33] Assume now that the emitter-base junction is biased beyond the second resonance of the quantum well so that the electron injection energies exceed E_2. The injected electrons that approach the DB ballistically have their energy associated mostly with perpendicular motion so that $E_\perp > E_2$. These electrons therefore cannot tunnel resonantly into the QW. On the other hand, the part of the electron distribution that evolves following scattering has electrons with reduced $E_\perp$. This part contains electrons with $E_\perp$ equal to E_1 and E_2, so that they are still able to tunnel resonantly into the QW when incident on the DB. This provides an increasing background to the collector current as V_{EB} increases. We therefore have a rather broad peak region and a small peak-to-

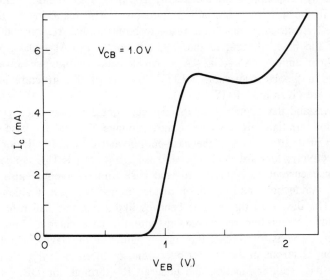

Fig. 8 Transfer characteristics of the resonant-tunneling transistor of Fig. 7b at 10 K.

valley ratio in Fig. 8. Since the DB region in our device has no intentional dopant impurities, the possible effects of elastic scattering centers in the QW on the peak-to-valley ratio[34] are relatively unimportant in this case.

It is interesting that we have not observed a peak corresponding to the first resonance of the well ($E_1 = 48$ meV). This is primarily because the peak current associated with the first resonance is smaller than the one through the resonance by the ratio (≈ 10) of the transmission coefficients of the individual barrier at the two resonant energies (see Eq. 2), so that the first peak is masked by the rapidly rising emitter current for $V_{EB} > V_{bi}$; in addition, the effects of scattering on the incident distribution will reduce the peak-to-valley ratio and broaden the peak, as previously discussed.

8.3.3 Thermionic Injection RTBTs Operating at Room Temperature

The first operating RTBT was demonstrated in 1986 and was designed to have minority electrons thermally injected into the DB.[18] This made the design of the device much less critical and the structure implemented in the AlGaAs system operated at room temperature. The band diagram of this transistor is shown in Figs. 9a and 9b under operating conditions. The alloy composition of the region adjacent to the emitter is adjusted in such a way that the conduction band in this region lines up with or is slightly below the bottom of the ground-state subband of the QW. For the 74-Å well with 21.5-Å AlAs barriers the first quantized energy level is $E_1 = 65$ meV. The Al mole fraction was chosen to be $x = 0.07$ (corresponding to $E_g = 1.521$ eV) so that $\Delta E_c \approx E_1$. The QW is undoped; nevertheless, it is easy to show that there is a high concentration ($\approx 7 \times 10^{11} \text{cm}^{-2}$) two-dimensional hole gas in the well. These holes have transferred by tunneling from the nearby $Al_{0.07}Ga_{0.93}As$ region. This reduces scattering in the well by essentially eliminating elastic scattering by the doping impurities.[34] Electrical contact was made to both the well region and the GaAs portion of the base adjacent to the DB, but not to the $Al_{0.07}Ga_{0.93}As$ region. The wide-gap emitter ($Al_{0.25}Ga_{0.75}As$) provides the well-known advantages of heterojunction bipolar transistors (HBT). Details of the structure and of the processing are given in Ref. 18.

To understand the operation of the device, consider a common emitter bias configuration. Initially the collector-emitter voltage V_{CE} and the base current I_B are chosen in such a way that the base-emitter and the base-collector junctions arer, respectively, forward- and reverse-biased, Fig. 9a. If V_{CE} is kept constant and the base current I_B is increased, the base-emitter potential also increases until a flat conduction-band condition in the emitter-base p-n junction region is reached. The device in this regime behaves like a conventional transistor with the collector current linearly increasing with the base current, Fig. 9c. The slope of this curve is, of course, the current gain β of the device. In this region of operation, electrons in the emitter overcome, by thermionic injection, the barrier of the base-emitter junction and undergo RT through the DB. If the base current is further increased above the value corresponding to the flat-band con-

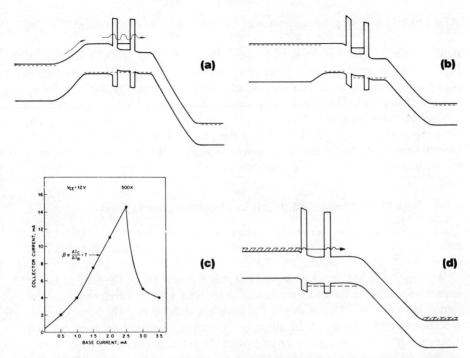

Fig. 9 Energy-band diagrams of the RTBT with thermal injection for different base currents I_B at a fixed collector emitter voltage V_{CE} (not to scale). As I_B is increased, the device first behaves as a conventional bipolar transistor with current gain (a), until nearly flat-band conditions in the emitter are achieved. For $I_B > I_{Bth}$ a potential difference develops across the AlAs barrier between the contacted and uncontacted regions of the base. This raises the conduction band edge in the emitter above the first resonance of the well, thus quenching resonant tunneling and the collector current (b). The collector current versus base current in the common emitter configuration, at room temperature, is shown in (c). The line connecting the data points is drawn only to guide the eye. An alternative RTBT design is shown in (d).

dition, the additional potential difference drops primarily across the first semi-insulating AlAs barrier (Fig. 9b), between the contacted and uncontacted portions of the base, since the highly doped emitter is now fully conducting. This pushes the conduction band edge in the $Al_{0.07}Ga_{0.93}As$ above the first energy level of the well, thus quenching the RT. The net effect is that the base transport factor and the current gain are greatly reduced. This causes an abrupt drop of the collector current as the base current exceeds the threshold value (Fig. 9c). Thus, the device has negative transconductance.

It should be clear that this device is not equivalent to a series combination of an RT diode and a bipolar transistor since electrical contact is made to the QW. The base-emitter voltage, therefore, directly modulates the energy difference between the states of the well and the emitter quasi-Fermi level. Recently an

RTBT based on this operating principle, but with the base layer restricted to the GaAs quantum well,[15] has been reported.[35] An RTBT with the DB between the base and the emitter has been made that exhibits negative transconductance at liquid nitrogen temperature.[17] Since the quantum well is not contacted and is placed out of the base, this device, unlike the ones of Refs. 18 and 35, can be thought of as a monolithic series integration of a DB and a bipolar transistor.

Several alternative RTBT designs are possible; one is shown in Fig. 9d. Here the p region between the DB and the emitter is eliminated. The well is heavily doped with low diffusion acceptors (e.g., C). The operating principle is the same as that of the device of Fig. 9a.

8.3.4 Speed and Threshold Uniformity Considerations in RTBTs

The insertion of a DB in an HBT structure offers new, interesting circuit opportunities but also raises questions concerning its effect on speed and threshold uniformity.

With regard to the speed issue, the introduction of a DB in the base or in the emitter will increase the emitter-collector delay time τ_{ec} and therefore reduce the cutoff frequency f_T. This is due to the tunneling delay time, which, in general, is a complicated function of the shape of the incident perpendicular energy distribution. If the latter is much broader than the resonance width (Γ) and nearly centered on one of the resonances, it can be shown that the transit time across the RT structures (τ_T) is approximately given by[10]

$$\tau_T = \frac{d}{v_G} + \frac{2\hbar}{\Gamma}, \tag{3}$$

where d is the width of the RT structure and v_G is the electron group velocity. The first term represents the semi-classical transit time across the structure and is ≤ 0.1 ps for the RT DBs of interest here. The second term is the so-called phase time. In the RT transistor structures with potential practical impact (e.g., Fig. 9), the first resonance width is much smaller than the quasi-Fermi energy in the emitter, thus satisfying the first assumption underlying Eq. 3. The condition that the tunneling wave packet be nearly centered on the resonance is only partially valid, thus making Eq. 3 good for an approximate estimate of the delay time associated with RT. It is clear from this expression that, to minimize τ_T, the resonant width Γ, which depends exponentially on the barrier thickness, must be maximized. Consider an RTBT structure of the type previously discussed (Fig. 9). For a 17-Å AlAs barrier thickness and a 45-Å GaAs well, tunneling resonance calculations give $E_1 = 0.136$ eV for the first energy level and $2\hbar/\Gamma = 0.45$ ps.[10] The first term in Eq. 3 is 0.08 ps (assuming a drift velocity $\geq 10^7$ cm/s, since overshoot effects following injection in the DB are possible).[10] Thus, $\tau_T \approx 0.5$ ps. It is well known that AlGaAs/GaAs HBT without an RT DB and uniform composition in the base can achieve values of $f_T > 50$ GHz. The introduction of this DB in an HBT with $f_T = 50$ GHz will increase τ_{ec} by 0.5 ps, giving $f_T \approx 43$ GHz. This example shows that RTBTs with suitably designed DBs should have cutoff frequencies and overall speeds of response comparable with that of state-of-the-art HBTs. A $Ga_{0.47}In_{0.53}As$ HBTs with an $Al_{0.48}In_{0.52}As$ (44 Å)/$Ga_{0.47}In_{0.53}As$ (38 Å) DB in the emitter having an f_T of 12.5 GHz has recently been reported.[17] The microwave performance of multistate RTBTs will be discussed in the next section.

Concerning the threshold (V_{Bth}) uniformity, let us recall that a conventional HBT has excellent uniformity (a few mV) both on the same wafer and from wafer to wafer, since V_{Bth} is given by the base-emitter built-in voltage. The latter is proportional to the bandgap and weakly (logarithmically) dependent on doping. The introduction of a DB in an HBT will induce greater fluctuations in V_{Bth}. To estimate V_{Bth} consider the case of an RTBT with a DB in either the emitter or the base. The voltage position of the collector current peak (transistor fully on) is given approximately by $V_{Bth} + 2E_1/q$, where E_1 is the energy of the first resonance of the well. E_1 can fluctuate across a wafer primarily as a result of in-plane thickness fluctuations, ΔL. Problem 1 provides an upper limit for the fluctuations in E_1; the corresponding fluctuation in the peak position is

$$\Delta V_p = \Delta V_{Bth} + \frac{4E_1}{q}\frac{\Delta L}{L}. \tag{4}$$

Thickness fluctuations in state-of-the-art MBE material are of the order of one monolayer, that is, $\Delta L = 2.5$ Å. For an RTBT with the double barrier considered in this section, Eq. 4 gives $\Delta V_p = 30$ mV. One obtains $\Delta V_p = 21.5$ mV for an RTBT containing AlInAs (25 Å)/GaInAs (50 Å) DB. Values ≤ 100 mV are adequate for the circuits envisioned with this technology.

8.4   DEVICES WITH MULTIPLE-PEAK *I-V* CHARACTERISTICS AND MULTIPLE-STATE RTBTS

A simple approach to realize multiple-peak *I-V* characteristics is the integration of a number of RT diodes. In this method, a single resonance of different quantum wells is used to generate the multiple peaks. Hence, they occur at almost the same current level and exhibit similar peak-to-valley ratios as those required by the circuit applications that will be discussed. However, these devices do not have the gain and input-output isolation of three-terminal devices.

There are two different ways to integrate RT diodes to achieve this characteristic. One, is to integrate them horizontally[36,37] so that the diodes are in parallel in the equivalent circuit. The other, is to integrate them vertically[38,39] so that they are in series.

The rest of this section is devoted to RTBTs with multiple peaks in their transfer characteristics and their digital and analog circuit applications.

8.4.1   Horizontal Integration of RT Diodes

The equivalent circuit of the horizontally integrated RT diodes is shown in Fig. 10a. This scheme can obviously be extended to more than two RT diodes, in which case the resistance shown in Fig. 10a provides the useful function of a monolithically integrated voltage divider.[36]

In operation the substrate current—that is, the current through the terminal S—is measured as a function of positive bias applied between the terminals S and A (which is grounded) for different values of the potential difference V_{BA} applied between B and A. The current through S is the sum of the two currents flowing through the two RT diodes. For zero potential difference V_{BA}, the structure behaves as a conventional RT diode and the I-V curve displays only one peak (Fig. 10b, $V = 0$) that corresponds to the quenching of the RT through the two DBs simultaneously. With a positive voltage (V_{BA}) applied to B, RT through the diode B is quenched at a higher substrate bias than through diode A, leading to the presence of an additional peak at a higher voltage in the characteristics (Fig. 10b, $V = 0.5$ V, 1 V). Note that the position of this new peak moves toward higher voltage with increasing V_{BA} while the other remains fixed. The separation between the peaks is equal to the applied V_{BA}.

This structure was the first to demonstrate multiple-peak I-V characteristics and was used to implement many circuits with reduced complexity.[36] However, the requirement of an additional power supply for the bias V_{BA} limits its applicability. A variation of the same structure is to add a resistance in series with one of the diodes[37] so that the ohmic drop across the same serves to bias this diode at a different voltage with respect to the other. This eliminates the need for the additional bias supply.

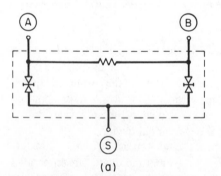

(a)

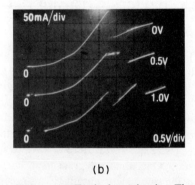

(b)

Fig. 10 Parallel integration of resonant tunneling diodes. (a) Equivalent circuit. The choice of the circuit symbol for the RT diode (two back-to-back tunnel diodes) is motivated by the symmetry of the current-voltage characteristic of the RT diode. (b) Current through terminal S versus positive V_{SA} at 100 K with positive potential difference (V_{BA}) between terminals B and A as the parameter. Terminal A is grounded.

8.4.2   Vertical Integration of RT Diodes

Vertical integration of RT structures is achieved by stacking a number of DBs in series, separated by heavily doped cladding layers to quantum mechanically decouple the adjacent DBs from each other.[38,39] The DBs are designed so that the ground state in the QW is substantially above the Fermi level in the adjacent cladding layers.   The band diagram of the structure under bias is shown in Fig. 11.   When bias is applied, the electric field is higher at the anode end of the device (Fig. 11a) because of charge accumulated in the QWs under bias. Quenching of RT is thus initiated across the DB adjacent to the anode and then propagates sequentially to the other end, as the high-field region widens with increasing applied voltage, as shown in Figs. 11a and 11b.   Once RT has been suppressed across a DB, the voltage drop across it quickly increases with bias because of the increased resistance.   The non-RT component through this DB provides continuity for the RT current through the other DBs on the cathode side.   An NDR region is obtained in the *I-V* characteristics, corresponding to the quenching of RT through each DB.   Thus, with *n* diodes, *n* peaks are present in the *I-V* curve.

The technique for generating multiple peaks by combining tunnel diodes in series is well known.[40] However, the mechanism in that arrangement is differ-

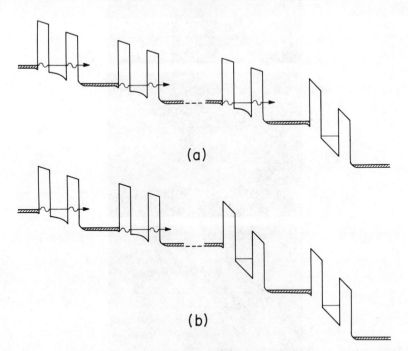

Fig. 11   Vertical Integration of resonant-tunneling diodes.   Band diagram under applied bias (a) with RT quenched through the DB adjacent to the anode and (b) after expansion of the high-field region to the adjacent DB with increasing bias. The arrows indicate the RT component of the current.

ent. The tunnel diodes used in such a combination must have different characteristics with successively increasing peak currents, so that each of them can go into the NDR region only when its corresponding current level is reached.[40] Besides, structures using RT diodes have significant advantages over ones using tunnel diodes.[36]

We have tested devices consisting of two, three, and five RT $Al_{0.48}In_{0.52}As$ (50 Å)/$Ga_{0.47}In_{0.53}As$ (50 Å) DBs in series, separated by a 1000-Å-thick n^+-$Ga_{0.47}In_{0.53}As$ region. The resulting I-V characteristics taken in both polarities of the applied voltage at room temperature are shown in Figs. 12a, b, and c, respectively. Positive polarity here refers to the top of the mesa being biased positively with respect to the bottom. Note that in this polarity, the devices show two, three, and five peaks in the I-V characteristics, as expected. In the negative polarity, the third peak is not observed in the device with three DBs because of rapidly increasing background current. This is believed to be due to structural asymmetry unintentionally introduced during growth.

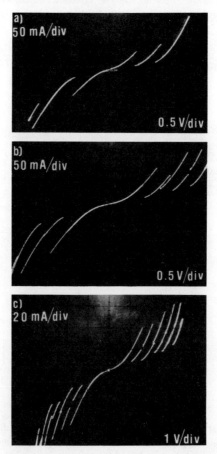

Fig. 12 Current-voltage characteristics of the devices with (a) two, (b) three and (c) five vertically integrated RT double barriers, taken for both bias polarities at 300 K.

8.4.3   Multi-State RTBTs

The stacked RT structure discussed above was used to design an RTBT exhibiting multiple NDR and negative transconductance characteristics.[21,22] A schematic of this transistor is shown in Fig. 13. The device essentially consists of a $Ga_{0.47}In_{0.53}As/Al_{0.48}In_{0.52}As$ *n-p-n* transistor with a stack of two $Ga_{0.47}In_{0.53}As$ (50 Å)/$Al_{0.48}In_{0.52}As$ (50 Å) RT DBs, as discussed before, embedded in the emitter. Details of the structure (doping and layer thicknesses) are given in Ref. 21. The operation of the transistor can be understood from the band diagrams in the common-emitter configuration shown in Fig. 14. The collector-emitter bias (V_{CE}) is kept fixed and the base-emitter voltage (V_{BE}) is increased. For V_{BE} smaller than the built-in voltage ($V_{bi} \approx 0.7$ eV at 300 K) of the $Ga_{0.47}In_{0.53}As$ *p-n* junction, most of the bias voltage falls across this junction (Fig. 14a), since its impedance is much greater than that of the two DBs in series, both of which are conducting via RT. The device in this region behaves as a conventional bipolar transistor with the emitter and the collector current increasing with V_{BE} (Fig. 15) until the base-emitter junction reaches the flat-band condition. Beyond flat-band, most of the additional increase in V_{BE} will fall across the DBs (Fig. 14b), and as RT through these quenches sequentially by the mechanism of Fig. 11, abrupt drops in the emitter and, hence, the collector current are observed (Fig. 15). The highest peak-to-valley ratio in the transfer characteristics at room temperature is 4:1, but increases to about 20:1 at 77 K.

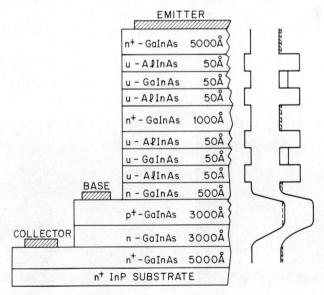

Fig. 13   Schematic structure of the multiple-state RTBT and its equilibrium conduction-band diagram.

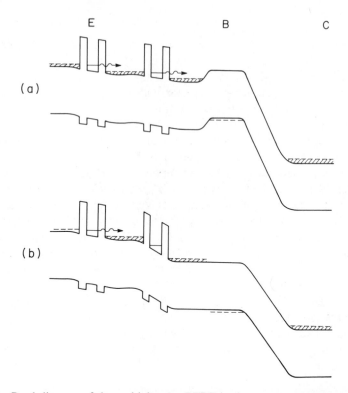

Fig. 14 Band diagram of the multiple-state RTBT in the common-emitter configuration for different base-emitter bias conditions. (a) Electrons resonantly tunnel through both DBs; in this regime the transistor operates as a conventional bipolar. (b) Quenching of RT through the DB adjacent to the *p-n* junction gives rise to a negative differential resistance region in the collector current. Quenching of RT through the other DB produces a second peak in the *I-V* curve.

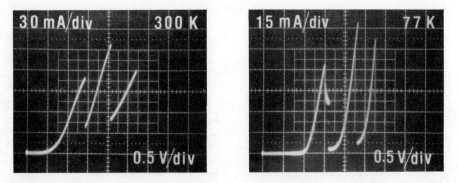

Fig. 15 Collector current versus base-emitter voltage in the common-emitter configuration for V_{CE} = 2.5 V at 300 K and 77 K.

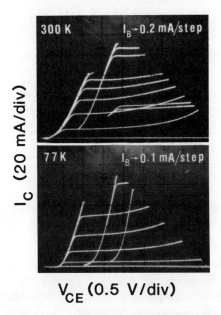

Fig. 16 Common-emitter output characteristics of the multiple-state RTBT. Collector current versus collector-emitter voltage for different base currents at 300 K (top) and 77 K (bottom).

Figure 16 shows the common-emitter output characteristics of the transistor (I_C versus V_{CE} at different I_B) at room temperature and 77 K. At low base currents I_B (and hence low base-emitter voltages, V_{BE}), the device behaves as a conventional bipolar transistor, as discussed before, with a large current gain (200 at 77 K and 70 at 300 K). With increasing I_B (V_{BE}) beyond the flat-band condition, the excess applied voltage V_{BE} starts appearing across the series of DBs in the emitter. As RT through them quenches sequentially, at threshold base currents I_{Bth1} and I_{Bth2}, the electron current across the base-emitter junction drops abruptly, while the hole current, flowing by thermionic emission, continues to increase. This results in sudden quenching of the current gain at these threshold base currents, consequently, the collector current I_C also quenches, giving rise to two NDR regions (Fig. 16). The highest peak-to-valley ratios observed are 6:1 at room temperature and 22:1 at 77 K. Note that the small-signal current gain of the transistor at room temperature in its second (1.2 mA $< I_B <$ 1.6 mA) and third ($I_B >$ 1.6 mA) operation regions are reduced to 40 and 20, respectively. This is expected, since the hole current flowing from the base toward the emitter increases with increasing V_{BE}, thus reducing the injection efficiency. This reduction of the current gain is less pronounced at 77 K, since the thermionic flow of holes is much lower at this temperature.

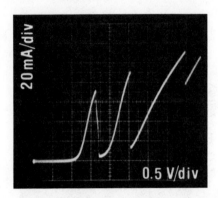

Fig. 17 Common-emitter transfer characteristic of a multiple-state RTBT with three DBs in the emitter at 77 K. I_C versus V_{BE} is shown for $V_{CE} = 4.75$ V.

Figure 17 shows the common-emitter transfer characteristic of a similar transistor with three DBs in the emitter at 77 K. The third peak is shifted out to a significantly higher voltage compared to the other two. Our systematic studies also indicated large hysteresis associated with it. Such behavior is not uncommon in RT devices whenever there is large parasitic resistance.[41] When three DBs are put in series, the parasitics also add up and enhance the effect. The structure with three peaks must be optimized to minimize these effects.

To minimize the flow of holes from the base to the emitter, an n^+-$Al_{0.48}In_{0.52}As$ layer can be inserted between the stack of DBs and the base, similar to the use of a wide-gap emitter in HBTs. Preliminary results with an $n = 1 \times 10^{18} cm^{-3}$ 500-Å AlInAs layer, followed by the growth of an equally thick and doped AlInGaAs grading layer before the DBs, give a β of 4000 at 77 K.

8.4.4 Microwave Performance of Multi-State RTBTs

In this section we discuss the high-frequency operation of multi-state RTBTs.[42] The device structure, grown lattice matched to an InP substrate by MBE, is very similar to the one discussed in Section 8.4.3. For microwave evaluation, the present structure was grown on a semi-insulating InP substrate instead of an n^+ substrate. Furthermore, the base layer thickness was reduced by a factor of two, down to 1500 Å, and the doping was doubled ($4 \times 10^{18} cm^{-3}$) to reduce the base transit time without increasing the base resistance.

The emitter-up transistor in a mesa configuration was obtained by successive steps of photolithography and wet-chemical etching to expose base and collector layers. After planarization of the structures with dielectric deposition, a nonalloyed metallization (Ti/Au) was deposited for the contacts. The emitter area is 42 μm^2. The I-V characteristics of the devices are virtually identical to those of Figs. 15 and 16. The only difference is that the collector currents are considerably smaller due to the scaled-down area.

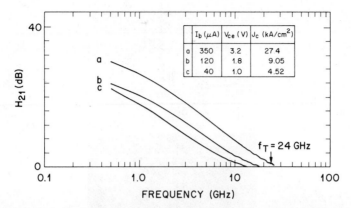

Fig. 18 Current gain (h_{21}) as a function of frequency for different bias points in the common-emitter configuration. The corresponding collector current density J_C is also indicated.

Scattering (S) parameter measurements were performed in the frequency range from 0.5 to 26.5 GHz with a wafer prober used in conjunction with an automatic network analyzer. Figure 18 displays the current gain (h_{21}) as a function of frequency for different bias conditions. Curve a refers to an operating point after the second peak in the common-emitter characteristics. The f_T obtained by extrapolation using a –20 dB/decade straight line is 24 GHz, which is the highest ever achieved in an RTBT. Previously, a comparable cutoff frequency was obtained in an RHET with a single collector current peak.[43] Curve b refers to a bias point between the two peaks. For curve c the base current $(40\ \mu A)$ is such that no NDR appears in the corresponding common-emitter characteristic.

8.5 CIRCUIT APPLICATIONS

8.5.1 Frequency Multiplier

The transfer characteristics of Fig. 15 were used to design the frequency-multiplier circuit shown in Fig. 19a. As the input voltage is increased, the collector current increases resulting in a decrease in the collector voltage until the device reaches the negative-transconductance regions, where sudden drops in the collector current and increases in the output voltage are observed. Under suitable bias (V_{BB}), such that the base-emitter junction is biased between the two peaks of the common-emitter transfer characteristic, triangular input waves will be multiplied by a factor of three and sine waves by a factor of five.[22] Unlike two-terminal multipliers, the output signal in this case is ground referenced and is also isolated from the input. These advantages are obtained in this circuit because the multiple peaks are present in the transfer characteristic of a transis-

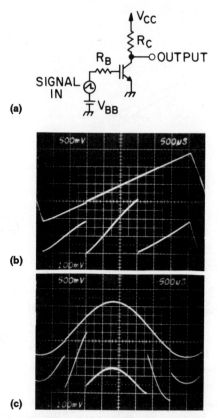

(a)

(b)

(c)

Fig. 19 Frequency multiplier using the multiple-state RTBT (a), the experimental results of multiplying sawtooth (b), and sine-wave (c) input signals at room temperature.

tor rather than the *I-V* curve of a two-terminal device as in Ref. 36. It should be noted that in applications where two-terminal devices cannot be used, conventional frequency-independent multipliers require the use of a phase-lock loop and a digital frequency divider.

The gain of the circuit is determined by the transconductance of the transistor and the collector resistance R_C. However, too large a value for R_C will lead to saturation of the device at large input voltages. The saturation can, of course, be avoided by the choice of a larger supply voltage V_{CC}, but the maximum usable V_{CC} is limited by the collector-base breakdown of the device. In our circuits we had $V_{CC} = 3.0$ V, $V_{BB} = 1.8$ V, $R_C = 5$ Ω and $R_B = 50$ Ω. Figure 19b and c shows the experimental results of multiplying the frequency of triangular and sine wave inputs, respectively. The polarity of the output signals (bottom traces) are inverted in the display for clarity of presentation.

For frequency multiplication at high frequencies, the devices described in Section 8.4.4 were biased in the common-emitter configuration with

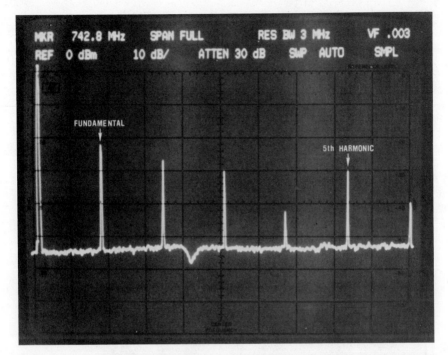

Fig. 20 Power output spectral response of the RTBT used as a frequency multiplier. Vertical scale is –10 dB/div. measured from the top horizontal line (0 dBm reference). The frequency span is 1.8 GHz (180 MHz/div.).

$V_{CE} = 3.2$ V and the characteristic impedance of the 50Ω line as the load. The base-emitter junction was dc biased at 2.0 V via a bias tee. A 350-MHz sine wave was applied to the base. The amplitude was adjusted to achieve a base-emitter voltage swing large enough to bring the device into the negative-transconductance regions of the transfer characteristic. The output power versus frequency was displayed on a spectrum analyzer (Fig. 20). Note that the amplitude of the fifth harmonic is much larger than that of the fourth and the sixth. The efficiency of the multiplier (power ratio of the fifth harmonic to the fundamental) is $\approx 15\%$, which is close to the maximum achievable (20%) in a resistive multiplier ($= 1/n$ where n is the harmonic under consideration).

8.5.2 Parity Generator

Figure 21a shows a four-bit parity generator circuit employing the previously discussed multiple-state RTBT.[44] The voltages of the four input bits of the digital word are added at the base node of the transistor by the resistive network to generate a step-like waveform. The quiescent bias of the transistor, adjusted by the resistance R_{B1}, and the values of the resistances R_0 are chosen to select the

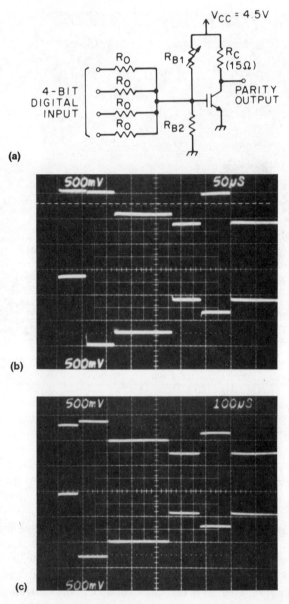

Fig. 21 (a) Four-bit parity generator circuit using an RTBT ($R_0 = 15$ kΩ, $R_{B1} = 6.9$ kΩ, $R_{B2} = 2.4$ kΩ, and $R_C = 15$ Ω). (b) Collector (top trace) and base (bottom trace) waveforms in parity generator circuit at 77 K. (c) Collector (top trace) and base (bottom trace) waveforms at 300 K.

operating points of the transistor alternately at low and high collector-current levels (i.e., valleys and peaks of the transfer characteristics) at the successive steps of the summed up voltage. In this circuit, $R_0 = 15\ k\Omega$, $R_{B1} \approx 6.9\ k\Omega$, $R_{B2} = 2.4\ k\Omega$, $R_C = 15\ \Omega$, and $V_{CC} = 4.5$ V. The output voltage at the collector would thus be high or low, depending on the number of input bits set high being even or odd, respectively. Thus for this four-bit parity generator, only one transistor is used, compared to 24 bits needed in an optimized conventional circuit using three exclusive ORs. Also note that the four-bit binary data are first converted to a multi-state signal that is then processed by the device. This is equivalent to processing all four bits in parallel, which results in improved speed compared with conventional sequential processing of binary logic. Such multistate processing elements thus show potential in replacing clusters of circuits in existing binary logic systems. Parity generators using horizontally[36] and vertically integrated[45] RT diodes were demonstrated before. The advantage of the present circuit is that a separate summing amplifier is not required, resulting in further reduction in complexity.

To test the circuit, a pseudo-random sequence of four-bit binary words was used rather than a monotonically increasing staircase waveform,[45] since the latter does not take into account the effect of hysteresis in the I-V characteristics. The train of input data produced both positive and negative steps at the base of the transistor. Experimental results at 77 and 300 K are shown in Figs. 21b and 21c, respectively, where the top traces show the output waveforms and the bottom traces the base waveforms of the transistor. Considering the dotted line in the upper trace as a logic threshold level, we find that the output is low for the second and the fourth voltage levels at the base, while it is high for the others. Also note that, at room temperature, the differential transconductance of the device decreases appreciably at higher voltages, making the design of the circuit more critical.

8.5.3 Multi-State Memory

A suitable load line drawn on an I-V characteristics curve with n peaks will intersect the curve at $n+1$ points in the positive slope part, as illustrated in Fig. 22 in the case of a curve with two peaks.[36] Thus, the circuit shown in the inset of Fig. 22 will have $n+1$ stable operating points and can be used as a memory element in $n+1$ state logic systems. Even in a binary computer, the storage system could be built around an $n+1$ logic to increase the packing density and the data converted to and from binary at the input/output interface. This scheme has been demonstrated with the horizontally integrated RT structures exhibiting two peaks in the I-V curve.[36] With a supply voltage $V_{SS} = 16$ V, load resistance $R_L = 215\ \Omega$ and the device biased to $V_{BA} = 0.7$ V, the three stable states were measured to be at 3.0, 3.6, and 4.3 V. The corresponding load line drawn on the measured characteristic of the device at $V_{BA} = 0.7$ V intersects at 2.8, 3.4, and 4.1 V, respectively, which are in close agreement with the measured values

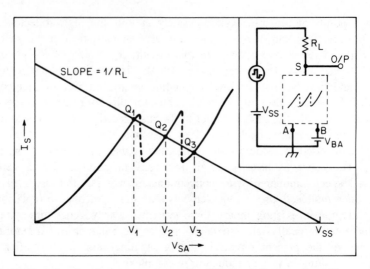

Fig. 22 Schematic of a 3-state memory cell. The load line on the *I-V* curve shows three stable operating points Q_1, Q_2, and Q_3.

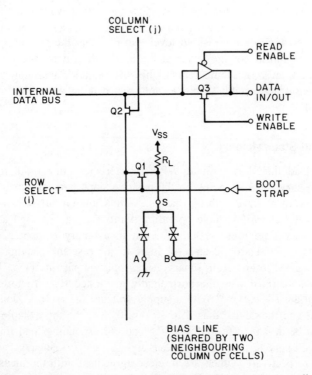

Fig. 23 Typical layout of an IC using the 3-state memory cells.

of the three stable operating points. Similar memory cells utilizing triple-well resonant tunneling diodes have been demonstrated.[46]

This three-state memory cell is also suitable for integration in memory integrated circuits (ICs) with Read/Write and Decoding network laid out as shown in Fig. 23. The memory cells are placed in a matrix array and a particular element in the array is addressed by activating the corresponding row and column select lines. A row select connects each device in that row to the corresponding column lines. The column select finally connects the selected column to the data bus. Consider the element (i,j) of the memory matrix shown in Fig. 23. When the row select line is activated, it turns on the driving switch Q_1. It also turns on the switches for every element in the ith row. The column select logic now connects the jth column only to the data bus. The ternary identity cell T acts as the buffer between the memory element and the external circuit for reading data. For reading data from the memory, the identity cell is activated with the Read Enable line, and data from the element number (i,j) in the matrix goes, via the data bus, to the In/Out pin of the IC. When the Write Enable line is activated, data from the external circuit is connected to the data bus and is subsequently forced on the (i,j)th element in the array and written there.

8.5.4 Analog-to-Digital Converter

Among other circuit applications of the multi-state RTBT, the analog-to-digital converter, briefly mentioned in Ref. 14 and shown in Fig. 24, is potentially the most significant. The analog input is simultaneously applied to an array of RTBT circuits having different voltage scaling networks. To understand the

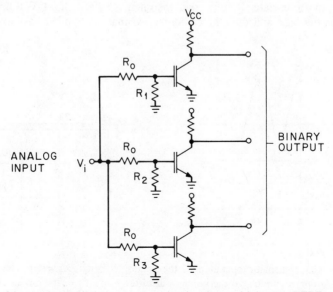

Fig. 24 Analog-to-digital converter circuit using multiple-state RTBTs.

operation of the circuit, consider the simplest system comprising only the two transistors Q_1 and Q_2. The voltages at different points of this circuit are shown in Fig. 25a for various input voltages V_i. Consider that the resistances R_0, R_1, and R_2 are so chosen that the base voltages V_{B1} and V_{B2} of the transistors Q_1 and Q_2 vary with V_i according to the curves V_{B1} and V_{B2}, respectively. With the input voltage at V_1, the output of both the transistors will be at the operating point P_1 (high state). With the input changing to V_2, the output of Q_1 will become low (P_2), while that of Q_2 will remain high (closer to P_1). Applying this logic to the input voltages V_3 and V_4, it can be easily shown that this circuit follows the truth table of Fig. 25b. The outputs of the RTBT array thus constitute a binary code representing the quantized analog input level. The system can be extended to more bits with larger numbers of peaks in the I-V curve. Note that this is a flash converter requiring only n transistors for n-bit conversion, as compared to 2^n analog comparators in conventional flash converters. Furthermore, the RTBTs not only work as the comparators, but also give the digital output directly, eliminating the 2^n-to-n bit decoder needed in conventional circuits. This further reduces the circuit complexity and enhances the speed of operation. However, it is very difficult to implement this circuit with present RTBTs. It should be noted that successful operation of the circuit relies on transfer characteristics where the current remains at a high or low level for a significant span of the base-emitter voltage. In the multi-state RTBTs implemented so far, the current gradually increases with the input voltage and then suddenly drops, followed by another gradual rise.

It should be mentioned that in all these circuits the minimum allowable collector voltage is determined by the maximum input-signal voltage applied at the base terminal, which is higher than that in a normal bipolar transistor. In fact, in the multiple-state RTBTs demonstrated so far, the QWs are positioned between the base and the emitter contact so that the applied base-to-emitter volt-

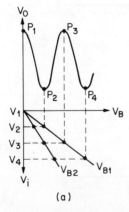

TRUTH TABLE

INPUT	OUTPUT	
V_i	Q_2	Q_1
V_1	1	1
V_2	1	0
V_3	0	1
V_4	0	0

(a)　　　　　　　(b)

Fig. 25 The schematic operation of the analog-to-digital converter circuit of Fig. 24, involving only two bits: (a) the voltages at different points of the circuit at various input voltages; (b) the truth table.

age is used to bias the emitter base *p-n* junction *and* the QWs. The base potential is then elevated to a relatively high value under operating conditions in the common emitter mode. As a result, the quiescent collector bias must be large to allow for sufficient output signal swing without forward biasing the collector-base junction. This requires proper care on the part of the circuit designer and careful device design so as to achieve a sufficiently high breakdown voltage.

8.6   UNIPOLAR RESONANT TUNNELING TRANSISTORS

Several unipolar three-terminal devices have been proposed and implemented that utilize the RT structure as electron injectors to generate voltage-tunable NDR and negative-transconductance characteristics.

The schematic band diagram of the resonant-tunneling hot-electron transistor (RHET) is shown in Fig. 26.[17] The structure consisted of an RT double barrier placed between GaAs base and emitter layers. The RT structure was made of a 56-Å-thick GaAs quantum well sandwiched between two 50-Å-thick $Al_{0.33}Ga_{0.67}As$ barriers. The RT double barrier between the base and the emitter

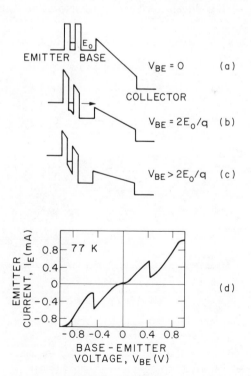

Fig. 26 The band diagrams of the RHET at (a) $V_{BE} = 0$, (b) $V_{BE} = 2E_0/q$ (maximum RT current), (c) $V_{BE} > 2E_0/q$ (RT quenched), illustrating the operating principle of the device, and (d) base-emitter current-voltage characteristic measured at 77 K.

simply served the purpose of injecting, through RT, high-energy electrons into the base region. The high-energy electrons are transported ballistically through the 1000-Å-thick n^+ base region before being collected at the 3000-Å-thick $Al_{0.20}Ga_{0.80}As$ collector barrier. The barriers and the quantum well were undoped, whereas the emitter, base, and the collector layers were n-type, doped to $1 \times 10^{18} cm^{-3}$.

The operation of the device in the common-emitter configuration with a fixed collector-emitter voltage V_{CE} is schematically shown in the band diagrams of Fig. 26. When the base-emitter voltage V_{BE} is zero (Fig. 26a), there is no electron injection; hence, the emitter and collector currents are zero even with a positive V_{CE}. A peak in the emitter and the collector current occurs when V_{BE} is equal to $2E_0/q$ [where E_0 is the energy of the first resonant state in the quantum well (Fig. 26b)]. With further increase in V_{BE}, RT is quenched (Fig. 26c) with a corresponding drop in the collector current.[17]

This device could in principle be used for the same applications discussed in Section 8.5. The Fujitsu group has demonstrated its application as an exclusive NOR gate,[17] which is essentially the parity generator circuit of Fig. 21 with two inputs.

In 1985, the quantum-wire transistor was proposed.[47] In this device the resonant tunneling is of two-dimensional electrons into a one-dimensional quantum

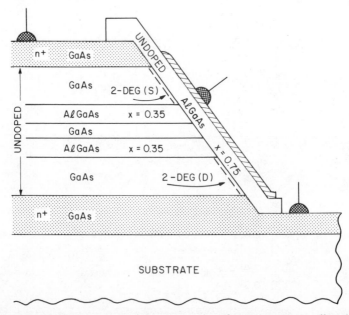

Fig. 27 Schematic cross section of the proposed surface resonant-tunneling device, the quantum-wire transistor structure. A "V-groove" implementation of the quantum wire is assumed. Thicknesses of the two undoped GaAs layers outside the double barrier region should be sufficiently large ($\geq$ 1000 Å) to prevent the creation of a parallel conduction path by the conventional (bulk) RT.

well. The idea of the quantum-wire transistor is illustrated in Fig. 27, where a GaAs/AlGaAs heterostructure implementation is assumed. The device consists of an epitaxially grown undoped planar quantum well and a double AlGaAs barrier sandwiched between two undoped GaAs layers and heavily doped GaAs contact layers. The working surface defined by a V-groove etching is subsequently overgrown epitaxially with a thin AlGaAs layer and gated. The thickness of the gate barrier layer ($d \geq 100$ Å) and the Al content in this layer ($x \geq 0.5$) should be chosen so as to minimize gate leakage. The thicknesses of the quantum-well barrier layers are chosen so that their projection on the slanted surface should be ≤ 50 Å each. The Al content in these layers should be typically $x \leq 0.45$. Application of a positive gate voltage V_G induces 2D electron gases at the two interfaces with the edges of undoped GaAs layers outside the quantum well. These gases will act as the source (S) and drain (D) electrodes. At the same time, there is a range of V_G in which electrons are not yet induced in the quantum-wire region (which is the edge of the quantum-well layer) because of the additional dimensional quantization.

To understand the operation of the device, consider first the band diagram in the absence of a source-to-drain voltage, $V_{DS} = 0$ (Fig. 28a). The diagram is drawn along the x direction (from S to D parallel to the surface channel). The y direction is defined as the one normal to the gate and the z direction as that along the quantum wire. Dimensional quantization induced by the gate results in a zero-point energy of electronic motion in the y direction, represented by the bottom E_0 of a 2D subband that corresponds to the free motion in the x and z directions. The thicknesses of the undoped S and D layers are assumed to be large enough (≥ 1000 Å) that the electronic motion in the x direction in these layers can be considered free. On the other hand, in the quantum-well region of the surface channel, there is an additional dimensional quantization along the x direction that defines the quantum wire. Let t be the x projection of the quantum-well layer thickness; then the additional zero-point energy is approximately given by

$$E_0' - E_0 = \frac{\pi^2 \hbar^2}{2m^* t^2}. \tag{5}$$

This approximation is good only when the barrier height substantially exceeds E_0'.

Application of a gate voltage V_{GS} moves the 2D subband E_0 with respect to the (classical) bottom of the conduction band E_C and the Fermi level E_F. The operating regime of this device with respect to V_{GS} at $V_{GS} = 0$ corresponds to the situation where E_F lies in the gap $E_0' - E_0$. A resonant-tunneling condition is started by the application of a positive V_{DS}, as illustrated in Fig. 28b. In this situation, the energy of certain electrons in S matches unoccupied levels in the quantum wire (Fig. 28c). Compare this with the tunneling of 3D electrons into 2D density of states discussed in Chapter 2. In the present case, the dimensionality of both the emitter and the base is reduced by one. Hence, the emitter Fermi sea of Fig. 20 (Chapter 2) has become a disk in this case and the Fermi

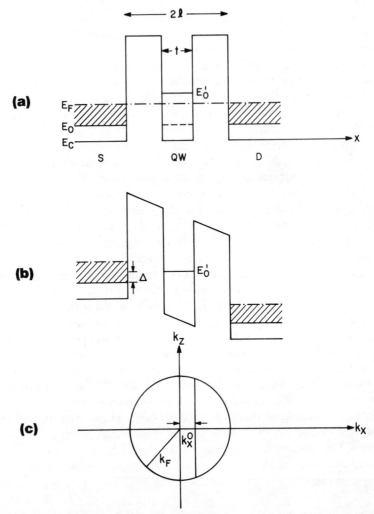

Fig. 28 Illustration of the quantum-wire transistor operation. (a) Band diagram along the channel in "equilibrium," i.e., in the absence of a drain bias. (b) Band diagram for an applied bias V_{DS}, when the energy of certain electrons in the source (S) matches unoccupied levels of the lowest 1D subband E_0' in the quantum wire. (c) Fermi disk corresponding to the 2D degenerate electron gas in the source electrode. Vertical chord at $k_x = k_x^0$ indicates the momenta of electrons that can tunnel into the quantum wire while conserving their momentum k_z along the wire.

disk is replaced by a resonant segment, as shown in Fig. 28c. Since both k_x and k_y are quantized in the quantum wire, RT requires conservation of energy and the lateral momentum k_z. This is true only for those electrons whose momenta lie in the segment $k_x = k_x^0$ (Fig. 28), where

$$\frac{\hbar^2(k_x^0)^2}{2m^*} = \Delta. \tag{6}$$

It should be noted that the energies of all electrons in this segment ($k_x = k_x^0$) lie in the band $E_0 + \Delta \leq E \leq E_F$. However, only those electrons in this energy band that satisfy the momentum conservation condition are resonant. As V_{DS} is increased, the resonant segment moves to the left (Fig. 28c), toward the vertical diameter $k_x = 0$ of the Fermi disk, and the number of tunneling electrons grows, reaching a maximum $[2m^*(E_F - E_0)]^{1/2}/\pi\hbar$ per unit length in the z direction when $\Delta = 0$. At higher V_{DS}, when $\Delta < 0$, there are no electrons in the source that can tunnel into the quantum wire while conserving their lateral momentum. This gives rise to the NDR in the drain circuit.

In the present device, additional flexibility is achieved through the gate electrode. The gate voltage in this structure not only determines the number of electrons available for conduction but also controls the position of the E_0' level in the quantum wire with respect to E_0 in the source. This latter control is affected by the fringing electric fields and gives rise to the interesting possibility of negative transconductance, as in the resonant-tunneling field-effect transistor (RT-FET). The corresponding electrostatic problem has been solved by suitable conformal mappings.[47] In the operating regime of the device, an increasing $V_{GS} > 0$ lowers the electrostatic potential energy in the base (quantum wire) with respect to the emitter (source), nearly as effectively as does an increasing V_{DS}.

Resonant-tunneling gate field effect transistors (RT-FET)[48-50] have also been developed. In addition the integration of RT diodes and FETs[51-53] and their circuit applications[54] has been demonstrated.

Bandgap engineering allows the utilization of RT in other transistor structures. One such structure is the gated QW transistor.[55] This is the first transistor in which negative transconductance is achieved by directly controlling the potential of the QW.

The structure was proposed under the name Stark-effect transistor.[56] The key ideas of the Stark-effect transistor were the use of a QW collector and the inverted sequence of layers in which the controlling electrode (here referred to as *gate*[55]) was placed "behind" the collector layer. It was predicted that the gate field would modify the position of the collector subbands with respect to the emitter Fermi level and thus modulate the tunneling current. As demonstrated in Ref. 55, the structure offers additional advantages, namely NDR and negative transconductance. Moreover, the operation of the device is only partly governed by the Stark effect. In fact, another mechanism, *the quantum capacitance*,[57] is essential for understanding its operation.

The device grown by MBE in the AlGaAs material system consisted of an undoped quantum-well collector 120 Å thick to which contact was provided. This layer was separated from an n^+-doped emitter by a 40-Å-thick undoped AlAs tunneling barrier. On the other side of the collector a 1200-Å-thick undoped AlAs barrier was followed by the n^+ gate. The doping of the 5000-Å-thick n^+ layers was nominally $2 \times 10^{18} \text{cm}^{-3}$. The energy diagram of the device is sketched in Fig. 29.

The emitter-collector I-V characteristics of the device are expected to peak at biases that maximize the RT of the emitter electrons into the 2D collector sub-

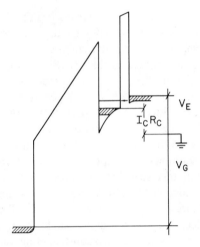

Fig. 29 Band diagram of the gated quantum-well resonant-tunneling transistor with the collector at reference and the biases $V_G > 0$ and $V_E < 0$ corresponding to peak resonant tunneling of emitter electrons into the second subband of the well.

bands. Transistor action in the structure is obtained through the influence of the gate field on the alignment of the 2D electron gas energy levels relative to the emitter Fermi level. This occurs, as anticipated above, for the combined action of the generalized Stark effect (see Problem 6) and the quantum capacitance effect. The contribution to the capacitance, not present in a classical metal, arises from the energy that has to be spent to raise the Fermi energy in the well as the carrier concentration is increased by the increasing gate field. This causes the gate field to penetrate beyond the 2D metal in the quantum well and induce charges on the emitter electrode.[57]

Figure 29 shows the band diagram of the device in the common-collector configuration with applied biases $V_G > 0$ and $V_E < 0$ such that the bottom of the conduction band in the emitter is in resonance with the second collector subband; this corresponds to a peak in the current. The RT current can be subsequently reduced by increasing either V_E (in modulus) or V_G. The former leads to the observation of NDR, the latter to the observation of negative transconductance.

Figure 30 shows experimental data at cryogenic temperatures. The expected features are indeed present and were observed, although less pronounced, up to liquid-nitrogen temperature. In particular, the data show, for the transconductance, a value of the order of ≈ 1 mS. The operation of the device can be modeled quantitatively with great accuracy taking into account the two mechanisms governing this structure.[55] This device has the advantage of a negligible gate current (it is always several orders of magnitude smaller than the emitter current), which gives a large current-transfer ratio but suffers from the drawback of a relatively small transconductance.

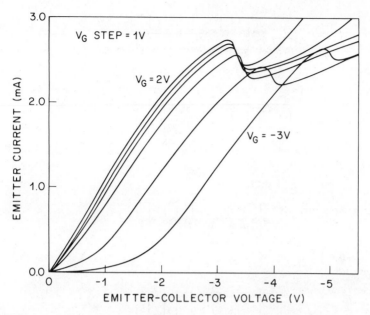

Fig. 30 Common-collector characteristics of the resonant-tunneling transistor of Fig. 29 at various V_G (2, 1, 0, -1, -2, -3 V). The measurements were performed at 7 K.

8.7 SUPERLATTICE-BASE TRANSISTORS

Negative transconductance also can be obtained using suitably designed minibands in the superlattice base of a transistor. The emitter is degenerately doped so that electrons can be injected by tunneling into the miniband (Fig. 31a). When the base-emitter voltage exceeds the bias required to line up the bottom of the conduction band in the emitter with the top of the miniband (Fig. 31b), the collector current is expected to drop. This negative transconductance arises in a straightforward manner from the conservation of lateral momentum and energy during tunneling into the miniband, similar to the situation in RT DBs. This effect has recently been observed in an InP/GaInAs superlattice HBT.[58]

The structure is grown by chemical beam epitaxy on an n^+-400-InP substrate. An $n = 5 \times 10^{17} cm^{-3}$ 5000-Å $Ga_{0.47}In_{0.53}As$ buffer layer is followed by an undoped n-type $Ga_{0.47}In_{0.53}As$ 1.8-μm-thick collector. The base consists of a p^+- ($2 \times 10^{18} cm^{-3}$) $Ga_{0.47}In_{0.53}As$ 500-Å-thick region, adjacent to the collector layer, followed by a 20-period $Ga_{0.47}In_{0.53}As$ (70 Å)/InP(20Å) superlattice. The barrier layers are undoped, while all the GaInAs wells are heavily doped ($2 \times 10^{18} cm^{-3}$) p-type. A 20-Å undoped InP doping set-back layer separates the superlattice from the 5000-Å-thick n^+- ($\approx 2 \times 10^{18} cm^{-3}$) InP emitter. This superlattice design ensures the formation of relatively wide minibands, which guarantees Bloch conduction of injected electrons through the base. The

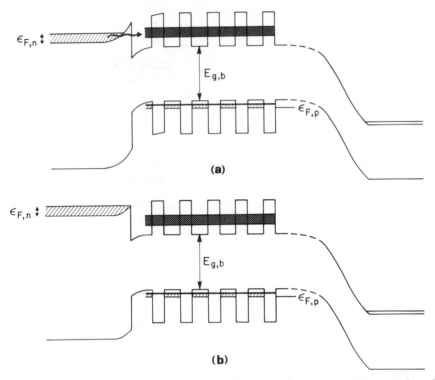

Fig. 31 Band diagram of superlattice base HBT under injection conditions: (a) into the miniband and (b) at the suppression of injection in the miniband.

calculations show that the ground-state electron miniband extends from 41 to 96 meV, while the heavy-hole miniband extends from 11.9 to 12 meV. Energies are measured from the classical bottom of the conduction- and valence-band wells, respectively. Conduction-band non-parabolicities were included in these envelope function calculations; $\Delta E_c = 0.23$ eV and $\Delta E_v = 0.39$ eV were used for the band discontinuities.

Figure 32 shows the common-base transfer characteristics at 7 and 77 K. Consider first the collector current at 7 K. At these low temperatures thermionic emission from the emitter is completely negligible, and injection from the emitter is dominated by tunneling. From the band diagram (Fig. 31) it is clear that only a small (theoretically zero at 0 K in the absence of tail states) electron current can flow from the emitter to the collector until the quasi-Fermi energy in the emitter is lined up with the bottom of the miniband. This requires an emitter-base bias given by

$$V_{BE}^{th} = \frac{E_{F,p} + E_{1,hh} + E_{g,b} + E_{1,e}}{q} = 0.88 \text{ V} , \qquad (7)$$

where $E_{F,p}$ is the Fermi energy in the base ($= 15$ meV), $E_{1,hh}$ is the bottom of

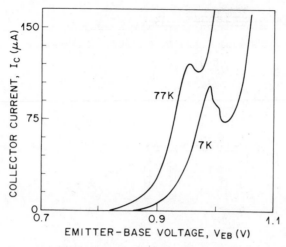

Fig. 32 Transfer characteristic of the superlattice-base HBT in the common-base configuration at two different temperatures. For both curves $V_{CB} = 2$ V.

the heavy-hole miniband ($= 12$ meV), $E_{g,b}$ is the GaInAs bulk bandgap ($= 0.812$ eV), and $E_{1,e}$ is the bottom of the first electron miniband ($= 41$ meV). The data of Fig. 32 indeed show that the collector current rapidly increases for $V_{BE} \geq 0.87$ V. The suppression of injection into the miniband requires increasing the emitter-base voltage by more than $\Delta_1 + E_{F,n} = 118.6$ meV, where $\Delta_1 = 55$ meV is the width of the first electron miniband and $E_{F,n} = 80$ meV is the quasi-Fermi energy in the emitter. Thus, a minimum in the current is expected at

$$V_{EB}^P = V_{EB}^{th} + \frac{\Delta_1 + E_{F,n}}{q} = 1.015 \text{ V} , \qquad (8)$$

in excellent agreement with the experimental value (≈ 1.00 V). Following this the collector current rises rapidly for $V \geq 1.02$ V. This is expected, since, at a bias $\approx E_g + \Delta E_c/q \approx 1.03$ V, the conduction band edge in the emitter becomes flat, leading to a steep increase in the injection efficiency. Note that at 77 K the peak shifts to a lower voltage. The shift (≈ 25 mV) is close, as expected, to the GaInAs band gap lowering (≈ 30 meV) as the temperature is varied from 7 to 77 K.

The common-base characteristics I_C, I_E versus V_{EB} showed a maximum common-base current gain $\alpha = I_C/I_E = 0.75$ at $V_{BE} \approx 0.98$ V. This value is consistent with the maximum gain $\beta = I_C/I_B \approx 3.2$ measured in the common-emitter configuration. These values of α and β, although far from optimal, indicate that transport in the base is via miniband conduction rather than by hopping. Previously an HBT with an AlGaAs/GaAs superlattice base had been reported.[59] Although miniband conduction in the base was demonstrated, no negative transconductance was shown, since the structure did not use a tunneling emitter for injection.

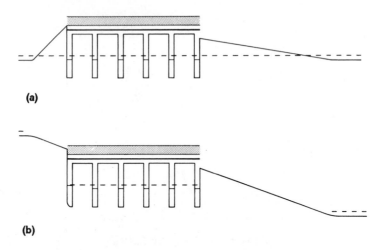

(a)

(b)

Fig. 33 Conduction-band energy diagram of the superlattice-base transistor. (a) At equilibrium, and (b) in the common-base configuration near the peak of the current-voltage characteristic. Further increase in the negative emitter-base bias will suppress injection in the base because of quantum reflections by the minigap.

Negative transconductance can also be achieved by controlling injection into minibands above the top of the barriers. Recently, a tunneling emitter transistor was proposed in which hot electrons transfer through the base by miniband conduction in a continuum state.[60] Here we present the operation of a new superlattice-base unipolar transistor in which electrons are injected into a miniband in the classical continuum.[61]

The structure, whose equilibrium conduction-band energy diagram is sketched in Fig. 33a, was grown by MBE. It consists of a 8000-Å n^+-collector followed by an undoped $Al_xGa_{1-x}As$ layer 5000 Å thick with x varying from 0 to 0.25. On top of these the SL base was grown, consisting of 5.5 periods of 40-Å n^+-GaAs/200Å undoped $Al_{0.31}Ga_{0.69}As$. An undoped $Al_xGa_{1-x}As$ injector layer 500 Å thick followed, with x varying from 0 to 0.33, corresponding to a band discontinuity $\Delta E_c = 273$ meV, roughly at the bottom of the chosen miniband. Finally, an n^+-emitter layer 3000 Å thick concluded the growth. In all the doped layers $n = 2 \times 10^{18} cm^{-3}$.

The operation of the device is easily understood with the help of Fig. 33, where the common-base operation mode is illustrated. At a fixed positive collector-base bias, the negative emitter-base bias (V_{EB}) is increased and the collector current is measured. By appropriately tailoring the compositionally graded emitter barrier, electrons are injected into the third miniband. The energy dispersion of the latter was calculated in the envelope function approximation, taking into account band non-parabolicities, and is large enough (≈ 23 meV) to guarantee miniband conduction. Increasing V_{EB} will further flatten the triangular injector and increase the injection current. However, part of the bias will appear in a depletion region in the base, thus shifting the top of the band discontinuity with respect to the miniband. When this shift is larger

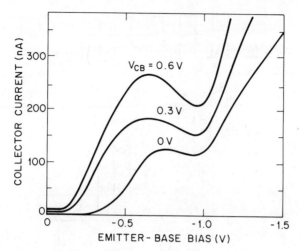

Fig. 34 Collector current as a function of emitter-base bias at fixed collector-base (V_{CB}) bias at 30 K. The curves shown are relative to $V_{CB} = 0, 0.3, 0.6$ V.

than the miniband width, the injected electrons will not be able to satisfy the energy and lateral momentum-conservation conditions at the interface and will experience strong quantum-mechanical reflections. Consequently, the injection efficiency will drop together with the collector current and the *I-V* characteristics will exhibit negative transconductance. This effect is shown in the experimental curves shown of Fig. 34. In fact, a 23-meV shift ($\approx$ miniband width) of the top of the injector band discontinuity is required to suppress electron injection. This corresponds to an almost total depletion of the first well of the base and gives, by a simple electrostatic computation, an emitter-base bias of ≈ 0.6 V, in agreement with the peak position for $V_{CB} > 0$ (Fig. 34).

In the present structure the base transport factor is $\ll 1$, leading to a small I_C. The design must be optimized and stray leakage paths eliminated in order to enhance α.

The structures discussed in this section are of interest, primarily from a physics point of view, as tools to investigate transport in two-dimensional systems and superlattices. Their operation so far has only been demonstrated at cryogenic temperatures. These devices have several performance shortcomings in comparison with the much more advanced RTBTs described in the preceding sections.

8.8 SUMMARY AND FUTURE TRENDS

8.8.1 Negative Differential Resistance in Superlattices, Bloch Oscillators, and Quantum Interference Transistors

In this section we discuss briefly other potentially useful quantum effects and their device applications; these structures, however, have been investigated far

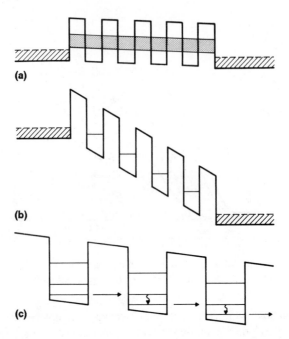

Fig. 35 (a) Conduction-band diagram of superlattice at low applied bias, the shaded region shows a miniband. (b) Stark ladder quantization in a superlattice. (c) Sequential resonant tunneling in a superlattice.

less, both from an experimental and theoretical point of view, than RT devices, and most of them are still at the proposal stage.

In the original proposal[62] of the superlattice concept, it was shown that, under the action of a suitable electric field, electrons can occupy the negative-mass region of the first minizone (i.e., where the $E(k)$ dispersion curvature changes sign) thereby decelerating and giving rise to negative differential resistance (Fig. 35a). In a superlattice with dispersion relationship of the type

$$E(k) = \frac{\Delta}{2} [1 - \cos ka] , \tag{9}$$

where Δ is the miniband width and a is the superlattice period, the field dependence of the drift velocity is found to be[62]

$$v_d(E) = \frac{qE\tau}{m^*_{||}} \frac{1}{1+(\omega_B\tau)^2} . \tag{10}$$

In Eq. 10, $m^*_{||}$ is the band-edge effective mass along the direction of the superlattice and is given by $2\hbar^2/\Delta a^2$, τ is the momentum relaxation time, and ω_B is the Bloch frequency, expressed as

$$\omega_B = \frac{qEa}{\hbar} . \tag{11}$$

From Eq. 10 it is clear that at low fields the velocity increases linearly with E, with a mobility $q\tau/m^{*}_{||}$. The threshold field for the onset of NDR is given by the condition $\omega_{B}\tau = 1$ and is $\approx 10^{3}$ V/cm for typical values of the superlattice period d (≤ 100 Å) and the relaxation time $\tau (\leq 10^{-12}$s). The use of more complex dispersion curves $E(k)$ alters the value of the threshold field only slightly.[62]

If the scattering time is sufficiently long, that is, the condition $\omega_{B}\tau > 2\pi$ is satisfied, electrons can undergo coherent rf oscillations with an angular frequency ω_{B}, the so-called Bloch or Zener oscillations, due to Bragg reflections at the minizone boundaries at $k = \pm \pi/a$.[62] This can be easily verified by a simple integration of the equation of motion, $\hbar dk/dt = qE$. It should be clear that the observation of these oscillations is much more difficult than that of NDR because of the different threshold conditions. Neither of the two effects has been observed so far. It should be noted that to build an actual Bloch oscillator, that is, a device capable of extracting power from Bloch oscillations, the electrons should oscillate in phase. For this to occur, electrons must be injected into the field region in phase, that is, with the same wave vector k.[63] This phase initialization will ensure coherence, which will be maintained, since the condition $\omega_{B}\tau > 2\pi$ is assumed satisfied.

The difficulty of observing Bloch oscillations also stems from the fact that, in order to achieve $\omega\tau > 2\pi$, the field must be substantially increased above that required for NDR. Thus, transitions into the second minizone by tunneling, when electrons reach the first boundary of the minizone, cannot be neglected. This effect can suppress Bloch oscillations unless the tunneling probability is made negligible. This is difficult to do for fields approaching 10^{4} V/cm, because typical minigaps are a few tens of meV wide. The Zener tunneling can be conveniently suppressed in a lateral (or surface) superlattice, for example, a two-dimensional array of dots with spacings less than the de Broglie wavelength. The additional periodicity in such structures can be used to advantage to increase the minigap widths to values ≥ 200 meV.

Bloch transistors that utilize lateral superlattices to achieve NDR and Bloch oscillations have been proposed.[64] For example, the Bloch-FET consists of a doubly periodic array of metal dots embedded in the metal-oxide semiconductor FET (MOSFET) oxide. These dots induce a two-dimensional superlattice potential in the inversion layer. The minigaps can be tuned by varying the electron concentration in the inversion layer by means of the gate voltage. Obviously such structures require a demanding electron-beam technology (we are not yet capable of producing 100-Å dot spacings), and the use of small-effective-mass materials such as InSb. For a detailed quantitative analysis of Bloch oscillators, consult Ref. 63.

Other methods have been proposed to achieve NDR in a superlattice. One of these is based on the phenomenon of wave function localization in a superlattice.[65-67] In an electric field, the electronic wavefunctions extend over a number of periods of the order of Δ/qaE and are separated in energy by qaE (the so called Wannier-Stark ladder). As the field is increased, the wavefunctions become increasingly localized in space, up to the extreme point where they are confined to one well (Fig. 35b). Complete localization leading to Stark-ladder quantization was recently demonstrated through optical-absorption experiments.[68]

Conduction properties are deeply affected by this localization, since the increasing confinement of the electronic states causes a decreasing overlap between them and therefore a decreasing transition probability. This behavior was recently demonstrated experimentally.[69] Moreover this localization-induced NDR and Bragg-diffraction induced mechanism presented at the beginning of this section were shown to be physically equivalent.[69]

If the field is further increased so that the ground state of a well is brought into resonance with the excited states of the neighboring well (Fig. 35c), a series of peaks in the *I-V* characteristic due to sequential resonant tunneling is expected, as discussed in Chapter 2. This negative differential conductance has been clearly observed in an AlInAs/GaInAs superlattice.[70]

Periodic negative conductance has been observed in multi-layer structures of weakly coupled doped wells, due to the non-uniform electric fields, as sequential resonant tunneling extends from one period to next, with increasing bias.[71,72]

An interesting NDR device, which exploits resonant tunneling between minibands, has been demonstrated.[73] This structure, which could be called a miniband tunnel diode, consists of two superlattices separated by a thick barrier. When a bias is applied, the voltage drops primarily across the barrier, and NDR is achieved by first bringing into and then out of resonance the ground-state miniband of one of the superlattices with the first excited miniband of the other superlattice. (See Section 9.2.2.)

Finally, we briefly discuss another recently proposed method of achieving negative transconductance in a transistor. The underlying physics can form the basis for a new class of devices based on quantum interference.[74,75] Electrons in two parallel high-mobility channels, obtained, for example, by modulation doping and separated by a barrier, can be made to interfere constructively (at cryogenic temperatures) if the phase-coherence length is longer than the channel length. Note that ballistic mean free paths in modulation-doped GaAs can exceed 10 μm at temperatures $\approx$ 4 K and in small electric fields. A voltage difference is then applied between the two channels by means of a suitable gate. This induces a corresponding phase difference. If this difference is π, the drain current will reach a minimum as a result of destructive interference, and a large negative transconductance is expected. Clearly the exploitation of wave phenomena in the lateral direction in semiconductors is in its infancy, and many exciting developments in this area of physics can be expected.

8.8.2 Summary

In this chapter, we have reviewed recent developments in quantum-effect devices. These structures can be viewed both as exciting tools to investigate the physics of transport over short distances and as potentially interesting devices for digital and analog circuit applications. Certainly, among these structures, the RTBT emerges as one of the most promising devices in light of the large negative transconductance and its ability to operate on multiple states. In fact, the latter characteristic might be the most significant in light of its potentially far reaching implications for circuits and in particular new computer architectures based on multiple-valued logic. Of course, a word of caution is necessary in this context. It is difficult to see how a technology based on quantum-effect devices alone could substitute conventional architectures based on silicon two-state devices. III-V materials are "light-years" away from silicon in terms of the level of integration and reliability. For quantum-effect devices to have large-scale impact it would be necessary to demonstrate their operation in silicon-based materials such as silicon-germanium alloys. Work on RT in such heterostructures is in its infancy and one cannot yet draw conclusions on the usefulness of such heterstructures for quantum-effect circuits and devices. A more realistic view sees quantum-effect devices as having future impact in certain niche applications such as the ones discussed in this chapter (multipliers, parity generators, etc.) and for oscillators in the millimeter-wave region, provided that enough power can be generated.

ACKNOWLEDGMENT

The authors are grateful to A. Y. Cho, A. L. Hutchinson, R. Kiehl, L. Lunardi, S. Luryi, R. J. Malik, P. R. Smith, M. Shoji, D. Sivco, W. T. Tsang, and A. S. Vengurlekar for collaborations and useful discussions.

PROBLEMS

1. Molecular beam epitaxy interfaces are typically abrupt to within one or two monolayers (one monolayer ≈ 2.8 Å in GaInAs), due to terrace formation in the growth plane. Estimate the energy level broadening for the ground and first excited electron states of a 150-Å GaInAs quantum well bound by thick AlInAs barriers. (*Hint*: Assume the case of a two-monolayer thickness fluctuation and an infinitely deep quantum well. The electron effective mass in GaInAs is $0.0427m_0$).

2. a) Derive the formula for the energy levels of the compositionally graded parabolic well of Fig. P1 in term of ΔE_c and L. Ignore the finiteness of the well height and the dependence of the effective mass m^* on the alloy composition.

   b) What is the energy-level spacing for a semiparabolic well? [*Hint*: A simple physical argument based on symmetry can be used to obtain (b) from the solution of (a)].

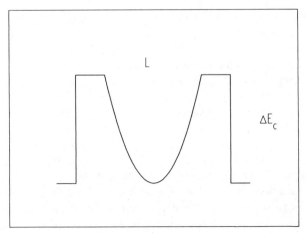

Fig. P1

3. Show that the application of an electric field to a parabolic well of infinite width does not change the energy-level spacing.

4. Consider a rectangular resonant-tunneling double barrier. Show that in the WKB limit the width of the nth resonance is approximately given by

$$\Gamma_n \approx E_n \exp\left[-\frac{2L_B}{\hbar}\sqrt{2m^*(\Delta E_c - E_n)}\right], \qquad (P1)$$

where m^* is the effective mass in the barrier of length L_B and ΔE_c is the barrier height.

5. Consider a resonant-tunneling double barrier carrying a current density $J = 10^5$ A/cm^2 through the first resonance assumed of width $\Gamma = 0.1$ meV. Calculate the electron density in the quantum well.

6. Consider the gated quantum-well transistor of Section 8.6 biased approximately as shown in Fig. 29. Assume for the ground subband the function $\psi_1 = (b^3/2)^{1/2} z e^{-bz/2}$ and for the first excited subband the second sine solution of the corresponding square well. Show that the first-order variation in the first excited quantum-well state E_2 in response to a variation δV_G in the gate potential is given by

$$\delta E_2 = \frac{6q}{L_G L_W b^2}\left[1 - \frac{\alpha^4 + 3\alpha^2 + 6}{6(1+\alpha^2)^3}\right]\delta V_G, \qquad (P2)$$

where L_G and L_W are the thicknesses of the gate-to-drain barrier and of the quantum-well drain layer, $\alpha = 4\pi/bL_W$, and b was defined above. (*Hint:* Use first-order perturbation theory to calculate δE_2 in response to a variation in the quantum-well potential $\delta\phi$. The latter can be obtained by integration of the Poisson equation, given ψ_1.)

7. Demonstrate that in a superlattice the Bloch mobility can be expressed as

$$\mu = \frac{q \, \Delta d^2 \tau}{2 \hbar^2} ,$$ (P3)

where Δ is the miniband width, d is the superlattice period, and τ is the relaxation time. [*Hint*: Assume a band structure along the superlattice of the form $E = \Delta/2(1 - \cos(k_{||}d))$ where $k_{||}$ is the wave vector parallel to the superlattice axis.]

8. Show that in a superlattice-base transistor of the type of Fig. 31 the base transit time is given by

$$t_B = \frac{\hbar^2}{d^2} \frac{W_B^2}{\tau k_B T \Delta} ,$$ (P4)

where W_B is the base thickness, T the lattice temperature, k_B the Boltzman constant, and the other quantities are as defined in the previous problems. (*Hint*: The transport in the base is diffusive.)

9. Calculate the energy dispersion relation in the superlattice of Fig. 33. (*Hint*: This problem is conveniently solved numerically by calculating the transmissivity of the structure. At every interface use as boundary conditions $\psi^{\text{left}} = \psi^{\text{right}}$ and

$$\frac{1}{m^{*\text{left}}} \frac{d\psi^{\text{left}}}{dz} = \frac{1}{m^{*\text{right}}} \frac{d\psi^{\text{right}}}{dz}.$$

To obtain the results mentioned in Section 8.7, energy-band non-parabolicity should be taken into account.)

REFERENCES

1. J. A. Morton, "From Physics to Function," *IEEE Spect.*, p. 62, September 1965.
2. L. Esaki, "New Phenomenon in Narrow Germanium *p-n* junctions," *Phys. Rev.* **109**, 603 (1958).
3. W. S. Boyle and G. E. Smith, "Charged-Coupled Semiconductor Devices," *Bell Syst. Tech. J.* **49**, 587 (1970).
4. F. Capasso, "Graded-Gap and Superlattice Devices by Bandgap Engineering," in *Semiconductors and Semimetals*, Vol. 24, R. K. Willardson and A. C. Beers, Eds. Academic Press, New York, 1987, p. 319.
5. D. A. B. Miller, D. S. Chemla, T. C. Damen, A. C. Gossard, W. Wiegmann, T. H. Wood, and C. A. Burrus, "Novel Hybrid Optically Bistable Switch," *Appl. Phys. Lett.* **45**, 13 (1984).
6. L. L. Chang, L. Esaki, and R. Tsu, "Resonant Tunneling in Semiconductor Double Barriers," *Appl. Phys. Lett.* **24,** 593 (1974).

7. T. C. L. G. Sollner, W. D. Goodhue, P. E. Tannenwald, C. D. Parker, and D. D. Peck, "Resonant Tunneling through Quantum Wells at Frequencies Up to 2.5 THz," *Appl. Phys. Lett.* **43**, 588 (1983).

8. E. R. Brown, T. C. L. G. Sollner, C. D. Parker, W. D. Goodhue, and C. L. Chen, "Oscillations Up to 420 GHz in GaAs/AlAs Resonant Tunneling Diodes," *Appl. Phys. Lett.* **55**, 1777 (1989).

9. T. P. E. Broekaert, W. Lee, and C. G. Fonstad, "Pseudomorphic $In_{0.53}Ga_{0.47}As/AlAs/InAs$ Resonant Tunneling Diodes with Peak-to-Valley Current Ratios of 30 at Room Temperature," *Appl. Phys. Lett.* **53**, 1545 (1988).

10. E. R. Brown, T. C. L. G. Sollner, W. D. Goodhue, and C. L. Chen, "High Speed Resonant Tunneling Diodes," *Proc. SPIE* **943**, 2 (1989).

11. F. Capasso, F. Beltram, S. Sen, and A. Y. Cho, "Physics and Device Applications of Resonant Tunneling," in *Physics of Quantum Electron Devices*, F. Capasso, Ed., Springer-Verlag, Heidelberg, 1990.

12. R. H. Davis, and H. H. Hosack, "Double Barrier in Thin Film Triodes," *J. Appl. Phys.* **34**, 864 (1963).

13. L. V. Iogansen, "The Possibility of Resonance Transmission of Electrons in Crystals through a System of Barriers," *Zh. Eksp. Teor. Fiz.* **45**, 207 (1963) [English transl. *Sov. Phys. JETP* **18**, 46 (1964)].

14. F. Capasso and R. A. Kiehl, "Resonant Tunneling Transistor with Quantum Well Base and High-Energy Injection: A New Negative Differential Resistance Device," *J. Appl. Phys.* **58**, 1366 (1985).

15. B. Riccò and P. M. Solomon, "Tunable Resonant Tunneling Semiconductor Emitter Structure," *IBM Tech. Disclos. Bull.* **27**, 3053 (1984).

16. R. T. Bate, G. A. Frazier, W. R. Frensley, J. K. Lee, and M. A. Reed, "Prospects for Quantum Integrated Circuits," *Proc. SPIE* **792**, 26 (1987).

17. N. Yokoyama, K. Imamura, S. Muto, S. Hiyamizu, and H. Nishi, "A New Functional Resonant Tunneling Hot Electron Transistor (RHET)," *Jpn. J. Appl. Phys.* **24**, L-853 (1985). For a comprehensive review of RHETs, see N. Yokoyama, H. Ohnishi, T. Futatsugi, S. Muto, T. Mori, K. Imamura, and A. Shibatomi, "Resonant-Tunneling Transistors Using InGaAs-based Materials," *Proc. SPIE* **943**, 14 (1988).

18. F. Capasso, S. Sen, A. C. Gossard, A. L. Hutchinson, and J. H. English, "Quantum Well Resonant Tunneling Bipolar Transistor Operating at Room Temperature," *IEEE Electron Dev. Lett.* **EDL-7**, 573 (1986).

19. T. Futatsugi, Y. Yamaguchi, K. Ishii, K. Imamura, S. Muto, N. Yokoyama, and A. Shibatomi, "A Resonant Tunneling Bipolar Transistor (RBT): A Proposal and Demonstration for New Functional Device with High Current Gains," *Tech. Dig. IEDM*, p. 286 (December 1986).

20. T. Inata, S. Muto, Y. Nakata, T. Fujii, H. Ohnishi, and S. Hiyamizu, "Excellent Negative Differential Resistance of InAlAs/InGaAs Resonant Tunneling Barrier Structure Grown by MBE," *Jpn. J. Appl. Phys.* **25**, 983 (1986).

21. F. Capasso, S. Sen, A. Y. Cho, and D. L. Sivco, "Multiple Negative Transconductance and Differential Conductance in a Bipolar Transistor by Sequential Quenching of Resonant Tunneling," *Appl. Phys. Lett.* **53,** 1056 (1988).

22. S. Sen, F. Capasso, A. Y. Cho, and D. L. Sivco, "Multiple State Resonant Tunneling Bipolar Transistor Operating at Room Temperature and Its Application as a Frequency Multiplier," *IEEE Electron Dev. Lett.* **9,** 533 (1988).

23. F. Capasso, K. Mohammed, and A. Y. Cho, "Resonant Tunneling through Double Barriers, Perpendicular Quantum Transport Phenomena in Superlattices, and Their Device Applications," *IEEE J. Quant. Electron.* **QE-22,** 1853 (1986).

24. S. Sen, F. Capasso, A. L. Hutchinson, and A. Y. Cho, "Room Temperature Operation of $Ga_{0.47}In_{0.53}As/Al_{0.48}In_{0.52}As$ Resonant Tunneling Diodes," *Electron. Lett.* **23,** 1229 (1988).

25. S. Sen, F. Capasso, A. C. Gossard, R. A Spah, A. L. Hutchinson, and S. N. G. Chu, "Observation of Resonant Tunneling through a Compositionally Graded Parabolic Quantum Well," *Appl. Phys. Lett.* **51,** 1428 (1987).

26. F. Capasso, S. Sen, A. Y. Cho, and A. L. Hutchinson, "Resonant Tunneling Spectroscopy of Hot Minority Electrons Injected in Gallium Arsenide Quantum Wells," *Appl. Phys. Lett.* **50,** 930 (1987).

27. J. R. Hayes and A. F. J. Levi, "Dynamics of Extreme Nonequilibrium Transport in GaAs," *IEEE J. Quant. Electron.* **QE-22,** 1744 (1986).

28. C. Rine, Ed., *Computer Science and Multiple Valued Logic*, North-Holland, Amsterdam, 1977, p. 101.

29. A. F. J. Levi, "Scaling Ballistic Heterojunction Bipolar Transistors," *Electron. Lett.* **24,** 1273 (1988).

30. T. Weil and B. Vinter, "Equivalence between Resonant Tunneling and Sequential Tunneling in Double-Barrier Diodes," *Appl. Phys. Lett.* **50,** 1281 (1987).

31. R. J. Malik, R. Nottenburg, E. F. Schubert, J. F. Walker, and R. W. Ryan, "Carbon Doping in Molecular Beam Epitaxy of GaAs from a Heated Graphite Filament," *Appl. Phys. Lett.* **53,** 2661 (1988).

32. A. F. J. Levi, S. L. McCall, and P. M. Platzman, "Nonrandom Doping and Elastic Scattering of Carriers in Semiconductors," *Appl. Phys. Lett.* **54,** 940 (1989).

33. A. F. J. Levi and Y. Yafet, "Nonequilibrium Electron Transport in Bipolar Devices," *Appl. Phys. Lett.* **51,** 42 (1987).

34. E. Wolak, K. L. Lear, P. M. Pitner, E. S. Hellman, B. G. Park, T. Weil, and J. S. Harris, Jr., "Elastic Scattering Centers in Resonant Tunneling Diodes," *Appl. Phys. Lett.* **53,** 201 (1988).

35. M. A. Reed, W. F. Frensley, R. J. Matyi, J. N. Randall, and A. C. Seabaugh, "Realization of a Three Terminal Resonant Tunneling Device: The Bipolar Quantum Resonant Tunneling Transistor," *Appl. Phys. Lett.* **54,** 1034 (1989).

36. S. Sen, F. Capasso, A. Y. Cho, and D. Sivco, "Resonant Tunneling Device with Multiple Negative Differential Resistance: Digital and Signal Processing Applications with Reduced Circuit Complexity," *IEEE Trans. Electron Dev.* **ED-34,** 2185 (1987).

37. J. Söderström and T. G. Andersson, "A Multiple-State Memory Cell Based on the Resonant Tunneling Diode," *IEEE Electron Dev. Lett.* **9,** 200 (1988).

38. R. C. Potter, A. A. Lakhani, D. Beyea, E. Hempling, and A. Fathimulla, "Three-Dimensional Integration of Resonant Tunneling Structures for Signal Processing and Three-State Logic," *Appl. Phys. Lett.* **52,** 2163 (1988).

39. S. Sen, F. Capasso, D. Sivco, and A. Y. Cho, "New Resonant Tunneling Devices with Multiple Negative Resistance Regions and High Room Temperature Peak to Valley Ratio," *IEEE Electron Dev. Lett.* **9,** 402 (1988).

40. *General Electric Tunnel Diode Manual,* 1st ed., Vol. 66 (General Electric, 1961).

41. M. Tsuchiya, H. Sakaki, and J. Yoshino, "Room Temperature Observation of Differential Negative Resistance in an AlAs/GaAs/AlAs Resonant Tunneling Diode," *Jpn. J. Appl. Phys.* **24,** L466 (1985).

42. L. M. Lunardi, S. Sen, F. Capasso, P. R. Smith, D. L. Sivco, and A. Y. Cho, "Microwave Multiple-State Resonant Tunneling Bipolar Transistor," *IEEE Electron Dev. Lett.* **10,** 219 (1989).

43. T. Mori, K Imamura, H. Ohnishi, Y. Minami, S. Muto, and N. Yokoyama, "Microwave Analysis of Resonant Tunneling Hot Electron Transistor at Room Temperature," *Ext. Absts. 20th Conf. on Solid State Dev. and Mater.* **507,** (1988).

44. S. Sen, F. Capasso, A. Y. Cho, and D. L. Sivco, "Parity Generator Circuit Using a Multi-State Resonant Tunneling Bipolar Transistor," *Electron. Lett.* **24,** 1506 (1988).

45. A. A. Lakhani, R. C. Potter, and H. S. Hier, "Eleven-Bit Parity Generator with a Single, Vertically Integrated Resonant Tunneling Device," *Electron. Lett.* **24,** 681 (1988).

46. T. Tanoue, H. Mizuta, and S. Takahashi, "A Triple Well Resonant Tunneling Diode for Multiple Valued Logic Applications," *IEEE Electron Dev. Lett.* **9,** 365 (1988).

47. S. Luryi and F. Capasso, "Resonant Tunneling of Two Dimensional Electrons through a Quantum Wire: A Negative Transconductance Device," *Appl. Phys. Lett.* **47,** 1347 (1985); also Erratum, *Appl. Phys. Lett.* **48,** 1693 (1986).

48. F. Capasso, S. Sen, F. Beltram, and A. Y. Cho, "Resonant Tunneling Gate Field-Effect Transistor," *Electron. Lett.* **23,** 225 (1987).

49. S. Sen, F. Capasso, F. Beltram, and A. Y. Cho, "The Resonant Tunneling Field-Effect Transistor: A New Negative Transconductance Device," *IEEE Trans. Electron Dev.* **ED-34,** 1768 (1987).

50. F. Capasso, S. Sen, and A. Y. Cho, "Negative Transconductance Resonant Tunneling Field Effect Transistor," *Appl. Phys. Lett.* **51,** 526 (1987).

51. A. R. Bonnefoi, T. C. McGill, and R. D. Burnham, "Resonant Tunneling Transistors with Controllable Negative Differential Resistance," *IEEE Electron Dev. Lett.* **EDL-6,** 636 (1985).

52. T. K. Woodward, T. C. McGill, and R. D. Burnham, "Experimental Realization of a Resonant Tunneling Transistor," *Appl. Phys. Lett.* **50,** 451 (1987).

53. T. K. Woodward, T. C. McGill, H. F. Chung, and R. D. Burnham, "Integration of a Resonant-Tunneling Structure with a Metal-Semiconductor Field-Effect Transistor," *Appl. Phys. Lett.* **51,** 1542 (1987).

54. T. K. Woodward, T. C. McGill, H. F. Chung, and R. D. Burnham, "Applications of Resonant-Tunneling Field-Effect Transistors," *IEEE Electron Dev. Lett.* **EDL-9,** 122 (1988).

55. F. Beltram, F. Capasso, S. Luryi, S. N. G. Chu, and A. Y. Cho, "Negative Transconductance Via Gating of the Quantum Well Subbands in a Resonant Tunneling Transistor," *Appl. Phys. Lett.* **53,** 219 (1988).

56. A. R. Bonnefoi, D. H. Chow, and T. C. McGill, "Inverted Base-Collector Tunnel Transistors," *Appl. Phys. Lett.* **47,** 888 (1985).

57. S. Luryi, "Quantum Capacitance Devices," *Appl. Phys. Lett.* **52,** 501 (1988).

58. F. Capasso, A. S. Vengurlekar, A. L. Hutchinson, and W. T. Tsang, "A Negative Transconductance Superlattice Base Bipolar Transistor," *Electron. Lett.* **25,** 1117 (1989).

59. J. F. Palmier, C. Minot, J. L. Lievin, F. Alexandre, J. C. Harmand, J. Dangla, C. Dubon-Chevallier, and D. Ankri, "Observation of Bloch Conduction Perpendicular to the Interfaces in a Superlattice Bipolar Transistor," *Appl. Phys. Lett.* **49,** 1260 (1986).

60. C. S. Lent, "The Resonant Hot Electron Transfer Amplifier: A Continuum Resonance Device," *Superlat. and Microstruct.* **3,** 387 (1987).

61. F. Beltram, F. Capasso, A. L. Hutchinson, and R. J. Malik, "Injection in a Continuum Miniband: Observation of Negative Transconductance in a Superlattice-Base Transistor," *Appl. Phys. Lett.* **55,** 1534 (1989).

62. L. Esaki and R. Tsu, "Superlattice and Negative Differential Conductivity in Semiconductors," *IBM J. Res. Devel.* **14,** 61 (1970).

63. P. Roblin and M. W. Muller, "Time-dependent Tunneling and the Injection of Coherent Zener Oscillations," *Semicond. Sci. Technol.* **1,** 218 (1986).

64. "R. K. Reich, R. O. Grondin, D. K. Ferry, and G. J. Iafrate, "The Bloch-FET: A Lateral Surface Superlattice Device," *IEEE Electron Dev. Lett.* **EDL-3,** 381 (1982).

65. G. H. Dohler, R. Tsu, and L. Esaki, "A New Mechanism for Negative Differential Resistance in Superlattices," *Solid State Commun.* **17,** 317 (1975).

66. R. Tsu and G. Dohler, "Hopping Conduction in a Superlattice," *Phys. Rev. B* **12**, 680 (1975).

67. J. Bleuse, G. Bastard, and P. Voisin, "Electric Field Induced Localization and Oscillatory Electro-Optical Properties of Semiconductor Superlattices," *Phys. Rev. Lett.* **60**, 220 (1988).

68. E. E. Mendez, F. Agulló-Rueda, and J. M. Hong, "Stark Localization in GaAs/AlGaAs Superlattices under an Electric Field," *Phys. Rev. Lett.* **60**, 2426 (1989).

69. F. Beltram, F. Capasso, D. L. Sivco, A. L. Hutchinson, S.-N. G. Chu, and A. Y. Cho, "Scattering Controlled Transmission Resonances and Negative Differential Conductance by Field-Induced Localization in Superlattices," *Phys. Rev. Lett.*, June 4, 1990.

70. F. Capasso, K. Mohammed, and A. Y. Cho, "Sequential Resonant Tunneling through a Multiquantum-Well Superlattice," *Appl. Phys. Lett.* **48**, 478 (1986).

71. L. Esaki and L. L. Chang, "New Transport Phenomenon in a Semiconductor Superlattice," *Phys. Rev. Lett.* **33**, 495 (1974).

72. K. K. Choi, B. F. Levine, R. J. Malik, J. Walker, and C. G. Bethea, "Periodic Negative Conductance by Sequential Resonant Tunneling through an Expanding High Field-Superlattice Domain," *Phys. Rev. B* **35**, 4172 (1987).

73. M. J. Kelly, R. A. Davies, A. P. Long, N. R. Couch, P. H. Beton, and T. M. Kerr, "Vertical Transport in Multilayer Semiconductor Structures," *Superlat. Microstruct.* **2**, 313 (1986).

74. S. Datta, M. R. Melloch, S. Bandopadhyay, R. Noren, M. Vaziri, M. Miller, and R. Reifenberger, "Novel Interference Effects between Parallel Quantum Wells," *Phys. Rev. Lett.* **55**, 2344 (1985).

75. S. Datta, "Quantum Interference Devices," in *Physics of Quantum Electron Devices*, F. Capasso, Ed., Springer-Verlag, Heidelberg, 1990.

9 Microwave Diodes

S. M. Sze

AT&T Bell Laboratories
Murray Hill, New Jersey

9.1 INTRODUCTION

In this chapter we consider some special two-terminal microwave semiconductor devices. The basic principles of microwave diodes have been discussed in Chapters 9, 10, and 11 in Ref. 1. The microwave frequencies cover the range from about 0.1 GHz (10^8 Hz) to about 1000 GHz with corresponding wavelengths from 300 cm to 0.3 mm. The microwave frequency range is usually grouped into different bands.[2] The bands and the corresponding frequency ranges as designated by the Institute of Electrical and Electronics Engineers (IEEE) are listed in Table 1. For frequencies between 30 and 300 GHz, we

TABLE 1 IEEE Microwave Frequency Bands

Designation	Frequency Range (GHz)	Wavelength (cm)
VHF	0.1–0.3	300.00–100.00
UHF	0.3–1.0	100.00–30.00
L band	1.0–2.0	30.00–15.00
S band	2.0–4.0	15.00–7.50
C band	4.0–8.0	7.50–3.75
X band	8.0–13.0	3.75–2.31
Ku band	13.0–18.0	2.31–1.67
K band	18.0–28.0	1.67–1.07
Ka band	28.0–40.0	1.07–0.75
Millimeter	30.0–300.0	1.00–0.10
Submillimeter	300.0–3000.0	0.10–0.01

High-Speed Semiconductor Devices, Edited by S.M. Sze. ISBN 0-471-62307-5
© 1990 John Wiley & Sons, Inc.

have the millimeter-wave band because the wavelength is between 10 and 1 mm. For even higher frequencies, we have the submillimeter-wave band. There are two other microwave band classifications adopted by the U.S. Department of Defense, which are listed in Tables 2 and 3. To avoid confusion it is recommended that both the band and the corresponding frequency range be used when referring to microwave devices.

For lower microwave frequencies (< 30 GHz), the transistors discussed in previous chapters can provide high-efficiency, high-power operations with low noise. However, it is difficult to extend operation of these transistors into the millimeter-wave band, because of their inherent structural complexities. Therefore, two-terminal microwave diodes are key devices for generation, amplification, and detection applications in the millimeter-wave band and the submillimeter-wave band.

Millimeter-wave technology offers many advantages for communications and radar systems such as radio astronomy, clear-air turbulence detection, nuclear

TABLE 2 Department of Defense Microwave Frequency Bands

Designation	Frequency Range (GHz)
P band	0.225–0.390
L band	0.390–1.550
S band	1.550–3.900
C band	3.900–6.200
X band	6.200–10.900
K band	10.900–36.000
Q band	36.000–46.000
V band	46.000–56.000
W band	56.000–100.000

TABLE 3 Department of Defense New Microwave Frequency Bands

Designation	Frequency Range (GHz)	Designation	Frequency Range (GHz)
A band	0.100–0.250	H band	6.000–8.000
B band	0.250–0.500	I band	8.000–10.000
C band	0.500–1.000	J band	10.000–20.000
D band	1.000–2.000	K band	20.000–40.000
E band	2.000–3.000	L band	40.000–60.000
F band	3.000–4.000	M band	60.000–100.000
G band	4.000–6.000	N band	100.000–200.000
		O band	200.000–300.000

spectroscopy, air-traffic-control beacons, and weather radars. The advantages of millimeter waves over lower microwave and infrared systems include light-weight, small-size, broad bandwidths (several GHz), operation in adverse weather (e.g., transmitting through fog, haze, smoke, and dust), and narrow beamwidths with high resolution. The principal frequencies of interest in the millimeter-wave band are centered around 35, 60 94, 140 and 220 GHz.[3] The reason for these specific frequencies is mainly because of the atmospheric absorption of horizontally propagated millimeter waves as shown in Fig. 1. The atmospheric "windows" where absorption reaches a local minimum are found at about 35, 94, 140, and 220 GHz. The absorption peak due to O_2 at 60 GHz can be used for secure communication systems.

The developments of new materials and new technologies [e.g., superlattice and molecular beam epitaxy (MBE) as discussed in Chapter 1] and new device building blocks (e.g., planar-doped barrier and quantum well as discussed in Chapter 2) have been adopted to create new structures for microwave diodes and to extend microwave performance into the millimeter-wave band. We will consider tunnel devices, IMPATT and related transit-time devices, and transferred-electron devices in the subsequent sections and present their state-of-the-art performances.

9.2 TUNNEL DEVICES

We consider two-terminal devices associated with quantum tunneling phenomena in this section. In 1958, while studying the internal field emission in a heavily doped p-n junction, Esaki discovered an "anomalous" current-voltage characteristic in the forward direction, that is, a negative-resistance region over part of the forward characteristic. He explained this anomaly by the quantum tunneling

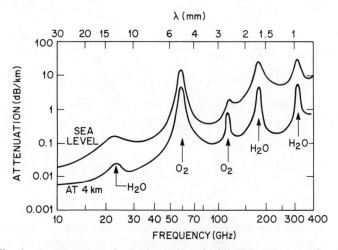

Fig. 1 Average atmospheric absorption of millimeter waves. (After Ref. 3)

concept and obtained reasonable agreement between tunneling theory and the experimental results.[4] This device is referred to as the tunnel diode or the Esaki diode.

In 1974, Chang et al. observed the resonant-tunneling in double-barrier structures.[5] A resonant-tunneling diode (RTD) can respond to electrical impulses in subpicoseconds. Therefore the RTD can provide a basis for novel electronic devices that operate at even higher frequencies than tunnel diodes.

9.2.1 Tunnel Diode

The impact of the tunnel diode on the physics of semiconductors has been large, leading to important developments such as tunneling spectroscopy, and to increased understanding of tunneling phenomenon in solids. Because of its mature technology, the tunnel diode is used in special low-power microwave applications, such as local oscillators for satellite communications and high-speed sampling. It has also been used as a highly conductive interconnect for multiple-junction cascade solar cells, as high-precision pressure and temperature sensors, and as components for high-density static random-access memory cells.

Current-Voltage Characteristic. A tunnel diode consists of a simple p-n junction in which both p and n sides are very heavily doped with impurities. Figure 2 shows a typical static current-voltage characteristic of a tunnel diode along with four band diagrams corresponding to four different bias conditions.

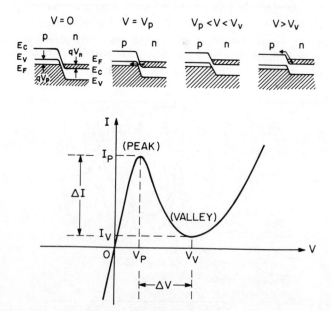

Fig. 2 Static current-voltage characteristics of a typical tunnel diode. I_p and V_p are the peak current and peak voltage, respectively. I_V and V_V are the valley current and valley voltage, respectively. The insert shows the band diagrams of the device at different bias voltages.

When there is no voltage applied to the diode, it is in thermal equilibrium (refer to the band diagram for $V = 0$). Because of the high dopings, the Fermi level is located within the allowed bands themselves. The amount of degeneracy, qV_p and qV_n is typically 50 to 200 meV, and the depletion layer width is of the order of 100 Å or less.

When a forward bias is applied, a band of energies exists for which there are filled states on the n side corresponding to available and unoccupied states on the p side. The electrons can tunnel from the n side to the p side. When the applied bias equals approximately $(V_p + V_n)/3$, the tunneling current reaches its peak value I_P, and the corresponding voltage is called the peak voltage V_P. When the forward voltage is further increased, there are fewer available unoccupied states on the p side ($V_P < V < V_V$ where V_V is the valley voltage), and the current decreases. Eventually, the band is "uncrossed," and at this point the tunneling current can no longer flow. With still further increase of the voltage the normal thermal current will flow (for $V > V_V$).

Therefore the static current-voltage characteristic is the result of the tunneling current and the thermal current. In addition, there is an excess current due to tunneling via states in the bandgap. The complete static characteristic is given by the sum of the three current components[1]:

$$I = I_t + I_x + I_{th} \tag{1}$$

where

$$I_t = \text{tunneling current} \approx I_P \left[\frac{V}{V_P} \right] \exp \left[1 - \frac{V}{V_P} \right] \tag{2}$$

$$I_x = \text{excess current} = I_V \exp[\gamma (V - V_V)] \tag{3}$$

$$I_{th} = \text{thermal current} = I_0 \exp \left[\frac{qV}{\eta kT} \right] \tag{4}$$

In these equations, γ is a constant, I_0 is the saturation current, and η is the ideality factor.

The peak current is related to the tunneling probability T_t, which can be written as[1]

$$T_t \sim \exp \left[- \frac{\alpha \sqrt{m^*} \, E_g^{3/2}}{3q\hbar \mathcal{E}} \right] , \tag{5}$$

where α is $4\sqrt{2}/3 = 1.88$ for a triangular energy barrier, and $\pi/2\sqrt{2} = 1.11$ for a parabolic energy barrier, m^* is the tunneling effective mass, E_g is the bandgap, and $\mathcal{E}$ the electric field. The electric field for an abrupt junction is

$$\mathcal{E} \approx \sqrt{\frac{2q}{\epsilon_s} \frac{N_A N_D}{N_A + N_D} (V + V_{bi})} , \tag{6}$$

where ϵ_s is semiconductor permittivity, N_A and N_D are the doping concentrations on the p side and n side, respectively, and V_{bi} is the built-in potential. From Eqs. 5 and 6, we note that higher tunneling probability (and larger tunneling current) can be obtained for higher doping concentrations, lower effective mass, and lower bandgap.

Figure 3 shows the I-V characteristics of a Ge tunnel diode at three temperatures.[6] The three current components are also shown for the case at 77 K. As expected, the tunneling-current component is essentially independent of temperature, since the tunneling probability, Eq. 5, is only slightly dependent on temperature. The excess current is a weak function of temperature; however to minimize I_V, we must reduce the excess current. The thermal current is the most temperature-sensitive component, and the ideality factor η is larger than 1 $(1 < \eta < 2)$ due to recombination of carriers in the depletion regions, which are heavily doped.

Microwave Performance.   Most tunnel diodes are made using the alloy or planar process. In the alloy process, a small dot (25 μm diameter of n-type material) is alloyed to a p-type semiconductor to form an abrupt junction, and the junction is etched back to form a mesa structure with a junction diameter of the order of a few microns to tens of microns.[7] The planar tunnel diodes are fabricated using planar technology, including oxidation, lithography, etching, and controlled alloy. The planar process generally provides more reliable devices. A typical planar tunnel diode[8] is shown in Fig. 4a.

The equivalent circuit of a tunnel diode is shown in Fig. 4b, which consists of four elements: the series inductance L_s, the series resistance R_s, the diode capacitance, and the negative diode resistance, $-R$ (which is given approximately as $-2V_P/I_P$). The real part of input impedance R_{in} of the equivalent cir-

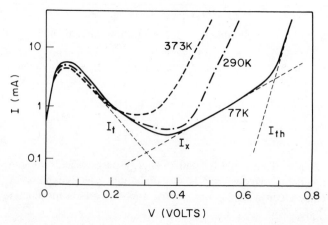

Fig. 3 Current-voltage characteristics of a Ge tunnel diode at three temperatures. The three current components are shown for the case at 77 K. (After Ref. 6)

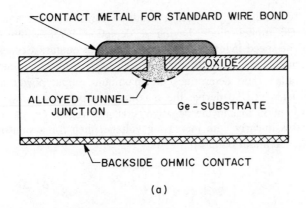

CONTACT METAL FOR STANDARD WIRE BOND

OXIDE

ALLOYED TUNNEL JUNCTION

Ge - SUBSTRATE

BACKSIDE OHMIC CONTACT

(a)

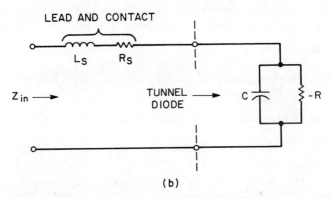

LEAD AND CONTACT

L_S R_S

Z_{in} →

TUNNEL DIODE →

C   -R

(b)

Fig. 4 (a) Cross-section view of a planar tunnel diode. (After Ref. 8) (b) Equivalent circuit of tunnel diode.

cuit is given by

$$R_{in} = R_s + \frac{-R}{1 + (\omega RC)^2}.$$ (7)

From Eq. 7, we see that R_{in} will be zero at a certain frequency, and we denote that frequency as the resistive cutoff frequency f_r at which the diode no longer exhibits negative resistance:

$$f_r \equiv \frac{1}{2\pi RC} \sqrt{\frac{R}{R_s} - 1}.$$ (8)

In most applications, it is desirable to have f_r much larger than the operating frequency.

A figure of merit for tunnel diodes is the speed index,[1] which is defined as the ratio of the peak current to the capacitance at the valley voltage, I_P/C_j. It is a measure of the current available for charging the device capacitance. Since the negative resistance is inversely proportional to the peak current, a large speed index (or small RC product) is required for fast switching. The speed index for a Ge tunnel diode having a depletion layer width of 55 Å is 40 mA/pF. (The speed index is also related to power output. See Problem 1.)

Another figure of merit is the ratio of peak current to valley current, I_P/I_V. This peak-to-valley ratio determines the maximum output power and dc-to-ac conversion efficiency. To derive the maximum output power, we shall use a simplified assumption that in the negative-resistance region ($V_P \leq V \leq V_V$) the current decreases linearly with increasing voltage; then for a sinusoidal voltage waveform, the maximum output power is

$$P_{max} = \frac{1}{2\pi} \left[\frac{\Delta I}{2} \right] \left[\frac{\Delta V}{2} \right] \int_0^{2\pi} \sin^2\theta \, d\theta = \frac{1}{8} (\Delta I \, \Delta V) , \qquad (9)$$

where ΔI and ΔV are the current and voltage ranges of the negative-resistance region (Fig. 2). A more accurate result based on a realistic I-V characteristic gives[9]

$$P_{max} = \frac{3}{16} (\Delta I \, \Delta V) = \frac{3}{16} I_P V_P \left[1 - \frac{I_V}{I_P} \right] \left[\frac{V_V}{V_P} - 1 \right] . \qquad (10)$$

The maximum dc-to-ac conversion efficiency is given by

$$\eta \equiv \frac{P_{max}}{P_{dc}} = \frac{3/16 \, (\Delta I \, \Delta V)}{(I_P - \Delta I/2) \, (V_P + \Delta V/2)} = \frac{3}{4} \frac{(\gamma_1 - 1)(\gamma_2 - 1)}{(\gamma_1 + 1)(\gamma_2 + 1)} , \qquad (11)$$

where $\gamma_1 \equiv I_P/I_V$, and $\gamma_2 \equiv V_V/V_P$. Therefore to increase both the maximum power output and the conversion efficiency, the peak-to-valley ratio γ_1 must be maximized.

Typical peak-to-valley ratios are 2:1 for SiC and InAs, 4:1 for Si, 5:1 for InP, 10:1 for Ge, and 12:1 for GaSb, GaAs, and Ga$_{0.7}$Al$_{0.3}$As. In general, the ratio for a given semiconductor can be improved by increasing the doping concentrations on both n and p sides. The ultimate limitation on the ratio depends on the peak current, which increases with decreasing tunneling mass and bandgap, and on the valley current, which decreases with decreasing concentration of energy levels in the bandgap.

In a tunnel diode, very high doping concentrations are required on both sides of the junction (N_A, $N_D \geq 10^{19}$ cm^{-3}) to achieve a high tunneling current. These high dopings result in relatively large capacitance, which limits the cutoff frequency f_r. The achieved maximum experimental oscillation frequency (from a GaAs tunnel diode)[10] is 103 GHz with a power output of a few tenths μW. Higher power outputs have been obtained at lower frequencies: about 1 mW at 10 GHz and 0.2 mW at 50 GHz.

9.2.2 Superlattice Tunnel Diode

A superlattice tunnel diode (STD) consists of two semiconductor superlattices separated by a moderately thin tunneling barrier.[11] As described previously in

Chapter 2, a superlattice is an artificial semiconductor formed by compositionally modulated multi-layer structures (e.g., GaAs/AlGaAs system) in which the thickness of each layer is less than the de Broglie wavelength (defined as $\lambda = \hbar/p$ where $\hbar$ is the reduced Planck's constant and p is the carriers momentum). For most semiconductors, λ is of the order of 200 Å at 300 K. Because of the strong quantum-mechanical effects in a superlattice, minibands are formed with typical bandwidths and separations between minibands of tens of meV.

The current-voltage characteristic of a STD at room temperature is shown in Fig. 5. It is quite similar to that of the classic tunnel (Esaki) diode. The structure is made of the GaAs/AlGaAs material system that is grown in a molecular beam epitaxial system, with silicon as the n-type dopant, on a semi-insulating (100) GaAs substrate. In order of growth, the layers are (1) 0.5 μm of GaAs (doped to 4×10^{18} cm^{-3}) as a buffer layer, (2) 20 Å of undoped GaAs as a spacer layer to act as a barrier to defect diffusion, (3) a superlattice section consisting of three 25-Å undoped Al$_{0.25}$Ga$_{0.75}$As layers alternating with three 60-Å GaAs layers (doped to 5×10^{17} cm^{-3}), (4) the central barrier layer of 40-Å undoped AlGaAs layer, (5) a second superlattice section consisting of three 60-Å GaAs layers (doped to 5×10^{17} cm^{-3}) alternating with three 25-Å undoped AlGaAs layers, (6) an undoped 20-Å GaAs spacer layer, and (7) an 0.2-μm GaAs capping layer (doped to 4×10^{18} cm^{-3}). Note the total thickness of the active region is only 550 Å.

Since the average doping in the superlattice is relatively high $(3.5 \times 10^{17}$ cm$^{-3})$, the superlattice has high conductivity similar to a "metallic" system with the Fermi level in the lowest miniband. The band diagram of

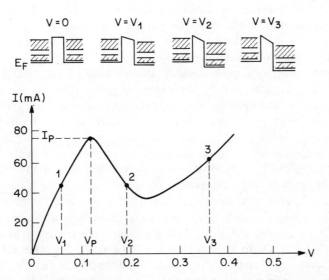

Fig. 5 Current-voltage characteristics of a superlattice tunnel diode at room temperature. The insert shows the band diagrams of the diode at different bias voltages. (After Ref. 11)

the STD is therefore similar to that of a metal-insulator-metal tunnel barrier. However, the superlattice miniband structure is imposed on the initial and final energy states for the tunneling process. The band diagram in Fig. 5 for $V = 0$ is for the thermal equilibrium condition. We have shown two minibands for the left-side and two for the right-side superlattice. The necessary conditions for tunneling in a STD are the same as for the tunnel diode: we must have occupied energy states on the left-side superlattice, and unoccupied energy states at the same energy on the right-side superlattice.

When a voltage is applied to the STD, most of it will be dropped across the central barrier. At low bias voltages (the band diagram for $V = V_1$, in Fig. 5) the current increases with the bias voltage, since carriers can tunnel from the first miniband on the left side to the first miniband on the right side. When the voltage drop across the tunnel barrier exceeds the width of the first miniband, there are no available states on the right side and the current decreases (the band diagram for $V = V_2$). The current will increase again when the voltage is increased by an amount equal to the separation of the first and the second minibands, where carriers can tunnel from the first miniband on the left side to the second miniband on the right side (the band diagram for $V = V_3$).

A STD also has three current components, similar to that of a tunnel diode. The tunneling current component has already been discussed. The thermal current component results from the thermally activated carrier injection across the top of the central barrier. The excess current component is from conduction of electrons that are thermally excited into the second miniband, where the effective tunnel barrier is lower, or from scattering in the tunneling barrier, which produces transitions that are not allowed by total energy or transverse momentum conservation between the initial and final states. To minimize the excess current, one can increase the miniband gap (i.e., by changing the parameters of the superlattices) or by reducing the thickness of the tunneling barrier (i.e., by reducing the number of scattering events).

The room-temperature peak-to-valley current ratio of the STD (Fig. 5) is about 2:1. It is expected that by improving the growth process, by minimizing the excess current, and by optimizing the material parameters, the peak-to-valley current ratio along with other device characteristics can be substantially improved.

The STD has certain intrinsic advantages over the tunnel diode (TD): the high capacitance can be reduced by vertically stacking several STDs by repeating the MBE growth cycles (stacking TDs is very difficult to implement), and the absence of impurity atoms in the high-field region in STD eliminates one source of TD failure due to field-assisted migration of these impurities.

It should be pointed out that STDs are quite different from the resonant-tunneling diode, RTD (to be considered in Section 9.2.3). One difference is that STDs tunneling is nonresonant. A second difference is that the small miniband width reduces the voltage required to achieve negative differential resistance (e.g., the peak voltage is 0.12 V for the STD as compared to 0.3 to 2 V for the RTD).

9.2.3 Resonant-Tunneling Diode

The resonant tunneling in a semiconductor double-barrier structure results from spatial quantization; that is, for a carrier de Broglie wavelength comparable to the barrier thickness, the carrier momentum in the longitudinal direction is quantized giving rise to a set of discrete energy levels. The existence of discrete energy levels in a double-barrier structure was first observed in 1974 in a quantum-well structure.[5] The resonant-tunneling experiments at terahertz frequencies (10^{12} Hz) using a double-barrier structure in 1983 has stimulated renewed interest in negative-resistance devices.[12] Because of the improvements in fabrication technology (especially the MBE process), peak-to-valley ratios of 30:1 at 300 K and oscillating frequencies in excess of 400 GHz have been obtained.[13,14] It is expected that the resonant tunneling diode will be a key device for analog applications in the submillimeter-wave region, and a major building block for multi-terminal high-speed logic devices (refer to Chapters 2 and 8).

Transmission Coefficient.   A schematic double-barrier structure RTD is shown in Fig. 6a where the two barriers are assumed to be identical (a symmetric RTD).[15a] The basic device parameters are the barrier height E_0 (~ 1 eV for GaAs/AlAs/GaAs barrier), the barrier width L_B (typically 20 to 50 Å), and the quantum-well width L_W (typically 50 to 100 Å). Also shown in Fig. 6a are the quantized discrete energy levels (i.e., E_1, E_2, etc.).

When an incident electron has a longitudinal energy E that exactly equals one of the discrete energy levels, it will transit through the symmetric double barrier with a unity transmission coefficient (i.e., the resonant-tunneling phenomenon). However, the transmission coefficient decreases rapidly as the energy deviates from the discrete energy levels. For example, an electron with an energy 10 meV higher or lower than the level E_1 will result in 10^5 times reduction in the transmission coefficients as depicted in Fig. 6b.

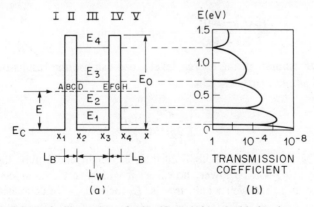

Fig. 6 (a) Schematic illustration of AlAs/GaAs/AlAs double-barrier structure (resonant tunneling diode) with 25-Å barriers and 70-Å well. (b) Transmission coefficient versus electron energy for the structure. (After Ref. 15a)

To calculate the transmission coefficient, we consider Fig. 6a where the five regions (I, II, III, IV, V) are specified by the coordinates (x_1, x_2, x_3, x_4). The Schrödinger equation for the electron in any region can be written as

$$-\frac{\hbar^2}{2m_i^*}\left[\frac{d^2\psi_i}{dx^2}\right] + V_i\psi_i = E\psi_i, \qquad i = 1,2,3,4,5 \qquad (12)$$

where $\hbar$ is the reduced Planck constant, m_i^* the effective mass in the ith region, E the incident energy, and V_i and ψ_i are the potential energy and the wave function in the ith region, respectively. The wavefunction ψ_i can be expressed as

$$\psi_i(x) = A_i \exp(jk_ix) + B_i \exp(-jk_ix) , \qquad (13)$$

where A_i and B_i are constants to be determined from the boundary conditions, and $k_i = \sqrt{2m_i^*(E - V_i)}/\hbar$. Since the wave functions and their first derivatives (i.e., $\psi_i/m_i^* = \psi_{i+1}'/m_{i+1}^*$) at each potential discontinuity must be continuous, we obtain the transmission coefficient (for identical effective mass across the five regions)[16]

$$T_t = \frac{1}{1 + E_0^2(\sinh^2\beta L_B)H^2/[4E^2(E_0-E)^2]} , \qquad (14)$$

where

$$H \equiv 2[E(E_0-E)]^{1/2}\cosh\beta L_B \cos kL_W - (2E - E_0)\sinh \beta L_B \sin kL_W$$

and

$$\beta \equiv \frac{\sqrt{2m^*(E_0-E)}}{\hbar}, \qquad k = \frac{\sqrt{2m^* E}}{\hbar} .$$

The resonant condition occurs when $H = 0$, and thus $T_t = 1$. The resonant-tunneling energy levels E_n can be calculated by solving the transcendental equation:

$$\frac{2[E(E_0-E)]^{1/2}}{(2E - E_0)} = \tan kL_W \tanh \beta L_B . \qquad (15)$$

As a first-order estimate of the energy levels, one can use the results of a quantum well with infinite barrier height:

$$E_n \approx \left[\frac{\pi^2\hbar^2}{2m^*L_W^2}\right]n^2 . \qquad (16)$$

For a double-barrier structure with finite barrier height and width, the energy level (for a given n) will be lower; however, it will have a similar dependence on the effective mass and well width, that is, E_n increases with decreasing m^* or L_W.

The calculated energy levels, E_n, at which the transmission coefficient exhibits the first and second resonant peaks in GaAs/AlAs double-barrier structure are

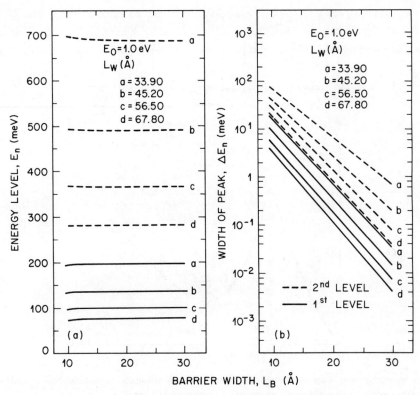

Fig. 7 (a) Calculated energy of electrons at which the transmission coefficient shows the resonant peak in AlAs/GaAs/AlAs structures as function of barrier width L_B for various well widths L_W. (b) Full-width at half-maximum (FWHM) versus barrier width for the first and second resonant peaks. (After Ref. 15b)

shown in Fig. 7a as a function of barrier width L_B with the well width L_W as a parameter.[15b] It is apparent that E_n is essentially independent of L_B, but is dependent on L_W as pointed out previously (Eq. 16). The calculated widths of the peak, ΔE_n (i.e., the full-width at half-maximum point where $T_t = 0.5$) are shown in Fig. 7b as functions of L_B and L_W. For a given L_W, the width ΔE_n decreases exponentially with L_B.

The resonant-tunneling current I_t is related to the integrated flux of electrons whose energy is in the range where the transmission coefficient is large. Thus, I_t is proportional to the width ΔE_n, and sufficiently thin barriers are required to achieve a high current density.

Current-Voltage Characteristics. Figure 8 shows a schematic current-voltage characteristic along with electron energy diagrams for various dc voltages.[12] At thermal equilibrium, $V = 0$, the energy diagram is similar to that in Fig. 6a (here only the lowest energy level E_1 is shown). As we increase the applied

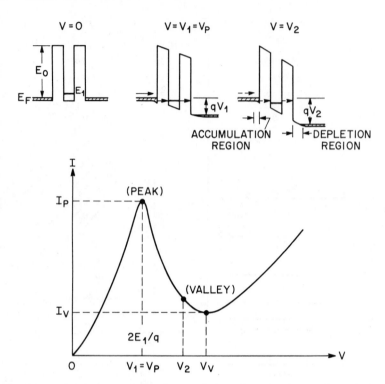

Fig. 8 Schematic current-voltage characteristics of resonant tunneling diode. Also shown are the energy diagrams at various dc voltages. (After Ref. 12)

voltage, the electrons (i.e., occupied energy states) near the Fermi level to the left side of the first barrier tunnel into the quantum well, and subsequently tunnel through the second barrier into the unoccupied states in the right side. Resonance occurs when the energy of the injected electrons becomes approximately equal to the energy level E_1 where the transmission probability is maximum. (Under bias, the transmission through a symmetric RTD cannot be unity because of the field-induced asymmetry. However, it can be very close to unity if the barriers are very thin). This is illustrated by the energy diagram for $V = V_1 = V_P$ where the conduction band edge on the left side is lined up with E_1. The magnitude of the peak voltage must be at least $2E_1/q$, but is usually larger because of additional voltage drops in the accumulation and depletion regions:

$$V_P > \frac{2E_1}{q}. \tag{17}$$

When the voltage is further increased, that is, at $V = V_2$, the conduction band edge is above E_1 and the number of electrons that can tunnel decreases, resulting in a small current. The valley current I_V is due mainly to the excess current components I_x, such as electrons that tunnel via an upper valley (e.g., X-valley in AlAs) in the barrier. At room temperature and higher, there are other components due to inelastic tunneling current, either phonon assisted or impurity assisted. To minimize the valley current, we must improve the quality of the heterojunction interfaces and eliminate impurities in the barrier and well regions. For even higher applied voltages, $V > V_V$, we have the thermionic current component I_{th}, due to electrons injected through higher discrete energy levels in the well or thermionically injected over the barriers. The current I_{th} increases monotonically with increasing voltage similar to that of a tunnel diode. To reduce I_{th}, we should increase the barrier height and design a diode that operates at relatively low bias voltages.

The measured I-V characteristics of a mesa-type RTD is shown[17] in Fig. 9. The cross section of the RTD is also shown in the insert. The alternating GaAs/AlAs layers are grown sequentially by molecular beam epitaxy on an n^+-GaAs substrate. The barrier widths are 17 Å and the well width is 45 Å. Assuming a 1-eV barrier height, the first energy level E_n ($n = 1$) occurs at about 140 meV and ΔE_1 is about 2 meV. The active regions are defined with ohmic contacts. The top contact is used as a mask to isolate the region under the contact by etching mesas. The peak-to-valley ratio is 4:1 for this device at room temperature.

An RTD suitable for microwave integrated-circuit applications[18] is shown in the insert of Fig. 10. The device material is grown by MBE on semi-insulating GaAs substrate. A 0.9-μm n^+ layer (Si-doped to 10^{18} cm^{-3}) is grown to provide the bottom ohmic contact. A 700-Å undoped GaAs spacer layer is grown on top of this, followed by the double barriers. The latter consists of a 40-Å GaAs well sandwiched between two 15-Å AlAs barriers. This is followed by a 100-Å undoped spacer layer and a 0.45-μm top ohmic-contact layer (doped to 10^{18} cm^{-3}). A four-step masking process is employed to fabricate the RTD

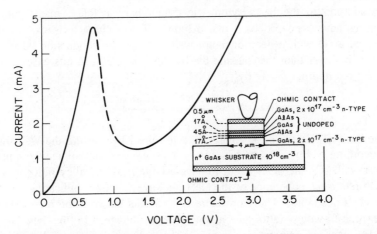

Fig. 9 Measured current-voltage characteristics of a mesa-type resonant tunneling diode. The insert shown the cross section of the diode. (After Ref. 17)

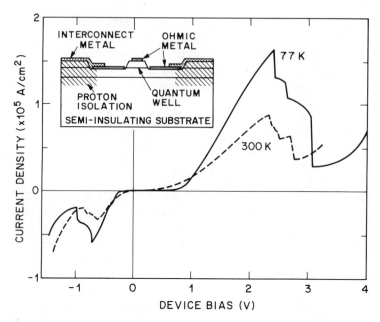

Fig. 10 Static current-voltage characteristics of a proton-implanted, microwave-compatible resonant-tunneling diode. The insert shown the device cross section. (After Ref. 18)

structure and to form ohmic contacts. Device isolation is achieved by proton implantation. The device exhibits peak current densities of 0.9×10^5 and $1.7 \times 10^5 \, A/cm^2$ at 300 and 77 K, respectively, with corresponding peak-to-valley ratios of 2.5 and 6.1. The average specific negative resistance is found to be about $-650 \, \Omega\text{-cm}^2$.

In addition to the basic symmetric double-barrier RTD, many other RTD structures have been proposed that exhibit different characteristics. Figure 11a shows an RTD with a deep quantum well, that is, the conduction band edge of the well is lower than that outside the barriers by ΔE_0. In this case, peak voltage is given approximately by[19]

$$V_P \approx \frac{2(E_1 - \Delta E_0)}{q}. \tag{18}$$

In the basic RTD, the peak voltage is determined by varying the well thickness L_W according to Eqs. 16 and 17. In order to reduce peak voltage we must use a thicker well. On the other hand, to have a large peak-to-valley ratio, it is necessary to retain the coherence of the electron wave function, indicating that a thinner well is desirable. Therefore there is a conflict between the requirement of a large peak-to-valley ratio and a small peak voltage. In the new structure, Fig. 11a, we can employ a thinner well to realize a large peak-to-valley ratio, and at the same time adjust ΔE_0 for a small peak voltage. One approach is to

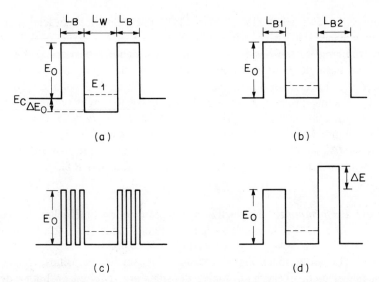

Fig. 11 Band diagrams of four resonant-tunneling diodes: (a) deeper quantum well, (b) thicker right-side barrier, (c) superlattice barrier, and (d) low-high barrier.

use an $In_yGa_{1-y}As$ quantum well in the double barrier structure with $Al_xGa_{1-x}As$ and GaAs emitter and collector. The deep quantum well results from the fact that $In_yGa_{1-y}As$ has a narrower bandgap than GaAs. Although $In_yGa_{1-y}As$ is not lattice-matched to the $Al_xGa_{1-x}As$ barrier, it is thin enough (< 100 Å) to form a dislocation-free strained-layer (also called a pseudo-morphic) quantum well. For an RTD with 23.2 Å AlAs (with $x = 1$) barriers and 41.5 Å $In_{0.7}Ga_{0.3}As$ well layer, a peak-to-valley ratio of 51:1 has been obtained at 77 K.[13]

In an RTD that is symmetric in both composition and doping concentration (e.g., Fig. 6a), the *I-V* characteristics are antisymmetric about the origin. If we increase the right-side barrier thickness ($L_{B2} > L_{B1}$), we obtain an asymmetric RTD as shown[20] in Fig. 11b. The *I-V* characteristic of such an RTD is also asymmetric, that is, there is a negative differential resistance region for positive biases (+V on right side) and none for negative biases. Many different *I-V* curves can be generated by varying the relative thicknesses of the two barriers.

By replacing the single hetero-barriers with thin, short-period superlattices, one can form a superlattice RTD as shown[21] in Fig. 11c. The structure can reduce the roughness at the heterojunction interfaces by superlattice smoothing, and the barrier height is determined by the coupled states with the superlattice. A symmetric superlattice RTD has been fabricated. It has a quantum well of 45 Å undoped GaAs, sandwiched between two superlattice barriers consisting of three 7.3-Å AlAs layers (3 monolayers) alternating with three 7.3-Å GaAs layers (3 monolayers). The device has an antisymmetric *I-V* characteristic. The peak-to-valley ratio is 3:1 at 300 K and 17:1 at 80 K.

Figure 11d shows a low-high barrier RTD. Such a device can lead to a substantially enhanced negative differential resistance and a large peak-to-valley ratio as compared to the basic RTD having two identical barriers.[22] It is clear that for the basic RTD at resonant-tunneling condition (Fig. 8, $V = V_P$), the right barrier is lower than the left barrier due to the applied voltage. To maximize the tunneling process, it is desirable to make the right barrier higher so that at resonance the effective barrier heights E_0 and E'_0 are equal (see insert of Fig. 12). Figure 12 shows the calculated I-V characteristic of a low-high RTD ($Al_{0.3}Ga_{0.7}As/Al_{0.45}Ga_{0.55}As$). The peak-to-valley ratio in the forward direction is 11:1, while a symmetric $Al_{0.3}Ga_{0.7}As/GaAs/Al_{0.3}Ga_{0.7}As$ RTD has a ratio of only 3.5:1.

Microwave Performance. Neglecting resonant-tunneling time delay, the equivalent circuit of a RTD is identical to that of a tunnel diode (Fig. 4b). The cutoff frequency due to RC delays is given by the same expression, Eq. 8. However, f_r for the RTD can be much higher because of its smaller parasitics. In the p-n junction tunnel diode, very high doping densities are required on both sides of the junction ($\geq 10^{19}$ cm^{-3}), and therefore the capacitance is relatively large. For an RTD, the main contribution to the capacitance is from the depletion region (refer to the band diagram for $V = V_2$ in Fig. 8). Since the doping density there can be much lower ($\approx 10^{17}$ cm^{-3}), the depletion capacitance is relatively lower. For the RTD in Fig. 9, the series resistance is 15 Ω, the

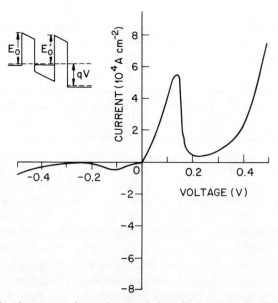

Fig. 12 Calculated current-voltage characteristics of a low-high barrier resonant tunneling diode. Insert shows the band diagram of a low-high barrier under resonant condition. (After Ref. 22)

capacitance is 20 fF, and the negative resistance is $-77 \, \Omega$. Therefore, the maximum cutoff frequency[23] is about 200 GHz.

In addition to the limitation due to circuit components considered above, the maximum oscillation frequency can also be limited by the transit time τ_{depl} through the depletion layer, and the lifetime τ_w of an electron in the quantum well. The frequency limit due to τ_{depl} is given roughly by

$$f_{depl} \approx \frac{1}{2\pi \, \tau_{depl}} \qquad (19)$$

and for a depletion layer thickness of 700 Å (insert in Fig. 9) and a carrier velocity of 10^7 cm/s, f_{depl} is over 220 GHz. The lifetime is given by the uncertainty principle, $\tau = \hbar/\Delta E_n$, and the frequency limit corresponding to the lifetime can be approximated by[24]

$$f_{\Delta E_n} \approx \frac{1}{2\pi(\hbar/\Delta E_n)} \, . \qquad (20)$$

The value of ΔE_1 for the RTD in Fig. 9 is about 2 meV, so that $f_{\Delta E_n}$ is about 300 GHz. Therefore the most important frequency limitation of this RTD is due to the circuit that limits the maximum cutoff frequency to about 200 GHz. The maximum observed fundamental oscillation frequency for this device is, in fact, 201 GHz.[25]

By minimizing the circuit parasitics, substantially higher oscillation frequencies up to several hundred gigahertz are expected. For example, an RTD with very thin (11 Å) AlAs barriers has yielded oscillations up to 420 GHz at room temperature.[14]

For the fastest Ge and GaAs p-n junction tunnel diodes, the speed indices are 40 and 70 mA/pF, respectively. For the RTD (Fig. 9), the peak current density is 4×10^4 A/cm^2 and the specific capacitance is about 1×10^5 pF/cm^2. Thus, the speed index of the RTD is 400 mA/pF, which is a factor of 5 to 10 better than tunnel diodes.[26]

Neglecting the resonant-tunneling time and transit time across the depletion layer, the power output and the conversion efficiency for RTDs are given by the same expressions as for the p-n junction tunnel diodes, Eqs. 10 and 11. To increase the output power and efficiency, the peak-to-valley ratio must be maximized. Output powers of 200 μW at 20 GHz, 60 μW at 56 GHz, 2 μW at 227 GHz, and about 0.2 μW at 420 GHz have been obtained.[14,25,26]

9.3   IMPATT AND RELATED TRANSIT-TIME DIODES

The basic operational principles of IMPATT (*imp*act ionization *a*valanche *t*ransit *t*ime) diodes have been considered in Ref. 1. Figure 13 shows a schematic diagram of an idealized transit-time diode that consists of two regions—the injection region and the drift region. For an IMPATT diode, charge carriers are generated in the injection region by impact ionization under high electric field

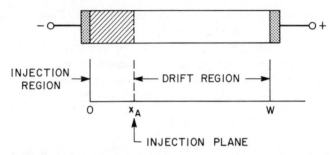

Fig. 13 Idealized transit-time diode with an injection region and a drift region. Carriers are injected into the drift region at $x = x_A$.

and subsequently injected into the drift region. Thus there are two delays—the "avalanche delay" due to the finite built-up time of the avalanche current and the "transit-time delay" due to the finite time for carriers to cross the drift region. These two delays cause the current to lag behind the voltage and give rise to a negative resistance. The IMPATT diode was proposed by Read[27] in 1958, and the first observation of IMPATT oscillation was obtained by Johnston et al.[28] in 1965.

When the drift region in Fig. 13 consists of two segments in which the velocity of carriers in one segment is significantly less than that in the second segment, higher conversion efficiency is expected. This device is called a DOVATT (double-velocity avalanche transit time) diode.[29]

When the electric field in the injection region is sufficiently high such that in addition to impact ionization, some carriers are injected into the drift region by tunneling processes, then we have the MITATT (mixed tunnel-avalanche transit time) mode.[30] If the electric field in the injection region becomes so high that only tunneling can occur, we have the TUNNETT (tunnel transit time) mode.[31]

Two related transit-time diodes are the BARITT (barrier injection transit time) diode and the QWITT (quantum-well injection transit time) diode. For the BARITT diode, carriers are injected into the drift region by thermionic emission over a potential barrier.[32] In a QWITT diode, the injection region consists of a quantum well; when the carrier energy lines up with the resonant energy level in the quantum well, the carriers will be injected into the drift region.[33]

IMPATT diode is one of the most powerful solid-state sources of microwave power. It can generate the highest cw (continuous wave) power at millimeter-wave frequencies, and is most extensively used in that frequency range (i.e., 30 to 300 GHz). Other transit-time devices may complement IMPATT diodes, because they have different characteristics and properties such as lower noise (BARITT diode), higher efficiency (DOVATT and QWITT diodes), or higher frequency of operation (MITATT and TUNNETT diodes).

9.3.1 Static Characteristics

The basic members of the IMPATT diode family are the single-drift devices and the double-drift devices. Figure 14 shows the single-drift IMPATT diodes in

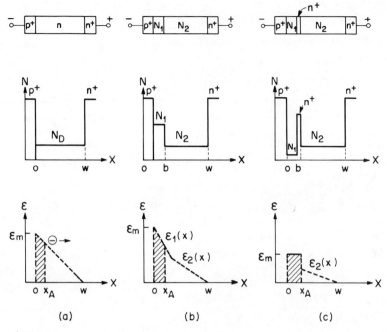

Fig. 14 Doping profiles and electric field distributions at avalanche breakdown condition of three single-drift IMPATT diodes: (a) one-sided abrupt p-n junction, (b) hi-lo structure, and (c) lo-hi-lo structure.

which only one type of charge carriers (i.e., electrons) is traversing the drift region.

Figure 14a shows the doping profile and electric field distribution at avalanche breakdown condition of a one-sided abrupt p-n junctions. Because of the strong dependence of the ionization rates on an electric field, most of the avalanche multiplication processes occurs in a narrow region near the highest field between 0 and x_A (cross-hatched area) where x_A is the width of the avalanche region (i.e., the distance over which 95% of the contribution to the ionization integrand is obtained).[1]

Figure 14b shows the hi-lo structure, in which a high doping N_1 region is followed by a lower doping N_2 region. With proper choices of the doping N_1 and its thickness b, the avalanche region can be confined within the N_1 region. In the limit when N_2 becomes intrinsic, we have the Read diode (p^+-n-i-n^+). However, we generally refer Fig. 14b as the Read diode even though N_2 is not intrinsic. Figure 14c is the lo-hi-lo structure, in which a "clump" of donor atoms is located at $x = b$. Since a nearly uniform high-field region exists from $x = 0$ to $x = b$, the avalanche region x_A is equal to b, and the maximum field can be much lower than that for a Read diode.

Figure 15 shows the double-drift devices in which both electrons and holes participate in device operation over two separate drift regions. The double-drift devices have higher conversion efficiency and higher output power than single-

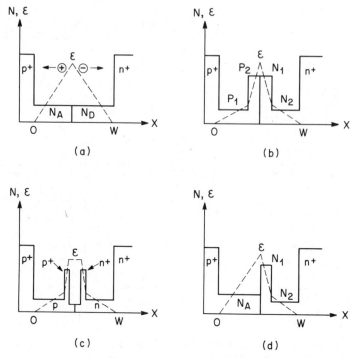

Fig. 15 Doping profiles and electric-field distributions of four double-drift IMPATT diodes: (a) flat-profile, (b) hi-lo structure or double-drift Read diode, (c) lo-hi-lo structure, and (d) hybrid Read structure in which the *p*-side has a flat-profile and the *n*-side has a hi-lo profile.

drift devices. Figure 15a describes the doping profile and electric-field distribution of a two-sided abrupt *p-n* junction. This structure is also called a double-drift flat-profile device (DDF). The avalanche region is located near the center of the depletion layer.

Figure 15b shows a double-drift hi-lo structure or a double-drift Read diode (DDR) that consists of a lo-hi structure on the *p* side and a hi-lo structure on the *n* side. Figure 15c shows a double-drift lo-hi-lo structure, the avalanche region is given by the distance between the p^+ clump and the n^+ clump. Figure 15d shows the double-drift hybrid Read structure (DDH) in which the *p* side has a flat doping profile but the *n* side has a Read-type hi-lo profile.

The selection of a particular device structure depends on many factors, such as the operating frequency, the dc-to-ac conversion efficiency, power output and ease of fabrication. The double-drift lo-hi-lo structure (Fig. 15c) is expected to have the highest efficiency, but it is also the most difficult to fabricate. The double-drift hybrid Read diode (Fig. 15d) is a good compromise, since it has a good efficiency and is relatively easier to make. Of course, the simplest structure is the single-drift *p-n* junction (Fig. 14a).

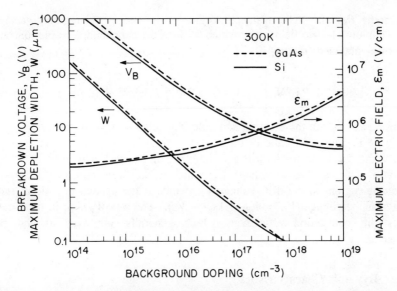

Fig. 16 Breakdown voltage, maximum depletion width, and maximum electric field at breakdown for GaAs and Si one-sided abrupt *p-n* junctions. (After Ref. 34)

Breakdown Voltage. The breakdown voltage V_B (including the built-in potential V_{bi}) is given by the area underneath the electric-field versus distance plot (Figs. 14 and 15). For the single-drift one-sided abrupt junction (Fig. 14a), V_B is simply given by $\mathcal{E}_m W/2$. The results[34] for GaAs and Si are shown in Fig. 16. Note that as the doping increases both the breakdown voltage and the depletion width decrease. However, the maximum electric field increases with increasing doping. At dopings near or above 10^{18} cm^{-3}, tunneling mechanism also contributes to the breakdown process and eventually dominates.

For the hi-lo diode and the lo-hi-lo diode, the breakdown voltages are given, respectively, by[1]

$$V_B(\text{hi-lo}) = \left[\mathcal{E}_m - \frac{qN_1b}{2\epsilon_s}\right]b - \frac{1}{2}\left[\mathcal{E}_m - \frac{qN_1b}{\epsilon_s}\right](W - b) \qquad (21)$$

$$V_B(\text{lo-hi-lo}) = \mathcal{E}_m b + \frac{1}{2}\left[\mathcal{E}_m - \frac{qQ}{\epsilon_s}\right](W - b) \qquad (22)$$

where Q is the number of impurities/cm^2 in the clump. The maximum field at breakdown for a hi-lo diode with a given N_1 is the same as the value of the one-sided abrupt junction with the same N_1. The maximum field of a lo-hi-lo structure can be calculated from the position-independent ionization coefficient. Similar approaches can be used to obtain breakdown voltages for various double-drift diodes.

Avalanche Region.   The avalanche region is defined as the region adjacent to the maximum electric field, over which 95% of the contribution to the ionization integral is obtained, that is

$$\int_0^{x_A} \alpha_n \exp\left[-\int_0^x (\alpha_p - \alpha_n)dx'\right] dx = 0.95 , \qquad (23)$$

where α_n and α_p are the electron and hole ionization rate, respectively.

Figure 17 shows the ratio of the avalanche region width to the total depletion layer width for some single-drift and double-drift diodes.[35] This ratio determines the ratio of the avalanche voltage V_A, which is the dc voltage drop across the avalanche region, to the drift voltage V_D, which is the applied dc voltage minus the avalanche voltage ($V_D = V_B - V_{bi} - V_A$). The ratio V_A/V_D, in turn, determines the conversion efficiency, which generally improves as the ratio decreases.

9.3.2 Dynamic Characteristics

We now consider the injection delay and transit-time effect of the idealized device shown in Fig. 13. Assume that a conduction current pulse is injected at $x = x_A$ with a given phase angle ϕ with respect to the total current; also assume that the applied dc voltage across the diode causes the injected carriers to travel at the saturation velocity v_s in the drift region; the ac resistance R is given by[36]

$$R = \frac{\cos \phi - \cos(\phi + \theta)}{\omega C \theta} . \qquad (24)$$

The angle θ is the drift-region transit angle $\omega(W - x_A)/v_s$, and C is the diode capacitance per unit area $\epsilon_s/(W - x_A)$ where ϵ_s is the dielectric permittivity.

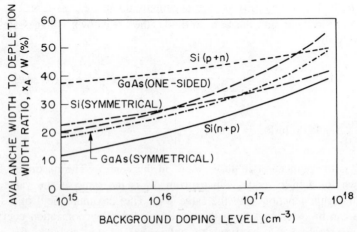

Fig. 17   Ratio of avalanche region width to total depletion-layer width as a function of doping for GaAs and Si diodes. (After Ref. 35)

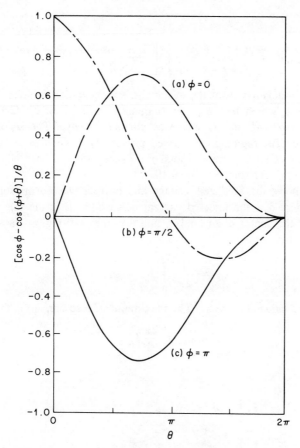

Fig. 18 Normalized ac resistance versus transit angle for three different injection phase delays: (a) $\phi = 0$, (b) $\phi = \pi/2$, and (c) $\phi = \pi$.

When ϕ equals to zero (no injection phase delay), the resistance is proportional to $(1 - \cos \theta)/\theta$, which is always greater or equal to zero as shown in Fig. 18a, that is, there is no negative resistance. Therefore, in the structure, transit-time effect alone cannot give rise to negative resistance. However, for any non-zero ϕ, the resistance is negative for certain transit angles. For example, at $\phi = \pi/2$, the largest negative resistance occurs near $\theta = 3\pi/2$ (Fig. 18b). This is the basic operation of the BARITT diode and TUNNETT diode. For $\phi = \pi$, the largest negative resistance occurs near $\theta = 3\pi/4$ (Fig. 18c). This corresponds to the IMPATT operation. This analysis confirms the importance of the injection delay. The task of finding active transit-time devices has thus been reduced to finding a means to delay the injection of conduction current into the drift region.

For IMPATT diodes, small-signal analysis indicates that the total resistance is the sum of the resistances of the avalanche region, the drift region and a series resistance R_s:

$$R = \left[\frac{1}{1 - \omega^2/\omega_r^2} \right] \left[\frac{1 - \cos\theta}{\omega C\theta} \right] + R_s , \qquad (25)$$

where the first bracket is from the avalanche region, which behaves like an LC parallel circuit in which the resonant frequency ω_r is given by $(2\alpha' v_s J_0/\epsilon_s)^{1/2}$ and where $\alpha' \equiv d\alpha/d\mathcal{E}$ and J_0 is the dc current density. The second bracket comes from the drift region as discussed previously. When $\omega > \omega_r$, the first bracket becomes negative, and the total resistance also becomes negative, provided that the series resistance R_s is small.

Figure 19 shows the idealized voltage and current waveforms for a transit-time diode, where δ is the injected current pulsewidth, ϕ is the injection phase delay and is at the center of the pulse, and θ is the drift-region transit angle.[30] The dc current is

$$I_{dc} = \frac{1}{2\pi} \int_0^{2\pi} I_{\text{ind}} \, d(\omega t) , \qquad (26)$$

where I_{ind} is the induced current. The maximum induced current is then

$$I_{\max} = \frac{2\pi}{\theta} I_{dc} . \qquad (27)$$

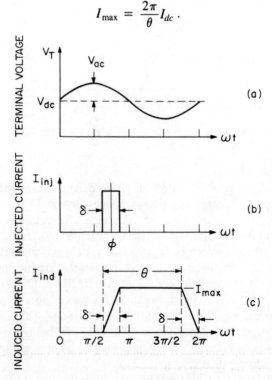

Fig. 19 Idealized voltage and current waveforms for a transit-time diode. (After Ref. 30)

The dc power input is

$$P_{dc} = V_{dc} I_{dc} .$$ (28)

The ac power output is

$$P_{ac} = \frac{1}{2\pi} \int_0^{2\pi} I_{ind}(\omega t) V_{ac} \sin \omega t \, d(\omega t) .$$ (29)

The dc-to-ac conversion efficiency is obtainable from Eqs. 28 and 29:

$$\eta \equiv \frac{P_{ac}}{P_{dc}} = \frac{V_{ac}}{V_{dc}} \frac{\sin(\delta/2)}{\delta/2} \frac{\cos\phi - \cos(\phi + \theta)}{\theta} .$$ (30)

For IMPATT operation, $\phi \approx \pi$, and Eq. 30 reduces to

$$\eta = \frac{V_{ac}}{V_{dc}} \frac{\sin(\delta/2)}{(\delta/2)} \frac{\cos\theta - 1}{\theta} .$$ (31)

To improve η, the pulsewidth δ should be reduced. For a given θ, the best efficiency can be obtained under the sharp-pulse condition (i.e., $\delta = 0$):

$$\eta = \frac{V_{ac}}{V_{dc}} \frac{\cos\theta - 1}{\theta} .$$ (32)

The maximum efficiency is obtained when $\theta = 0.74\pi$ where

$$\eta = \frac{2.27}{\pi} \frac{V_{ac}}{V_{dc}} .$$ (33)

If $V_{ac}/V_{dc} = 0.5$, the maximum efficiency is about 36%.

9.3.3   Power and Efficiency Limitations

At lower frequencies, the cw performance of an IMPATT diode is limited by thermal considerations, that is, by the power that can be dissipated in a semiconductor chip. The major contributions to the thermal resistance R_T are (1) the semiconductor layer, $d_s/A\varkappa_s$ where d_s is the layer thickness, A the device area, and $\varkappa_s$ the semiconductor thermal conductivity, and (2) the thermal spreading resistance of the heat sink, $(4\varkappa_c)^{-1} (\pi/A)^{1/2}$, where $\varkappa_c$ is the thermal conductivity of the heat sink ($\varkappa_c$ is 3.9 W/cm-K for copper and 20 W/cm-K for diamond). Clearly, the thermal resistance decreases as the semiconductor thickness decreases and as the diode area increases. Also a diamond heat sink can substantially reduce the thermal resistance. Under thermal limitations,[1] the power output will decrease as $1/f$.

Figure 20 shows the calculated maximum cw power output versus frequency for mesa structures on copper and diamond heat sinks for millimeter-wave Si and GaAs double-drift hybrid diodes.[37] The use of a diamond heat sink can increase the power output by a factor of 3 or more.

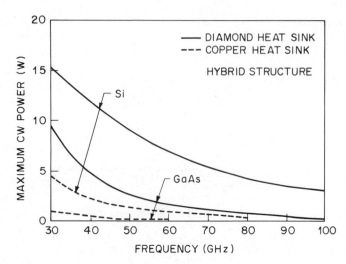

Fig. 20 Calculated maximum cw power output versus frequency for mesa structure on copper and diamond heat sinks for millimeter-wave GaAs and Si double-drift hybrid diodes. (After Ref. 37)

At higher frequencies, the power output is limited by the semiconductor material properties (such as the maximum field $\mathscr{E}_m$, and the saturation velocity v_s) and the minimum impedance levels in the microwave circuitry. Under the material and circuit limitations (also referred to as the electronic limitations), the power output will decrease as $1/f^2$.

The power output obtainable from IMPATT diodes can be increased if they are stacked in series. At lower frequencies, the power output is thermal limited. The expected increase of power from stacked devices with a copper heat sink[38] are relatively smaller, as shown by the solid lines in Fig. 21. The solid lines can move upward by a factor of 3 if a diamond heat sink is used. The maximum power output is limited by the maximum allowable temperature rise. Consequently, even if the number of layers is increased, there is no appreciable increase in power output. At higher frequencies the power-frequency limitation due to material properties is given by[38]

$$ P_m f^2 \approx \left(\frac{n^2 \mathscr{E}_m^2 v_s^2}{4\pi X_c} \right) \eta , \qquad (34) $$

where P_m is the maximum power, n is the number of stacked devices, X_c is the reactance $(1/2\pi f C)$, and η is the efficiency. It is apparent that for a given operating frequency, the power output increases as n^2, as shown by the dotted lines in Fig. 21.

At even higher frequencies, the power output will decrease more rapidly than $1/f^2$ due to reduced efficiency.[39] The efficiency expression can be written as[1]

$$ \eta \approx \frac{2}{\pi} \frac{V_{ac}}{V_{dc}} \frac{1}{1 + V_A/V_D} , \qquad (35) $$

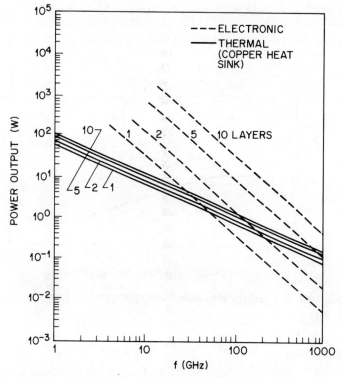

Fig. 21 Calculated cw power output versus frequency for multi-layered IMPATT diodes. (After Ref. 38)

where V_A and V_D are the dc voltage drop across the avalanche region and across the drift region, respectively. To improve efficiency one must increase the ac voltage modulation (i.e., V_{ac}/V_{dc}), and reduce the V_A/V_D ratio. However, V_A must be high enough to initiate the avalanche process rapidly.

In practical IMPATT diodes, many factors reduce efficiency. These factors include the space-charge effect, reverse saturation current, series resistance, the skin effect, the tunneling effect, and minority carrier storage effects.[1] In addition, at very high frequency operation, the field required for impact ionization becomes very high. The ionization rate will vary slowly at high field, causing a broadening of the injected current pulsewidth δ, so that the term $[\sin(\delta/2)/(\delta/2)]$ in Eq. 31 decreases, resulting in reduced efficiency.[30]

Another factor that limits performance is the finite response time of impact ionization when an electric field is applied. The avalanche response times[40] for Si and GaAs are shown in Fig. 22. For Si this time is about 10^{-13} s, which is small compared to the transit time in the submillimeter-wave region. Therefore, Si IMPATT diode are expected to be efficient up to 100 GHz or higher frequencies. For GaAs, however, the response time is longer. This is the major reason that GaAs IMPATT diodes have lower efficiencies and lower power output for frequencies near or above 100 GHz.

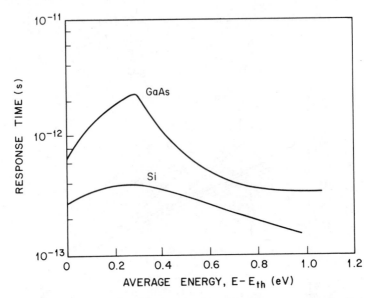

Fig. 22 Response time (energy relaxation time) versus average energy in GaAs and Si. (After Ref. 40)

9.3.4 Noise Behavior

The noise in an IMPATT diode arises mainly from the statistical nature of the generation rates of electron-hole pairs in the avalanche region. An important expression for the noise behavior is the noise measure, which is defined as

$$M \equiv \frac{\overline{V}_n^2}{4kT_0(-R)B_1}, \tag{36}$$

where $\overline{V}_n^2$ is the mean-square noise voltage, $-R$ is the negative resistance, k is Boltzmann's constant, $T_0 = 290\ K$, and B_1 is the noise bandwidth. The mean-square noise voltage $\overline{V}_n^2$ varies as $(W/x_A)^2/J_0$; therefore the noise measure is expected to decrease as the ratio (W/x_A) decreases or the dc current density J_0 increases.

Figure 23 shows the experimental noise measures of a Si IMPATT diode versus percent of maximum power output.[41] We note that the noise measure remains constant up to about 75% of the maximum power output, then increases rapidly as power increases. Therefore, a suitable trade-off is to operate the device at about 75% maximum power for the minimum noise measure. Such power level will also provide reasonable junction temperature for reliable long-term operation.

Figure 24 shows the noise measure of IMPATT diodes and transferred-electron devices (TEDs) for equal power output at X-band and W-band[42] (TED will be discussed in Section 9.4). At low frequencies (X-band), the noise measure of IMPATT diodes is much higher (40 to 50 dB) compared to the noise

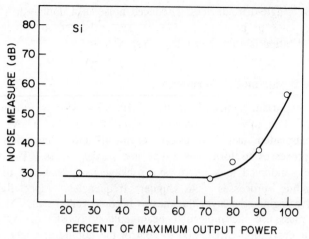

Fig. 23 Noise measure versus percent of maximum power output for a Si IMPATT diode. (After Ref. 41)

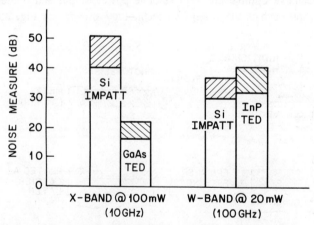

Fig. 24 Noise measures (the cross-hatched areas) of IMPATT diodes and TEDs for equal power output at X band and W band. (After Ref. 42)

measure of TEDs (18 to 22 dB) as shown by the cross-hatched areas. However, as the frequency increases, the noise measure of IMPATT diodes decreases while that of TEDs increases, and at 100 GHz the noise measures of both devices become comparable (~ 30 to 40 dB).

The reduced noise measure in Si IMPATT diodes at higher frequencies is due to two reasons. First, the increased peak field leads to closer equality for electron and hole ionization rates; second, the saturation current is increased (due to reduced lifetime in highly doped semiconductor), thus reducing the carrier multiplication required. The noise measure is found to be lower in double-drift devices than in single-drift devices. Reduced noise in a double-drift device is due to the presence of both carriers in the avalanche region, which leads to reduced multiplication factors, hence less noise.

9.3.5   Device Design and Performances

The two most important semiconductors for IMPATT diodes are silicon and gallium arsenide. For lower-frequency operation, the Si and GaAs IMPATT diodes are fabricated using diffusion, chemical-vapor deposition or ion-implantation processes to form the n-type and p-type layers. The fabricated diode is usually mounted in a microwave package using standard die-bonding and wire-bonding processes. At higher frequencies, especially in the millimeter-wave region, the layer thicknesses become very small. At these frequencies we must use the molecular beam epitaxy (MBE) or metal-organic chemical-vapor deposition (MOCVD) to control the doping and layer thickness precisely.

The conventional packaging methods give rise to large variation of device performance in millimeter-wave region. New packaging approaches have been developed to achieve optimal and reproducible power output and operating frequency. One approach is the beam-lead technology shown[43] in Fig. 25. Heat-

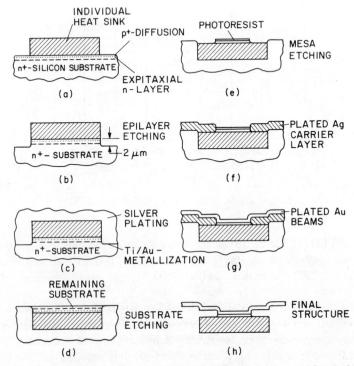

Fig. 25 Beam lead technology for IMPATT diode fabrication. (After Ref. 43)

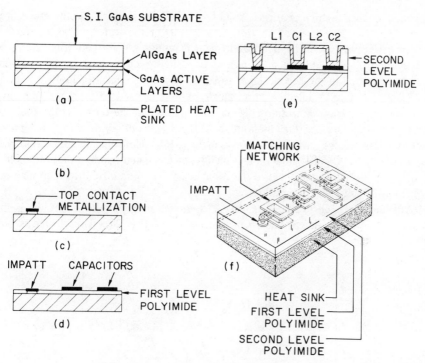

Fig. 26 (a)–(e), Process sequence of monolithic IMPATT oscillator. (f) Monolithic IMPATT oscillator with lumped-element impedance matching circuits. (After Ref. 44)

sink areas are electroplated and defined on the epitaxial layer (Fig. 25a). The heat sink serves as an etch mask to remove the semiconductor material to a predetermined depth (Fig. 25b). A Ti/Au layer is evaporated, followed by electroplating of silver on the whole surface (Fig. 25c). The semiconductor surface is thinned to reduce series resistance and to minimize the skin effect (Fig. 25d). Single mesa diodes are formed by lithography and etching processes (Fig. 25e). A second silver plating is used as a carrier layer for the beam leads (Fig. 25f). The beam leads are formed by a gold plating onto the silver layer and defined by an etching step (Fig. 25g). The diodes are separated by removing the two silver layers and the intermediate gold foil (Fig. 25h). The advantage of this approach lies in the reproducibility of the diode thickness and diameter as well as the beam leads.

For millimeter-wave integrated circuits, it is desirable to integrate IMPATT diode and impedance-matching circuits on the same chip without requiring individual device packaging. Figure 26 shows one approach to fabricating a monolithic IMPATT oscillator.[44] The semiconductor layers are formed on a semiinsulating substrate, then an electroplating step is used to form the heat sink (Fig. 26a). An etching step is used to remove all semiconductor materials except the active layer (Fig. 26b). The top-contact metal of the device is deposited and defined, which in turn is used as the etch mask to define the diode area (Fig. 26c). A thin layer of polyimide is spun over the wafer. The top surface of the polyimide layer is metallized and capacitors are formed (Fig. 26d). A second and thicker layer of polyimide is deposited over the first layer, and windows are etched in the second layer to reveal the top contacts of IMPATT diode and the capacitors. Selective area electroplating is used to fabricate inductive circuit element and connections between the diode and the circuit (Fig. 26e). A schematic diagram of the monolithic IMPATT diode with lumped-element, impedance-matching circuits is shown in Fig. 26f. Note that in this monolithic approach, the IMPATT diode is completely encapsulated inside a durable dielectric, and is directly connected to the impedance-matching circuit to minimize the device-to-circuit transition parasitics.

Another approach to integrate IMPATT diodes into monolithic circuits is the lateral IMPATT diode, as shown in the insert[45] of Fig. 27. This device is pla-

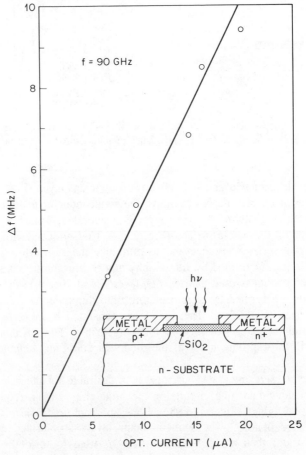

Fig. 27 Measured optical tuning characteristic for W-band IMPATT diode. Insert shows a planar lateral IMPATT diode. (After Refs. 45 and 46)

nar, and its contacts and drift region are all adjacent to the wafer surface. The lateral diode can be used for optical switching (i.e., to drive the device into or out of oscillation by altering the negative-resistance properties of the device), or to induce optical tuning of the oscillator frequency. The tuning sensitivity (frequency shift Δf) at low optical intensity is proportional to the optically generated current as shown[46] in Fig. 27.

The state-of-the-art conversion efficiency of cw IMPATT diodes is shown in Fig. 28. At low frequencies, both single-drift and double-drift GaAs diodes have attained 36% efficiency. Such high efficiency is due mainly to the small V_A/V_D ratio in GaAs devices. However, the efficiency of GaAs diodes drops rapidly above 60 GHz, mainly caused by the relatively long avalanche response time. The Si IMPATT diodes have lower efficiencies ($\sim 12\%$ for single-drift devices and 15% for double-drift devices), but the devices can maintain a constant efficiency up to about 100 GHz before substantial reduction occurs.

A summary[47,48] of the state-of-the-art of solitary mesa-type IMPATT power output is given in Fig. 29. (Also shown are results for tunnel devices discussed in Section 9.2 and related transit-time devices to be considered). At lower frequencies, the power output is thermal-limited and varies as f^{-1}; at higher frequencies (> 60 GHz for GaAs and > 100 GHz for Si) the power varies approximately as f^{-3}. The rapid decrease of power with frequency is due to the combined effect of material-property limitation and the reduction of conversion efficiency.

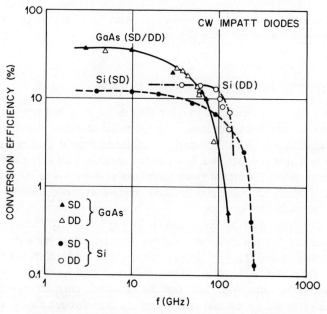

Fig. 28 State-of-the-art IMPATT conversion efficiency.

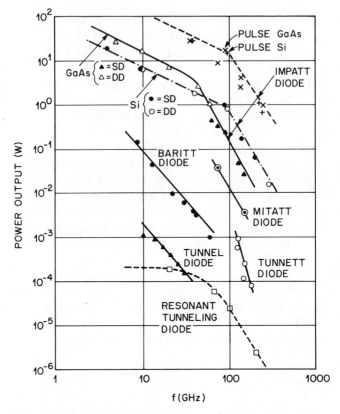

Fig. 29 State-of-the-art IMPATT performances. Also shown are the power output for related transit-time devices and tunnel devices.

9.3.6 DOVATT, TUNNETT, and MITATT Diodes

To achieve conversion efficiencies higher than those described previously, many novel structures have been proposed. One potential candidate is the DOVATT[29] diode as shown in Fig. 30a. The device is different from a conventional single-drift diode in that it consists of a low-velocity material (with saturation velocity v_1) and a high-velocity material (with saturation velocity v_2) in the drift region. The device is identical to the conventional single-drift diode in the limit where $\beta \equiv v_1/v_2 = 1$.

The electric-field distribution is shown in Fig. 30b. By choosing a lower bandgap material for the avalanche region, we can lower both the maximum field at breakdown and the V_A/V_D ratio. Because of the heterojunction, there is a discontinuity in the field profile. The terminal voltage and current waveforms (using sharp-pulse approximation for the injected current) are shown in Fig. 30c. During the period that charge carriers drift through the N_1 region, we have $I_1 = qnv_1A$ where n is the average electron density and A is the diode

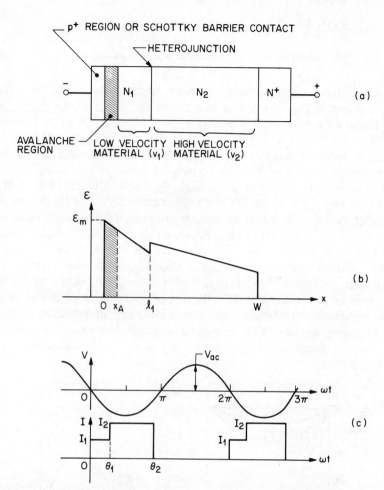

Fig. 30 (a) Schematic diagram of a DOVATT diode. (b) Electric-field distribution. (c) Idealized current and voltage waveforms at the terminals of the DOVATT diode. (After Ref. 29)

area. After the transit angle θ_1 the charge carriers enter the N_2 region with velocity v_2. The current becomes $I_2 = qnv_2A$. The electron pulse is collected at an overall transit angle θ_2 and the current drops to zero. The transit angles θ_1 and θ_2 can be varied by varying the lengths of N_1 and N_2 regions.

The efficiency for the waveforms of Fig. 30c is given by

$$\eta \equiv \frac{P_{ac}}{P_{dc}} = \frac{V_{ac}}{V_{dc}} \frac{(\beta - 1)\cos \theta_1 - (\beta - \cos \theta_2)}{(\beta - 1)\theta_1 + \theta_2} , \qquad (37)$$

where (V_{ac}/V_{dc}) is the ratio of peak ac voltage to dc applied voltage. The overall transit angle θ_2 can be determined by maximizing η by setting $d\eta/d\theta_2 = 0$, and we obtain

$$\sin \theta_2 = \frac{(1 - \beta)\cos \theta_1 + \beta - \cos \theta_2}{(\beta - 1)\theta_1 + \theta_2}.$$ (38)

The optimum transit angle θ_2 versus the transit angle θ_1 through the low-velocity region is shown in Fig. 31a with β as a parameter. We note that the optimum transit angle for conventional diodes is $\theta_2 = 133°$ (for $\beta = 1$) as considered previously. For values of β less than unity, the optimum values of θ_2 become smaller. In the limit that $\beta = 0$, θ_2 approaches 90° corresponding to a delta-function current pulse at a 90° overall transit angle.

The efficiency for optimum transit angle θ_2 is shown in Fig. 31b. For the conventional single-drift diode the efficiency is about 36% (for $V_{ac}/V_{dc} = 0.5$). For smaller values of β, higher efficiencies are expected. For each value of β, there exists an optimum value of θ_1. For example, if $\beta = 0.25$, the optimum θ_1 is 60° and the θ_2 is 121° (from Fig. 31a), and the efficiency can reach 43%. DOVATT diodes have been made using $Ga_{0.6}Al_{0.4}As$ for the N_1 region and GaAs for the N_2 region.[49] Initial results show a cw power output of 2.2 W at 9 GHz with an efficiency of 11%. By further optimization of the device and material parameters, higher power and higher efficiency are expected.

By extending the DOVATT concept to a double-drift device, we have the double DOVATT diode.[50] The avalanche region will be located at the central

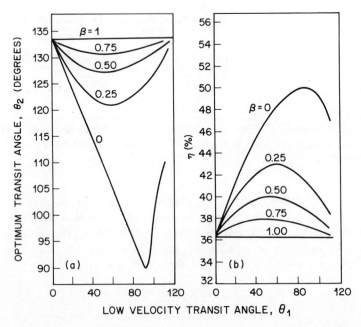

Fig. 31 (a) Optimum transit angle θ_2 versus the transit angle θ_1 through the low-velocity region with β as a parameter where β is the velocity ratio (v_1/v_2). (b) Efficiency for optimum transit angle θ_2. (After Ref. 29)

high-field region, and there are two drift regions (one for electrons and one for holes). Each drift region will have a low-velocity segment and a high-velocity segment. Higher power and higher efficiency are expected from the double DOVATT diodes than the single-drift DOVATT diodes, especially in the millimeter-wave range.

An interesting limiting case of the DOVATT diode is the heterojunction IMPATT diode in which the width of the avalanche region is equal to the length of the first segment (i.e., $x_A = l_1$ as shown in Fig. 30b). Typically a low-bandgap semiconductor is used for the avalanche region with a large-bandgap semiconductor for the drift region. Since the avalanche voltage V_A will be low for low-bandgap semiconductor, the ratio of the avalanche voltage to the drift voltage V_A/V_D can be reduced, thus improve the conversion efficiency (Eq. 35). A p^+-GaInAs/n-GaInAs/n-InP/n^+-InP heterojunction IMPATT structure is considered.[51] The avalanche region consists of a 0.12-μm n-Ga$_{0.47}$In$_{0.53}$As layer ($E_g = 0.78$ eV) and the drift region consists of a 4.38-μm n-InP layer ($E_g = 1.35$ eV). The calculated performances for 10-GHz operation show a maximum efficiency of 45% and a cw power of 4 W.

When the injection region becomes sufficiently small such that only tunneling current is present (no avalanche current), we have the TUNNETT (*tunnel transit time*) diode. The potential advantages of TUNNETT diodes as compared to IMPATT diodes are high-frequency operation, low noise and low bias voltage. For TUNNETT operation, the maximum injection occurs at the point where the ac voltage reaches its maximum. With reference to Fig. 19, this means that the phase angle ϕ for the injected current is equal to $\pi/2$. Equation 30 for $\delta \approx 0$ becomes

$$\eta = \frac{V_{ac}}{V_{dc}} \frac{\sin\theta}{\theta}. \tag{39}$$

The optimum efficiency is obtained for $\theta = 3\pi/2$ where

$$\eta = \frac{2}{3\pi} \frac{V_{ac}}{V_{dc}}. \tag{40}$$

If $V_{ac}/V_{dc} = 0.5$, then $\eta \approx 10\%$. The TUNNETT diode will therefore have lower efficiency than the IMPATT diode. However, because of the tunneling injection process, the noise will be much lower.

The insert in Fig. 32 shows the doping profile of a TUNNETT diode.[31,52] The hyperabrupt junction is used to give high electric field for tunneling. As an example, for $N_0 = 10^{16}$ cm^{-3}, $N_t = 10^{19}$ cm^{-3}, and $L = 100$ Å, the voltage across the diode for 300-GHz operation is 2.17 V and the maximum field is 1.75×10^6 V/cm. The electric field drops to 1.75×10^5 V/cm over a distance of less than 300 Å. The oscillation frequency and the power output as a function of dc input current density are shown in Fig. 32. For this device, the oscillation frequency is between 180 and 190 GHz, and the maximum power output is 80 μW. Devices with maximum cw power of 1 mW has been obtained

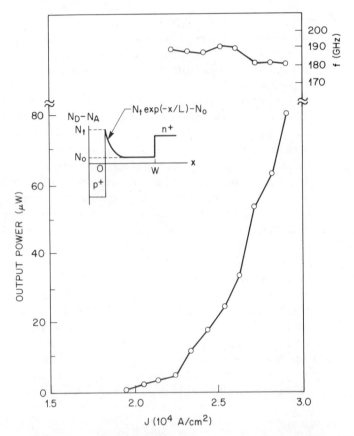

Fig. 32 Power output and frequency versus dc input-current density for a TUNNETT diode. Insert shows the doping profile of the diode. (After Refs. 31 and 52)

at 128 GHz. The performance of present TUNNETT diode is shown in Fig. 29. Higher powers and higher frequencies of operating (up to 1000 GHz) are expected for optimized TUNNETT diodes.

When both tunneling and avalanche breakdown exist in the injection region, we have the MITATT (*mi*xed *t*unneling *a*valanche *t*ransit *t*ime) diode.[30] For MITATT operation, the injection phase delay is between the IMPATT operation ($\phi = \pi$) and the TUNNETT operation ($\phi = \pi/2$), that is, $\pi/2 < \phi$(MITATT) $< \pi$. Therefore the efficiency will lie between the IMPATT and TUNNETT diodes and so would the noise performance.

The insert in Fig. 33 shows a MITATT structure[53] that consists of a p^+-GaAs/n^+-GaAs tunnel injection region, the maximum field for an applied voltage of 8.5 V is 1.24×10^6 V/cm. The drift region consists of a 400-Å $Ga_{0.6}Al_{0.4}As$ layer and a 2600-Å GaAs layer. The structure is similar to a DOVATT diode, however, the injected current contains both avalanche and tunneling components. Figure 33 shows that the oscillation frequency is around

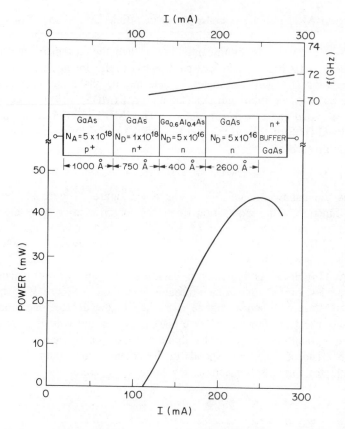

Fig. 33 Power output and frequency versus dc input current for a MITATT diode. Insert shows the MITATT structure. (After Ref. 53)

72 GHz and the peak power output is 43 mW. The power versus frequency plot for the MITATT diodes is shown in Fig. 29. As expected, for a given frequency the MITATT power lies between that of the IMPATT and TUNNETT diodes.

9.3.7 BARITT Diode

As discussed in Section 9.3.2, all transit-time devices contain a moderately high-field region in which carriers travel at saturation velocity and develop a terminal current that contributes to a phase delay between the ac voltage and current. However, this delay alone cannot give rise to a negative resistance. Therefore, we need an additional delay to achieve a negative resistance.

In BARITT (*bar*rier *i*njection *t*ransit *t*ime) diode, the injection mechanism is provided by thermionic emission over a barrier.[32] Since the maximum injection occurs at the peak ac voltage, $\phi = \pi/2$, the optimum efficiency is given by

Eq. 40. The BARITT diode operates at lower efficiency and lower power than the IMPATT diode. On the other hand, the noise associated with thermionic injection over a barrier is significantly lower than the avalanche noise in an IMPATT diode. BARITT diodes are particularly useful for applications in self-mixing oscillators. A key parameter in determining the sensitivity of a self-mixing oscillator is the minimum detectable signal (MDS) power level, that is, how weak a signal the device can detect. This MDS is affected by both the noise and bandwidth of the system and the mixing characteristics of the device. The MDS is given by[54]

$$\text{MDS} = kTB \cdot \text{NF} , \qquad (41)$$

where B is the bandwidth and NF is the noise figure. Figure 34 shows the MDS as a function of the modulation frequency for three microwave diodes, all operated at the same power level of 10 mW. Note that over the frequency range from 30 Hz to 50 kHz, the BARITT diode has the lowest MDS and therefore the highest sensitivity.

The BARITT diode is basically a back-to-back pair of p-n junctions or Schottky diodes biased into reach-through condition. Figure 35a shows a p^+-n-p^+ structure.[55] When a voltage is applied to the device, one junction is forward biased and the other is reverse biased. When the voltage is above the reach-through condition, the BARITT diode has the electric-field profile shown in Fig. 35b. The point x_R corresponds to the potential maximum for minority-carrier (hole for this case) injection; the point x_s separates the low-field drift

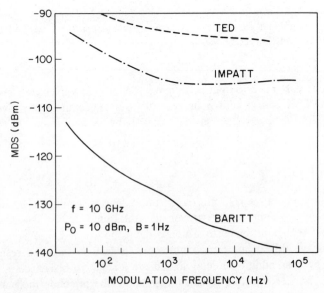

Fig. 34 Minimum detectable signal (MDS) versus modulation frequency for BARITT diode, IMPATT diode, and TED. (After Ref. 54)

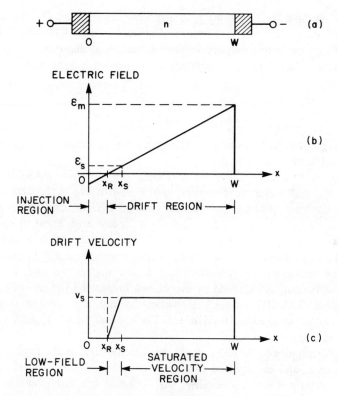

Fig. 35 Device cross section, field distribution, and carrier drift velocity of a BARITT diode. (After Ref. 55)

region from the saturation-velocity drift region, that is, for $\mathcal{E} > \mathcal{E}_s$, $v = v_s$ as shown in Fig. 35c. The reach-through voltage V_{RT} for a uniformly doped n region is

$$V_{RT} \approx \frac{qN_DW^2}{2\epsilon_s} - \left[\frac{2qN_DV_{bi}}{\epsilon_s} \right]^{1/2} W , \qquad (42)$$

where N_D is the doping concentration, W the length of the n region, ϵ_s the dielectric permittivity, and V_{bi} the built-in potential.

The transit time in the drift region ($x_R < x < W$) is given by[56]

$$\tau_d = \int_{x_R}^{x_s} \frac{dx}{\mu_n \mathcal{E}(x)} + \int_{x_s}^{W} \frac{dx}{v_s}$$

$$\approx \frac{3.75\epsilon_s}{q\mu_n N_D} + \frac{W - x_s}{v_s} , \qquad (43)$$

where μ_n is the low-field mobility. In the limit that $x_s \ll W$, τ_d becomes W/v_s. As discussed previously, the external current in a BARITT diode travels three-fourths of a cycle to reach the negative terminal, so that $\theta = \omega\tau_d = 3\pi/2$, or $f = 3/(4\tau_d)$. Therefore, the frequency of oscillation is approximately given by

$$f \approx \frac{3v_s}{4W}. \qquad (44)$$

More accurate values for the optimum frequency can be obtained by substituting Eq. 43 into the expression $f = 3/(4\tau_d)$.

Figure 36 shows a design plot for uniformly doped silicon BARITT diodes operated in the millimeter-wave region.[57] Also shown are some experimental results (in solid dots). For example, to design a 60-GHz device, we can choose n-type Si layer with a doping of 2.3×10^{16} cm^{-3}, and a thickness of 1.05 μm. The corresponding reach-through voltage is 20 V. For the same frequency of operation, we can choose a higher doping with a correspondingly thicker layer. However, the maximum allowed thickness is 1.1 μm, above that, avalanche breakdown will occur as indicated by the dashed line in the figure. Microwave performance of BARITT diodes[58] is shown in Fig. 29. Power output of 100 mW has been obtained at 10 GHz, and 1 mW at 60 GHz. Typical efficiencies are in the range of 0.5 to 2%.

In the uniformly doped BARITT diode, the electric field at the injection point ($x = x_R$) is zero, and the injected carriers must travel a certain distance (from x_R to x_s) to reach the saturation velocity. An optimum structure would be one in

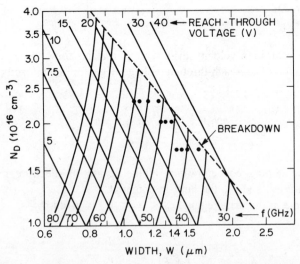

Fig. 36 Doping density versus depletion layer width for uniformly doped BARITT diode with reach-through voltage and frequency as parameters. The breakdown condition is also shown. The dots represent results of fabricated diodes. (After Ref. 57)

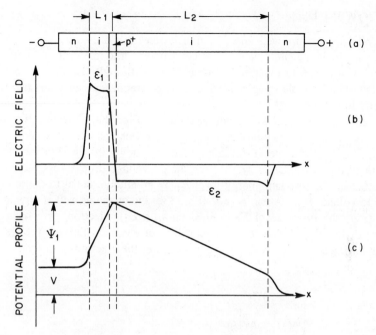

Fig. 37 (a) Schematic diagram of a triangle-BARITT diode. (b) Electric field profile and (c) electrostatic potential distribution under a dc bias. (After Ref. 59)

which the carriers are injected from a high field region and the injected carriers immediately attain their saturation velocity in the drift region. Such a structure is shown in Fig. 37a, which is a planar-doped n-i-p^+-i-n structure where the p^+ region is a ultra-thin charge sheet[59] (see Chapter 2 on planar-doped devices). The electric field and potential distributions are shown in Figs. 37b and c, respectively. All acceptors in the p^+ region are ionized forming a negatively charged sheet, which gives rise to a potential barrier of triangular shape. As can be seen we have a high field $\mathcal{E}_1$ in the injection region and a uniform and moderately high field $\mathcal{E}_2$ in the drift region resulting in velocity saturation over the entire drift region.

The maximum negative resistance for the optimum BARITT diode is given by

$$R = \frac{1}{8\pi\epsilon_s\omega}\left[L_2 - 2\pi\sqrt{3}\left[L_1 - \frac{2kT}{q\mathcal{E}_1}\right]\right] \qquad (45)$$

for a phase delay of $\phi = \pi/3$ with $(\phi + \theta) = 2\pi$. The estimated power output per unit area is 5×10^3 W/cm^2, which is about an order of magnitude greater than in the conventional BARITT diodes.

9.3.8 QWITT Diode

The QWITT (*quantum-well injection transit-time*) diode uses resonant tunneling through a single quantum well to inject carriers into the drift region of the device.[33] Compared to other transit-time devices, the QWITT diode can have a more favorable injection angle, a superior high-frequency characteristics, and a relatively low noise.

The structure of a QWITT diode is shown in Fig. 38a. The injection region is represented by the resonant-tunneling diode, which has been discussed in Section 9.2. Consider a special biasing condition, shown in Fig. 38b for the band diagram of the device at $\omega t = 0$, when the dc voltage drop across the quantum well is below the resonant energy level E_1 by an amount equal to the amplitude of the ac voltage V_{ac}. At $\omega t = \pi/2$ (Fig. 38c) the quantum well is now at resonance (i.e., the Fermi level E_F is lined up with E_1), electrons will be injected into the drift region. The resonant tunneling of electrons through the well will peak at an injection angle of $\pi/2$ in the ac cycle; the electrons will then traverse

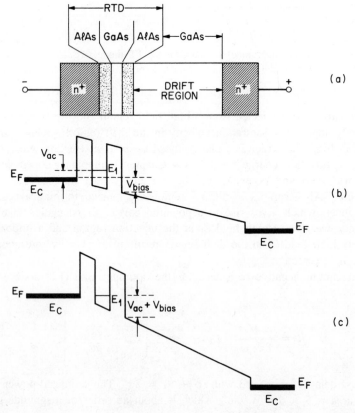

Fig. 38 (a) Schematic diagram of a QWITT diode. (b) Band diagram of the device at $\omega t = 0$. (c) Band diagram and $\omega t = \pi/2$. (After Ref. 33)

the drift region at their saturation velocity, giving rise to a dynamic negative resistance, similar to that of BARITT and TUNNETT operations.

The QWITT diode can also be biased to move the Fermi level in the n^+ region above the resonant energy level so that current injection can peak at $3\pi/2$ in the negative half of the ac voltage cycle ($\phi = 3\pi/2$). The optimum transit angle θ for such a case is $\pi/2$. The efficiency as calculated from Eq. 32 is 47% for $V_{ac}/V_{dc} = 0.5$. Thus, the device can have higher efficiencies than other conventional transit-time devices. Because of the ultra-short tunneling time in the injection region, the QWITT diode is expected to extend the normal frequency limit associated with transit-time devices.

9.4   TRANSFERRED-ELECTRON DEVICES

The transferred-electron device (TED) and IMPATT diode are the two most important microwave devices, especially for the millimeter-wave applications. The power output and efficiency of TEDs are generally lower than that of IMPATT diodes. However, TEDs have lower noise, lower operating voltages, and relatively easier circuit designs.

The microwave oscillation due to transferred-electron effect was discovered[60] by Gunn in 1963. Later Kroemer[61] pointed out that the observed microwave oscillations were consistent with a theory of negative differential resistance proposed by Ridley and Watkins[62] and by Hilsum.[63] This effect is referred to as the Ridley-Watkins-Hilsum effect or as the Gunn effect, and the most important semiconductors for such effect are gallium arsenide (GaAs) and indium phosphide (InP).

The transferred-electron effect is the transfer of conduction electrons from a high-mobility energy valley to low-mobility, higher-energy satellite valleys. The energy-momentum (E-k diagrams for GaAs and InP have been shown in Chapter 1. For both semiconductors, the bottom of the conduction band is located at $k = 0$ (Γ point). The first higher subband is located along the $<100>$ axis (X axis). Simplified E-k plots for GaAs and InP are shown in the insert of Fig. 39. Note that the separation ΔE between the Γ and X points is 0.31 eV for GaAs and 0.69 eV for InP.

The measured room-temperature velocity-field characteristics for GaAs and InP are shown[42] in Fig. 39. The threshold field $\mathscr{E}_T$, defining the onset of negative differential resistance (NDR), is 3.5 kV/cm for GaAs and 10.5 kV/cm for InP. The peak velocity v_p at room temperature is 2.1×10^7 cm/s for GaAs and 2.6×10^7 cm/s for InP, both having a donor concentration of 10^{16} cm^{-3}. The peak velocity[64] decreases linearly with temperature or log(N_D). The maximum negative differential mobility μ_- is about –2400 cm^2/V-s for GaAs and –2000 cm^2/V-s for InP.

The basic requirements for the electron-transfer mechanism to produce NDR are (1) the lattice temperature must be low enough that, in the absence of a bias field, most electrons are in the lower conduction-band minimum (Γ point), or

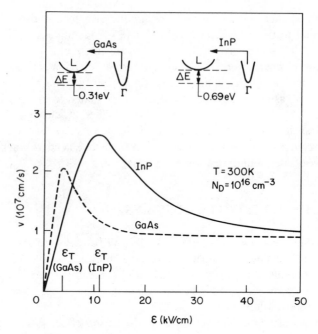

Fig. 39 Velocity-field characteristics of GaAs and InP. Insert shows the two-valley model for electron transfer. (After Ref. 42)

$kT < \Delta E$; (2) in the lower conduction-band minimum, the electrons must have high mobility, small effective mass, and low density of states; whereas in the upper satellite valleys, the electrons must have low mobility, large effective mass, and high density of states; and (3) the energy separation between the two valleys must be smaller than the semiconductor bandgap, so that avalanche breakdown does not set in before electrons are transferred into the upper valleys. Some important properties of GaAs and InP are listed in Table 4.

9.4.1 Modes of Operation

The three most important modes of operation for TEDs are the accumulation layer mode, the transit-time mode, and the quenched-domain (dipolar-layer) mode.[65] The various modes of operation are determined by the cathode-contact property, type of circuit used, and the operating bias voltage.

In a uniformly doped semiconductor bulk material with an equilibrium carrier concentration n_0, when there is a small local fluctuation of carriers, the locally created space-charge density is $(n - n_0)$. The time rate of the early-stage space-charge growth is given by

$$n - n_0 = (n - n_0)_{t=0} \exp\left[\frac{t}{|\tau_R|}\right], \qquad (46)$$

TABLE 4 Properties of GaAs and InP

Properties	GaAs	InP
E_g (eV)	1.42	1.35
$\Delta E(\Gamma - L)$ (eV)	0.31	0.69
Intervalley relaxation time (ps)	1.5	0.75
v_p/v_v ratio	2.4	4.0
Threshold field (kV/cm)	3.2	10.5
Temperature dependence of efficiency ($°C^{-1}$)	0.2%	0.05%
Diffusion coefficient (cm^2 /s)	142	72
Maximum negative differential mobility (cm^2/V-s)	−2400	−2000
Thermal conductivity (W/cm-°C)	0.54	0.68
High frequency limit (GHz)	~ 100	~ 200

where τ_R is the dielectric relaxation time

$$|\tau_R| \equiv \frac{\epsilon_s}{qn_0|\mu_-|} , \qquad (47)$$

where ϵ_s is the dielectric permittivity, and μ_- is the negative differential mobility. If Eq. 46 remains valid throughout the entire transit time of the space-charge layer, the maximum growth factor would be $\exp(L/v|\tau_R|)$, where L is the length of the active region and v is the average drift velocity of the space-charge layer. For large space-charge growth, this growth factor must be greater than unity, making $L/v|\tau_R| > 1$ or

$$n_0L > \frac{\epsilon_s v}{q|\mu_-|} . \qquad (48)$$

For n-type GaAs and InP, the right-hand side is about 10^{12} cm^{-2}. Hence an important boundary that separates various modes of operation is the (carrier concentration) × (device length) product, $n_0L = 10^{12}$ cm^{-2}.

An important figure of merit for microwave devices is the dc-to-ac conversion efficiency, which is the ratio of the rf power output to the dc power input. The efficiency is proportional[65,66] to $(\gamma - 1)/(\gamma + 1)$ where γ is the ratio of the peak velocity to valley velocity. Since γ is 4 in InP and 2.4 in GaAs (refer to Table 4), the efficiency of InP is expected to be ~50% higher than GaAs.

Accumulation Layer Mode. When a TED with subcritical n_0L product (i.e., $n_0L < 10^{12}$ cm^{-2}) is connected to a resonant circuit, it may oscillate in the accumulation layer mode. Figure 40a shows the electric field versus distance near the peak of the rf cycle and the voltage and current waveforms versus time.

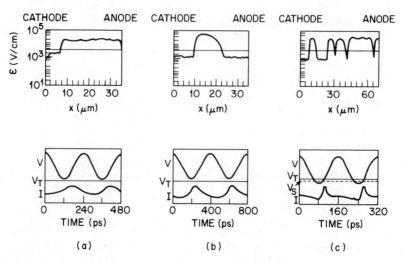

Fig. 40 Electric field profiles and voltage and current waveforms for various modes of operation in TEDs. (a) Accumulation layer mode. (b) Transit-time mode. (c) Quenched-domain mode. (After Ref. 65)

The voltage is always above the threshold value ($V > V_T \equiv \mathcal{E}_T L$). These waveforms are far from ideal, and the dc-to-ac conversion efficiency is about 5% for GaAs and 7.5% for InP.

Transit-Time Mode. When the $n_0 L$ product is greater than 10^{12} cm^{-2}, the space-charge perturbations in the semiconductor increase exponentially in space and time to form dipole layers (or domains) that propagate to the anode. The dipole is usually formed near the cathode contact, since the largest doping fluctuation and space-charge perturbation exists there. The cyclic formation and subsequent disappearance of the fully developed domains at the anode give rise to the experimentally observed Gunn oscillations. Figure 40b shows the electric-field profile at the peak voltage and the voltage and current wave forms, for a device with $n_0 L$ product of 2.1×10^{12} cm^{-2}. The transit time across the device is simply L/v and the corresponding frequency of oscillation is v/L. The maximum conversion efficiency for this mode is about 10% for GaAs and 15% for InP.

Quenched-Domain Mode. In the transit-time mode, most of the voltage across the device is dropped across the high-field domain itself. As the bias voltage is reduced, the width of the domain is also reduced, eventually the domain width shrinks to zero. The bias voltage at which this occurs is V_s. Thus, the domain is "quenched" when the bias voltage across the device is reduced below V_s. When the bias voltage swings back above threshold, a new domain is nucleated, and the process repeats. Therefore, a TED in a resonant circuit can operate at frequencies higher than the transit-time frequency if the domain is quenched before it reaches the anode. Figure 40c shows the multiple-domain formation and the voltage and current waveforms for a device operated in quenched-domain mode. Multiple domains form because one domain does not have enough time to readjust and absorb the voltage of the other domains. The $n_0 L$ product is 4.2×10^{12} cm^{-2} and the oscillation frequency is about four times the transit-time frequency. The efficiency of quenched-domain mode can reach 13% for GaAs and 20% for InP.

9.4.2 Device Structures

High-performance TEDs require several key device technologies: (1) a high-quality, highly doped, n-type substrate material, (2) the epitaxial growth of high-mobility layers for buffer and active regions, (3) a high-efficiency cathode contact, and (4) a substrate thickness that should be less than one skin depth at the operating frequency [the skin depth is given by $(\pi f \mu \sigma)^{-1/2}$ where μ is the permeability and σ is the conductivity]. At 100 GHz, the skin depth for a highly doped substrate is about 10 μm. Another important factor is the heat sink. An integrated heat sink, or diamond heat sink similar to those used for IMPATT diodes, should be used to reduce thermal resistance and to increase power output.

Graded Doping Profile.[67]   A graded doping profile for a TED is shown in Fig. 41a where the doping concentration in the active region is graded exponen-

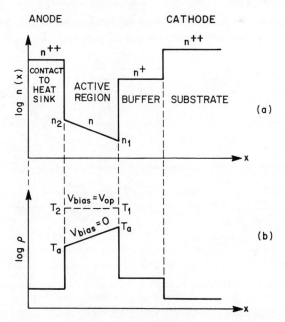

Fig. 41 GaAs graded doping TED. (a) Doping profile with graded doping. (b) Device resistivity profile showing the effect of self-heating. (After Ref. 67)

tially from n_2 to n_1 where the n_2 side (anode) is positively biased with respect to the substrate (cathode). The reason for such grading is to compensate for the variation of resistivity produced by the temperature gradient. For a given donor concentration, the resistivity $\rho = (q\mu_n n)^{-1}$ is essentially dependent on the mobility μ_n, which varies approximately as T^{-1}. When a TED is biased, a mobility gradient will develop as a result of heat generation in the active region. Since the temperature decreases exponentially from the anode to cathode, the mobility will increase exponentially toward the cathode. Therefore the graded doping profile will result in a constant resistivity across the active region under operating condition as shown by the dashed curve in Fig. 41b. Constant resistivity results in higher efficiency and higher power than conventional TEDs with uniform dopings profiles. A computer-aided MBE or MOCVD system is required to control precisely the doping concentration and doping profile during epitaxial growth.

Current-Limiting Contact.[68,69] Using a Schottky barrier with a low barrier height (Fig. 42, insert a) or a low-doping notch (Fig. 42, insert b) as the cathode contact can produce a strong current-limiting effect that, in turn, can help to create a uniform electric field across the active region. The degree of current limiting can be described by the current-limiting ratio, which is the ratio of bias current at the operating point to the saturation current.

Figure 42 shows the effect of the current-limiting cathode on device efficiency. It is clear that the efficiency increases as the current-limiting ratio decreases. The maximum efficiency is generally achieved with a ratio of 0.2. For smaller ratios, the cathode contact approaches an ohmic contact that will cause non-uniform field distribution, which reduces the efficiency.

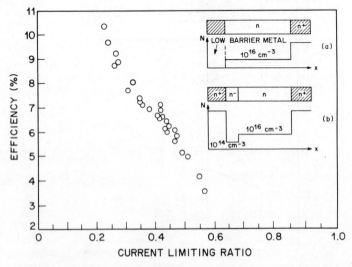

Fig. 42 Efficiency versus current-limiting ratio. Insert shows the doping profiles for low-barrier Schottky contact and low-doping notch structure. (After Refs. 68 and 69)

Injection Cathodes.   For conventional TEDs, the transfer of electrons from the Γ minimum into the subsidiary L valleys is achieved by a large applied electric field.   The electrons acquire the energy ΔE over a distance of 1 μm or more. The transfer is a noisy process because many other scattering mechanisms may take energy from the accelerated electrons and delay the precise position at which the transfer occurs.   By using an injection cathode, we can accelerate the electrons over much shorter distances and eliminate much of the noise associated with the other scattering mechanisms.

Figure 43 shows three injection cathodes.   Figure 43a is the planar-doped barrier (PDB) cathode[70] (refer to Chapter 2 for a discussion on PDB).   The PDB

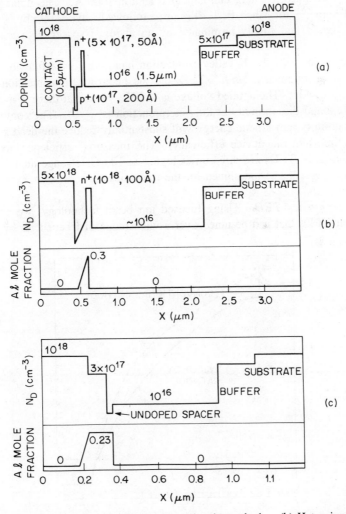

Fig. 43 Injection cathodes. (a) Planar-doped barrier cathode. (b) Heterojunction cathode with n^+ clump. (c) Heterojunction cathode with undoped spacer. (After Refs. 70 and 71)

doping profile is asymmetric to provide a long acceleration region. The 50-Å n^+-doped clump with a charge density of 2.5×10^{11} cm^{-2} located between the injection cathode and the n-type drift region will eliminate the "dead zone" or an excessively wide depletion region in the active region. GaAs TEDs with PDB injection cathodes have produced the highest cw power at 94 GHz.

Figure 43b shows a heterojunction cathode of Al$_x$Ga$_{1-x}$As (linearly graded composition with $x = 0$ to $x = 0.3$ over a distance of 500 Å) on GaAs. This device has a minimal dead zone, and an power output of 40 mW with 4% efficiency at 100 GHz is predicted. Figure 43c shows another heterojunction injection cathode.[71] The Al mole fraction is graded linearly from $x = 0$ to 0.23 within 500 Å and then held constant at 0.23 for 1000 Å. The doping density was 10^{18} cm^{-3} for the first 0.2 μm, followed by 0.09 μm doped to 3×10^{17} cm^{-3} and finally terminated by an undoped spacer with a length of 100 Å.

Figure 44 shows[71] the Γ valley conduction band edge as a function of position for a conventional n^+-n-n^+ TED and the heterojunction injection cathode shown in Fig. 43c. The applied voltage is 1.5 V. Note that there is a 200-meV conduction-band discontinuity at the heterojunction ($x = 0$). The conversion of this discontinuity into kinetic energy will substantially reduce the dead zone, and therefore improve the device efficiency. The measured efficiency and power output from such TEDs are three times higher at 100 GHz range than from conventional n^+-n-n^+ TEDs designed for the frequency.

Tunable-Frequency TED.   Using focused ion beam technology, we can fabricate planar TEDs that can be tuned over a wide range of frequencies by varying

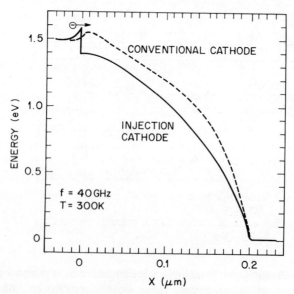

Fig. 44 Γ-valley conduction band edge as a function of position for conventional TED and heterojunction TED. (After Ref. 71)

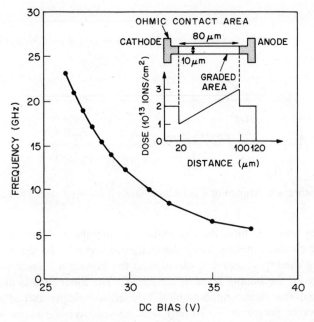

Fig. 45 Frequency versus bias voltage of a planar TED fabricated by ion beam technology. Insert shows the device geometry and doping profile. (After Ref. 72)

the dc bias across the device. The insert in Fig. 45 shows the device geometry and the doping profile.[72] The dose is varied linearly from 10^{13} to 3×10^{13} cm^{-2} over the 80 μm length of the active region. This profile is obtained by vectoring the beam with increasing dwell time from one ohmic contact to the other.

In a TED in which the doping level increases from cathode to anode, the applied electric field decreases from cathode to anode. The length over which the electric field exceeds the domain sustaining value ($\mathcal{E}_s \approx 2$ kV/cm) defines the length of repeated travel of the domain. When a domain reaches the point at which the field has dropped to $\mathcal{E}_s$, it is quenched, and a new domain is nucleated at the cathode. By varying the bias voltage, this point can be shifted along the device active region, thus varying the transit-time frequency. As the bias voltage increases, the transit length and transit time increase and the frequency of oscillation decreases. This is shown in Fig. 45. By varying the dc bias from 26 to 37 V, the domain propagation length increases from 4 to 17 μm, and the corresponding frequency decreases from 23 to 6 GHz. This TED is useful as a wide-band voltage-controlled oscillator and is suitable for integration into monolithic microwave circuits.

Monolithic TED Oscillator. Monolithic integrated-circuit techniques are being developed to obtain low-cost, compact-size, and light-weight microwave systems. Figure 46 shows a monolithic GaAs TED oscillator.[73] The TED layers are first grown on semi-insulating substrate. An ohmic metal is selectively

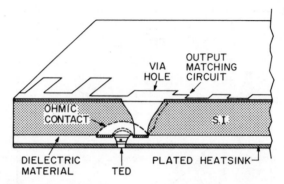

Fig. 46 A schematic diagram of a monolithic GaAs TED oscillator. (After Ref. 73)

plated onto the wafer to form the top contact for the diode. The device geometry is defined by mesa etching away the unwanted top n^+ and the n active layers. A ring geometry of metal is plated onto the bottom n^+ layer to form the ohmic contact for the bottom layer of the diode. The diode area is protected by photoresist and the remaining bottom n^+ layer completely etched away. A dielectric layer of polyimide ($\sim 10~\mu$m) is deposited over the entire wafer surface, then a via hole is opened in the dielectric layer above the first top ohmic contact. A thick gold layer is electroplated on the polyimide to form the heat sink. The substrate is thinned to 100 μm and the wafer is flipped over for the backside processing. Another via hole is formed above the second top ohmic metal to enable the device to be connected to other circuit components. The metallization for the via hole and the oscillator biasing circuit are formed by selective e-beam and gold electroplating processes. This initial circuit achieves an power output of 4 mW at 42 GHz.

9.4.3 Microwave Performances

The state-of-the-art efficiency versus frequency is shown in Fig. 47 for GaAs and InP TEDs. Because of the higher ratio of peak velocity to valley velocity in InP, the efficiencies in InP devices are indeed higher than that of GaAs devices. It is interesting to compare the results with the efficiencies of IMPATT diodes. We note that for $f < 50$ GHz, TED efficiencies are less than that of GaAs IMPATT diodes, and for $f > 50$ GHz, TED efficiencies are less than that of silicon IMPATT diodes.

Figure 48 shows the state-of-the-art microwave power output versus frequency. For lower frequencies, the power varies as $1/f$ due to thermal limitations. At high frequencies, the power varies as $1/f^2$ due to electronic limitations. The millimeter-wave cw performances of InP TEDs are higher than those of GaAs devices, mainly due to the higher efficiency for InP. However, the pulse results of GaAs devices are higher, presumably due to more advanced GaAs technology.

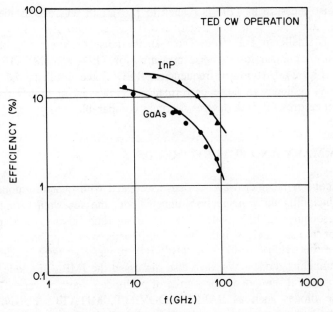

Fig. 47 State-of-the-art efficiency versus frequency for InP and GaAs TEDs.

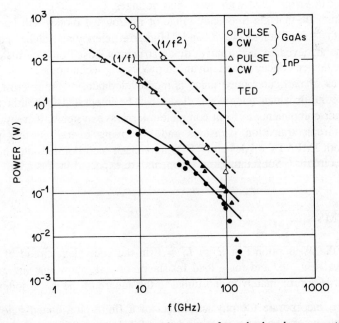

Fig. 48 Output microwave power versus frequency for pulsed and cw-operated InP and GaAs TEDs.

The noise measure in TED is related to $D/|\mu_-|$, where D is the diffusion coefficient and $|\mu_-|$ is the negative differential mobility. Since the ratio $D/|\mu_-|$ is smaller in InP (from Table 4), the noise is expected to be lower in InP TEDs. A comparison of noise measures for TEDs and IMPATT diodes is shown in Fig. 24. At lower frequencies, TEDs have a substantial advantage over IMPATT diodes in noise performance. However, at higher frequencies, the noise measures for both devices becomes comparable.

9.5   SUMMARY AND FUTURE TRENDS

We have considered microwave devices associated with quantum-tunneling phenomena, including the *p-n* junction tunnel diode, the superlattice tunnel diode, and the resonant-tunneling diode. The tunneling time is very short, permitting the use of tunnel devices into the submillimeter-wave region (>300 GHz). Microwave detection up to THz (10^{12} Hz) range has been made using a resonant-tunneling diode. We have also discussed the IMPATT diode, which is the most powerful semiconductor source of millimeter-wave power. The related transit-time diodes such as BARITT, DOVETT, MITATT, TUNNETT, and QWITT diodes can complement the IMPATT diode for special lower-power applications. The TEDs have been extensively used as local oscillators and amplifiers, covering the frequency range from 1 to 100 GHz. Although TEDs have lower power and lower efficiency than IMPATT diodes, they have lower noise and can be operated with lower bias voltages.

We anticipate that the frequency range of microwave diodes will be extended well into the submillimeter-wave band. Higher efficiency and higher power output will be achieved by using novel diode structures based on new material systems, new device concepts, and advanced processing technologies.

One area of particular importance is the development of the microwave integrated circuits in which microwave diodes can be integrated monolithically with other circuit components to form complete microwave systems, thereby reducing device-to-circuit transition parasitics and improving overall system performances. Both IMPATT diodes and TEDs have been incorporated into microwave integrated circuits. Substantial improvements are expected in this area.

PROBLEMS

1. With the assumption that $R_s = L_s = 0$ in the equivalent circuit of a tunnel devices, Fig. 4b, and for a load resistance of 1 Ω, show that the cw power output is approximately proportional to the square of the speed index.

2. a) To incorporate the physical effect of a finite quasibound-state lifetime into an equivalent-current model for RTD, a modified equivalent circuit has been proposed. The circuit is similar to the of Fig. 4b except that

that $-R$ branch is replaced by a conductance $-G (\equiv -1/R)$ in series with an inductance L, which is given by τ_N/G where τ_N is the Nth quasibound-state lifetime. Find the cutoff frequency f_{RCL}, the frequency at which the real part of the impedance is zero, assuming that the lead inductance L_s is negligibly small.

b) Calculate f_{RCL} for an RTD with $\tau_1 = 6$ ps, $C = 77$ fF, $G = -0.05$ mhos, and $R_s = 5 \, \Omega$; and compare this frequency with f_{RC} from Eq. 8.

3. Figure 19 shows the idealized voltage and current waveforms for a transit-time diode, where δ is the injected pulsewidth, ϕ is the phase delay and is at the center of the pulse, and θ is the drift-region angle. The dc current is

$$I_{dc} = \frac{1}{2\pi} \int_0^{2\pi} I_{\text{ind}} \, d(\omega t) \, ,$$

where I_{ind} is the induced current. Verify that

a) $I_{\max} = \dfrac{2\pi}{\theta} I_{dc}$.

b) $\eta = \dfrac{P_{ac}}{P_{dc}} = \dfrac{V_{ac}}{V_{dc}} \dfrac{\sin(\delta/2)}{(\delta/2)} \dfrac{\cos \phi - \cos(\phi + \theta)}{\theta}$.

c) For IMPATT operation, $\phi \approx \pi$,

$$\eta = \frac{V_{ac}}{V_{dc}} \frac{\sin(\delta/2)}{(\delta/2)} \frac{\cos \theta - 1}{\theta} \, .$$

4. a) Derive the transmission coefficient for a symmetric double-barrier resonant-tunneling diode, assuming that the effective mass is constant throughout the double-barrier structure (Fig. 6a).

b) Find the lowest four resonant energy levels for a symmetric double-barrier structure with $L_B = 20$ Å, $L_W = 20$ Å, $E_0 = 3.1$ eV, and $m^* = 0.42 \, m_0$.

5. Estimate the maximum cw power output of a silicon IMPATT diode operated at 140 GHz. The diode capacitance is 0.03 pF. Assume the power output is limited by material properties (with a maximum electric field of 3×10^5 V/cm and a saturation velocity of 10^7 cm/s).

6. Consider a GaAs double drift lo-hi-lo IMPATT diode shown in Fig. 15c with an avalanche region width (where the electric field is constant) of 0.4 μm and a total depletion width of 3 μm. The n^+ or p^+ clump has a charge Q of 1.5×10^{12}/cm^2. (a) Find the breakdown voltage of the diode and the maximum field at breakdown. (b) Is the field in the drift region high enough to maintain the velocity saturation of electrons? (c) Find the operating frequency.

7. Design a silicon single-drift lo-hi-lo IMPATT diode operated at 94 GHz.

8. For the BARITT diode shown in Fig. 35, the width of the n-type silicon, W, is 1 μm and the doping is $2.5 \times 10^{16} \, \mathrm{cm}^{-3}$. (a) Find the reach-through voltage, V_{RT}, at which the reverse-biased depletion region reaches through to the forward-biased depletion region. (b) Find the flat-band voltage at which the electric field is zero at the forward-biased metal-semiconductor contact. (c) Find the frequency of oscillation when the device is operated at 400 K. (d) Estimate the power conversion efficiency of the BARITT diode, assuming $V_{ac} = V_{RT}/2$ where V_{ac} is peak ac voltage.

9. An InP TED is 1 μm long with a cross-sectional area of $10^{-4} \, \mathrm{cm}^2$ and is operated in the transit-time mode. (a) Find the minimum electron density n_0 required for transit-time mode. (b) Find the time between current pulses. (c) Calculate the power dissipated in the device, if it is biased at one half the threshold.

10. In a transferred-electron device, if a domain is suddenly quenched during transit so that the excess domain voltage is changed from V_{ex} to zero in a time that is short compared to the transit time, the change in total current through the device during this time, integrated with respect to time, should give a measure of the charge stored in the domain, Q_0. Relate this charge Q_0 to the domain excess voltage V_{ex} for a triangle electric field distribution, i.e., the field increases linearly from $\mathscr{E}_r$ to $\mathscr{E}_{dom}$ over a distance x_A for the accumulation layer and decreases linearly from $\mathscr{E}_{dom}$ to $\mathscr{E}_r$ over a distance x_D for the depletion layer (assuming that the charge in each layer is uniform).

REFERENCES

1. S. M. Sze, *Physics of Semiconductor Devices*, 2nd ed., Wiley, New York, 1981.

2. S. Y. Liao, *Microwave Solid State Devices*, Prentice-Hall, Englewood, NJ, 1985.

3. G. R. Thorn, "Advanced Applications and Solid-State Power Sources for Millimeter-Wave Systems," *Proc. Soc. Photo-Optical Inst. Opt. Eng. (SPIE)*, **544**, 2 (1985).

4. L. Esaki, "New Phenomenon in Narrow Germanium *p-n* Junctions," *Phys. Rev.*, **109**, 603 (1958).

5. L. L. Chang, L. Esaki, and R. Tsu, "Resonant Tunneling in Semiconductor Double-Barriers," *Appl. Phys. Lett.*, **24**, 593 (1974).

6. L. Kirkup and S. Wallace, "Temperature Dependence of Germanium Tunnel Diode *I-V* Characteristics," *Eur. J. Phys.*, **8**, 93 (1987).

7. R. S. Virk, "Using Tunnel Diode Technology," *MSN CT*, p. 71, November 1987.

8. J. Tantum and K. Hinton, "Tunnel Diodes Complement High-Performance Detectors," *Microwave RF,* **24**, 115 (1985).

9. C. S. Kim and A. Brandli, "High-Frequency High-Power Operations of Tunnel Diodes," *IRE Trans. Circuit Theory,* **CT8**, 416 (1961).

10. C. A. Burrus, "Gallium Arsenide Esaki Diodes for High-Frequency Applications," *J. Appl. Phys.,* **32**, 1031 (1961).

11. R. A. Davies, M. J. Kelly, and T. M. Kerr, "Tailoring the *I-V* Characteristics of a Superlattice Tunnel Diode," *Electron. Lett.,* **23**, 90 (1987).

12. T. C. L. G. Sollner, W. D. Goodhue, P. E. Tannenwald, C. D. Parker, and D. D. Peak, "Resonant Tunneling through Quantum Wells at Frequencies up to 2.5 THz," *Appl. Phys. Lett.,* **43**, 588 (1983).

13. I. Mehdi, G. I. Haddad, and R. K. Mains, "Performance Criteria for Resonant Tunneling Diodes as Millimeter-Wave Power Sources," *Microwave Opt. Technol. Lett.,* **2,** 172 (1989).

14. E. R. Brown, T. C. L. G. Sollner, C. D. Parker, W. D. Goodhue, and C. L. Chen, "Oscillations Up to 420 GHz in GaAs/AlAs Resonant-Tunneling Diodes," *Appl. Phys. Lett.* **55**, 1777 (1989).

15. (a) M. Tsuchiya, H. Sakaki, and J. Yashino, "Room Temperature Observation of Differential Negative Resistance in an AlAs/GaAs/AlAs Resonant Tunneling Diode," *Jpn. J. Appl. Phys.,* **24**, L466 (1985); (b) E. R. Brown, MIT Lincoln Laboratory, private communication.

16. H. Yamamoto, "Resonant Tunneling Condition and Transmission Coefficient in a Symmetrical One-Dimensional Rectangular Double-Barrier System," *Appl. Phys.,* **A42**, 245 (1987).

17. E. R. Brown, T. C. L. G. Sollner, W. D. Goodhue, and C. L. Chen, in "High Speed Resonant Tunneling Diodes," *Quantum Well and Superlattice Physics II*, F. Capasso, G. H. Dohler, and J. N. Schulman, Eds., *Proc. Soc. Photo-Optical Inst. Eng. (SPIE)*, **943**, 2 (1988).

18. S. K. Diamond, E. Ozbay, M. J. W. Rodwell, D. M. Bloom, Y., C. Pao, E. Wolak, and J. S. Harris, "Fabrication of 200 GHz f_{max} Resonant Tunneling Diodes for Integrated Circuit and Microwave Applications," *IEEE Electron Dev. Lett.,* **10**, 104 (1989).

19. H. Toyoshima, Y. Ando, A. Okamoto, and T. Itoh, "New Resonant Tunneling Diode with a Deep Quantum Well," *Jpn. J. Appl. Phys.,* **25**, L786 (1986).

20. W. D. Goodhue, T. C. L. G. Sollner, H. Q. Le, E. R. Brown, and B. A. Vojak, "Large Room-Temperature Effects from Resonant Tunneling through AlAs Barriers," *Appl. Phys. Lett.,* **49**, 1086 (1986).

21. J. W. Lee and M. A. Reed, "Molecular Beam Epitaxial Growth of AlGaAs/(In, Ga)As Resonant Tunneling Structures," *J. Vac. Sci. Technol.,* **B5**, 771 (1987).

22. J. Zou, J. Xu, and M. Sweeny, "Effects of Asymmetric Barriers on Resonant Tunneling Current," *Semicond. Sci. Technol.,* **3**, 819 (1988).

23. T. C. L. G. Sollner, E. R. Brown, W. D. Goodhue, and H. Q. Le, "Observation of Millimeter-Wave Oscillations from Resonant Tunneling Diodes and Some Theoretical Considerations of Ultimate Frequency Limits," *Appl. Phys. Lett.*, **50**, 332 (1987).

24. H. C. Liu, "Tunneling Time through Heterojunction Double-Barrier Diodes," *Superlat. Microstruct.*, **3**, 379 (1987).

25. E. R. Brown, W. D. Goodhue, and T. C. L. G. Sollner, "Fundamental Oscillations up to 200 GHz in RTD and New Estimates of Their Maximum Oscillation Frequency from Stationary-State Tunneling Theory," *J. Appl. Phys.*, **64**, 1519 (1988).

26. T. C. L. G. Sollner, E. R. Brown, and H. Q. Le, "Microwave and Millimeter-Wave Resonant-Tunneling Devices," *Lincoln Lab. J.*, **1**, 89 (1988).

27. W. T. Read, "A Proposed High Frequency Negative Resistance Diode," *Bell Syst. Tech. J.*, **37**, 401 (1958).

28. R. L. Johnston, B. C. Deloach, Jr., and B. Cohen, "A Silicon Diode Oscillator," *Bell Syst. Tech. J.*, **44**, 369 (1965).

29. M. G. Adleretein and H. Statz, "Double-Velocity IMPATT Diodes," *IEEE Trans. Electron Dev.*, **ED-26**, 817 (1979).

30. M. E. Elta and G. I. Haddad, "High-Frequency Limitations of IMPATT, MITATT and TUNNETT Mode Devices," *IEEE Trans. Micro. Theo. Tech.*, **MTT-27**, 442 (1979).

31. K. Motoya and J. Nishizawa, "TUNNETT," *Int. J. Infrared Millimeter Wave*, **6**, 483 (1985).

32. D. J. Coleman, Jr. and S. M. Sze, "The Baritt Diode—A New Low Noise Microwave Oscillator," IEEE Device Res. Conf., Ann Arbor, June 28, 1971; "A Low-Noise Metal-Semiconductor-Metal (MSM) Microwave Oscillator," *Bell Syst. Tech. J.*, **50**, 1695 (1971).

33. V. P. Kesan, D. P. Neikirk, B. G. Streetman, and P. A. Blakey, "A New Transit-Time Device Using Quantum Well Injection," *IEEE Electron Dev. Lett.*, **EDL-8**, 129 (1987).

34. Y. Okuto and C. R. Crowell, "Threshold Energy Effect on Avalanche Breakdown Voltage in Semiconductor Junctions," *Solid-State Electron.*, **18**, 161 (1975).

35. W. F. Schroeder and G. I. Haddad, "Avalanche Region Width in Various Structures of IMPATT Diodes," *Proc. IEEE*, **59**, 1245 (1971).

36. P. Weissglas, "Avalanche and Barrier Injection Devices," in *Microwave Devices—Device Circuit Interactions*, M. J. Howes and D. V. Morgan, Eds., Wiley, New York, 1976, Chap. 3.

37. R. K. Mains and G. I. Haddad, "Properties and Capabilities of Millimete—Wave IMPATT Diodes," in *Infrared and Millimeter Waves*, Vol. 10, K. J. Button, Ed., Academic, New York, 1983.

38. B. Jogai, K. L. Wang, G. P. Li, and R. Abdeshaah, "Power-Frequency Scaling of Multi-Layer Microwave Devices," *Superlat. Microstruct.*, **1**, 299 (1985).

39. G. Salmer, "Physical Frequency Limitations of 2-Terminal Devices," *IEE Proc.*, **130**, 80 (1983).

40. D. Lippens, J. L. Nieruchalski, C. Dalle, and P. A. Rolland, "Comparative Studies of Si, GaAs and InP Millimeter-Wave IMPATT Diodes," *Int. J. Infrared Millimeter Waves*, **7**, 771 (1986).

41. D. M. Brookbands, A. M. Howard, and M. R. B. Jones, "Si IMPATTs Exhibit Low Noise at mm-Waves," *Microwaves RF*, **22**, 68 (1983).

42. W. Harth, "Microwave Semiconductor Devices: Status and Trends," *Mikrow. Mag.*, **14**, 106 (1988).

43. J. Freyer and R. Pierzina, "Encapsulation Techniques for Millimeter-Wave IMPATT Diodes," *Arch. Elektron. Uebertragung.*, **40**, 321 (1986).

44. B. Bayraktaroglu and H. D. Shih, "High Efficiency Millimeter Wave Monolithic IMPATT Oscillators," Digest of Microwave and Millimeter Wave Monolithic Circuit Symposium, p. 82 (1985).

45. P. J. Stabile and B. Lalevic, "Lateral IMPATT Diodes," *IEEE Electron Dev. Lett.*, **10**, p. 249 (1989).

46. A. J. Seeds, J. F. Singleton, S. P. Brunt, and J. R. Forrest, "The Optical Control of IMPATT Oscillators," *J. Lightwave Tech.*, **LT-5**, 403 (1987).

47. G. Jerinic, J. Fines, M. Cobb, and M. Schindler, "Ka/Q Band GaAs IMPATT Amplifier Technology," *Int. J. Infrared Millimeter Waves*, **6**, 79 (1985).

48. W. Harth, M. Claassen, and J. Freyer, "Si- and GaAs-IMPATT Diodes for Millimeter Waves," *Mikrow. Mag.*, **13**, 318 (1987).

49. W. E. Hoke and R. Traczewski, "Fabrication and Evaluation of GaAs-GaAlAs Double-Velocity IMPATT Diodes," *Tech. Dig. IEEE IEDM*, December 8, 1980, p. 452.

50. B. B. Pal, R. U. Khan, and P. Chakrabarti, "A New Solid-State Device as a Source of Power in mm-Wave," *J. Inst. Electron. Telecommun. Eng.*, **32**, 397 (1986).

51. J. C. DeJaeger, R. Kozlowski, and G. Salmer, "High Efficiency GaInAs/InP Heterojunction IMPATT Diodes," *IEEE Trans. Electron Dev.*, **ED-30**, 790 (1983).

52. S. Shimizu, "A Study of TUNNETT Diode," *Rec. Elect. Commun. Eng. Conversazione Tokohu U.*, **55**, 102 (1987).

53. N. S. Dogan, J. R. East, M. E. Elta, and G. I. Haddad, "Millimeter-Wave Heterojunction MITATT Diodes," *IEEE Trans. Micro. Theo. Tech.*, **MTT-35**, 1308 (1987).

54. J. R. East, H. Nguyen-Ba, and G. I. Haddad, "Design, Fabrication, and Evaluation of Baritt Devices for Doppler System Applications," *IEEE Trans. Micro. Theo. and Tech.*, **MTT-24**, 943 (1976).

55. J. L. Chu and S. M. Sze, "Microwave Oscillation in *pnp* Reach-Through Baritt Diodes," *Solid State Electron.*, **16**, 85 (1973).

56. H. Nguyen-Ba and G. I. Haddad, "Effects of Doping Profile on the Performance of Baritt Diodes," *IEEE Trans. Electron Devices,* **ED-24**, 1154 (1977).

57. U. Guttich, "Baritt-Dioden für das V-Band," *Mikrow. Magaz.*, **1**, 37 (1987).

58. U. Guttich and J. Freyer, "BARITT Diodes for Millimeter-Wave Frequencies," *Int. J. Electron.*, **59**, 625 (1985).

59. S. Luryi and R. F. Kazarinov, "Optimum Baritt Structure," *Solid State Electron.*, **25**, 943 (1982).

60. J. G. Gunn, "Microwave Oscillation of Current in III–V Semiconductors," *Solid State Commun.*, **1**, 88 (1963).

61. H. Kroemer, "Theory of the Gunn Effect," *Proc. IEEE,* **52**, 1736 (1964).

62. B. K. Ridley and T. B. Watkins, "The Possibility of Negative Resistance Effects in Semiconductors," *Proc. Phys. Soc. Lond.*, **78**, 293 (1961).

63. C. Hilsum, "Transferred Electron Amplifiers and Oscillators," *Proc. IRE,* **50**, 185 (1962).

64. W. Kowalsky and A. Schlachetzki, "Transferred-Electron Effect in InGaAsP Alloys Lattice-Matched to InP," *Solid State Electron.*, **28**, 299 (1985).

65. H. W. Thim, "Solid State Microwave Sources," in *Handbook on Semiconductors*, Vol. 4, *Device Physics*, C. Hilsum, Ed., North-Holland, Amsterdam, 1980.

66. I. G. Eddison, "Indium Phosphide and Gallium Arsenide Transferred—Electron-Devices," *Infrared and Millimeter Waves*, Vol. 11, *Millimeter Components and Techniques, Part III*, Academic, Orlando, 1984, p. 1.

67. J. Ondria and R. L. Ross, "Improved Performance of Fundamental and Second Harmonic MMW Oscillators through Active Device Doping Concentration Contouring," *1989 IEEE MTT-S Int. Microwave Symp. Dig.*, **2**, 977 (1987).

68. B. Fank, J. Crowley, and C. Hang, "InP Gunn Diode Sources," SPICE Millimeter Wave Technology III, *Proc. Soc. Photo-Optical Inst. Eng.*, (*SPIE*), **544**, 22 (1985).

69. F. B. Fank, J. D. Crowley, M. C. Buswell, C. Hang, P. H. Wolfert, P. Tringali, and L. C. Ching, "High Efficiency InP Millimeter-Wave Oscillators and Amplifiers," Proc. 14th European Microwave Conference, p. 575 (1984).

70. J. Ondria, "Latest Advances in MMW Oscillators," IEEE Colloquium on Solid State Components for Radar, p. 1 (1988).

71. A. Al-Omar, J. P. Krusius, Z. Greenwald, D. Woodard, A. R. Calawa, and L. F. Eastman, "Space Charge Effects on Heterojunction Cathode

AlGaAs Gunn Oscillators," Proc. IEEE/Cornell Conference on Adv. Concepts in High Speed Semiconductor Devices and Circuits, p. 365 (1987).

72. H. J. Lezec, K. Ismail, L. J. Mahoney, M. I. Shepard, D. A. Antoniadis, and J. Melngailis, "A Tunable-Frequency Gunn Diode Fabricated by Focused Ion-Beam Implantation," *IEEE Electron Dev. Lett.*, **9**, 476 (1988).

73. J. C. Chen, C. K. Pao, and D. W. Wong, "Millimeter-Wave Monolithic Gunn Oscillators," IEEE 1987 Microwave and Millimeter-Wave Monolithic Circuit Symposium Digest, p. 11 (1987).

10 High-Speed Photonic Devices

W. T. Tsang

AT&T Bell Laboratories
Murray Hill, New Jersey

10.1 INTRODUCTION

Semiconductor light-emitting diodes (LED), lasers, and photodetectors are critically important devices in optical fiber communications systems[1] because of their small size and high reliability. Figure 1 shows a chart of some possible important applications of these devices. As sources, LEDs and lasers are easy to modulate to encode information. As optical receivers, semiconductor photodetectors have high quantum efficiency in converting the input optical signal back into the original electrical format. Further, unlike solid-state or gas lasers, the current injection crystalline semiconductor LED and laser are potentially very low cost because they are fabricated using planar processing. With more advanced technologies, they can be integrated monolithically with other optical and electronic devices to form a complex circuit. This technology is generally called opto-electronic integrated circuits (OEIC). Such advanced technology will substantially lower the cost and enhance the performance of optical circuits and equipment.

This chapter will focus on the high-speed aspects of semiconductor photonic devices. Further, because the silica fibers have the least loss and chromatic dispersion in the wavelength range of 1.0 to 1.6 μm, the devices considered here will be mostly built out of the quarternary III-V. InGaAsP alloy systems that span this wavelength range with InP (0.9 μm) and $In_{0.53}Ga_{0.47}As$ (1.67 μm) at their end compositions. InP is a binary compound semiconductor and is used as the substrate. Thin crystalline layers of InGaAsP having different compositions are grown epitaxially over the substrate. The lattice constants of these layers are kept the same as the InP substrate underneath. Different composition results in different energy bandgap and hence emit photons with different energy.

High-Speed Semiconductor Devices, Edited by S.M. Sze. ISBN 0-471-62307-5
© 1990 John Wiley & Sons, Inc.

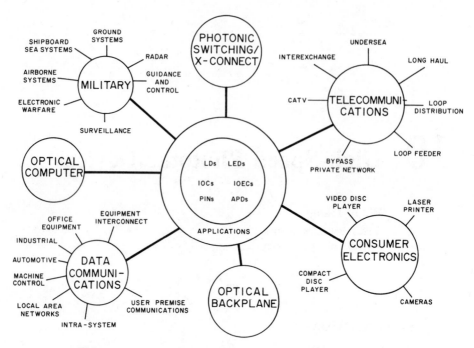

Fig. 1   Possible applications of LEDs, laser, and photodetectors.

10.2   LIGHT-EMITTING DIODES

Though current injection semiconductor lasers are the preferred optical sources in lightwave systems, LED sources can provide a reliable, inexpensive alternative. LEDs utilize relatively simple driving circuits that do not need feedback to control power output, operate over a wide range of temperatures, and have projected lifetimes one to two orders of magnitude longer than those of laser diodes made of the same materials. Long lifetime is essential for the reliability of the system. In general, LEDs are much less temperature-sensitive than lasers. However, LEDs have substantially lower output power and speed. Long-wavelength InGaAsP LEDs emitting near 1.3-μm wavelength, where silica fibers have low attentuation and minimum dispersion, can have repeater spacings of tens of kilometers and data rates of a few hundred Mb/s. Thus, the primary applications are for inexpensive moderate-speed data links. Data rates[2] of 560 Mb/s have been demonstrated with 1.3-μm LED systems.

10.2.1   Device Structures

There are principally three different types of LEDs as shown in Fig. 2 for the surface-emitting diodes, the edge-emitting diodes, and the superluminescent diodes, repetively. In the surface-emitting diodes,[3] the output beam emits per-

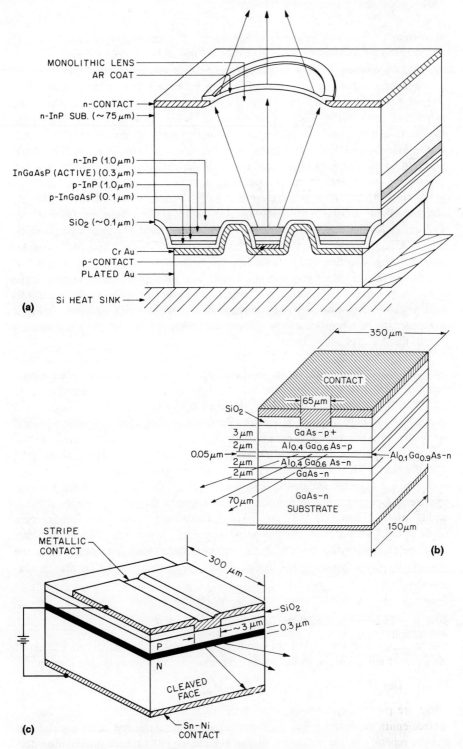

Fig. 2 (a) Small-area mesa etched InGaAsP/InP surface-emitting LED structure. (b) GaAlAs stripe geometry edge emitter. (c) Superluminescent diode with absorbing region.

pendicular to the plane of the layer structure. The emitting area is typically 25 μm in diameter as shown in this example by the etched circular mesa. Current injection is limited to this mesa by the SiO$_2$ masking layer. The InGaAsP active-layer composition in these LEDs can be varied to provide output wavelengths from about 1 μm to nearly 1.6 μm. However, because the chromatic dispersion of the silica fibers is minimum at ~1.3 μm and the emissions from LEDs are broadbanded (~40 nm to 100 nm) an integral lens is usually formed at the exit face of InP substrate to improve coupling efficiency to the fibers. Since InP is a wider bandgap material than the InGaAsP active layer, the InP substrate is transparent to the emitted beam.

In Fig. 2b, an efficient edge-emitter LED[4,5] emits its beam in a relatively directed beam parallel to the active layer as in a semiconductor laser diode. This improves the coupling efficiency into a fiber, especially into a single-mode fiber. A single-mode fiber refers to an optical fiber designed so that it guides only the lowest order optical mode in its core. To avoid lasing one end is reflective and the other end (emitting end) has an antireflection coating. The injection of current is limited to the narrow stripe by the SiO$_2$ masking. In this example, AlGaAs alloys are used in the forming of the LED. AlGaAs emits at a wavelength between 0.9 and ~0.7 μm dependent on the Al and Ga composition in the AlGaAs active layer.

A superluminescent diode,[6] Fig. 2c, has one end of the active stripe made optically lossy to prevent reflection and thus suppressing lasing. The output beam is emitted from the opposite end. In operation, the current passed through the diode is increased until stimulated emission and amplification occur, but because of the high loss at one end (no current injection applied in this portion), no feedback exists and no oscillation occurs. Therefore, in the current-injection region, there is gain and output increases rapidly with current due to single-pass amplification. Due to self-absorption in the unpumped region, the spectral width of the output decreases to less than 10 nm. Devices have been made to operate in a pulsed mode to provide a peak output of 60 mW in an optical bandwidth[6] of 6 to 8 nm at 0.87 μm. The disadvantage of this superluminescent diode is the high injection current needed (typically three times higher than that of a laser). With the cost of semiconductor lasers droping rapidly and the demand for Gb/s systems even in data links, the demand for LEDs may gradually decrease.

10.2.2 Modulation Characteristics

The modulation of the LED light output is accomplished by direct modulation of the injected current through the diode according to the following relationship[7,8]:

$$|I(\omega)| = \frac{I(0)}{\sqrt{1 + (\omega\tau)^2}}, \tag{1}$$

where I(ω) is the intensity of the light output being modulated at an angular frequency ω, I(0) is the intensity of the optical power at zero modulation frequency, and τ is the effective carrier lifetime inside the active medium where photons are generated. The modulation bandwidth, usually defined as the frequency where the response has fallen from 3 dB below its dc value ($f_{3\,dB}$), is given by

$$f_{3\,dB} = \frac{\Delta\omega}{2\pi} = \frac{1}{2\pi\tau}. \tag{2}$$

In the lightly doped active region and when injection of both electrons and holes occurs, the injected-carrier density Δn is proportional to the current density J and inversely proportional to the thickness of the active layer W, that is,

$$\Delta n = \frac{J\tau_r}{qW}, \tag{3}$$

Where $\tau_r = 1/B\,\Delta n$ is the radiative recombination lifetime, B is the radiative recombination probability, and q is a unit of electrical charge. Using the fact that $\tau \approx \tau_r$, the modulation bandwidth becomes

$$f_{3\,dB} = \frac{1}{2\pi} \left[\frac{BJ}{qW} \right]^{1/2}. \tag{4}$$

Thus, the bandwidth of a double-heterostructure (DH) LED with a lightly doped ($\sim 5 \times 10^{17} \text{cm}^{-3}$) active layer and operated in the bi-molecular recombination region, in the electron and hole concentrations are roughly equal, increases as $(J/W)^{1/2}$. Figure 3 confirms this relationship[9] in InGaAsP LEDs.

For heavy active-layer dopings, the modulation bandwidth becomes independent of the injected current density because the doped concentration is higher than the injected carriers. As a result, $f_{3\,dB}$ is inversely proportional to the doping concentration in the active layer. Modulation bandwidths greater than 1 GHz

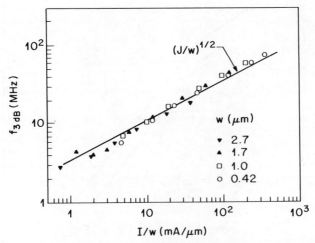

Fig. 3 The modulation bandwidth as a function of (J/w) for InGaAsP LEDs. (After Ref. 9)

have been demonstrated[10] with acceptor doping concentration $N_A \approx$ $1.5 \times 10^{19} \text{cm}^{-3}$.

In practical devices, parasitic capacitance can impose a limitation on the device modulation bandwidth. The capacitance C is comprised of a space-charge capacitance associated with the *p-n* junction area (the entire diode chip) and a diffusion capacitance that is related to the carrier lifetime in the small light-emitting area.

Figure 4 shows the results on GaAlAs and InGaAsP LEDs, where the reciprocal relationship between modulation bandwidth and output power is apparent. These results indicate that the output power P can be related to the bandwidth $f_{3\,dB}$ by $P \sim (f_{3\,dB})^{-\gamma}$. For $f_{3\,dB} < 100 \text{ MHz}$, $\gamma \approx 2/3$ and for $f_{3\,dB} > 100 \text{ MHz}$, $\gamma \approx 4/3$. The output power of the GaAlAs LEDs is about a factor of two higher than that of the InGaAsP LEDs at all bandwidths. This is mostly because the photon energy at 0.85 μm is larger by a factor of 1.53 than that at 1.3-μm wavelength.

At present, LED-based lightwave systems at the wavelength of minimal dispersion (near 1.3 μm) of silica graded-index fibers can have many applications especially in data links that offer advantages of simplicity, reliability, and economy over the laser-based systems.[11,12] Data rates up to a few hundred megabits per second and repeater spacings over tens of kilometers are achievable with

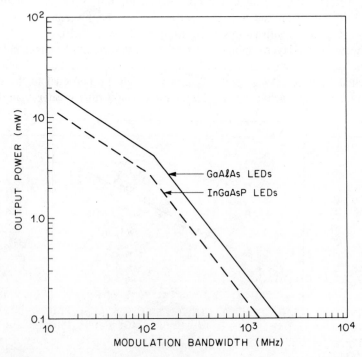

Fig. 4 Output power and bandwidth of LEDs at 0.8 and 1.3 μm wavelengths. (After Refs. 11 and 12)

LEDs and PIN photodetectors for both single-mode and multi-mode fiber systems.

The limitations of LED-based systems are the available power at high modulation rates (close to 1 Gb/s), the wide spectral width, and the difficulty in modulating beyond, at most, 2 Gb/s. The wide spectral width of the LED limits its application to the 1.3-μm region where the chromatic dispersion in the optical fiber is minimum. At other wavelengths, the chromatic dispersion causes the optical pulses to spread as it travels down the length of the fiber because different wavelengths travel at a different speed in the fiber.

10.3   SEMICONDUCTOR LASERS

10.3.1   Laser Structures

For high-speed (beyond 1 Gb/s) lightwave applications, semiconductor lasers are the natural choice. The modulation speed depends both on the intrinsic and extrinsic properties of the lasers. The extrinsic dependence is related to the particular laser structure and geometry employed. Figure 5 shows several of the important stripe-geometry laser structures[13] in their end-view cross sections.

Figure 5a is an inverted-rib laser structure. The wide-gap waveguide layer underneath the active layer has a rib-like structure running lengthwise inside the laser cavity. The difference in thickness produces a larger effective index of refraction in the rib region than on both sides of the rib. This results in waveguiding along the rib structure. The injection of current through the active layer is limited to the narrow stripe opening in the SiO$_2$ masking layer. In this structure, the active layer is planar. The ridge-waveguide laser structure shown in Fig. 5b employs the same principle for waveguiding in the lateral dimension. Waveguiding occurs underneath the ridge.

The etched-mesa buried heterostructure laser shown in Fig. 5c is quite different from the last two laser structures. In this structure, the active layer is first etched into a narrow stripe, InP is then regrown to bury the active stripe. Since the surrounding InP has a lower refractive index than the GaInAsP active stripe, strong waveguiding occurs. At the same time, the wider energy gap of InP than the GaInAsP also confines the injected carriers to the active stripe. This results in more efficient use of injected carriers and leads to lower lasing threshold. In the structure shown here, current injection is confined by the SiO$_2$ mask on top and the reversed p-n junction on both sides of the active stripe. This junction becomes reverse biased when the laser diode is forward biased. In forming this structure, two epitaxial growth sequences are needed, one for the growth of the base layer and the other for the regrowth after mesa-etching. In Fig. 5d, a V-groove is first etched into the substrate, then by liquid-phase epitaxial growth, the active layer and cladding layers are grown. Liquid-phase epitaxy is a growth technique in which epitaxial layers are deposited on the single crystalline substrate in direct contact with molten solutions composed of the constituent ele-

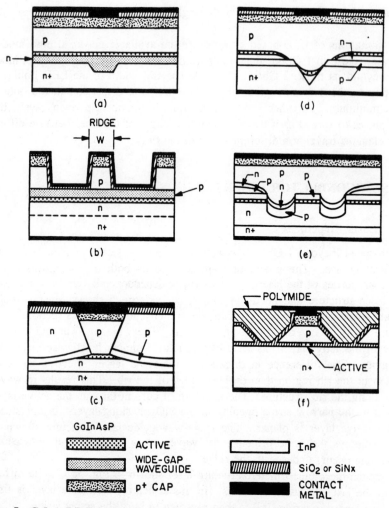

Fig. 5 GaInAsP/InP laser structures: (a) inverted-rib, (b) ridge-waveguide, (c) etched-mesa buried heterostructure, (d) channeled substrate buried heterostructure, (e) double-channel planar buried heterostructure, (f) constricted mesa. (After Ref. 13)

ments. A crescent stripe is formed directly above the V-groove. This forms the active stripe that is completely buried in the wide bandgap InP. In this structure, there is both optical and carrier confinement.

Figure 5e shows the cross-sectional view of a double-channel planar buried heterostructure laser. The two etched channels on both sides of the mesa containing the active stripe defines the width of the active stripe. Current confinement is achieved by reverse-biased *p-n* junctions similar to the structure described in Fig. 5c. The constricted mesa laser shown in Fig. 5f has the narrow active stripe buried by SiO_2, and the structure planarized by using polymide. These are some commonly used laser structures.

In conventional double heterostructures, the active layers are typically ~ 0.1 μm thick. If the thickness is reduced to typically less than 20 nm, quantum size effects are observed. Lasers, known as quantum-well (QW) lasers, constructed out of such active layers have characteristics quite different from conventional DH lasers.

In a quantum-well structure (Fig. 6a), the confinement of electrons and holes in one dimension causes a quantization in the allowed energy levels.[14] The density of states changes from the parabolic dependence in DH to a steplike structure. If the carriers are confined in two dimensions (quantum wire, Fig. 6b) or even in three dimensions (quantum box, Fig. 6c), the peak density of states becomes even larger. Such modifications in the density of states of the QW lasers result in reduced threshold-temperature dependence, lower threshold currents, higher modulation frequency, and narrower laser linewidth. These superior characteristics in QW lasers over DH lasers are much more pronounced in AlGaAs lasers than in InGaAsP lasers. A detailed discussion on QW lasers can be found in Ref. 46.

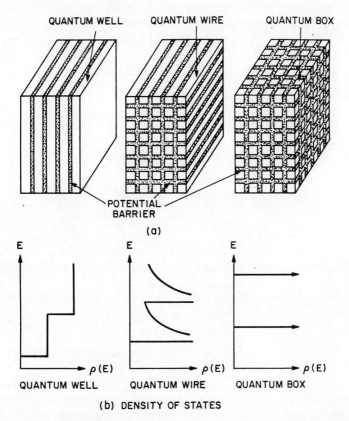

Fig. 6 Schematic diagram and density of states for quantum-well, quantum-wire and quantum-box laser. (After Ref. 14)

However, when the active layer becomes very thin, and with a refractive index step of ~0.3 between the cladding and active layers, most of the optical field does not overlap with the active QW. As a result, the threshold increases substantially. Two solutions can be implemented. One is to have an active region consisting of multiple QWs so that the total QW and barrier thicknesses are thick enough for efficient optical overlap. The other solution is to provide a separate means of confining the carriers and the optical field. A very efficient example is the graded-index waveguide, separate-confinement heterostrucutre[15] (GRIN-SCH) with one or multiple QWs shown in Fig. 7. This GRIN-SCH not only improves the optical-confinement efficiency but also improves the carrier-collecting efficiency. The optical confinement factor is improved due to the parabolic refractive index profile. This concentrates more optical energy to the active quantum well. The carrier collecting efficiency is improved due to the change in density of states in the graded layers are reduced. As a result, less carriers are wasted. This structure provides the lowest current thresholds ever achieved in any semiconductor laser structures. If multiple QWs are employed with barriers in between them, the barriers should not have too high an energy or the carriers cannot be injected uniformly across all the QWs.[16]

The structures just described provide control of the laser's transverse modes, laterally and vertically. Several longitudinal modes resulting from the Fabry-Perot resonances of the cavity formed by the cleaved laser mirrors generally have sufficient gain to reach threshold and oscillate simultaneously. In semiconductor lasers, the reflecting mirrors are formed by cleaving the crystal along

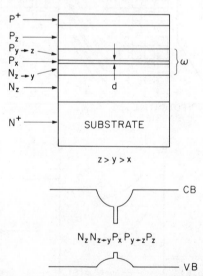

Fig. 7 A schematic diagram of a graded-index waveguide separate confinement heterostructure (GRIN-SCH) laser. Energy-band diagram of a GRIN-SCH laser: $N_i P_i$ stand for $N, P - Al_i Ga_{1-i}As$; n,p stand for $n,p - GaAs$; CB, conduction band; VB, valence band; P_{y-z} represent the AlAs concentration in $P - AlGaAs$ is varied from $Al_y Ga_{1-y}As$ to $Al_z Ga_{1-z}As$, and likewise for N_{z-y}.

certain crystal planes. This forms two parallel facets of the resonator. Such multi-mode operations, especially under pulse-code modulation, leads to pulse spreading during propagation through dispersive fibers and to mode partition noise. To avoid these effects, much effort has gone into development of single-frequency lasers, which remain stable under high-speed modulation.

The key to achieving single longitudinal-mode operation is to provide adequate gain or loss discrimination between the single desired mode and all the unwanted modes of the laser resonator. There are several ways of accomplishing this, such as using a short cavity,[17] adding an additional in-line, passive-coupled optical cavity,[18] using two independently adjustable active-coupled cavities,[19] using the distributed feedback (DFB) and the distributed Bragg reflector (DBR) structures.[20,21] The cleaved-coupled-cavity (C³) laser also offers the capability of electronically tuning the wavelength of the single longitudinal mode[19] (Fig. 8). This first demonstration in 1983 advanced semiconductor laser research from fixed single-frequency laser into electronically tunable single-frequency laser. In C³ lasers, the two sections are driven separately. If only one section is driven above threshold, the other can be used as a current-controlled tuning element that makes use of the free carrier dependence of the refractive index and hence changes the effective resonator length. As the effective resonator length changes, the lasing wavelength changes. Since the two coupled cavities have slightly different longitudinal resonant-mode spacings and they are optical coupled, only the longitudinal modes that are common to both (close in frequency) are enhanced. As a result, the effective longitudinal-mode spacing of the coupled-cavity resonator is significantly increased. This increases the gain discrimination of the side modes, which are further away from the gain peak of the active material. Dynamic single-mode operation of C³ lasers with side-mode suppression ratios of several thousand under high-speed modulation has been achieved.

Recently, this wavelength tuning concept was applied to multi-section DFB and DBR lasers. The grating in these lasers fixes the frequency around a predetermined frequency.

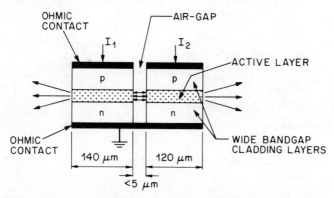

Fig. 8 Multiple element resonators, a 4-mirror configuration, in which the gap is formed by etching, or by cleaving to form a cleaved-coupled cavity (C³) laser.

An elegant approach to fixed single-frequency operation is the integration of wavelength selectivity directly into the semiconductor laser structure using a distributed Bragg reflector employing a surface grating etched into the epilayer, as shown in Fig. 9. In these lasers, the longitudinal modes are spaced symmetrically (for lasers without facet or other uniformities) around λ_B at wavelengths given by

$$\lambda = \lambda_B \pm \frac{(m + \frac{1}{2})\, \lambda_B^2}{2 \bar{n}_e L_e} , \qquad (5)$$

where m is the mode index, L_e is the effective grating length, $\lambda_B = 2 \bar{n}_e \Lambda / \mathcal{L}$ is the Bragg wavelength, $\bar{n}_e$ is the effective waveguide index, and Λ and $\mathcal{L}$ are the period and length of the grating. There are two equivalent lowest-order modes ($m = 0$) and oscillation on at least two frequencies is expected. However, because actual DFB are not perfectly symmetrical, single-frequency operation on one of the modes is often the result. By intentionally introducing a quarter-wave phase shift in the grating region, enhanced single-frequency DFB operation can be obtained. By incorporating two or more independent electrical sections in DFB or DBR lasers, as in the C^3 structure, frequency tuning can be accomplished.

10.3.2 Modulation Characteristics

The dynamics of semiconductor lasers have been extensively described in the literature.[22,23] Here, we present a simple but accurate analysis for most high-speed lasers. The rate equations governing the carrier and photon dynamics in a semiconductor laser can be expressed as follows:

$$\frac{dN}{dt} = -\frac{g_0(N - N_t)S}{1 + \epsilon S} + \frac{I}{qV} - \frac{N}{\tau_n} , \qquad (6)$$

$$\frac{dS}{dt} = \Gamma g_0 \frac{(N - N_t)S}{1 + \epsilon S} - \frac{S}{\tau_p} + \frac{\beta \Gamma N}{\tau_n} , \qquad (7)$$

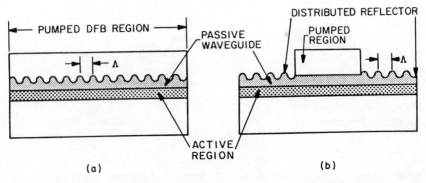

Fig. 9 Schematic representation of (a) DFB laser, (b) DBR laser.

where N and S are the electron and photon densities, g_0 is the differential gain, N_t is the carrier density for transparency, τ_n and τ_p are the spontaneous electron lifetime and the photon lifetime, Γ is the optical confinement factor, ϵ is a parameter characterizing the non-linear gain, β is the fraction of spontaneous emission coupled into the lasing mode, I is the current through the active layer, q is the electron charge, and V is the volume of the active layer.

Under small-signal modulation, an analytic expression for the intensity modulation response can be obtained by assuming a sinusoidal variation for $I = I_0 + ie^{jwt}$ and similarly for N and S. I_0, N_0, and S_0 represent the steady-state values; j is $\sqrt{-1}$. Substituting these into Eqs. 5 and 6, we have

$$\frac{S}{j} = \frac{A}{\omega_0^2 - \omega^2 + j\Gamma\omega} , \qquad (8)$$

where A, Γ and ω^2 are real

$$A = \frac{1}{qV}\left[\frac{\Gamma g_0 S_0}{1+\epsilon S_0} + \frac{\Gamma\beta}{\tau_n}\right] , \qquad (9)$$

$$\Gamma = \frac{g_0 S_0}{1+\epsilon S_0} + \frac{1}{\tau_n} - \frac{\Gamma g_0(N_0 - N_t)}{(1+\epsilon S_0)^2} + \frac{1}{\tau_p} , \qquad (10)$$

$$\omega_0^2 = -\left[\frac{g_0 S_0}{1+\epsilon S_0} + \frac{1}{\tau_n}\right]\left[\frac{\Gamma g_0(N_0 - N_t)}{(1+\epsilon S_0)^2} - \frac{1}{\tau_p}\right] \qquad (11)$$

$$+ \frac{g_0(N_0 - N_t)}{(1+\epsilon S_0)^2}\left[\frac{\Gamma g_0 S_0}{1+\epsilon S_0} + \frac{\Gamma\beta}{\tau_n}\right] ,$$

$$N_0 - N_t = \frac{S_0/\tau_p - \Gamma\beta N_t/\tau_n}{\Gamma g_0 S_0/(1+\epsilon S_0) + \Gamma\beta/\tau_n} . \qquad (12)$$

This transfer function is plotted in Fig. 10 for several values of ϵ. Typical values for the variables for 1.3-μm InGaAsP lasers were used in this calculation. From the poles of Eq. 8, we have

$$\omega = \frac{j\Gamma}{2} \pm \sqrt{\omega_0^2 - \frac{\Gamma^2}{4}} . \qquad (13)$$

The first term is the damping of the response. The second term is roughly the resonance frequency at

$$\omega_p = \sqrt{\omega_0^2 - \frac{\Gamma^2}{2}} . \qquad (14)$$

ω_p is usually called the resonance frequency (not resonant frequency) or relaxa-

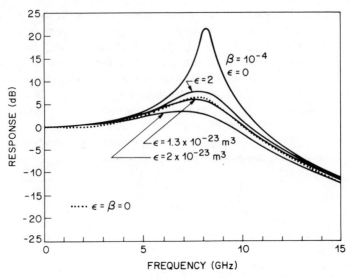

Fig. 10 Transfer function given in Eq. 8 for ($\cdots$) no damping ($\beta = \epsilon = 0$), and (———) several levels of nonlinear gain. ($\epsilon = 0$, $\beta = 0$), the top curve is for spontaneous emission damping only ($\epsilon = 0, \beta = 10^{-4}$). Values for the variables in Eq. 8 are $t = 0.15\ \mu m$, $w = 1\ \mu m$, $\Gamma = 0.35$, $R = 0.35$, $\alpha_i = 20\ cm^{-1}$, $g_0 = 1.8 \times 10^{-12}\ m^3/s$, $N_T = 1 \times 10^{18}\ cm^{-3}$.

tion frequency. By noting that $N_0 - N_t \sim (g_0 \Gamma \tau_p)^{-1}$ and $g_0 S_0 \ll 1/\tau_p$, we have

$$\omega_p \approx \left[\frac{g_0 S_0}{\tau_p(1 + \epsilon S_0)} \right]^{1/2}. \tag{15}$$

The output power of a semiconductor laser is

$$P = \frac{SVh\nu\,\alpha_m}{2\Gamma\tau_p(\alpha_m + \alpha_i)}, \tag{16}$$

where α_m is the mirror loss, and $h\nu$ is the photon energy. Using $W_p = 2\pi f_p$, Eq. 15 can be rewritten as

$$f_p = D\sqrt{P} \tag{17}$$

where

$$D = \frac{1}{2\pi} \sqrt{\frac{2g_0\Gamma(\alpha_i + 1/\mathcal{L}\ln 1/R)}{h\nu Wt\,\ln(1/R)}}, \tag{18}$$

and W, t and $\mathcal{L}$ are the width, thickness, and length of the waveguide, and R the mirror reflectivity. As an experimental convention, the bandwidth of a semiconductor laser is usually defined as 3 dB below its dc value ($f_{3\,dB}$) or one half of its dc value ($f_{6\,dB}$). The expression for $f_{3\,dB}$ and $f_{6\,dB}$ are

$$f_{3\,dB} \approx \left[\sqrt{1 + \sqrt{2}}\right] f_p , \tag{19}$$

$$f_{6\,dB} \approx \sqrt{3}\, f_p . \tag{20}$$

Equations 19 and 20 are valid if parasitics are insignificant.

For the analysis of parasitics, we assume that the laser chip is connected to a microstrip line of impedance R_s without significant discontinuities. The bond wire has an inductance (L), for a 1-mm long wire, $L \approx 1$ nH. The bonding pad has a capacitance (C). For a typical chip size (500 $\times$250 μm^2) with a 1000-Å SiO$_2$ layer under the bonding pad, the capacitance is ~40 pF. The resistance of the p-contact is typically several ohms. Under a forward-bias current of 50 mA, the p-n junction itself has an impedance of less than 1 Ω. Under such a simple circuit model, the 3 dB bandwidth is

$$f_{3\,dB} = f_{\mathscr{L}} \left[1 - \frac{1}{2Q^2} + \left(2 - \frac{1}{Q^2} + \frac{1}{4Q^4}\right)^{1/2}\right]^{1/2} , \tag{21}$$

where

$$f_{\mathscr{L}} = \frac{1}{2\pi} \left[\frac{R_s + R}{L} RC\right]^{1/2} \tag{22}$$

and

$$Q = \frac{[LRC(R_s + R)]^{1/2}}{L + RR_sC} . \tag{23}$$

Figure 11 shows contours of constant parasitic-limited bandwidth for a laser with $R = 4\ \Omega$ operating from a source resistance of 50 Ω. Thus, to design a laser for a bandwidth of 30 GHz, the bonding pad capacitance must be kept small (< 2 pF) by limiting the bonding pad area and using a thick, low dielectric-constant layer under it, and by reducing the area of current-confining p-n junctions. A schematic diagram of such a structure, the constricted mesa laser,[24] is shown in Fig. 5f. The capacitance of this laser can be 1 pF or less. The other requirement for 30-GHz operation is to reduce the package effects and reduce L to less than 0.2 nH.

It should be noted that the small-signal modulation characteristics depend strongly on the bias current. Consequently, the large-signal impulse response of the laser is not simply the Fourier transform of the small-signal frequency response. The knowledge of the impulse response of the laser is important for ultra-high-speed communication systems and gain switching.

Most optical-communications systems transmit information using pulse code modulation of the laser intensity by directly modulating the current to the laser. The two most common formats are the non-return-to-zero (NRZ) modulation

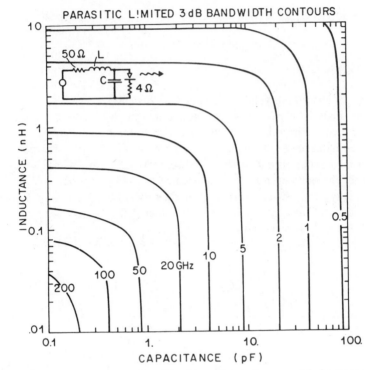

Fig. 11   3-dB bandwidth contours in the plane of capacitance and inductance assuming $R_s = 50$ and $R = 4\ \Omega$.

and the return-to-zero (RZ). In the former the current stays on during multiple ones, while in the latter, a "one" transmission consists of a transition from off to on to off again. The disadvantage of RZ transmission is that the bandwidth of the laser driver must be twice as large as the NRZ bandwidth, but the main advantage is easier clock recovery. The responses of a constricted mesa laser to NRZ and RZ modulations at bit rates from 2 to 16 Gb/s are shown in Fig. 12. The NRZ eye at 16 Gb/s is not much worse than RZ eyes at 8 Gb/s.

Current modulation results in the modulation of laser intensity. It modulates the electron density, which modulates the gain and the refractive index of the active layer. Consequently, the frequency of the laser is also modulated. This frequency modulation (FM) can have important applications in frequency-shift-keyed (FSK) coherent and incoherent optical communications systems, but it is a problem in high-speed direct amplitude-modulated systems operating at the dispersive regime of the silica fibers. This is because in dispersive fiber, different frequencies of light travel at different speeds. After a certain distance, the light pulse becomes smeared out in shape.

Starting with the rate equations and the relation between a frequency shift and an index shift $\Delta\nu/\nu = -\Delta\bar{n}/\bar{n}$, the frequency shift and power of the laser can

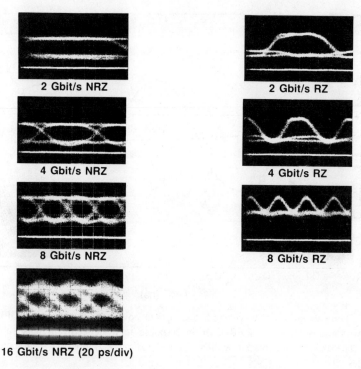

2 Gbit/s NRZ

2 Gbit/s RZ

4 Gbit/s NRZ

4 Gbit/s RZ

8 Gbit/s NRZ

8 Gbit/s RZ

16 Gbit/s NRZ (20 ps/div)

Fig. 12 Pseudorandom NRZ and RZ modulation of a high-speed 1.3-μm GaInAsP constricted mesa laser at bit rates of 2 Gb/s, 4 Gb/s, 8 Gb/s, and 16 Gb/s. (Courtesy J. E. Bowers)

be related[25]:

$$\Delta \nu = \frac{-\alpha}{4\pi} \left[\frac{dP/dt}{P} + \varkappa P \right]. \tag{24}$$

Where α is the change between the change in the real and imaginary parts of the active layer with a carrier density change, $\varkappa = 2\Gamma\epsilon(\alpha_m + \alpha_i)/(\mathcal{L}\alpha_m)$. The first term is independent of the laser structure, and leads to chirping during the relaxation oscillation of the modulation. The second term leads to a wavelength chirp between the high and low points in the optical waveform.[26] Two different examples are illustrated in Fig. 13. The ridge overgrown DFB laser has a large transient oscillation in power when the pulse is turned on. This results in a corresponding wavelength variation, $\Delta\lambda$, as a function of time. The DCPBH DFB laser has a smooth turn-on. The wavelength shift is also step-like. In lasers with small active volumes, such as the buried heterostructures, $\varkappa$ is large. The second term is large, but the laser is heavily damped. In weakly guided lasers, for example, the ridge waveguide structrues, $\varkappa$ is smaller and the laser intensity and frequency oscillate, but there is a smaller frequency shift between the on and off levels.

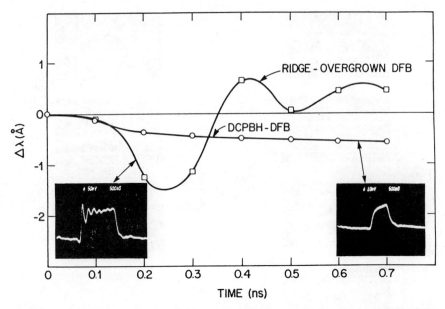

Fig. 13 Time variation of wavelength and optical power for a ridge type DFB (ridge overgrown) and a DCPBH DFB laser in response to a step current pulse applied at $t = 0$. The lasers were biased with an optical extinction ratio of 4:1 in both cases. (After Ref. 26)

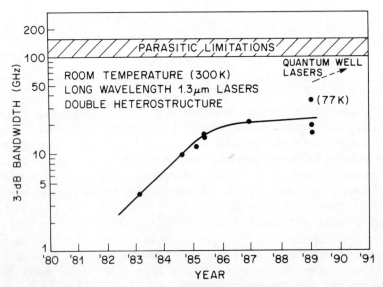

Fig. 14 Historical dependence of cw room temperature bandwidths of long-wavelength semiconductor lasers. The date indicated is the time of journal submission or conference presentation.

In Fig. 14 the typical modulation responses (intensity) of a number of long-wavelength laser structures are compared. These results represent the state-of-the-art but are not necessarily optimized. To make a very high frequency laser, the parasitics must be small and the resonance frequency must be sufficiently high. The maximum resonance frequency from Eq. 17 is $f_r^{max} = DP_{max}$, so D and P_{max} have to be maximized. As given in Eq. 18, D can be maximized by decreasing the waveguide width, increasing the gain coefficient by cooling the laser, doping the active layer, confining the carriers in quantum wells, "loading" the laser off its Fabry-Perot resonance, and decreasing the cavity length.

Figure 14 shows the increase in bandwidth of InGaAsP lasers since 1980. It is increasing rapidly. What is the ultimate bandwidth obtainable from a semiconductor laser? The limits will be set by parasitics, resonance frequency, and optical non-linearities. With a parasitic capacitance and inductance of ≥ 0.2 pF and ≤ 0.05 nH, a 100-GHz bandwidth is achievable. The present bandwidths are a long way from being inherent parasitic limited. The resonance frequency can be increased as discussed above to ≥ 45 GHz. At present, a bandwidth of 12 GHz (continuous wave) and 17 GHz (pulsed) can be obtained at laser output powers of 10 to 20 mW in long-wavelength lasers. With 100-mW capability, a further increase in bandwidth is realizable. Two potentially severe limitations to further increases in bandwidth are optical non-linearities such as spectral-hole burning or two-photon absorption. These can cause a saturation in bandwidth where no further increase in bandwidth with increased laser output power is possible. This maximum limit, based on estimates of spectral-hole burning, is between 25 and 60 GHz. Higher-frequency operation is possible by increasing the optical gain in reduced-dimension, quantum-confined lasers.

Among the methods of maximizing D, confining carriers in reduced dimensions in quantum wires and boxes is most interesting. Carriers in a quantum well have a modified density-of-state function (see Fig. 6) and a high differential gain. The optical gain is proportional to the number of carriers in the active medium. Since the density-of-state in QWs is modified, the carrier distributions, as a function of energy, are also modified from that of bulk material. This, in turn, resulted in a different spectral gain people. The gain coefficient in Eq. 18, g_0, is given by the slope of the gain profile at the lasing wavelength. Thus g_0 is also modified in QWs. Consequently, quantum-well lasers should have a larger slope D than conventional DH lasers.[27] This increase for GaAs lasers shows a factor of two improvement in D for QW with 50 to 200-Å wide wells. Even greater changes are predicted for quantum-wire and quantum-box lasers (Fig. 15).

10.4 PIN PHOTODETECTORS

High-speed and high-sensitivity photodetectors are required for high bit-rate optical communication systems. Two commonly used photodetectors used in these systems are the PIN and avalanche photodiodes (APDs). Because APD

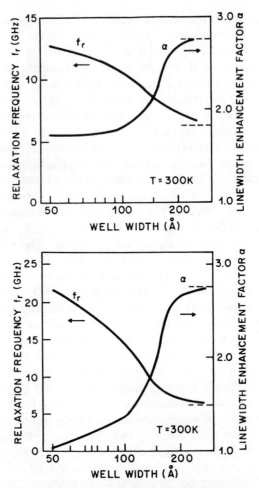

Fig. 15 (a) Resonance frequency and linewidth enhancement factor dependence on well width for a quantum-well laser. (After Ref. 27) (b) Same parameters as Fig. 14 for a quantum wire laser with width=thickness. (After Ref. 27)

has internal gain, it offers the most sensitive receivers at lower frequencies. But at the highest speed applications, for example, 16 Gb/s and above, the best sensitivity may be obtained with PIN/FET receivers because they have higher bandwidth capability and lower leakage current than do APDs. Since the primary wavelengths are 1.3 to 1.6 μm in optical communication systems, we will also discuss InGaAs photodetectors and limit the discussion to PIN and ADP photodiodes.

A PIN photodetector is structurally very simple. It consists of n-InP, n^--InGaAs, and p-InGaAs layers. The speed of a PIN depends on the time it takes for a carrier to drift across the depletion layer in the $\bar{n}$-InGaAs, the time it takes for the carrier to diffuse out of undepleted regions, the time it takes to charge and discharge the inherent capacitance of the diode plus the parasitic capacitance, and the trapping at the heterojunctions.

For high-speed operation, the intrinsic layer should be completely depleted and the field in this layer should be above 50 kV/cm (for InGaAs) so that the carriers travel at their saturation velocity. Figure 16 shows the dependence of velocity on electric field for GaAs and InGaAs lattice-matched to InP substrate.[28-32]

Figure 17 shows the calculated impulse response[47] of a 0.5-μm thick back-illuminated GaInAs/InP PIN for several different wavelengths, that is, different absorption coefficients α. For $\alpha L >> 1$, excitation is at one edge and results in a rectangular-shaped impulse response. For $\alpha L << 1$, the excitation is uniform and results in a triangular-shaped impulse response.

The frequency response has a $(\sin x)/x$ response for $\alpha L >> 1$. For arbitrary αL with optical excitation from the n-side the frequency response is given by[33]

$$\frac{i(\omega)}{i(0)} = \frac{1}{1 - e^{-\alpha L}} \left[- \frac{e^{j\omega\tau_n - \alpha L} - 1}{j\omega\tau_n + \alpha L} + e^{-\alpha L} \frac{e^{-j\omega\tau_n} - 1}{j\omega\tau_n} \right.$$

$$\left. - \frac{e^{j\omega\tau_p} - 1}{j\omega\tau_p} + e^{-\alpha L} \frac{1 - e^{\alpha L - j\omega\tau_p}}{\alpha L - j\omega\tau_p} \right]. \tag{25}$$

Where i is the detected current, ω is the angular modulation frequency, $\tau_n = (L/v_n)$ and $\tau_p = (L/v_p)$ are the electron and hole transit times, respectively, and v_n and v_p are their velocities. For p-side illumination, the p and n

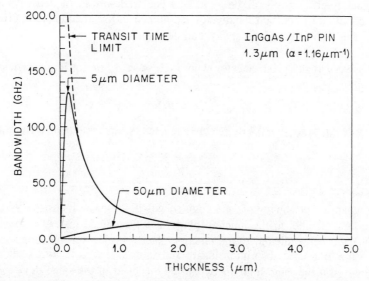

Fig. 16 Dependence of carrier velocity on electric field for GaInAs and GaAs. (After Refs. 28–32).

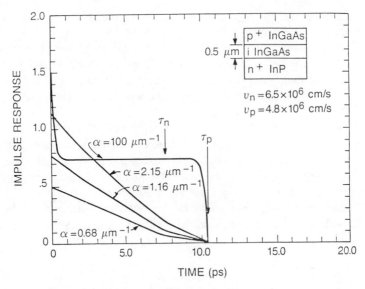

Fig. 17   Impulse response of a PIN detector for different value of α: $\alpha = 0.68\ \mu m^{-1}$ ($\lambda = 1.55\ \mu m$), $\alpha = 1.16\ \mu m^{-1}$ ($\lambda = 1.36\ \mu m$), $\alpha = 2.15\ \mu m^{-1}$ ($\lambda = 1.06\ \mu m$), ($v_p = 4.8 \times 10^6$ m/s, $v_n = 6.5 \times 10^6$ m/s, corresponding to GaInAs. (After Ref. 47)

subscripts in Eq. 25 are interchanged. For high-speed response, very thin intrinsic layers are needed to reduce the transit times. However, this sacrifices the quantum efficiency. In principle, bandwidths >200 GHz are possible, but the quantum efficiency will be $\sim 11\%$. The trade-off is plotted in Fig. 18, where contours of constant bandwidth are plotted in the area/quantum-efficiency plane. The assumed load resistance and detector series resistance is 50 Ω in this calculation.

For a very high speed detector where $\alpha L \ll 1$, the quantum efficiency is given by

$$\eta \approx (1 - R)\,\alpha L \ , \tag{26}$$

where R is the Fresnel reflectivity. The bandwidth of a thin detector is given by

$$f_{3dB} = \frac{0.45\,v}{L} \ , \tag{27}$$

where we assumed $v_n = v_p = v$.

The quantum efficiency of very high speed PINs can be increased by collecting the light parallel to the junction plane in a waveguide instead of having the light incident perpendicular to the junction plane. Such a scheme simultaneously realizes the long light-absorption length and short carrier-collecting path, so that both high efficiency and high speed can be achieved in the same device. However, such waveguide PIN detectors have poorer coupling efficiency, typically $\sim 50\%$. The overall expected bandwidth can be ~ 150 GHz.

THICKNESS (μm)

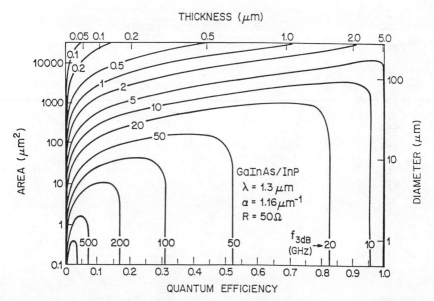

Fig. 18 Contours of constant 3-dB bandwidth in the detector area, depletion layer thickness plane. ($\alpha = 1.16 \, \mu m^{-1}$ for 1.3 μm wavelength, $v_n = 6.5 \times 10^6$ cm/s, $v_p = 4.8 \times 10^6$ cm/s, $\epsilon = 14.1$). (After Ref. 47)

Finally, trapping at the heterojunctions can be serious. This can be improved by incorporating a graded layer at the hetero-interfaces. A hetero-interface is the interface between two epilayers having different energy bandgaps. The leakage current in PIN devices comes from the generation-recombination current in the depleted region, diffusion current from the undepleted p and n sides of the junction, tunneling current, and surface leakage current. However, in a well-designed 4-Gb/s PIN-FET receiver, a leakage current of 10 μA causes a sensitivity penalty of less than 1 dB. At higher bit rates, the effect of leakage current is even less.

Figure 19 shows the rapid increase in bandwidth of long-wavelength photodetectors. A 200-GHz bandwidth is realizable.

10.5 AVALANCHE PHOTODIODES

In comparison with PINs, the internal gain of APDs can provide improved receiver sensitivity particularly at higher bit rates (≥ 500 Mb/s). In the InGaAsP system,[34] the basic layer structure shown in Fig. 20 is commonly used and is called the separate absorption and multiplication (SAM) APDs. The p-n junction and thus the high field region is located in the wide-gap material InP where tunneling is insignificant. Optical power is absorbed in the adjacent narrow-gap InGaAs layer. For this structure to operate properly three condi-

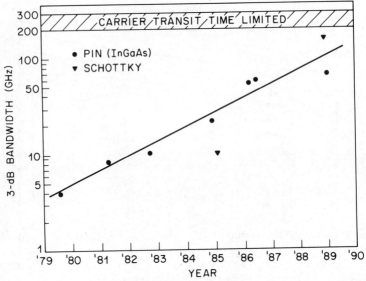

Fig. 19 Bandwidth of GaInAs photodetectors since their first demonstration in 1979. In some cases, the 3-dB bandwidth is inferred from time domain data; these points generally have the largest error bars.

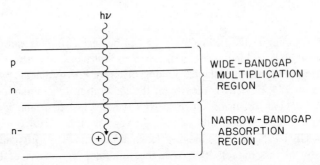

Fig. 20 Avalanche photodiode structure with separate absorption and multiplication regions (SAM-APD).

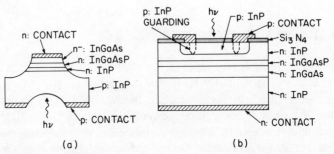

Fig. 21   (a) Mesa-structure and (b) planar-structure SAGM-APDs.

tions must be satisfied. (1) The electric field in the multiplication region must be high enough to produce useful gain. (2) The electric field at the absorbing layer interface must be low enough so that tunneling in the absorbing layer is negligible. (3) The depletion region must extend far enough into the absorbing region so that diffusion effects are eliminated.

Unfortunately, the frequency response of these SAM-APDs is usually poor due to carrier trapping at the hetero-interface.[35] The bandwidth is only a few hundred MHz. The pulse response of these SAM-APDs contains a fast and a slow component. The fast component has a time constant of <200 ps. The slow component decreases with increasing bias voltage and can be in the range of a few nanoseconds to microseconds.

To reduce carrier trapping in the SAM-APD, a graded interface consisting of one or more intermediate-bandgap InGaAsP layers is added between the InGaAs absorbing layer and the InP multiplication layer.[36-39] Figure 21 shows two different device structures of this type, which is the mesa type, the separate-absorption graded-multiplication (SAGM) ADP. The major advantage of the etched mesa type (Fig. 21a) is its fabrication simplicity. However, it is not planar and can have larger surface leakage current. Hence, for practical applications, planar guard-ring SAGM-APDs (Fig. 21b) are fabricated. In these structures, the InGaAs layer is the light-absorbing layer, the InGaAsP layer is the bandgap-grading layer, the n-InP layer is the carrier multiplication (avalanche) layer. The improvement in speed over the SAM-APD afforded by the SAGM-APD structure is illustrated in Fig. 22 for a device grown by chemical beam epitaxy[40] with three 200-Å intermediate-bandgap InGaAsP layers with corresponding wavelengths at 1.1, 1.3, and 1.55 μm. A gain-bandwidth product of 70 GHz was obtained at an incident wavelength of 1.3 μm, which is to be compared with a value of only a few hundred MHz for a SAM-APD.

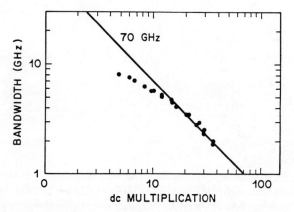

Fig. 22 The gain-bandwidth product of a SAGM-APD grown by chemical beam epitaxy, measured at λ = 1.3 μm.

The frequency response of these APDs can be approximated by[41]

$$\frac{M(\omega)}{M_0} = \frac{2\sin(\omega W/v)}{\omega W/v} \frac{1}{\sqrt{1+(\omega RC)^2}} \frac{1}{\sqrt{1+(\omega/e_h)^2}} \frac{1}{\sqrt{1+(\omega\tau_m M_0)^2}} \ , \quad (28)$$

where $M(\omega)$ and M_0 is the gain at ω and dc, respectively, $M_0\tau_m$ is the avalanche build-up time, and e_h is the hole-emission rate over the heterojunction barrier. The first term is an approximate expression for the frequency response of a transit-time-limited PIN having a depletion width W. The second term is the RC contribution, and the third term is due to the residual interface trapping. The regenerative nature of the avalanche process is covered in the last term. As the gain increases, this term eventually dominates giving rise to a constant gain-bandwidth product (see Fig. 22).

There have been a number of novel APD structures that, at least in theory, improve performance by "artificially" increasing the ratio of the ionization coefficients for holes and electrons. Figure 23 shows a structure incorporating a superlattice into the multiplication region.[42] The difference between the conduction and valence band discontinuities enhances the electron ionization rate relative to that of holes.

For the photodetector to be practical, it has to perform well in a lightwave receiver. Figure 24 is a summary of the measured receiver sensitivities that

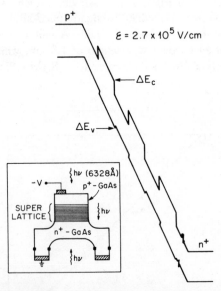

Fig. 23 A superlattice avalanche photodiode. The inset shows the layer strucutre of the device. The superlattice is composed of AlGaAs/GaAs thin layers ($\sim$ 200 Å). The electrons acquire more energy than the holes as they drop into the GaAs wells because of the larger bandgap discontinuities in conduction band edge than valence band edge. (After Ref. 42)

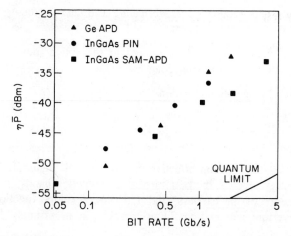

Fig. 24 The measured receiver sensitivities with InP/InGaAsP/InGaAs SAGM-APDs, Ge-APDs, and InGaAs PINs in the front-end circuits of long-wavelength receivers. (After Ref. 42)

have been achieved with InP/InGaAsP/InGaAs SAGM-APDs, Ge-APDs, and InGaAs PINs in the front-end circuits of long-wavelength receivers.[37,43-45] It should be noted that these comparisons between APDs and PINs are somewhat deceptive because the detectors were in different receivers. The PIN results were obtained with very low capacitance (<0.6 pF) hybrid front ends, whereas the capacitance of the APD front ends were 2 to 3 times larger. Nevertheless, it is clear that the APD has an edge over PIN. As pointed out in a previous section, PIN becomes advantageous at very high bit rates where PINs are not limited by the frequency response bandwidth as are APDs.

10.6 SUMMARY AND FUTURE TRENDS

Light-emitting diodes, semiconductor injection lasers, PIN photodetectors, and avalanche photodiodes are four very important devices in optical-fiber communications systems. Their performance, especially in the InGaAs alloys, has been optimized by device design, epitaxial growth, fabrication, and packaging in the last decade. It is seen that device parasitics are the major limitations to the ultimate speeds of both lasers and photodetectors. In InGaAsP lasers, a 100-GHz modulation bandwidth is potentially realizable, while in InGaAs photodetectors, a 200-GHz bandwidth is possible.

PROBLEMS

1. Describe the step sequence in fabricating the laser structure shown in Fig. 5a starting from InP substrate.

2. Describe the step sequence in fabricating the laser structure shown in Fig. 5e starting from InP substrate.

3. Explain how current injection is limited to the active mesa stripe in the DCPBH structure shown in Fig. 5e.

4. Using the Schrödinger-like equation

$$\left[-\frac{\hbar^2}{2m^*(z)} \frac{\partial^2}{\partial z^2} + V_c(z) \right] \chi_n(z) = \epsilon_n \chi_n(z) ,$$

where $m^*(z)$ is the electron effective mass of A or B material, $V_c(z)$ represents the energy level of the bottom of the conduction bands, ϵ_n is the confinement energy of the carriers, and $\chi_n(z)$ the envelope function, find the electron wave functions and confining energy ϵ_n in an infinitely deep well.

5. What are the continuity conditions at the interfaces of the QW?

6. Derive the density of state for a two-dimensional system, i.e., QWs.

7. How to maximize the resonance frequency of a semiconductor laser?

8. What is the physics behind the frequency chirping when a semiconductor laser is modulated?

9. If an optical fiber has a chromatic dispersion of 15 ps/km-nm, and a laser emitting pulses under modulation has a frequency chirping of 0.4 nm, what is the maximum distance that a 2-Gb/s system can have neglecting other system degrading effects.

10. If the effective grating period in a DFB laser changes due to changes in $\bar{n}_e$ because of carrier modulation during pulse-code modulation, how does the lasing wavelength change?

REFERENCES

1. W. T. Tsang, Ed., "Lightwave Communications Technologies," Vol. 22 A, B, C, D, and E in *Treatise of "Semiconductors and Semimetals,"* R. K. Willardson and A. C. Beer, Eds., Academic, New York, 1985.

2. J. L. Gimlett, M. Stern, L. Curtis, W. C. Young, Cheung, and P. W. Shumate, "Dispersion Penalties for Single-Mode-Fiber Transmission Using 1.3 and 1.5 μm LEDs," *Electron. Lett.*, **21**, 668 (1985).

3. T. Uji and J. Hayashi, "High-Power Single-Mode Optical-Fibre Coupling to InGaAsP 1.3 μm Mesa-Strucutre Surface-Emitting LEDs," *Electron. Lett.*, **21**, 418 (1985).

4. H. Kressel and M. Ettenberg, "A New Edge-emitting (AlGa)As Heterojunction LED for Fiber-Optic Communications," *Proc. IEEE*, **63**, 1360 (1975).

5. Y. Horikoshi, Y. Takanashi, and G. Iwane, "High Radiance Light-emitting Diodes," *Jpn. J. Appl. Phys.*, **15**, 485 (1976).

6. T. P. Lee, C. A. Burrus, Jr., and B. I. Miller, "A Stripe-Geometry Double-Heterostructure Amplified-Spontaneous-Emission (Superlumines-cent) Diode," *IEEE J. Quant. Electron.*, **QE-9**, 820 (1973).

7. Y. S. Liu and D. A. Smith, "The Frequency Response of an Amplitude-modulated GaAs Luminescence Diode," *Proc. IEEE (Lett.)*, **63**, 542 (1975).

8. H. Namizaki, M. Nagano, and S. Nakahara, "Frequency Response of $Ga_{1-x}Al_xAs$ Light-emitting Diodes," *IEEE Trans. Electron Dev.*, **21**, 688 (1974).

9. O. Wada, S. Yamakoshi, M. Abe, Y. Yishitoni, and T. Sakwai, "High Radiance InGaAsP/InP Lensed LEDs for Optical Communication Systems at 1.2-1.3 μm," *IEEE J. Quant. Electron.*, **QE-17**, 174 (1981).

10. J. Heinen, W. Huber, and W. Harth, "Light-emitting Diodes with a Modu-lation Bandwidth of More Than 1 GHz," *Electron. Lett.*, **12**, 553 (1976).

11. T. P. Lee, "LEDs and Photodetectors for Wavelength-Division-multiplexed Light-Wave Systems," *Opt. Laser Tech.*, **14**, 15 (1982).

12. R. H. Saul, "Recent Advances in the Performance and Reliability of InGAsP LEDs for Lightwave Communication Systems," *IEEE Trans. Elec-tron Dev.* **ED-30**, 285 (1983).

13. J. E. Bowers and M. A. Pollack, in *Optical Fiber Telecommunications II*, S. E. Miller and I. P. Kaminow, Eds., Academic, New York, 1988.

14. Y. Arakawa and A. Yariv, "Quantum Well Lasers-Gain, Spectra, Dynam-ics," *J. Quant. Electron.*, **QE-22**, 1887 (1986).

15. W. T. Tsang, "A Graded-Index Waveguide Separate-Confinement Laser with Very Low Threshold and a Narrow Gaussian Beam," *Appl. Phys. Lett.*, **39**, 134 (1981).

16. W. T. Tsang, "Extremely Low Threshold (AlGaAs)As Modified Multi-quantum Well Heterostructure Lasers Grown by Molecular-Beam Epitaxy," *Appl. Phys. Lett.*, **39**, 786 (1981).

17. T. P. Lee, C. A. Burrus, R. A. Linke, and R. J. Nelson, "Short-Cavity Single-Frequency InGaAsP Buried-Heterostructure Lasers," *Electron. Lett.* **19**, 82 (1983).

18. K. Y. Liou, C. A. Burrus, R. A. Linke, I. P. Kaminow, S. W. Granlund, C. B. Swan, and P. Besomi, "Single-Longitudinal-Mode Stabilized Graded-Index-Rod External Coupled-Cavity Laser," *Appl. Phys. Lett.*, **45**, 729 (1984).

19. W. T. Tsang, "The Cleaved-Coupled-Cavity (C^3) Laser," in *Semiconduc-tors and Semimetals: Lightwave Communications Technology*, R. K. Willardson and A. C. Beer, Eds, Vol. 22B, W. T. Tsang, vol. Ed., Chap. 5, Academic, London, 1985 pp. 257–373.

20. H. Kogelnik and C. V. Shank, "Stimulated Emission in a Periodic Structure," *Appl. Phys. Lett.*, **18**, 152 (1971).

21. H. Kogelnik and C. V. Shank, "Coupled-Wave Theory of Distributed Feedback Lasers," *J. Appl. Phys.*, **453**, 2327 (1972).

22. D. J. Channin, "Effect of Gain Saturation of Injection Laser Switching," *J. Appl. Phys.*, **50**, 3858 (1979).

23. J. Manning, R. Olshansky, D. M. Fye, and W. Powazinik, "Strong Influence of Nonlinear Gain on Spectral and Dynamic Characteristics of InGaAsP Lasers," *Electron. Lett.*, **21**, 496 (1985).

24. J. E. Bowers, B. R. Hemenway, A. H. Gnauck, and D. P. Wilt, "High-Speed InGaAsP Constricted-Mesa Lasers," *J. Quant. Electron.*, **QE-22**, 833 (1986).

25. T. L. Koch and J. E. Bowers, "Nature of Wavelength Chirping in Directly Modulated Semiconductor Lasers," *Electron. Lett.*, **20**, 1038 (1985).

26. T. L. Koch, and R. A. Linke, "Effect of Nonlinear Gain Reduction on Semiconductor Laser Wavelength Chirping," *Appl. Phys. Lett.*, **48**, 613 (1986).

27. Y. Arakawa and H. Skaki, "Multidimensional Quantum Well Laser and Temperature Dependence of Its Threshold Current," *Appl. Phys. Lett.*, **40**, 939 (1982).

28. T. H. Windhorn, L. W. Cook, and G. E. Stillman, "Temperature Dependent Electron Velocity-Field Characteristics for $In_{0.53}Ga_{0.47}As$ at High Electric Fields," *J. Electron Mat.* **11**, 1065 (1982).

29. T. P. Pearsall, Ed., *GaInAsP Alloy Semiconductors*, Wiley, New York, 1982.

30. P. Hill, J. Schlafer, W. Powazinik, M. Urban, W. Eichen, and R. Olshansky, "Measurement of Hole Velocity in *n*-type InGaAs," *Appl. Phys. Lett.*, **50**, 1260 (1987).

31. K.-H. Hellwege, Editor-in-Chief, *Landolt-Bornstein Numerical Data and Functional Relationships in Science and Technology*, Vol. 17, O. Madelung, Ed., Subvolume a, Physics of Group IV Elements and III–V Compounds, Springer-Verlag, Berlin, 1982, pp. 532–533.

32. S. M. Sze, *Physics of Semiconductor Devices*, Wiley, New York, 1981.

33. J. E. Bowers, C. A. Currus, and R. J. McCoy, "InGaAs PIN Photodetectors with Modulation Response to Millimetre Wavelengths," *Electron. Lett.* **21**, 812 (1985).

34. K. Nishida, K. Taguchi, and Y. Matsumoto, "InGaAsP Heterostructure Avalanche Photodiodes with High Avalanche Gain," *Appl. Phys. Lett.*, **35**, pp. 251 (1979).

35. S. R. Forrest, O. K. Kim, and R. G. Smith, "Optical Response Time of $In_{0.53}Ga_{0.47}AsInP$ Avalanche Photodiodes," *Appl. Phys. Lett.*, **41**, 95 (1982).

36. Y. Matsushima, A. Akiba, K. Sakai, Y. Kushiro, Y. Noda, and K. Utaka, "High-Speed-Response InGaAs/InP Heterostructure Avalanche Photodiode with InGaAsP Buffer Layers," *Electron. Lett.*, **18**, 945 (1982).

37. J. C. Campbell, A. G. Dentai, W. S. Holden, and B. L. Kasper, "High-Performance Avalanche Photodiode with Separate Absorption 'Grading' and Multiplication Regions," *Electron. Lett.*, **19**, 818 (1983).

38. K. Yasuda, T. Mikawa, Y. Kishi, and T. Kaneda, "Multiplication-dependent Frequency Responses of InP/InGaAs Avalanche Photodiode," *Electron. Lett.*, **20**, 373 (1984).

39. Y. Sugimoto, T. Torikai, K. Makita, H. Ishihara, K. Minemura, and K. Taguchi, "High-Speed Planar-Strucutre InP/InGaAsP/InGaAs Avalanche Photodiode Grown by VPE," *Electron. Lett.*, **20**, 653 (1984).

40. J. C. Campbell, W. T. Tsang, G. J. Qua, and B. C. Johnson, "High-Speed InP/InGaAsP/InGaAs Avalanche Photodiodes Grown by Chemical Beam Epitaxy," *IEEE J. Quant. Electron.*, **QE-24**, 496 (1988).

41. J. C. Campbell, W. S. Holden, G. J. Qua, and A. G. Dentai, "Frequency Response of InP/InGaAsP/InGaAs Avalanche Photodiodes with Separate Absorption, 'Grading', and Multiplication Regions," *IEEE J. Quant. Electron.*, **QE-21** (1985).

42. F. Capasso, W. T. Tsang, A. L. Hutchinson, and G. F. Williams, "Enhancement of Electron Impact Ionization in a Superlattice: A New Avalanche Photodiode with a Large Ionization Rate Ratio," *Appl. Phys. Lett.*, **40**, 38 (1982).

43. D. R. Smith, R. C. Hooper, P. P. Smyth, and D. Wake, "Experimental Comparison of a Germanium Avalanche Photodiode and InGaAs PINFET Receiver for Longer Wavelength Optical Communication Systems," *Electron. Lett.*, **18**, 453 (1982).

44. M. C. Brain, P. P. Smyth, D. R. Smith, B. R. White, and P. J. Chidgey, "PINFET Hybrid Optical Receivers for 1.2 Gbit/s Transmission Systems Operating at 1.3 and 1.55 μm Wavelength," *Electron. Lett.*, **20**, 894 (1984).

45. S. R. Forrest, G. F. Williams, O. K. Kim, and R. G. Smith, "Excess-Noise and Receiver Sensitivity Measurements of $In_{0.53}Ga_{0.47}As$/InP Avalanche Photodiodes," *Electron. Lett.*, **17**, 917 (1981).

46. W. T. Tsang, "Quantum Confinement Heterostructure Semiconductor Lasers," R. Dingle, Ed., Vol. 24, in *Treatise of "Semiconductors and Semimetals,"* R. K. Willardson and A. C. Beer, Eds., Academic, New York, 1987.

47. J. E. Bowers and C. A. Burrus, Jr., "Ultrawide-Band Long-Wavelength *p-i-n* Photodetectors," *J. Lightwave Technol.*, **LT-5**, 1339 (1987).

List of Symbols

Symbol	Description	Unit
a	Lattice constant	Å
$\mathcal{B}$	Magnetic induction	Wb/m^2
c	Speed of light in vacuum	cm/s
C	Capacitance	F
d	Oxide thickness	Å
$\mathcal{D}$	Electric displacement	C/cm^2
D	Diffusion coefficient	cm^2/s
E	Energy	eV
E_C	Bottom of conduction band	eV
E_F	Fermi energy level	eV
E_g	Energy bandgap	eV
E_V	Top of valence band	eV
$\mathcal{E}$	Electric field	V/cm
$\mathcal{E}_c$	Critical field	V/cm
$\mathcal{E}_m$	Maximum field	V/cm
f	Frequency	Hz
h	Planck's constant	J-s
$h\nu$	Photon energy	eV
I	Current	A
I_C	Collector current	A
J	Current density	A/cm^2
J_t	Threshold current density	A/cm^2
k	Boltzmann constant	J/K
kT	Thermal energy	eV
L	Length	cm or μm
m_0	Electron rest mass	kg
m^*	Effective mass	kg
$\bar{n}$	Refractive index	

Symbol	Description	Unit
n	Density of free electrons	cm^{-3}
n_i	Intrinsic carrier density	cm^{-3}
N	Doping concentration	cm^{-3}
N_A	Acceptor impurity density	cm^{-3}
N_C	Effective density of states in conduction band	cm^{-3}
N_D	Donor impurity density	cm^{-3}
N_V	Effective density of states in valence band	cm^{-3}
p	Density of free holes	cm^{-3}
P	Pressure	Pa
q	Magnitude of electronic charge	C
Q_{it}	Interface-trap density	charges/cm^2
R	Resistance	Ω
t	Time	s
T	Absolute temperature	K
v	Carrier velocity	cm/s
v_s	Saturation velocity	cm/s
v_{th}	Thermal velocity	cm/s
V	Voltage	V
V_{bi}	Built-in potential	V
V_{EB}	Emitter-base voltage	V
V_B	Breakdown voltage	V
W	Thickness	cm or μm
W_B	Base thickness	cm or μm
x	x direction	
∇	Differential operator	
∇T	Temperature gradient	K/cm
ϵ_0	Permittivity in vacuum	F/cm
ϵ_s	Semiconductor permittivity	F/cm
ϵ_i	Insulator permittivity	F/cm
ϵ_s/ϵ_0 or ϵ_i/ϵ_0	Dielectric constant	
τ	Lifetime or decay time	s
θ	Angle	rad
λ	Wavelength	μm or Å
ν	Frequency of light	Hz
μ_o	Permeability in vacuum	H/cm
μ_n	Electron mobility	cm^2/V-s
μ_p	Hole mobility	cm^2/V-s
ρ	Resistivity	Ω-cm
ϕ_{Bn}	Schottky barrier height on n-type semiconductor	V

Symbol	Description	Unit
ϕ_{Bp}	Schottky barrier height on p-type semiconductor	V
ϕ_m	Metal work function	V
ω	Angular frequency ($2\pi f$ or $2\pi\nu$)	Hz
Ω	Ohm	Ω

International System of Units

Quantity	Unit	Symbol	Dimension
Length	meter	m	
Mass	kilogram	kg	
Time	second	s	
Temperature	kelvin	K	
Current	ampere	A	
Frequency	hertz	Hz	$1/s$
Force	newton	N	$kg\text{-}m/s^2$
Pressure	pascal	Pa	N/m^2
Energy	joule	J	$N\text{-}m$
Power	watt	W	J/s
Electric charge	coulomb	C	$A\text{-}s$
Potential	volt	V	J/C
Conductance	siemens	S	A/V
Resistance	ohm	Ω	V/A
Capacitance	farad	F	C/V
Magnetic flux	weber	Wb	$V\text{-}s$
Magnetic induction	tesla	T	Wb/cm^2
Inductance	henry	H	Wb/A

APPENDIX C

Physical Constants

Quantity	Symbol/Unit	Value
Angstrom unit	Å	$1\ \text{Å} = 10^{-1}\ \text{nm} = 10^{-4} \mu\text{m}$ $= 10^{-8}\ \text{cm} = 10^{-10}\ \text{m}$
Avogadro constant	N_{AVO}	$6.02204 \times 10^{23}\ \text{mole}^{-1}$
Bohr radius	a_B	$0.52917\ \text{Å}$
Boltzmann constant	k	$1.38066 \times 10^{-23}\ \text{J/K}\ (R/N_{AVO})$
Elementary charge	q	$1.60218 \times 10^{-19}\ \text{C}$
Electron rest mass	m_0	$0.91095 \times 10^{-30}\ \text{kg}$
Electron volt	eV	$1\ \text{eV} = 1.60218 \times 10^{-19}\ \text{J}$ $= 23.053\ \text{kcal/mole}$
Gas constant	R	$1.98719\ \text{cal/mole} - \text{K}$
Permeability in vacuum	μ_0	$1.25663 \times 10^{-8}\ \text{H/cm}\ (4\pi \times 10^{-9})$
Permittivity in vacuum	ϵ_0	$8.85418 \times 10^{-14}\ \text{F/cm}\ (1/\mu_0 c^2)$
Planck constant	h	$6.62617 \times 10^{-34}\ \text{J-s}$
Reduced Planck constant	$\hbar$	$1.05458 \times 10^{-34}\ \text{J-s}\ (h/2\pi)$
Proton rest mass	M_p	$1.67264 \times 10^{-27}\ \text{kg}$
Speed of light in vacuum	c	$2.99792 \times 10^{10}\ \text{cm/s}$
Standard atmosphere		$1.01325 \times 10^5\ \text{Pa}$
Thermal voltage at 300 K	kT/q	$0.0259\ \text{V}$
Wavelength of 1-eV quantum	λ	$1.23977\ \mu\text{m}$

Index